The
World Atlas
of
Warfare

Military Innovations that Changed the Course of History

The *World Atlas* of *Warfare*

Military Innovations that Changed the Course of History

General Editor and main contributor:
Richard Holmes

VIKING
STUDIO
BOOKS

Contributors

Richard Holmes
Lt Col Richard Holmes is a Senior Lecturer at the
Royal Military Academy, Sandhurst, and a serving
officer in the Territorial Army. His publications include
The Little Field-Marshal (a biography of Sir John
French), *The Road to Sedan: The French Army 1866–70*,
and *Firing Line*. Among television documentaries he
was responsible for *Comrades in Arms? Dunkirk 1940*
and *Soldiers* (with John Keegan).

Matthew Bennett
Matthew Bennett MA is a Senior Lecturer at the
Department of Communications Studies, Royal Military
Academy Sandhurst. He has published a number of
articles and contributions to books on Classical and
Medieval warfare.

Anthony Clayton
Dr Anthony Clayton is a Senior Lecturer in history at
the Royal Military Academy Sandhurst. He is a
specialist in imperial history, and his two most recent
works are *The British Empire as a Superpower, 1919–39*
and *France, Soldiers and Africa*. The latter publication
is a major study of the French military presence in
Africa 1830–1962.

Eric Grove
Eric Grove is a widely published naval historian and
former lecturer at the Royal Naval College, Dartmouth.
In 1980–81 he was Exchange Professor at the US Naval
Academy, Annapolis and is currently Associate Director
of the Foundation for International Security, Oxford.
His major work to date is *Vanguard to Trident: British
Naval Policy since 1945*.

T A Heathcote
Dr T A Heathcote, since 1970 Curator of the Royal
Military Academy Sandhurst Collection, is a graduate
of the School of Oriental and African Studies,
University of London, and a Fellow of the Royal
Historical and Asiatic Societies, Previously on the staff
of the National Army Museum, his publications include
The Indian Army, 1822 and *The Afghan Wars 1839–1919*.
He served in the Territorial Army as an officer of the
Royal Artillery.

Ian Russell Lowell
The Revd Ian Russell Lowell, an Anglican team vicar,
became interested in the Late Bronze Age through Old
Testament studies. His main areas of research have
been the Hittites, the Sea Peoples and the Trojan War.
He has contributed numerous articles to the Society of
Ancients' magazine *Slingshot* and other journals.

John Pimlott
Dr John Pimlott, Senior Lecturer in the Department of
War Studies, Royal Military Academy Sandhurst, is the
author of numerous books on modern warfare, and
acted as Consultant Editor to the partwork *War in
Peace*. Among his recent works is a scholarly
publication on counterinsurgency.

John Sweetman
Dr John Sweetman is Head of Defence and
International Affairs, Royal Military Academy
Sandhurst. He is an authority on military aviation and
has written a number of books and articles on military
history and air power. His publications include
Schweinfurt: Disaster in the Skies, *The Ploesti Raid* and
Operation Chastise: The Dambusters Raid.

Charles Townshend
Professor Charles Townshend is Professor of Modern
History at Keele University, and a Fellow of the
National Humanities Center. He is the author of several
books on counterinsurgency, including *Britain's Civil
Wars: Counter Insurgency in the 20th Century* and
The British Campaign in Ireland 1919–1921.

Senior executive editor	James Hughes
Executive art editor	Paul Wilkinson
Specialist military editor	Richard O'Neill
Editor	Julia Gorton
Cartographic editor	Stephen Rogers
Design assistant	Rupert Chappell
Picture research	John and Diane Moore (Military Archive and Research Services) Anne-Marie Ehrlich
Production	Ted Timberlake
Artwork reconstructions	Stephen Biesty, Roy Huxley, Tony Gibbons, Richard Hook
Maps	Lovell Johns Ltd
Additional artwork	Hayward and Martin

Edited and designed by Mitchell Beazley International Ltd
Artists House, 14–15 Manette St, London W1V 5LB

VIKING STUDIO BOOKS
Published by the Penguin Group
Viking Penguin Inc., 40 West 23rd Street, New York, New
York 10010, U.S.A.
Penguin Books Ltd, 27 Wrights Lane, London W8 5TZ,
England
Penguin Books Australia Ltd, Ringwood, Victoria, Australia
Penguin Books Canada Ltd, 2801 John Street, Markham,
Ontario, Canada L3R 1B4
Penguin Books (N.Z.) Ltd, 182–190 Wairau Road, Auckland 10,
New Zealand

Penguin Books Ltd, Registered Offices: Harmondsworth,
Middlesex, England

First published in 1988 by Viking Penguin Inc.
Published simultaneously in Canada

Copyright © Mitchell Beazley International Limited, 1988
All rights reserved

Library of Congress Cataloging-in-Publication Data
Holmes, Richard, 1946–
    World atlas of Warfare.

    Bibliography: p.
    Includes index.
    1. Military art and science—History.   2. Military
history.   3. Technological innovations.   I. Title.
U27.H65   1988   355'.009   88-10678
ISBN   0-670-81967-0

Typeset by Servis Filmsetting Ltd
Origination by Scantrans (Singapore)
Printed in West Germany by Mohndruck GmbH, Gütersloh

Contents

Introduction

Military history is a particularly appropriate subject for an atlas, for war has shaped our world. International frontiers bear witness as much to the clash of armies as to the negotiations of diplomats. Indeed, diplomacy itself has often – some would say too often – served only to confirm the verdict of war. Alliances, whether short-lived associations between city-states in ancient Greece, or the superpower blocs of the modern world, have usually arisen to meet a perceived military threat and to devise a collective response. Trade may have been a major motive for the creation of many of the great empires of history, but it was soldiers and sailors who fought to establish and retain them, just as other warriors – tribesmen, soldiers in nationalist armies, or guerrillas – struggled to throw off the imperial yoke.

If war has played a leading role in the relationship between states, it has been scarcely less important within them. Almost all aspects of a state's corporate activities – political and social, economic and administrative – are affected by war itself, or by preparation for it. Many social systems have emphasized the connection between social status and military duty. The hoplite infantry of Greece were the political nation of their city-states in arms; the *junker* aristocracy of East Prussia paid for their social position on a hundred battlefields from Mollwitz to the Marne; and the *samurai* who dominated so much of Japanese history were a warrior caste whose status was both advertized and epitomized by the wearing of a pair of swords.

The army, the state, and the individual

Political institutions, too, have been shaped by military pressures. The rise of the nation state in early modern Europe was closely linked to the growth of standing armies. The burgeoning bureaucracies of the period were inspired largely by the need to raise troops and to levy taxes – taxes which were themselves needed to maintain fleets and armies. Effective central authority, as Richelieu recognized in 17th-century France, hinged upon disciplined armed forces which were controlled by the king, not by the great nobles. Modern state bureaucracies have functions far wider than those of Richelieu's day, but they still rank high amongst their duties the provision of money, men and equipment for defence. This close relationship between central authority and armed force underlines the importance of the army within the state. Armies provide governments with support against internal enemies who seek their violent overthrow, and, with their disciplined structures, reliable communications and specialist equipment they can also assist by maintaining essential services in the event of civil disturbance or natural disaster.

War is the most visible and dramatic function of armed force, and it, above all, is a time of supreme crisis for the human spirit. Soldiers themselves risk death, injury and capture, and their families and friends at home – even in those periods of history when marauding armies or droning bombers have not helped blur the distinction between soldier and civilian – suffer uncertainty and grief. Emotions and opinions are polarized: reasonable people do unreasonable things. Some, soldiers and civilians alike, find war to be the high point of their lives: they respond eagerly to its challenges, relish its comradeship and find stimulus in its danger. Others regard war as a supreme waste of time, resources and life: they bitterly resent not merely their own involvement in it, but also the very fact of its existence. For most, however, these essentially opposite emotions are often curiously intermingled, with terror, frustration and despair being set alongside excitement, enthusiasm and friendship.

The limits of the battlefield

Military history is, as we can already see, concerned with much more than battle. Yet the events of the battlefield have come to loom as large in the pages of military historians as they have in the memories of survivors. In part this is entirely justifiable: battle is a time of great personal risk for its participants, and, in a broader sense, it has what John Keegan calls "a central role" in war, standing in relation to it much as the market place does to commerce. But in part, too, this emphasis on battle is misleading. In the first place, battle is a comparatively rare event in war. The cliché which describes war as "90 percent boredom and 10 percent sheer terror" is not far from the truth: most soldiers spend far longer training, marching, camping, cooking, and waiting, waiting, waiting than they do fighting. Furthermore, as armies have become increasingly sophisticated technologically, so their supporting "tail" has thickened while their fighting "teeth" have grown fewer – albeit sharper. The modern soldier in combat may be supported by a dozen men whose primary function has little to do with the business of killing, and who may resemble civilian truck drivers, stores clerks or freight handlers more closely than they do the traditional fighting man.

Secondly, much of what happens on the battlefield

The Somme 1916: British troops go over the top

points unmistakably to the importance of events off it. The fighting qualities of soldiers owe much to the environments, physical and cultural, which produce them; to the training that has sharpened their military skills and bonded them into cohesive units; and to the self-esteem and discipline which help them to hold firm under stress. The weapons and equipment they use testify to the technical advancement of their society, and to the ability of its industry to supply arms and ammunition in the required quantity and quality. The skills of their commanders, and of the staff officers who serve them, may be decisive on the day of battle, but they too demand careful and lengthy preparation.

War and Technology

It is impossible to divorce war from the mechanical instruments with which it is fought, and grasping the fundamentally important role of technology is crucial to our understanding of military history. John H Morse, a former US Assistant Secretary of Defense, was not overstating the case when he declared, "It is more the march of technology than it is political decisions which largely drives the nature and structure of our societies, our strategy, the nature of military forces, their structure and the doctrine they develop." The evolution of military technology, and the first effective use of new weapons, tactics and equipment form the central thread within this book. Many authorities have pointed to a cyclical relationship, itself largely determined by technology, between mobility and firepower. The British military expert Major

German PzKpfw Mk IV medium tank, 1942

General J F C Fuller coined the terms "shock cycles" and "projectile cycles", while his contemporary Tom Wintringham preferred to think of these as armoured and unarmoured periods. The question of armoured shock versus defensive firepower was as important when French knights met English archers on the battlefield of Crécy in 1346 (see p. 39) as it was when Egyptian missile operators took on Israeli tanks in Sinai in 1973.

Yet we must also recognize that the effects of technology spill over the narrow confines of the battlefield. The development of strategic bombing doctrine and of aircraft to implement it gave technology a colossal strategic potential, a potential which the development of nuclear weapons has intensified to an unprecedented degree. At a low level, technology does not merely transform the weapons the soldier uses and the tactics he employs. It changes much else besides. It affects the transport that takes him into battle (even if it is as comparatively simple as the marching boot), the logistic system which supplies him with food and ammunition, the preventive medicine which protects today's soldiers from the diseases which ravaged the armies of history, and the surgery which repairs the damage done by bullet or shell splinter. And, no less important but perhaps more easily forgotten, it also revolutionizes the communications that permit commanders, from section commanders and squad leaders with their dozen men to generals whose theatre of operations embraces whole continents, to receive information and transmit instructions.

The human factor

Nonetheless, to see military history merely in terms of technology is, though, as misleading as to see war only in terms of battle. For, vital though technology is, it is not without limitations. It does not in itself bestow a Midas touch which will remedy bad generalship or poor tactics. There have been many occasions when the technological edge of one combatant was blunted by the sheer fighting muscle of the other. The superiority of the British Martini-Henry rifle over the Zulu *assegai* did not prevent the catastrophe of Isandhlwana in 1879 (see p. 120), any more than Argentine superiority in night-fighting equipment enabled the defenders of Tumbledown and Mount Longdon to withstand the attack of Scots Guards and Paras in 1982. And, much as technology may aid the process of command by furnishing reliable and secure communications, and speeding up the processing of data, it can only assist, not replace, the men who must ultimately make command decisions. Finally, both men and equipment alike must operate in what Karl von Clausewitz called "the realm of uncertainty." Danger, physical exhaustion, and the abrasive effects of terrain and climate combine to generate what he called "general friction." This obstructs the most carefully devised plans – as the fate of the Schlieffen plan in the summer of 1914 so effectively demonstrates – and wears out or breaks down the most elaborate of machines.

In the pages that follow we chart the history of the art of war and its impact upon our world. The book's central theme is the evolution of technology, but we are concerned with military history in its broadest sense, and lose sight neither of the interplay between military events and social, economic and political institutions, nor of the fact that wars are fought by individuals whose achievement, endurance and suffering must not be buried beneath operational narrative or statistical information. Any historian should strive to remain objective, but for the military historian objectivity is often an elusive goal. It is easy enough to agree with Henri Barbusse, himself an experienced combat veteran as well as an influential author, that "it would be a crime to show the good side of war . . . even if it had one."

However, an overwhelming concentration upon the dark side of war is itself distortive, and can too easily become ghoulish. Yet we must face the possibility that our examination of the wars of history may in itself imply that war is likely to retain its utility as a political instrument. If the evidence at our disposal suggests that war shows little sign of losing such utility, our emphasis upon the rise of military technology can only sharpen our concern for the future of a world already marked by war.

Richard Holmes

7

— chapter one —
The dawn of warfare

Ian Russell Lowell

The difference between a weapon and a tool is basically one of use. The true "dawn of warfare" may be said to be linked with the dawn of mankind itself, powerfully recreated in Stanley Kubrick's film *2001: A Space Odyssey*, when the man-ape knocks down his enemy and beats him to death with a femur. However, while acknowledging the primary achievement of this "Cain" of the Olduvai Gorge some 1.5 million years ago, this chapter will discuss events many thousands of years later, concentrating on the early development of warfare in the Middle East. All dates are BC.

Warfare is a corporate, social and political activity; therefore it is only when man makes the transition from hunter to farmer (c.9000), and from tribesman to city-dweller, as at Jericho (c.8000) or Çatal Hüyük (c.6000), that he develops militarily. The individual aggressive skills of the hunter are transformed into the collective defensive troop of citizen-soldiers. The developing technology of metalworking created the uniform weapon: the spear. From c.3000 the nucleated settlements of Sumer gave rise to efficient armies of spearmen, able to hold, advance, and fight in line through training and the invention of the bodyshield. Flanked by nobles on ass-drawn battlewagons, wars were won by discipline, stamina and morale.

Such checkerboard warfare continued between cities like Uruk, Ur, Larsa and Lagash for nearly 1,000 years, until the advent of Sargon, king of Agade or Akkad (c.2334–2279). The city-states of Sumer and Akkad had shared a common cultural heritage for centuries, but Sargon gave them a political unity which became the "classical" inheritance of the Middle East.

Sargon changed the face of warfare in many ways: he not only subjugated the other cities of Sumer, but he also pushed up the Euphrates to conquer desert tribes on the edge of the known world. His success came from utilizing hunting skills such as archery (known by the Sumerians, but not used in warfare) and mobility. He also had a corps of 5,400 "professional" troops who formed the backbone of his army, fleshed out with levies.

This army created the Akkadian empire, which stretched from the Persian Gulf (the "Lower Sea") to the Eastern Mediterranean (the "Upper Sea"), and crossed both the Amanus and the Taurus ranges into the Konya plain of Anatolia. Sargon's conquest stamped a political and imperial legacy on a geography previously touched fleetingly by regular trade routes.

It was a legacy that would outlast his memory, but it also set the pattern for early warfare in the Middle East: professional armies large enough to conquer and hold an empire, flexible enough to do battle over plains and mountains, against cities and tribes. It was the goal of kings and the death of kingdoms as they fought against each other and exhausted their resources. It would not end until 1,000 years later at the battle of Kadesh (1286).

The legacy of Sargon

The Akkadian empire collapsed after Sargon's grandson Naram-Sin (c.2254–2218) succumbed to pressure from the Amorites (*Amurru*, the "Westerners"), who created the city-states of Yamkhad (Aleppo), Mari and Babylon across Mesopotamia. From about 1900 to 1750, only the Assyrian trading colonies (*karu*) linked Anatolia to Mesopotamia, while the wave of Indo-European invaders, who were to form the states of Hatti and Mitanni, entered the Middle East from the north.

By c.1750, the Amorite Hammurabi, had conquered the other Mesopotamian cities to create the Babylonian empire. In the north, the Anatolian city-states were also conquered to form the Hittite kingdom under Hattusilis I (c.1650). In c.1595, his successor Mursilis I led a daring attack on Babylon itself. The Kassites established them

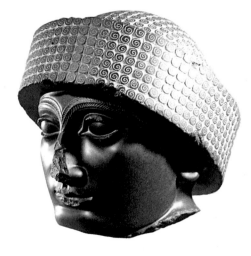

A very capable general, Sargon (c.2334–2279) merged the Akkadian and Sumerian cultures in a single empire, bequeathing a lasting imperial dream to the ancient world.

The stability of a settled culture permitted the technology of standardized weaponry – helmets, spears, throw-axes and bodyshields – to emerge in Sumeria during the 3rd millennium BC.

—TIME CHART—

c.8000BC	Jericho: walled city with moat and tower
c.6400	Çatal Hüyük: hunters armed with bows
c.5400	Hacilar: fortified "palace"
c.3000	Egypt: troops led to battle by standard
c.2900	Menes (Narmer) unifies Egypt
c.2800	Sumer: ass-drawn "battlewagons"
c.2500	Sumer: spearmen in shielded phalanx
2334–2279	Reign of Sargon of Agade: development of standing army
c.2300	Sumer: horse ("foreign ass") recorded
c.2300	Troy: fortified citadel
c.2000–1450	Minoan "thalassocracy" over Aegean
c.1792–1750	Hammurabi rules Babylonian kingdom
c.1700	Development of horse-drawn chariot
c.1650	Kingdom of Hatti founded by Hattusilis
1650–1540	Hyksos rule in Egypt
c.1600	Foundation of Mycenaean kingdoms
c.1600	Kingdom of Mitanni founded by Kirta
1595	Sack of Babylon by Mursilis of Hatti
1468	Battle of Megiddo

c.1460	Saustatar of Mitanni conquers Assyria
1375–1334	Reign of Suppiluliumas of Hatti
1340–1334	Mitannian campaign of Suppiluliumas
1332–31	Arzawan campaign of Mursilis II
1290–1224	Reign of Ramesses II of Egypt
1286	Battle of Kadesh (Kinza)
c.1270–1200	Hatti and Assyria in conflict
1269	Treaty between Hittites and Egyptians
c.1250	Sack of Troy (traditional date)
	Merenptah of Egypt defeats "Sea Peoples"
c.1200	Sack and destruction of Pylos
c.1190	Sack of citadel of Mycenae
	Ramesses III defeats "Sea Peoples"
c.1180	Destruction of Hattushash and Ugarit
c.1000	David rules kingdom of Israel
c. 800–640	Urartu and Assyria in conflict: development of cavalry
744–727	Reign of Tiglath-pileser III of Assyria: reformation of Assyrian army
605	Battle of Kargamis: Neo-Assyrian empire falls

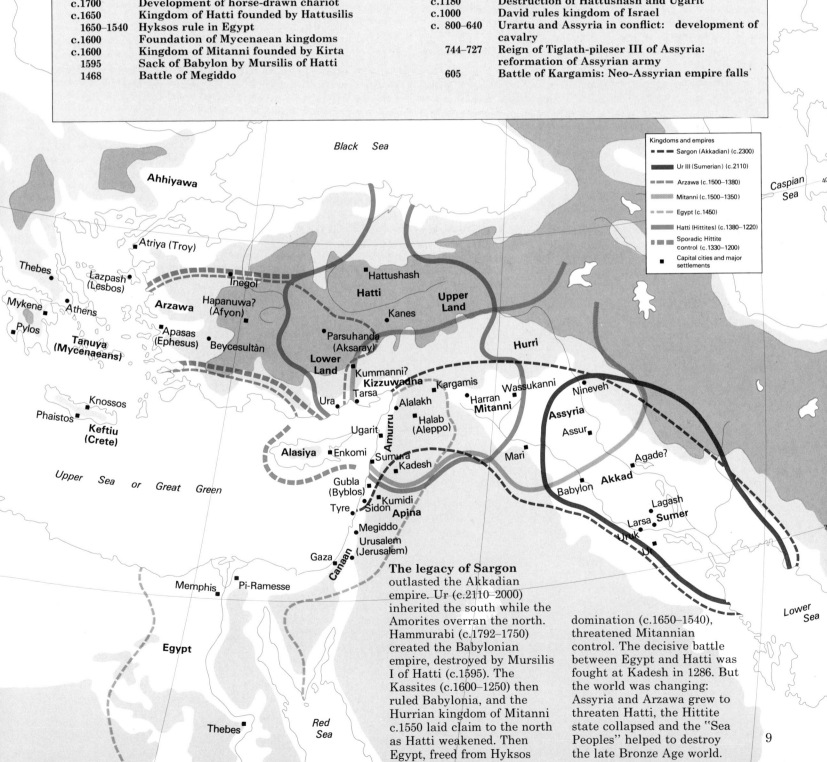

The legacy of Sargon outlasted the Akkadian empire. Ur (c.2110–2000) inherited the south while the Amorites overran the north. Hammurabi (c.1792–1750) created the Babylonian empire, destroyed by Mursilis I of Hatti (c.1595). The Kassites (c.1600–1250) then ruled Babylonia, and the Hurrian kingdom of Mitanni c.1550 laid claim to the north as Hatti weakened. Then Egypt, freed from Hyksos domination (c.1650–1540), threatened Mitannian control. The decisive battle between Egypt and Hatti was fought at Kadesh in 1286. But the world was changing: Assyria and Arzawa grew to threaten Hatti, the Hittite state collapsed and the "Sea Peoples" helped to destroy the late Bronze Age world.

9

selves as Hammurabi's successors under Agum II (c.1602–1585), but they had to contend with the growing strength of the Hurrians, who by c.1600 had formed the nucleus of the kingdom of Mitanni. This grew to control northern Mesopotamia and to push south towards Lebanon and Akkad.

The united kingdoms of Egypt, which from c.3000–1650 had remained largely unaffected by Sargon's legacy, were now invaded by the "Chiefs of Foreign Lands" (*Hikaukhoswet*). Known as the Hyksos, they ruled the Delta region of Egypt and influenced the whole country from c.1650–1540. From Lebanon and Canaan they brought with them both the legacy of Sargon and important new techniques of warfare.

Like the city-states of Sumer, Egypt had developed companies of citizen soldiers, but unlike Sumer these were based on the local district or *nome*, and also contained bowmen. By the 17th century, infantry armies had emerged, comprising companies of spearmen protected by bodyshields, companies of archers, and lightly armed auxiliaries such as Nubians and Libyans. Ranged against the Egyptian army, the invaders from Canaan had two new weapons. These were the light two-horse chariot and the composite bow.

Hurrian expertise had domesticated small wild horses, about 12 hands high, and trained them inspanned as a chariot team. The chariot, with its four-spoked (later six-spoked) wheels, could be used en masse as a shock attack weapon or as a mobile firing platform. Combined with the composite bow, with a range of 250yds (230m) far outshooting the Egyptian weapon, the Hyksos prevailed.

Over the next century, the Egyptians in the southern dynasty at Thebes learned their lesson. Under the leadership of Ahmose (c.1561–36), the Egyptians eventually expelled the Hyksos from their land and prepared to carry the war into the north.

Mitanni, Egypt and Hatti

Egypt was now reunited as a single state, although administratively divided into the two regions of Upper and Lower Egypt, and had a cultural and political stability greater than either Hatti or Mitanni. It also controlled Nubia to the south through a viceroy, and had some influence over the western Libyan tribes.

The Hittite core territory of Hatti lay within the bend of the River Marassantiya (classical Halys), with its capital at Hattushash (modern Bogazköy). Across the river to the east and south were the Upper and Lower Lands. Added to these were Anatolian client-states such as Kizzuwadna (Cataonia, Cilicia) and Tarhuntassa (Lycaonia), military marches like Arzawa (Lukka) and Gasga, and Mesopotamian provinces ruled by a viceroy at Kargamis.

Mitanni was situated between the Tigris and the Khabur rivers. Also known as Naharin or Hanigalbat, at its greatest it included the client kingdoms of Assyria and Arrapkha in the east, Nukhasse and Kizzuwadna in the west, and Kinza (Kadesh) in the south, where the prince

Control for Syria and its trade routes was a three-cornered fight between Mitanni, Egypt and Hatti for over 200 years (*left*). The Hittites eventually won by conquering the three important sites of Halab (ancient Yamkhad, Aleppo), Kargamis and Kinza (or Kadesh). Thus they controlled the important Syrian trade routes.

A heavy four-wheeled "battle-wagon" (*above*) depicted on the Standard of Ur (c.2500BC). Used by Sumerian kings and nobility, it was drawn by four onagers (wild asses) and armed with javelins.

acted as viceroy for Lebanon and Canaan.

Under the pharaohs Amunhotep I and Thothmes I (c.1536–1503), the Egyptians crossed the Lebanon and raided Naharin. However, the reign of the female pharaoh Hatshepsut (1490–1468) saw Mitanni regain control south of the Lebanon. Then Thothmes III (1468–1438) in his first campaign shattered the Canaanite confederation at the battle of Megiddo (1468).

In successive campaigns, Thothmes raided north of the

" . . . when the Prince of Kadesh sent out a mare, which was swift on her feet and which entered among the army, I ran after her on foot, carrying my dagger, and I [ripped] open her belly. I cut off her tail and set it before the king [Thothmes III]. . . . His majesty sent forth every valiant man . . . to breach the new wall which Kadesh had made. I was the one who breached it. . . . When I came out, I brought two men, maryanu, *as living prisoners. Then again my lord rewarded me. . . ."*

Amun-em-heb, Lieutenant of the Army, c.1458BC

Lebanon, hitting the stronghold of Kadesh, until he felt strong enough to cross into Naharin itself (1458). Mitanni had to come to terms with Egyptian interests north of the Lebanon, and a treaty was made. But under Saustatar (c.1460), it expanded east to dominate Assyria. Meanwhile, the Hittite kingdom had begun to recover, defeating

Anatolian alliances in the west and the Hurrian advances in the east. By the time of Suppiluliumas (1375–1335), the Hittites were ready to enter the Levant once again.

In c.1364, the Mitannian king Tusratta (c.1385–40) defeated Suppiluliumas and showed his loyalty to his southern allies by sending Pharaoh Amunhotep III spoils from the battle. However, by c.1350 the Hittites had reached the Euphrates, and after two campaigns Suppiluliumas, by c.1335, was in control of all the lands north of the Lebanon. The Hittite great-king had also sacked Kadesh, which had been in Egyptian hands since Thothmes III. Mitanni was crushed with the support of the Assyrians under Assur-uballit (c.1366–28).

As Egypt went into military decline under Akhenaten (1359–47) and his successors, the Hittites and Assyrians fought a vicarious war in support of rivals to the Mitannian throne. It was left to Suppiluliumas's son Mursilis II, to break Arzawa. Meanwhile Egypt had recovered, and the Hittites had to face an Egyptian threat from Ramesses II in the south. The place was the stronghold of Kadesh, the year 1286.

This was the high noon of early warfare. Ramesses won the battle of Kadesh, but he lost the campaign. The Egyptian army had to retreat south immediately, harried by the Hittites under the king's brother Hattusilis. The Lebanon proved to be a natural frontier, but hostilities did not cease until 1269, when Ramesses and Hattusilis (by then great-king) signed a pact of nonaggression and

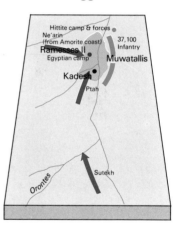

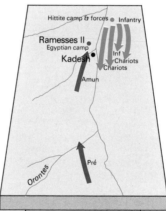

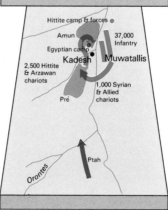

At Kadesh, Ramesses II and Amun brigade encamped in time to see Pre' brigade ambushed by 2,500 Hittite chariots. Looting the camp, the Hittite force was in turn ambushed by the Canaanite brigade from Amurru and Ptah brigade. A further 1,000 Hittite chariots failed to carry the day. Ramesses won the battle, but Muwatallis the war, as the Hittites pushed south as far as Damascus.

mutual aid. This treaty defined the two powers' spheres of influence, and as such merely affirmed the political and military reality of the previous 25 years.

More than a century of warfare stood between the two empires, evidenced by the exhaustion of their respective countries and the frontier states of the Levant. Social, economic, political and military force had been poured out and dissipated, while the respective spheres of influence of both states remained almost unchanged. But the treaty also noted a shift in affairs: new challenges from Assyria in the east and Arzawa in the west, and a threat from the "Sea Peoples" and Israelites in Canaan. The end was not immediate – the treaty remained in force for almost 100 years – but it was inevitable.

Hatti then suffered increasing attacks from Assyria, which defeated a Hittite army by the Euphrates in c.1262 and invaded Hittite territory two decades later. But Assyria failed to replace Hatti as the strong force in the north. Hittite power also crumbled in Anatolia before a resurgent Arzawa. Suppiluliumas II, the last Hittite king (c.1190), faced a disintegrating realm.

The city-state of Ugarit, allied to the Hittites, recorded

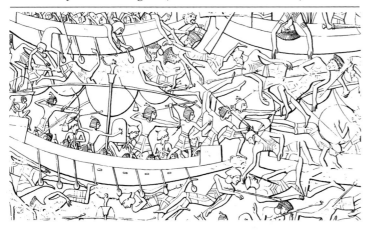

The *Keftiu* (Sea People) are defeated by Ramesses III.

an attack from hostile ships while its fleet was off the coast of Lukka and its army in Kizzuwadna. These were probably the ships of the "Sea Peoples", as the Egyptians called them; the Peleset (Philistines), Tjekker and Denyen, or else of the allied Sikala ("who live in ships" according to Hittite records). Their maritime power grew, through the weakness of Hatti and Egypt, to become another threat to

". . . the enemy ships are already here, they have set fire to my towns and have done very great damage in the country. . . . did you not know that all my troops were stationed in the Hittite country, and that all my ships are still stationed in the Lukka Lands and have not yet returned? So that my country is abandoned to itself . . . there are seven enemy ships that . . . have done very great damage. Now if there are more enemy ships let me know . . . so that I know the worst."

Hammurabi, King of Ugarit, to the King of Alašiya, c.1200BC

the stability of the Middle East. Ugarit and many major cities, including inland Hattushash, were sacked and deserted at this time. Others, like Kargamis, survived to become centres of the neo-Hittite culture, founded by

refugees from western Anatolia.

Egypt itself was also attacked. Ramesses' son Merenptah (1224–14) in c.1220 defeated a large force of Libyans and allies, including Shardana and Lukka (Arzawans), Ramesses III (1194–62) fought two wars against the Libyans and two battles against the "Sea Peoples": one off the Egyptian coast, the other inland on the Canaanite border. Among these invaders were Peleset (Philistines) and Tjekker who settled in Palestine and Denyen who settled in Kizzuwadna.

It took the Middle East nearly two centuries to recover from this collapse, and by then the antagonists had all but changed, consisting of a renewed and imperial Assyria, the city-states of the Greeks, and the new kingdoms of Lydia, Phrygia and Urartu. Warfare had also changed, as a result of new technology. The advances in iron-smelting meant cheaper and stronger weapons. The chariot was heavier and less manoeuvrable, its former tactical role being taken over by the cavalryman, and the foot soldier was better armoured. The Bronze Age had passed, yielding to the Iron Age of Greece and, later, Rome.

The martial state

Akkad, Babylon, Hatti, Mitanni and Egypt had created and kept their power through military strength: "professional" armies raised by company, regimented into battalions of foot soldiers and squadrons of chariots under the discipline and order of "career" officers. Military domination of the society sustained the imperial urge. The army, like the state, owed allegiance to the monarch and was in fact the private army of the king. Among the Hittites, all the troops, assembled by regiment and squadron under their colonels and captains, swore an oath of allegiance to the great-king, the royal family and his descendants at a special inspection before a campaign started.

In Egypt, the pharoah's role as martial leader is clearly portrayed in the temple reliefs of Ramesses II, and a similar role for the Hittites is indicated in the writings of Mursilis II, who recorded both the *Deeds* of his father Suppululiumas and the *Annals* of his own reign. Non-militaristic pharaohs like Hatshepsut and Akhenaten were exceptions, and weak great-kings like Tudkhaliyas the Younger and Urkhi-Tesub were overthrown. But the ruler also had divine attributes, and to break an oath to him was a sin.

The New Kingdom Egyptian army consisted of three, later four, brigades combining regiments of foot and chariot squadrons: Amun at Thebes, Pre' at Heliopolis, Ptah at Memphis, and Sutekh in the Delta. Added to these were companies of allies and mercenaries. The imperial Hittite army was made up of local regimented companies and squadrons together with allied brigades of horse and foot. The smallest tactical units were the company and the troop, each probably consisting of about 200 men and 10 chariots in the Egyptian army. The *Amarna Letters* and other sources indicate that two or more companies were organized into divisions for garrison duties and other purposes. An Egyptian regiment may have numbered about 5,000 men, with an attached squadron of 250 chariots. The Hittite regiment was nominally 1,000 men, with an attached squadron of 100 chariots. The ratio of horse to foot appears to be higher in Hittite brigades (1 to 10) than in Egyptian brigades (1 to 20).

On campaign, the amassed army could number many

Hittite chariotry at the battle of Kadesh, c.1286BC, was recorded as 3,500 in number. The Hittites could field a variety of chariots due to the nature of their empire,

but they looked to the Hurrians for military innovation, including siege warfare and horse-training. Egyptian records use the term *tuhiru* to describe Hittite charioteers and troops, but many of the allied charioteers at Kadesh are given the Hurrian term *mariyannu* – including those from Dardaniya, near Troy.

Canaanite chariot c.1450BC (*left*): the Canaanite chariot armies of the Hyksos conquered Egypt. Weaponry consisted of bow and javelins, with a rectangular shield, a quiver being attached to the car.

Egyptian chariot c.1190BC (*left*): The Egyptians realized the chariot's worth through the Hyksos invasion. Capturing enemy chariots and crew, they developed and refined the vehicle: adding extra quivers, strengthening the pole support, and eventually increasing the spokes to six, and then eight. Accompanying chariot squadrons were companies of runners, light infantry auxiliaries.

Mycenaean chariot c.1250BC (*left*, *above*): the Mycenaean chariot developed separately, retaining four-spoked wheels, and adding the characteristic "wings". At first a one-man vehicle (c.1600BC), it changed to a crew of two, the warrior armed with the long spear. In its final form, after c.1200BC, the car was stripped to just the frame.

thousands. Mursilis II claimed in his *Annals* to have captured 66,000 Arzawans during his two-year campaign against that country (1332–1). At the battle of Kadesh (1286) Ramesses II recorded on his temple reliefs a Hittite army of 3,500 chariots and 37,000 infantry. These were big battles, and many of the troops were auxiliaries and allies, raised for those particular campaigns. Ramesses listed at least 18 countries or states allied to the Hittites at Kadesh, and his own army had Shardana and Levantine auxiliaries, and included a separate brigade of Canaanites.

Each kingdom's military commitment was reflected in its ability to amass troops, and to feed and shelter them. Every soldier, whether noble charioteer or humble infantryman, was a professional, and the honour passed from father to son with a grant of land given in return for service. Officers were rewarded for bravery and discipline but severely punished for failure.

It took three years to train a chariot team, according to a Hurrian manual, and the grooms, woodworkers, and armourers needed for the chariot and its crew have a special mention in the Hittite laws. Specialized troops were trained for pioneering, siegecraft, scouting and intelligence work. Manuals and standing orders were written for garrison commanders and military governors. The skills of allied peoples and client states were used, such as the archery of the Lulakhi and Nubians, and Hurrian horsemanship. Prisoners and refugees might also be given land in return for military or agricultural service.

Geared to fulfilling the legacy of Sargon, Late Bronze Age society had become over-specialized and precariously balanced. Hattusilis III wrote that in a land ravaged by war it was not possible to sow corn for the next ten years. The dawn of warfare ends with a wasted Middle East.

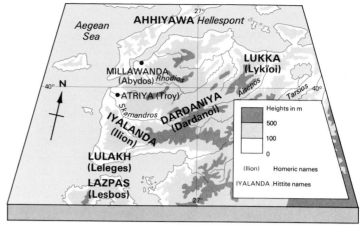

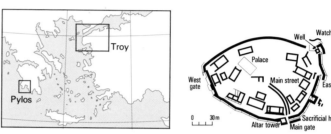

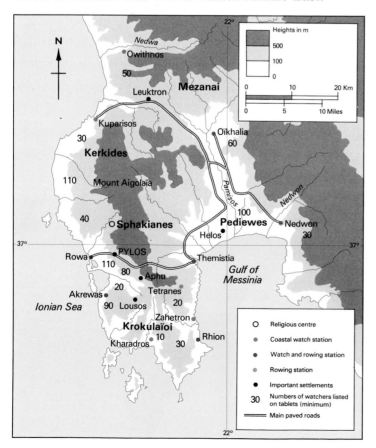

Pylos was destroyed c.1200BC, but from its archives we have a graphic picture of a small Late Bronze Age kingdom (*left*). Bronze was in small supply, agricultural output was low, and labourers' food strictly rationed. The defences of the realm were organized under two commands, the "Watchers" (Coast Guard) and "Rowers" (Fleet), which covered the whole coastline. There were ten companies of Watchers, totalling over 800 men. The Rowers numbered over 500. The records are damaged, so the totals can only be estimated, but probably an overall combined force of 2,000 men guarded the coast.

Homer's Troy (*above*), reconstructed from archaeological data, is probably the fortress of Atriya, mentioned in Hittite documents, in the land of Iyalanda (corrupted by later legend into Ilion). It was linked with Millawanda and the Lukka lands, whose people appear as allies of Muwatallis at Kadesh (1286BC) with the Dardaniya: the *Lykioi* and *Dardanoi* of Homer. It was strategically placed to guard the Hellespont and the tin route from Ahhiyawa (Thrace), and appears to have been destroyed either by siege and/or earthquake. However, it was only one, and not the earliest, Troy. Archaeology has revealed nine Trojan settlements, the earliest c.3000BC.

—THE EARLY IRON AGE—

The Hittite empire proved to be the keystone of the Late Bronze Age; when it weakened, almost all the structure collapsed. The Arzawans moved east into southern Anatolia and northern Mesopotamia to form Tabal and the Neo-Hittite states known in the Old Testament. The "Sea Peoples" settled in Kizzuwadna (Cilicia), Alasiya (Cyprus) and Palestine (named after the Philistines). The Assyrians initially expanded farther into northern Mesopotamia but failed to keep their hold on it.

In the south a weakened and divided Egypt eventually succumbed to a Libyan dynasty c.945BC. In Canaan, the Israelite tribes had united under King David by 1000BC, but became divided and weakened over the next 200 years as the Neo-Assyrian state grew under Adad-nirari III (810–783) and his successors. In the north, the Hurrian kingdom of Urartu, centred on Lake Van, expanded into northern Syria under Menuas (c.810–785), threatening Assyria itself.

A new political and military purpose came to Assyria with the accession of Tiglath-pileser III, who raised a national army from all lands under his influence from the Persian Gulf to the Taurus Mountains, Lake Van to the borders of Egypt. This territory was further extended by Assurbanipal (668–627) to include Egypt itself.

Meanwhile in the north, Rusas II of Urartu (c.685–645) faced increasing pressure from the nomadic Cimmerians and Scythians. In the west the kingdoms of Lydia and Phrygia took over the Anatolian lands that had once been Arzawan and Hittite; while Greek colonies were founded on the Aegean coastline and also on Cyprus. Furthermore, the overstretched Assyrian empire was to suffer and finally succumb to attack from the Medes, Persians and Babylonians. Assur-uballit II, the last Assyrian king, lost his kingdom at Kargamis in 605BC.

The development of iron-working meant stronger weapons could be cast and forged much more cheaply, even though iron needed to be smelted twice at a higher temperature than bronze (copper and tin). The long sword was soon adopted as the common weapon of the soldier and armour became available for the line infantryman – Tiglath-pileser's shock troops wore scale cuirasses.

The other major change was the development of the true cavalryman, sitting properly on the back of a horse and using a saddle. With the development of mounted archers and mounted spearmen, the chariot became heavier in design, carrying three or more crew, with an appropriately strengthened frame and wheels. The Assyrian army also perfected siege warfare, developing better battering rams, assault towers and engineering works.

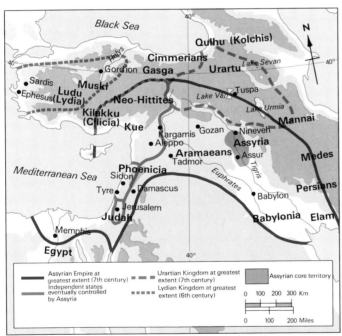

At the beginning of the early Iron Age (1st millennium BC), Assyria was weaker than her neighbours: the Neo-Hittites, Kassites (Kardunias) and Hurrians (Urartu). The new states of Israelites, Philistines and Aramaeans expanded as the old powers were eclipsed. More territory and plunder for their god Assur gave incentive to the Assyrian kings, and the Neo-Assyrian empire was created.

Iron Age technology gave greater protection to the ordinary soldier: Neo-Assyrian slingers of the army of Sennacherib (*above right*), deployed at the siege of Lachish (c.701BC), wear scale-armour baldrics. With the predominance of sieges in early Iron Age warfare, Neo-Assyrian armies developed powerful siege equipment (*right*). This six-wheeled battering ram was used against Dabigu (c.857BC).

chapter two

Warriors of Greece and Rome

Matthew Bennett

The image of warfare in the classical world is that it was dominated by the heavy infantrymen: the Greek hoplite, the Macedonian phalangite and the Roman legionary. True though this is at a tactical level, it should not blind us to the wide variety of troop types employed and the broad strategic vision of commanders. The ancient world was essentially a Mediterranean one, so control of the sea was crucial. Sieges played an important role, as did logistics, usually epitomized by the need for a grain supply.

We only possess the Greek version of the Persian invasions of 490BC and 480–79BC. It stresses how free men defeated oriental tyranny through their moral superiority, and has not been questioned. Certainly it was remarkable that Greece, divided into petty, squabbling, city-states, could withstand the military resources of a great empire. The Persians had sprung to world power over a century earlier, using a devastating combination of infantry archers and well-mounted cavalry. Their forces were more heterogeneous at the beginning of the 5th century BC, comprising many subject nations of doubtful enthusiasm. But the Medes and Persians, based on the 10,000-strong Immortals, still formed a redoubtable fighting machine.

King Darius' prestige had never quite recovered from his defeat by Scythian nomads – losing an army on the steppes – and this encouraged subject Ionian Greeks to rebel in 494 BC. They were crushed in a naval battle, and Darius determined to punish the mainland Greeks, and chiefly Athens, for backing the revolt. The invasion of 490 was a small expedition with that specific aim.

The Persians landed at Marathon Bay, but encountered a blocking force. Seeking to use their fleet to outflank the Greeks, they re-embarked their redoubtable cavalry. Seeing his opportunity, the Athenian general Miltiades ordered a charge. After a fierce struggle the heavily equipped hoplites routed the unarmoured Persians. Marching swiftly back to Athens, the victors prevented any further landing. This remarkable turnaround of events – the flight of the previously invincible Persians – had to be avenged. A decade later, Darius' son Xerxes led an army over a

bridge of boats across the Hellespont. Allegedly consisting of millions, this force may well have been 180,000-strong; half his empire's levy. He also used diplomacy to ensure that a Carthaginian attack prevented the Sicilian Greeks from sending aid. The position on the mainland seemed hopeless. The northerners – Thessalians and Boeotians – surrendered and joined the Great-King's forces. But the defence of the pass at Thermopylae, in which the Spartan king Leonidas and his small force fought to the death, while strategically insignificant, stiffened Greek morale. Athens was still taken and burnt. Even the cunning ambush of the Persian fleet at Salamis, while giving the Greeks naval dominance, did not expel the invader.

Xerxes left half his army under a capable general to mop up the opposition in the next campaigning season. The unwieldy hoplite phalanx was very vulnerable to the Persian cavalry, and the Greeks feared battle. The Persian, Mardonius, was hoping for political rifts in his enemies' ranks to win the war. At Plataea in 479 these unwilling opponents came to grips. It was a close run thing, but Mardonius was killed and his men fled.

The Peloponnesian War

Now opportunist Athens took the war onto Persian soil, liberating the Ionian Greeks and creating a federation of states based on the sacred island of Delos. Gradually this Delian League became an Athenian seaborne empire. Conservative Sparta did not show such far-flung ambitions, but generally ranged up against Athens in city-state coalitions. In 431BC, opposition became open war;

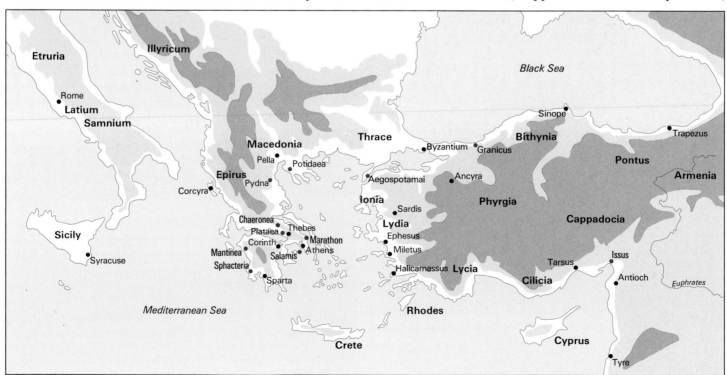

—TIME CHART—

499–448BC	Greco-Persian Wars
490	Greek hoplites defeat Persians at Marathon
480	Greek triremes gain victory over Persian navy at Salamis
479	Persians defeated at Plataea
432–404	Peloponnesian War
405	Athenian fleet destroyed at Aegospotamai
371	Spartans defeated by Theban deep phalanx tactics at Leuctra
338	Macedonian victory over Greeks at Chaeronea
336–323	Conquests of Alexander
331	Alexander's tactics defeat Persians at Gaugamela
323–281	Wars of the Diadochi (Alexander's successors)
281–272	War between Rome and Pyrrhus of Epirus
264–241	First Punic War: Carthage surrenders Sicily
218–201	Second Punic War
216	Romans routed at Cannae
202	Hannibal crushed at Zama
197	Roman legion overcomes Macedonian phalanx at Cynoscephalae
102	Marius defeats Teutones and Cimbri at Aquae Sextiae

101–100	Marius reforms the Roman legion
58–51	Caesar's Gallic Wars
52	Battle and siege of Alesia Roman Civil War
48 Aug 9	Pompey defeated at Pharsalus
44 Mar 15	Caesar assassinated
31 Sept 2	Anthony and Cleopatra defeated at Actium
AD42–60	Conquest of Britain
c.80	Danubian *limes* constructed
122–128	Hadrian's Wall built
284–305	Reign of Diocletian
c.300	Creation elite field armies; border soldier/farmers
312–337	Reign of Constantine
378	Gothic cavalry routs legions at Adrianople
by 400	Roman army "barbarized"; mostly German troops
410	Rome sacked by Goths under Alaric
451	Huns under Attila defeated by Romano-Goth army at Châlons
527–565	Reign of Justinian
568	Lombards invade Italy

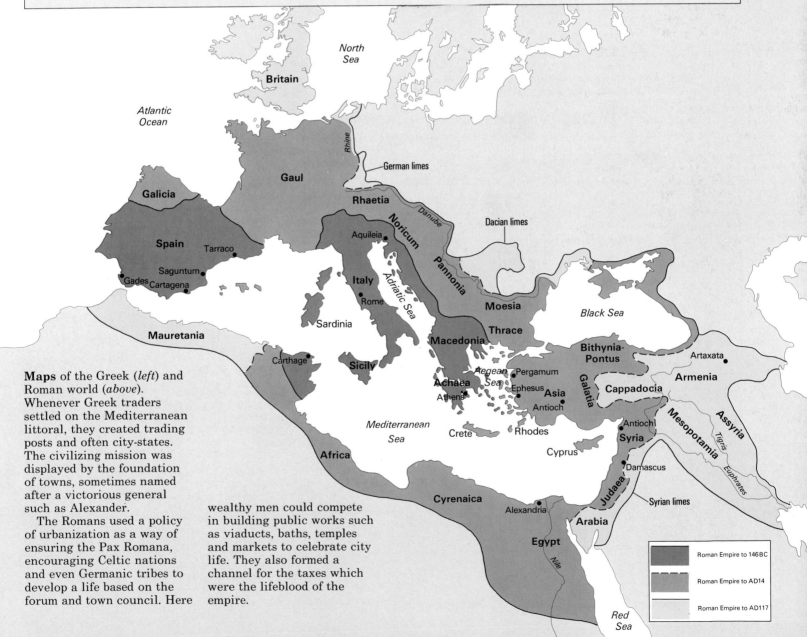

Maps of the Greek (*left*) and Roman world (*above*). Whenever Greek traders settled on the Mediterranean littoral, they created trading posts and often city-states. The civilizing mission was displayed by the foundation of towns, sometimes named after a victorious general such as Alexander.

The Romans used a policy of urbanization as a way of ensuring the Pax Romana, encouraging Celtic nations and even Germanic tribes to develop a life based on the forum and town council. Here wealthy men could compete in building public works such as viaducts, baths, temples and markets to celebrate city life. They also formed a channel for the taxes which were the lifeblood of the empire.

although neither side could seize the advantage.

Athens ruled the sea and Sparta the land. The Athenians dared not face the Spartans' hoplite army, which annually ravaged their territory, burning their crops. But the Athenian Long Walls, linking the city to the port of Piraeus, ensured that they could be safely supplied by sea. The Spartans were consistently defeated in naval operations. They proved their superiority on land at the battle of Mantinea in 418.

This stalemate was eventually broken by the Athenians overreaching themselves. A disastrous expedition to Syracuse, in Sicily (415–13BC) resulted in huge losses of men and ships. The Spartan admiral Lysander siezed this opportunity to challenge Athens at sea. Despite suffering several defeats he was eventually successful. In 405, he surprised the Athenian fleet while it lay beached at Aegospotomai, taking it entire and slaughtering the trained crews. Unable to resist any longer, Athens submitted, leading to a generation of Spartan hegemony.

Their domestic disputes settled for a moment, the Greeks turned once more to plunder Persia. First, Xenophon led 10,000 mercenaries to the assistance of Cyrus, a pretender to the throne. Following the battle of Cunaxa (401) in which the Greeks were victorious but Cyrus was killed, they had to march from Babylon to the Black Sea, entirely through hostile territory. Then, Agesilaus, the Spartan king, campaigned successfully in Asia Minor. Persian money stirred up opposition at home, but the Spartan marched his men back into Greece in only a month, and crushed the Corinthian alliance at Coronea (394).

Sparta was challenged and defeated by Thebes, chiefly through the tactical genius of Epaminondas, who perfected a deep phalanx to punch through the enemy line. At Leuctra in 371, he led a left-flank charge, killing the Spartan king. He lost his own life in a similarly brilliant victory at Mantinea in 362. Theban supremacy lasted as long as he lived. But hoplite warfare, although seemingly still dominant, was soon to be superseded.

Peltast and phalangite

Light skirmishers had inflicted a humiliating defeat on Spartan hoplites even during the Peloponnesian War, at Sphacteria (425BC). Although supreme in close order fighting, a Spartan force of over 400 men had been surrounded and harassed into surrender by peltasts landed from the Athenian fleet. Such lightly equipped archers, slingers and javelinmen also proved their worth during the March of the 10,000. By seizing high ground along the route and keeping the Persian cavalry at bay, they enabled the hoplites to march relatively undisturbed.

"When the two armies were close together, the Greek peltasts raised a shout and ran toward the enemy before anyone had given them the order. The enemy charged to meet them, both the cavalry and the Bithynians in close order, and they drove the peltasts back. But when the line of hoplites came up, moving at a quick pace, and at the same time the trumpet sounded and they sang the paean [battle chant] and then raised a shout as they brought their spears down for the attack, then the enemy stood their ground no longer, and took to flight."

Xenophon, on an incident during the March of the 10,000, 401BC

A hoplite (heavy infantryman) of c.480BC (*left*). The heavy armour includes a bronze helmet, greaves and large shield or *hoplon* (the method of carriage is clearly shown here) and a stiffened linen cuirass. Normal hoplite deployment (*above*) was a close-order phalanx, usually eight ranks deep, but less if covering a wider front. On good ground, peltasts (light armed troops) had little chance against such a formation, but in broken terrain they had the edge.

That the Spartans had learnt little about such innovations was apparent at Lechaeum in 390BC. There, when attacked on the march by peltasts under Iphicrates, the hoplites could offer no effective countermeasures. They were driven ignominiously from the coastal road into the sea. This imaginative Athenian general developed a new, lighter form of hoplite, suitable for more flexible battle conditions. He carried a longer spear, a smaller shield and wore lighter armour. These improvements may have influenced Philip of Macedon, who was hostage in Iphicrates' hands as a young man. Backward and remote, Macedonia had no tradition of hoplite fighting, but Philip succeeded in creating a new force of phalangite pikemen.

These carried a *sarissa* or long pike 15ft (4.5m) in length, a small buckler, and wore linen armour. By perfecting a deep phalanx, excellently drilled, Philip led his army, in only two decades, from petty victories over Illyrian barbarians to the conquest of all Greece (358–338BC). At Chaeronea, his phalanx, with the support of cavalry led by his son, Alexander, defeated an opposing hoplite army with ease. Having united the Greeks under his authority, Philip declared war on Persia; but he was assassinated and the task fell to Alexander.

Alexander's conquests and the Hellenistic world

Alexander's plan was to lead a small force of 40,000 men into the heart of the huge Persian empire. He had a firm grasp of strategy and logistics, and even arranged "forward dumps" on his intended line of march, using threats and diplomacy. He also attacked the bases of the Persian fleet, reducing its effectiveness against his own naval

Alexander the Great (356–323BC), King of Macedon, was born in a small and backward kingdom. He ended his life at Babylon, 33 years later, the ruler of a vast empire. This brief career was characterized by great personal bravery, combined with a sensitivity to the tactical opportunity that gave him continual victory.

Alexander's determination to carry through any task was implacable, whether he was reducing "impregnable" Tyre by land and sea (332), or pursuing Darius, or extending his conquests throughout the known world.

After his interview with the Egyptian priests at Siwah he believed himself divine, and his small force of devoted Macedonian pikemen and cavalry certainly treated him like a god. When they finally mutinied in the Punjab, after eight years and thousands of miles, he was deeply wounded; but he never lost faith in himself.

Alexander could massacre and enslave, as with the inhabitants of Thebes in 336; with equal single-mindedness he could strive to unite in brotherhood the peoples of Macedonia and Persia. Influenced by Aristotle (his tutor for three years), he was committed to the diffusion of Hellenistic civilization up to and beyond the boundaries of his extensive conquests, and left an enduring legacy.

In life, as in later repute, Alexander had a charisma that made him formidable in politics no less than in battle. He died at the height of his powers. For him there was no anticlimax.

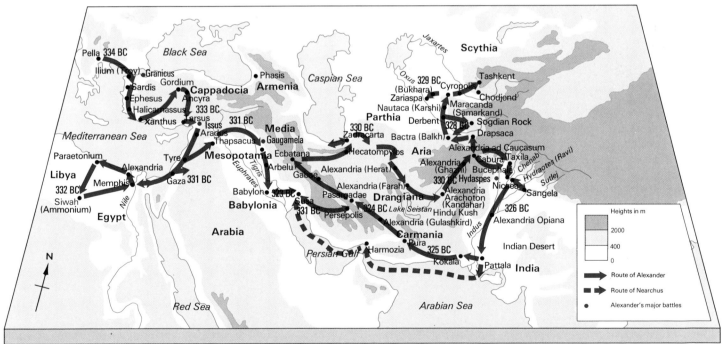

Alexander's conquests and empire. In a decade, Alexander conquered far beyond the boundaries of the world known to the Greeks. Within three years he had reached, and burned, the Persian capital of Persepolis – revenge for the destruction of Athens 150 years earlier.

He was drawn farther east to eliminate the last spark of Persian resistance, and to pursue a destiny which led him to victory on the Indus. At first Macedonian in composition, but increasingly heterogeneous as time went on, Alexander's armies triumphed over great physical barriers: the vast plains of Iran, the precipices of the Hindu Kush, and the arid wastelands of the Gedrosian desert (which almost destroyed them) on the long road back to Babylon – and Alexander's early death.

19

forces. In battle he proved a master. At the Granicus (334 BC), he forced a defended river crossing at the head of his Companion cavalry. The Persian cavalry were routed and their Greek mercenaries all killed or enslaved as traitors. At Issus (333), with his line of communications momentarily cut by Darius, the phalanx fought well, over difficult terrain, and Alexander's cavalry charge again won the day. In the final confrontation at Gaugamela (331), chariots and overwhelming numbers made no impact against Macedonian professionalism.

Irresistible in the field, no fortifications could defy him. He took "impregnable" Tyre by assault from land and sea. He led his army to the frontier of India, meeting and defeating elephants for the first time at the River Hydaspes in 326 (see p. 56). On his sudden death at Babylon in 323, at only 33 years old, he left no heir, and his "Successors" included many able generals who fought over the spoils.

Antigonus held Asia Minor and Syria, Ptolemy Egypt, and Seleucus, the eastern dominions of the empire (and each founded a dynasty named after them). Armies were still based on the phalanx, whose equipment gradually became heavier, and the formation more unwieldy. At sea too, larger vessels – quadriremes and quinqueremes – now ruled the waves. Antigonus used the advantage of his central position to try and unite the old empire. A confederation of other kings opposed him. Confident of his superiority in cavalry, he offered battle at Ipsus in 301BC. But he died amidst his shattered phalanx while the cavalry, under his dashing son Demetrius, took no part. They were too terrified by the strange sight, sounds and smells of the mighty elephants of Seleucus!

Third-century armies became increasingly heterogeneous and dependent on mercenaries. The opportunist actions of their generals are epitomized by Demetrius who, deprived of his inheritance, became a freebooter and raised fleets and armies to suit his immediate goal. His siege of Rhodes in 305BC was typical of the increased scale of warfare. He built a huge siege tower 130–160ft (45–50m) high in addition to other weapons. But he could not live up to his name (which was "Poliorkētēs" or city-taker) and the Rhodians erected their Colossus out of the proceeds of his discarded equipment.

Pyrrhus of Epirus was in the same mould: and it is his exploits which link the Greco-Macedonian style of warfare with new developments farther west.

The emergence of Rome

Sixth-century Rome fielded a hoplite army similar to the Greeks. There were five types of warrior ranging from the wealthy and heavily equipped to the slingers. The only cavalry was provided by a small number of nobles.

The Gallic invasion of c.390BC, which swept into the city, was associated with a reform of the military structure. This was the work of Camillus, who created the legion, a far more flexible battlefield instrument than the phalanx. There were three types of soldier: a first line of *hastati*, second of *principes*, both armed with heavy javelins (*pila*) and a third of spear-armed *triarii*. There were 10 maniples (*manipuli*) each of 2 centuries of each type, 120–160 strong for the first 2 lines and half that for the reserve. Together with an accompanying force of *velites* as skirmishers, a legion consisted of about 4,000 men. It was deployed in a

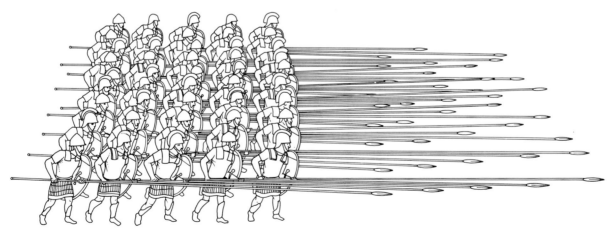

The Macedonian phalangist (*above*) was more lightly equipped than the hoplite, and carried a smaller shield. The *syntagma*, a unit of 256 men, 16 deep, was the basic tactical unit, and one which gave the phalanx great flexibility.

When attacked by stone-laden wagons in the Thracian mountains (335), or by Persian chariots at Gaugamela (331), the phalanx possessed the training and

discipline to make lanes between its ranks, through which the hostile forces passed harmlessly. At Chaeronea (338), the elite body of phalangites known as the Hypastists drew the Greeks from a defensive position and returned to the attack. The normal offensive formation was in echelon (units following each other), with cavalry and light infantry support to flank and rear.

A typical Republican legionary (*left*) of the 1st century BC, depicted on the Altar of Ahenobarbus. He wears a bronze helmet, mail tunic and sword, and carries an oval shield and javelins. This equipment was relatively cheap, extremely practical and ideally suited to the basic Roman tactic. This generally consisted of hurling javelins, followed by a charge with sword and shield.

checkerboard formation which allowed for a system of reliefs during combat. An army consisted of four legions with an equal number of allied troops from Italian cities, who provided much needed cavalry. The one weakness in the Roman system was the crucial one of command. Consuls, annually elected military amateurs, led in the field – often with disastrous consequences.

The legions are tested

Pyrrhus led his army of 25,000 men and 200 elephants into Italy at the request of the citizens of Tarentum, under pressure from Roman expansion. He was impressed by the discipline of the Roman infantry who soon showed their mettle in battle. At Heraclea in 280BC, the legions threatened to overwhelm the phalanx and were only defeated because Pyrrhus' small force of elephants panicked their allied cavalry, leaving the footmen exposed to flank attacks. The Romans suffered heavily but recovered to offer battle at Asculum. Here, on rough terrain unsuitable for elephants, they were eventually driven back into their camp. After these costly victories, the Greek tried his hand against the Carthaginians in Sicily. Here he achieved striking success, but quarrelled in victory with his employers and returned to Italy in 275. This time the Romans were ready, and at Beneventum managed to turn the elephants back into the enemy ranks. Pyrrhus' adventure was over, and the legions had been tested against one of the foremost generals of the age.

Rome against Carthage

Although in alliance during the invasion of Pyrrhus, it was inevitable that the powers of Rome and Carthage would eventually clash. Carthage was a mercantile state, seeking constantly expanding trade routes; Rome a land-based aggressor. Spheres of influence in the Mediterranean littoral overlapped in Sicily, Spain and Provence. Open war began in 264BC, resulting in Rome winning control of Sicily by 241. This was only possible through the creation of a fleet with which to challenge the Carthaginians at sea. The Romans constructed heavier vessels and fought actions which enabled them to board the enemy, rather than relying on manoeuvrability and ramming.

This defeat resulted in internal ructions at Carthage. Their land forces were largely mercenary. There was a body of citizen spearmen and cavalry, but Numidians, Gauls, Greeks and Spaniards fought for pay – and when they were not paid tried to seize power. Eventually they were overcome and Hamilcar Barca embarked on the conquest of Spain. On his death in 229BC he was succeeded by another talented general Hasdrubal. When Hasdrubal was assassinated in 221, Hamilcar's son took command. His name was Hannibal.

Only 26 years old at the time, Hannibal's brilliance almost brought Rome to her knees. Realizing that the Romans were winning the struggle in Spain, he conceived a bold stroke – the invasion of Italy. Marching through southern Gaul and across the Alps, a feat the Romans had not thought possible, he won a string of victories with a very mixed army. First he won a winter battle on the River Trebia, chiefly through a superiority in cavalry. His famous elephants were scattered by the skirmishing attack of the *velites*. The following year he achieved an army-sized

Ships played an important role in the struggle between Rome and Carthage, especially in the First Punic War (264–241BC). Carthage was the centre of a trade empire, and her twin harbour held more than 200 vessels. The reconstruction (*right*), based on recent archaeological work, shows galley pens from which the warships are emerging. However, the Romans developed boarding tactics (*above*) in the naval battles of Mylae, Encomus and the Aegatian Islands which enabled them to fight on sea as on land. Their ships carried a bridge on the bow, known as a *corvus* (raven), that was used for dropping onto an enemy vessel. They also carried artillery, both the crossbow-like *ballista* and the *onager* (literally "wild ass") shown above. Wooden castles, painted to look like stone, provided vantage points for missile men.

21

ambush at Lake Trasimene, destroying two legions. Hannibal then marched swiftly into southern Italy, where he hoped to raise opposition to Rome, but the results were disappointing. He had constant problems in recruiting to replace his losses.

Nevertheless, at Cannae in 216BC he achieved his greatest victory. By forming his centre of lightly armed Gauls, who were steadily pushed back by the legionaries, he drew the Romans into a trap. Enveloping their flanks and rear he surrounded and annihilated four legions. It was one of Rome's greatest defeats.

With his small forces Hannibal proved unable to take Rome, and in face of these defeats his enemy changed tactics. Under the direction of Fabius Maximus as Dictator, they now pursued a policy of avoiding battle in Italy while continuing the conquest of Spain. Hannibal found himself isolated and strategically outmanoeuvred. When his brother Hasdrubal crossed the Alps with reinforcements, he was defeated and killed on the Metaurus in 207.

The Roman general Scipio, who had conquered Spain by 206, proved a worthy opponent to Hannibal. Hoisting the Carthaginians on their own petard he led an army into Africa in 204. Hannibal was forced to return and in 202 the decisive battle was fought at Zama. This time the Romans had a cavalry superiority, provided by Carthage's erstwhile ally, the Numidian king, Massinissa. While the legions opened lanes in the ranks, as they had been trained to do, down which the elephants charged uselessly, their allied cavalry drove off the opposition. Returning, they completed the encirclement of the Carthaginian foot. Scipio won his title "Africanus"; Hannibal escaped; but it

Hannibal (247–182BC) was a fine general who won battles for the Carthaginians but lost the war. In the autumn of 218 he crossed the Alps in just 15 days with men, horses and elephants, descending on the amazed Romans to win 3 crushing victories within 6 months. Marching into southern Italy he went on to surround and overwhelm an army of 8 legions at Cannae in 216. These achievements illustrate Hannibal's great qualities as a general: mastery of logistics, combined with strategic surprise and tactical genius, to defeat superior enemy

forces. For his armies – Gauls, Spaniards and a few native Carthaginians – lacked the training and the rigid discipline of the legions, and were held together by his successes and leadership qualities.

When he invaded Italy, Carthage was losing the war at sea and in Spain. The purpose of the invasion was to raise troops from Rome's unwilling allies, and seemed to be succeeding while he had opposing armies to beat. But when Fabius Maximus resorted to a battle-avoiding strategy, leading to 10 years of attrition, he could sustain the contest no longer. Eventually recalled to Africa to counter Scipio's invasion, he was comprehensively defeated at Zama (202). The remaining two decades of his life were spent as a mercenary for Hellenistic rulers. In 183 the Romans demanded his extradition from the Seleucid capital, and Hannibal committed suicide to avoid capture.

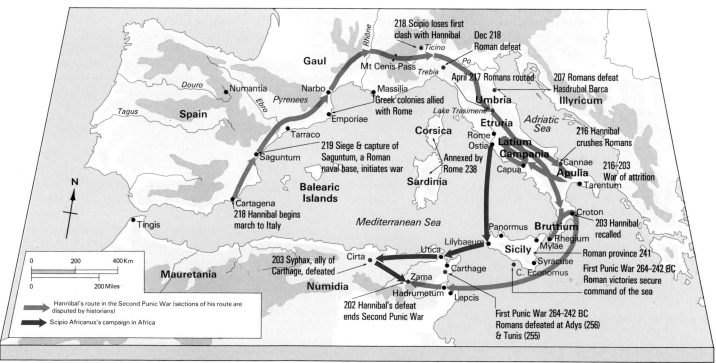

The struggle between **Rome and Carthage** began with the First Punic War (264–241BC), fought over Sicily. Rome, previously a land power, needed to win battles at sea, and achieved this with the construction of a large fleet of powerful craft.

The Second Punic War (218–201BC) opened with Hannibal's capture of Saguntum. His wholly unexpected crossing of the Alps in the autumn, and his tactical brilliance, brought him a string of victories in Italy, but he could not achieve his main aim of detaching Rome's allies. Avoiding battle in Italy, the Romans under Scipio recovered Spain and defeated Hasdrubal's relieving force.

When Scipio crossed to Africa, Hannibal was forced to return home. The Roman victory at Zama (202) was decisive.

was the end. Carthage surrendered and became a tributary state, allowed to linger on until she was vindictively destroyed on a legal pretext half a century later.

Legion and pilum against phalanx and sarissa

Meanwhile an opportunity for Roman expansion had opened up in the Hellenistic world. Ptolemaic Egypt had made gains along the coast of Asia Minor and Syria during the 3rd century, but on the death of Ptolemy IV, in 203BC, fell prey to the combined ambitions of Philip V of Macedon and Antiochus III, the Seleucid king. The independent states of Rhodes and Pergamum appealed to Rome for protection. The Senate already had a grudge against Philip for supporting Hannibal, and agreed to send its legions into Greece.

In 197 at Cynoscephalae, operating in broken ground, the flexible Roman formation outflanked the cumbersome phalanx. Soon it was the turn of Antiochus, whose forces were defeated at Thermopylae in 191. At the request of the Pergamenes, Roman forces carried the war into Asia. At Magnesia in 190, the Seleucid army was also overthrown. The Romans experienced an early cavalry reverse, but the tactic of stationing elephants in gaps in the phalanx proved disastrous. The legionaries knew how to terrify them into stampeding through the enemy lines, leaving the phalangites as easy meat for their swords.

The superiority of Roman tactics was exemplified against Perseus of Macedon at Pydna in 168. The heavy javelins of the Romans disrupted the phalanx's formation, allowing the legionary swordsmen to infiltrate enemy ranks. Hampered by their long weapons, the Greeks had no effective counter.

On a strategic level Rome was assuring her frontiers. Blockade by sea finally brought down Carthage in 146BC. In 133, the Romans were equally ruthless in Spain, starving Numantia into surrender by surrounding it with fortifications. The Jugurthine War 112–106 rounded off Rome's conquests in Africa by absorbing the Numidian kingdom. But Roman complacency following such a run of successes was rudely shaken by Germanic folk movements.

Marius and his army reforms

These migratory Cimbri and Teutones tribes inflicted a defeat on the Romans in southern Gaul in 112BC, and defied all attempts to expel them. In 105, they crushed another Roman army. The Senate turned to Marius, victor over Jugurtha, to defeat these fierce barbarians. The career of this man of ignoble birth was to set the scene for the political revolution in Rome of the following century.

This was no less a military revolution. The old, 3-line organization was abandoned in favour of 10 cohorts, each about 480 strong, uniformly armed in helmet, mail shirt, shield, sword and 2 *pila* (one light, one heavy). The 80-man century became the main tactical unit, and the centurions who commanded them were the backbone of the legions. Each man was expected to carry all his additional cooking and entrenching equipment, earning them the name of "Marius' Mules". The new army was toughened up through marches and camp construction, and, only when Marius considered it ready, led against the enemy. Their wild charge had overwhelmed many opponents over centuries; but now Rome had the answer. The *pila* were designed with flexible heads, which stuck in the

---

THE ROMAN CENTURION

Centurions formed the backbone of Rome's legions. Her rise to empire depended upon their courage and professionalism, as this declaration by a 2nd century BC veteran shows:

"I joined the army in the consulship of Publius Sulpicius and Gaius Aurelius [200BC] and I served two years in the ranks in the army which was taken across Macedonia, in the campaign against King Philip [of Macedonia]. In the third year Quinctus Flaminius promoted me, for my bravery, centurion of the tenth maniple of hastati. After . . . we had been brought back to Italy and demobilized, I immediately left for Spain as a volunteer with the consul Marcius Porcius. Of all living generals none has been a keener observer and judge of bravery than he . . . This general judged me worthy to be appointed centurion of the first century of hastati. I enlisted for the third time, again as a volunteer, in the army sent against the Aetolians and King Antiochus [191–188BC]; Marcus Acilius appointed me centurion of the first century of the principes . . . Twice after that I took part in campaigns in which the legions served for a year. Thereafter I saw two campaigns in Spain . . . Four times in the course of a few years I held the rank of chief centurion; thirty-four times I was rewarded for bravery by the generals; I have been given six civic crowns [for saving the life of a Roman citizen in battle]. I have completed twenty-two years of service in the army and I am now over fifty years old."

Livy, The History of Rome from its Foundation XLII 34

There were 59 centurions to each Roman legion. All except the 5 in the First Cohort commanded "centuries" of 80 men, consisting of *milites gregarii* (common soldiers), plus the *optio* (second-in-command) and *tessarius* (sergeant of the guard). Centurions rose through the ranks, being promoted by seniority and valour up the legion's command structure, which ran from the most junior centurion, *hastatus posterior* of the Tenth Cohort, to the *primus pilus* commanding the first century of the double-strength First Cohort. This would be an experienced veteran on whose advice the commanding officer or legate could rely.

The centurion's insignia included a vine staff (for keeping order), a transverse crest on the helmet, and large silver medals called *phalerae*, worn on a harness across the chest.

tribesmen's shields, dragging them from their hands. The legionaries followed up with a charge, pushing with their heavy shields and jabbing at the unprotected bodies of their opponents.

Marius defeated the tribes piecemeal, slaughtering the warriors and their families through a combination of ambush and hard fighting: the Teutones in 102 and the Cimbri the following year. In order to enable him to carry out his reforms and conduct his campaigns, he had been elected consul on six successive occasions. Such expedient crisis-management changed the nature of the army. The legions' loyalty was no longer to Rome – it had been transferred to their generals. This was apparent throughout the following century.

First, the pool of manpower available for recruitment increased dramatically. In 94BC, Italians living in Rome were granted citizenship. From 91–89 the Social War was fought over the issue of Italian allies winning this right – which made them eligible to serve in the legions. In the next year, Sulla, commanding in the Eastern provinces raised an army to fight Mithridates of Pontus. This also gave him power in Rome and the consulship. He cleared the motley Pontic forces out of Greece in two campaigns, the first in 87BC, the second in 85. Meanwhile, Lucullus raised a fleet, and in alliance with the professional Rhodians, won two naval battles.

Under the Senate, Rome had been satisfied with spheres of influence; now, ambitious individuals desired conquest. Sulla's protégé was Pompey. He made his name in Spain mopping up the remnants of support for Marius' Popular party. Meanwhile, in Italy, Crassus, equally ambitious,

defeated the slave revolt led by Spartacus, pinning the rebels behind massive entrenchments (73–71). Not to be outdone Pompey set out to destroy the pirates who were interfering with Rome's grain supplies. In a brilliant naval campaign he solved the problem by destroying their bases, but also resettling the survivors (67). He went on to finally defeat Mithridates who had been fighting Lucullus in the 70s. This wily character had reformed his army on the Roman model, which enabled him to maintain the contest over several decades.

The Age of Caesar

No man did as much to expand and define the Roman empire as Julius Caesar. Despite having no military experience when he became consul in 59BC, he seized the opportunity to win territory and glory. In less than a decade he conquered Gaul, invaded Britain and kept the Germans at bay. His campaigns were characterized by a mixture of cunning, determination and great personal bravery. He made the Roman legion an invincible tool of victory, drawing on all its resources of battlefield effectiveness and engineering skills.

He began, like Marius, by defeating migrating tribesmen, the Helvetii, with only five understrength legions. First he hemmed them in behind extensive earthworks, across their route at Lake Geneva; then he crushed them in battle. He followed up with an assault on the cause of the migration, the fearsome Germans of Ariovistus. Despite their size and reputation, legionary discipline overcame barbarian enthusiasm (58). Next Caesar defeated the Belgae and survived an ambush by their fiercest tribe, the

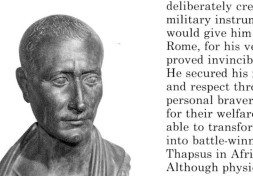

Julius Caesar (100–44BC) was already in his forties when he first took an army command. He is said to have wept when comparing his achievements with those of Alexander, but nevertheless he was to match the conquests of his hero in a whirlwind military career.

Caesar's campaigns were distinguished by their speed of movement and tenacity to the objective, and he quite consciously set out to be the first to cross the Rhine and to invade Britain. In his conquest of Gaul he

deliberately created the military instrument that would give him power in Rome, for his veteran legions proved invincible in the field. He secured his men's loyalty and respect through his personal bravery and concern for their welfare, and was able to transform raw legions into battle-winners, as at Thapsus in Africa (46BC). Although physically frail, he toughened himself, using "warfare as a tonic for his health" (Plutarch). He was also a great innovator and adaptor, shipping his troops across a Spanish river using coracles copied from those of the Britons.

Caesar fully deserved his success in warfare. He was capable of recovering from defeat swiftly, during the Gallic revolt of 52 and after Dyrrhachium in 48. And he always exploited his success to the maximum. He was assiduous in creating his own legend, but his talents justified it.

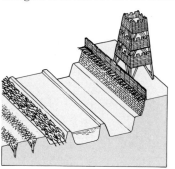

The showdown between the Romans and the Gauls, led by Vercingetorix (52BC) took place at the vast fortified plateau of Alesia in Burgundy. Caesar had just been repulsed at the siege of Gergovia and was determined to make no mistake. The Gauls were kept in with 11 miles of circumvallation, including ramparts, ditches and a 30-yard depth of mantraps, constructed by his 10 skilled and veteran legions. A similar fortification, 15 miles long, was built facing outwards when relieving forces endangered the Romans. The Gauls' desperate attacks

failed to break through, and Vercingetorix was forced to surrender. The operation emphasizes the importance of legionaries as builders as well as fighters.

Nervii. Only his inspiration and the steadiness of his veteran Tenth legion won the day (57). In 56 he won a naval victory over the larger vessels of the Veneti. In 55, he led a force across the Rhine to pre-empt a German attack. This required the construction of a bridge, which was an amazing feat of engineering, achieved in only ten days. Bolder still, he took ship with two legions to invade Britain. Dissatisfied with the result of this first expedition, he returned with five legions in 54, defeating the Britons

---

"As our soldiers were hesitating, mostly because of the depth of the water, the man who carried the eagle [standard] of the Tenth legion, after praying to the gods that his act would bring good luck to the legion, shouted out loudly: 'Jump down men, unless you want to betray your eagle to the enemy. I at any rate shall have done my duty to my country and my general!'

With these words he flung himself from the ship and began to carry the eagle towards the enemy. Then the soldiers jumped down from the ship all together, urging each other not to allow a disgrace like that to happen . . ."

Caesar's description of amphibious assault on Britain, 55BC

---

under Cassivellaunus.

Meanwhile, Gaul rose in revolt, inspired by the leadership of Vercingetorix. In a war of sieges, Caesar eventually pinned his opponent on the great hill of Alesia in Burgundy (52). Surrounding it with fortifications, he was forced to construct a similar line facing outwards, in order to beat off a relieving force. With the town's eventual surrender,

opposition was at an end.

Inevitably, these successes brought him into conflict with Pompey. In a lightning campaign he cleared Spain of Pompey's adherents (49), then turned to Illyricum to challenge his great rival. At Dyrrhachium in 48, there was an episode of trench warfare worthy of World War I, which resulted in a draw. Then Pompey gained the upper hand by cutting off Caesar's grain supply, but allowed himself to be drawn into battle. At Pharsalus, the veterans of Gaul made short work of their opponents. In 47, Caesar conquered Egypt; in 46 he consolidated northern Africa by defeating the last opposition to him at Thapsus. He returned to Rome with the intention of leading an expedition to the East, but was murdered by senatorial enemies, the victim of his own success (44).

The limits of empire

Rome's power had expanded continually for over two and a half centuries, and had still further to go; but by the beginning of the Christian era, the boundaries were beginning to become clearly defined. In the East, the Parthians had proved difficult opponents. Crassus' expedition of 53BC had ended in disaster. He attempted to march to Ctesiphon, the Parthian capital, but his legions were beaten by a combination of horse-archers and heavy, cataphract cavalry. Mark Anthony achieved little better in 36, just extracting his force by the skin of his teeth. He had been Octavian's ally at Philippi in 42, when Brutus and Cassius' attempt to restore the Republic had been crushed. He was to oppose Caesar's nephew at the sea battle of Actium in 31. Cleopatra's fleet of heavier ships was

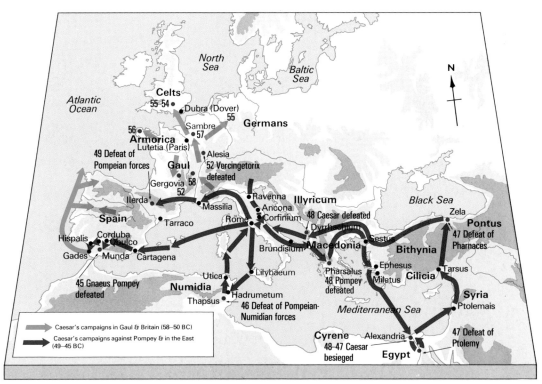

Julius Caesar's campaigns and conquests. A member with Pompey and Crassus of the First Triumvirate (Three-man rule) in 63BC, Caesar's conquest of Gaul, followed by the death of Crassus at Carrhae (53), led him to a showdown with Pompey. Despite initial setbacks he emerged the victor at Pharsalus (48), his rival being murdered while fleeing to Egypt. After campaigns in Egypt, Armenia, Africa and Spain he returned to Rome as undisputed ruler.

outmanoeuvred. Defeated, the lovers commited suicide, and Octavian was left as unchallenged ruler. This was a crucial moment in the development of Rome. Octavian took the title of prince, and the name Augustus. He was commander-in-chief – imperator – and he consolidated the Roman military achievement. He disbanded many of the Civil War legions, reducing them to 28 in number (and keeping one, the Praetorian Guard, with him in Rome). Specialist auxiliary forces providing cavalry, archers, slingers and javelinmen contributed a similar number. This force of 300,000 men served to defend a vast empire for almost three centuries.

Expansion was not quite over. Up until AD9 it seemed as if Germany must be absorbed. Augustus' talented nephews led armies and fleets as far as the Elbe in the last years of the first century BC. But the disaster of that year, when the Germans under Arminius destroyed three legions in the Teutoberger Wald, shocked the ageing emperor. The Romans were revenged in AD16, only to withdraw to the Rhine. Together with the Danube, these rivers delineated the northern frontier.

Rule and revolt

Despite the unsavoury reputation of some of her emperors, the legions of Rome were supreme. In AD42, Claudius invaded Britain, a producer of minerals and grain. It took just four legions to conquer most of the country by 47. The only serious opposition came from Caratacus, who conducted a guerrilla campaign in Wales. Harsh rule provoked a revolt in 60, led by the warrior queen Boudicca of the Iceni. Although Verulamium (St Albans) and London were burnt

A senior centurion in full regalia depicted on a 1st-century tombstone, one of the many soldiers who died when three legions under Varus were ambushed in AD9 by the German chief Arminius in the Teutoburger forest. He is wearing *phalerae* and a laurel wreath, and is carrying his vine staff. The inscription reads:

"Marcus Caelius, son of Titus, of the Lemonian tribe, born in Bolognia, Centurion of the first rank in the XVIIIth Legion. Aged 53 he fell in the Varian war. The bones of his freedmen are entitled to share his tomb. This monument was erected by his brother."

down, once the Britons were brought to battle they were utterly routed.

More serious was the Jewish revolt of 66. This, together with the collapse of Rome's grain supply, provoked a challenge to the Emperor Nero. First Galba, governor of Spain, seized power with his legions, but was murdered by Otho. Then Vitellius, commander on the Rhine, ousted him. Meanwhile, Vespasian, veteran of the British campaign, had been dealing harshly with the Jews. He marched his battle-tested legions back into Italy, winning the

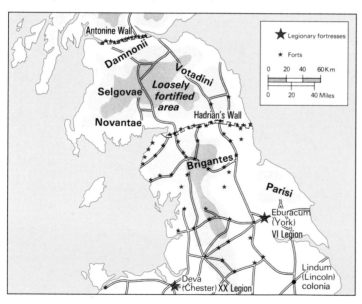

The emperor Hadrian's policy of consolidation included building walls "to separate the Romans from the barbarians." In northern Britain, a line of fortification stretching from the Solway to the Tyne still bears his name. This was mainly constructed between AD122–128 and supplemented in the following reign by the

Antonine Wall, running from the Forth to the Clyde estuaries. Between these lay a series of forts.

Hadrian's Wall became the permanent frontier, despite being breached on several occasions. It was rebuilt and re-garrisoned in the early 2nd, early 3rd and late 4th centuries, to be finally abandoned in the early 5th.

Standards were essential to the morale of the Roman army, and were treated with the reverence of holy objects. The reconstruction (*left*) of an early 3rd century auxiliary standard bearer comes from a tombstone at Carrawburgh on Hadrian's Wall.

Hadrian's Wall (*right*) stretched for 76 Roman miles, and the skilled masonry work was performed by the legionaries who left inscriptions at regular intervals. The western half (32 miles/50km) was made of a turf rampart 13ft (4m) high, whereas the remaining 44 miles (70km) consisted of a 15ft (4·5m) stone wall topped by a 6ft (1·8m) parapet. Besides milecastles and watchtowers, it contained 16 large forts.

Details of the soldiers' lives have been discovered from a mass of writing tablets discovered at Vindolanda, including the fact that they wore socks in the bitter British weather.

imperial dignity through his victory at Cremona. Thus AD69 was the year of the four emperors. Vespasian's son Titus captured Jerusalem in the following year, sacking the city, despoiling the Temple, and dispersing the Jews.

Defining the frontiers

In Titus' brief reign (79–81), and that of his brother, Domitian (murdered in 96), the Danubian frontier demanded attention. A strong state in Dacia (modern Hungary) and the raids of the nomadic Sarmatians, drew the ambition of the soldier-emperor Trajan (96–117). He mustered 10 legions for his invasion force, with an equal number of auxiliaries, the Danube flotilla and a mass of transport craft. This force can hardly have amounted to less than 100,000 men.

Unable to face him in the field, the Dacian king, Decebalus, relied on a fortress strategy for defence. But the Romans won a war of attrition, supported by their fleets (105–6). Trajan went on to campaign successfully in the east, conquering Parthia's western provinces (115–116). But he died the following year, and his successor, Hadrian, decided on retrenchment.

The *limes* or border zones were given stronger defences, often in stone. This was true on the Rhine, the Danube and in North Africa, with the construction of blockhouses, and in northern Britain Hadrian even built a wall from coast to coast. This secured the advances made by the province's governor, Agricola, between 78–84. Using a fleet to accompany his army, he defeated the Picts (of modern Scotland). Yet Rome did not attempt to hold the entire island of

Britain. This was for the same reason that the wilder parts of Germany were untenable: a small population and rugged terrain were inadequate to support Roman forces and Roman civilization. The Antonine Wall of Hadrian's successor, built farther north between the Forth and the Clyde estuaries, was short-lived.

The peace of the first half of the 2nd century was broken during the reign of Marcus Aurelius. Roman arms were successful against Parthia in 162–63, not least because her forces were more diversified than before. Heavy cavalry and horse-archers helped to meet the Easterners on their own terms. German pressure on the Danube was increasing, however, and the Marcomannic Wars demanded more troops to hold the frontier. After the interlude of incompetent Commodus, another soldier-emperor, Septimius Severus, fought a series of successful campaigns from Britain to Mesopotamia.

But the empire was increasingly on the defensive. The Severan dynasty proved unstable and short-lived, as the Praetorian Guard resumed its 1st century role of choosing the emperor. Meanwhile, more formidable enemies had appeared on the frontiers. The Sassanid Persians replaced the Parthians in 226, and sought to recover lands lost in the previous century. The German tribes had benefited from contact with Rome, and had begun to form dangerous, larger, combinations such as the Alemanni (All-Men) or the migratory Goths.

The 3rd and 4th centuries: collapse and recovery

The decline was as swift as it was unexpected. In AD 251,

the Emperor Decius was killed in battle against the Goths, who overran Greece. In 260 Valerian was captured by the Persians. Emperors rose and fell quickly. Secession took place in Britain and Palmyra. Not until 270 did Aurelian reconquer the latter, and he was assassinated in 275. The first man to bring about stability was Diocletian (284–305), who, realizing that the times required a commander on the spot, appointed Maximian as his colleague. Each emperor also had a junior, or Caesar, to aid in the military task. Under Constantine (312–37), who founded a new capital at Byzantium, the division of the empire into East and West was confirmed.

The army changed as much as the state. Instead of the 33 legions of the early 3rd century, stationed on a provincial basis, a new organization was set up. This consisted of a field army (in East and West) with a large cavalry component and elite legions, while the other forces sank to the level of soldier-farmers (*limitani*). The army was doubled in size to around 600,000 men, which put an increased strain on the empire's failing economic resources. Also, as every frontier was now suffering simultaneous attacks, the military demands were becoming overwhelming.

Sassanid Persia was held at bay until the Emperor Julian's expedition of 363. The campaign itself was successful, but it ended with the death of the Emperor in a skirmish. In 378 came a far more serious disaster. Emperor Valens, leading a large army against the Ostrogoths in Thrace, attacked their camp. While his forces were engaged the Gothic cavalry returned from foraging and fell

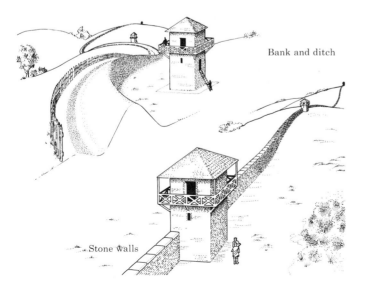

The *limes* (border defences) of Germany and the Danube lands stretched for 2,500 miles (4,000km). They varied from watch towers linked by beacons to bank and ditch or even stone walls. None of this work was on the scale of Hadrian's Wall, but there were large forts, naval bases, bridges and bridgeheads used for counterattacks to the north. Dacia was also part of the empire from c.AD106–275.

Sassanid Persia was for four centuries (c.AD226–650) Rome's greatest rival in the east. Roman armies always had difficulties with the terrain and the Persians' formidable combination of cataphract (armoured cavalry) and horse-archer. In this monumental rock carving from Naqsh-i-Rustam, Shapur I glorifies his victories over two 3rd-century emperors: Philip (249) and Valerian (260).

on his flank, driving the Romans into a helpless huddle. Valens was killed and his men massacred. It was Rome's greatest defeat since Cannae.

The Germanic invasions

The Ostrogoths had been fleeing the Huns, and they drove the Visigoths before them into Greece, and then c. AD400

"Some of our men fell without knowing who struck them down, others were buried beneath the weight of their assailants, and others were slain by the sword of a comrade . . . At the first coming of darkness the emperor, amid the common soldiers, fell mortally wounded by an arrow, and presently breathed his last; and he was never afterwards found anywhere. For since the foe were active in robbing the dead, no one among the fugitives or local inhabitants ventured to approach the spot."

Ammianus' account of Emperor Valens' death at Adrianople, AD378

---

to Italy. In 406, the Rhine froze over, allowing other Germanic tribes to pour into Gaul. The overstretched Roman defences cracked, and Imperial authority never recovered in the West. Rome was sacked by Alaric's Goths in 410, although by then Ravenna was the Emperor's military headquarters. The Alans and Vandals swept into Spain and, in 429, on to North Africa. Jutes, Angles and Saxons began their assault on Britain c. 440. Arthur – if he existed – only delayed their conquest by a generation. The Suevi settled in Spain, the Franks and Burgundians in

northern and southern Gaul respectively. Despite Aetius' victory over the Huns at Troyes in 451, the last Western Emperor, a child, was deposed in 479.

In contrast, the Eastern, Byzantine empire resumed the offensive in the next century. Emperor Justinian (527–65) was ambitious for reconquest, and he had a very talented general in Belisarius. He destroyed the Vandal state in Africa in a single campaign, secured the eastern frontier, and invaded Ostrogothic Italy. It took two decades of warfare to reconquer the country, however, a task completed by Narses when Belisarius fell from favour. In the battle of Taginae, 554, a Byzantine force of archers, spearmen, and dismounted cavalry lancers, defeated Goths relying on their horsemen alone. Such an army was far removed from the traditional Roman infantry and was largely composed of mercenaries.

Justinian's conquests were shortlived. In 568, the Lombards, a nation renowned for their cavalry, swept into Italy and set up a kingdom that was to endure for over a century. A war to the death with Persia that lasted until Heraclius' victory in 628, only left Byzantium prey to the Arabs, newly inspired by the Islamic faith. The Franks rose to be the greatest military power in the West, however. Their Carolingian dynasty established an empire based on military success: repelling the Arabs, conquering and converting the Germans, and absorbing Italy (and with it the Imperial title) in 800. The legions of Rome were long gone in all but name. Cavalry now formed the main strike force of any army; already the knight was riding onto the battlefields of Europe.

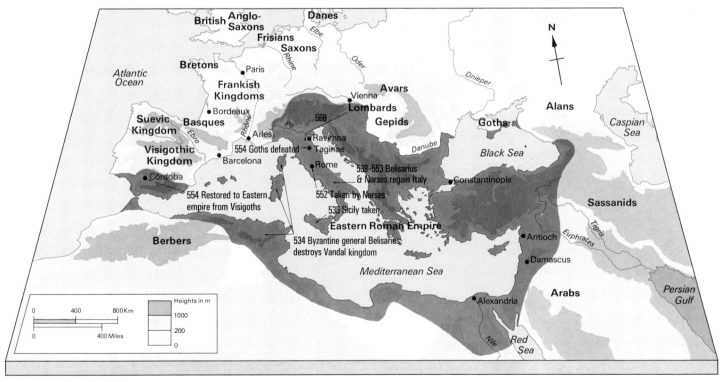

The sixth century saw the empire resurgent for a time under Justinian, with the movements of small but powerful cavalry forces, commanded by able generals, all over the Mediterranean.

This Reconquest was initially very successful in Africa and Spain, but the war in Italy dragged on, only resulting in the destruction of a stable Ostrogothic state and its replacement by the Lombards

in 568. Towards the end of his long reign, financial exhaustion, plague and barbarian attacks undid much of Justinian's achievement.

— chapter three —
Men of iron

Matthew Bennett

At the battle of Poitiers in AD732, a Frankish army defeated an invasion force of Spanish Muslims. This marked the high tide of Islamic expansion in the West. Charles Martel "The Hammer" led the Franks that day. He replaced the last of the Merovingians, the dynasty which had ruled Gaul since the conquests of Clovis c.500. His Carolingian successors, notably Charles the Great – Charlemagne – built a great empire that stretched from the Pyrenees to the Elbe and the upper Danube, and took in most of Italy.

This was only achieved by decades of warfare, mostly under Charlemagne. He was capable of lightning campaigns, such as the conquest of the Lombard kingdom in 774, or the destruction of the Avars in 796. The Carolingian army also sustained more than two decades of attrition in defeating and converting the Saxon tribes. Charlemagne was crowned Emperor of the Romans at Christmas 800, and this was no more than his due. His attempts to revive the administration and learning of the empire was also reflected in the army.

Essentially this was based on the military household of the ruler, expanded into a field force for campaign. Much has been made of the Carolingian use of cavalry; and it is true that armoured retinues of mounted warriors accompanied Charles on campaign, forming the strike weapon in battle. In effect these were knights – the warriors who personify the medieval period. But to overemphasize their role does disservice to the Frankish skill at logistics and siege warfare. The widespread use of the stirrup proved less of a military and social catalyst than some have claimed, for this riding-aid was not much in evidence before the mid-9th century – long after the period of greatest expansion.

Carolingian warriors were indeed men of iron. Cavalry and infantry alike wore mail coats and helmets (although the high cost of war horses already emphasized the

increasing social status of the mounted man). Charlemagne forbade the export of iron armour and weapons, and decreed what equipment his soldiers had to provide for service in the army. The Franks' only real competitors in warfare were the Byzantines, who were just emerging from a dark age of religiously inspired civil war, onslaughts from Arabs in the south, and attacks from Slavs and Bulgars in the north. The upstart Westerners proved too powerful for them, establishing Frankish authority in southern Italy and Moravia.

The Viking Age

When Charlemagne died in 814, Scandinavian pirates had already begun raiding along northern coasts. His empire was held together by his son, Louis the Pious, but inheritance disputes amongst his grandchildren weakened resistance to attack. A division into separate kingdoms took place in 843 and again, after a brief reunification, in 888. In 885–6 Paris was besieged by a Viking fleet, and only survived owing to the warlike energies of its bishop.

This was typical of what was happening: the empire and its successor kingdoms lacked the organization to defend their territories. Increasingly the responsibility fell upon leaders at a local level. A strong man with followers and a fortification – what became known as a castle – exercised effective authority. Levies of mounted warriors made up

Well-equipped horsemen such as these provided the punch for Carolingian field armies (adopting stirrups in the mid-9th century). They manoeuvred in units with banners at their head. But they formed only a small elite, the sworn followers of counts, dukes and kings. This was because horse, armour and weapons were so expensive. A mounted man needed the income from a landed estate of at least 400 acres to afford them.

Charlemagne (AD742–814). Only the greatest generals enter legend; Charlemagne's exploits became the background to some of the most popular stories in the Middle Ages. Yet the most famous poem concerning him, the *Chanson de Roland*, records one of his rare defeats, suffered in 778 after an unsuccessful expedition into Spain.

The Spanish frontier was the only one where the borders of the Frankish empire were not expanded during Charlemagne's rule. In 773–4 he conquered the Lombard kingdom of north

and central Italy in a single campaign; towards the end of his career as a field commander, in 796, he destroyed the Avar state in Hungary through a combination of diplomacy and offensive strategy. Having arranged for the Bulgar khan to attack in the east, he simultaneously mounted an offensive on several fronts, along parallel lines of approach. Piercing the earthworks surrounding the Avar territory, he drove to its centre, and sacked the capital where lay two centuries of booty.

By contrast, he directed a war of attrition against the Saxons, who were eventually converted to Christianity. Among his achievements was the construction of a pontoon bridge over the Rhine. In Hungary he emulated Trajan, in Germany Caesar. Charles was indeed worthy of the imperial title.

—TIME CHART—

from c.790	Viking raids against Western Europe and Britain
by c.870	Vikings reach Black Sea
911	Nucleus of Norman state established
from c.1000	Christian reconquest of Spain begins
1066	Hastings: Norman spearmen, bowmen and heavy cavalry defeat Anglo-Saxon axemen
c.1030–1090	Norman conquest of Sicily and southern Italy
from c.1050	Crossbows in use in Western Europe
1096–1291	Crusades: West European military expeditions to the Middle East
1097	Dorylaeum: Crusader heavy cavalry smashes Seljuk horse-archers
1098–9	Crusaders take Antioch and Jerusalem
from 1100	Construction of Crusader castles in the Levant
1106	Tinchebrai: dismounted knights prove effective against cavalry charges
1169	Ayyubid dynasty founded under Saladin
1195	Alarcos: resurgent Muslims defeat Castilians
from c.1200	Plate replaces mail armour for heavy cavalry
1212	Las Navas de Tolosa: Castile wins control of central Spain from Muslims
1282	Construction of concentric-style castles initiated by Edward I in Wales
1298	Falkirk: English longbowmen decimate Scottish pikemen
from 1320s	Gunpowder weapons in use in Europe
c.1335	Longbow adopted as English infantry weapon
1340	Sluys: archery barrage secures English victory
1346	Crécy: longbowmen slaughter mounted knights
1415	Agincourt: longbowmen slaughter dismounted knights
1419–36	Hussite Wars: offensive-defensive *wagenburg* (wagon-fort) tactics defeat armoured cavalry
by 1450	Swiss pike tactics fully developed
1525	Pavia: arquebusiers repulse armoured cavalry

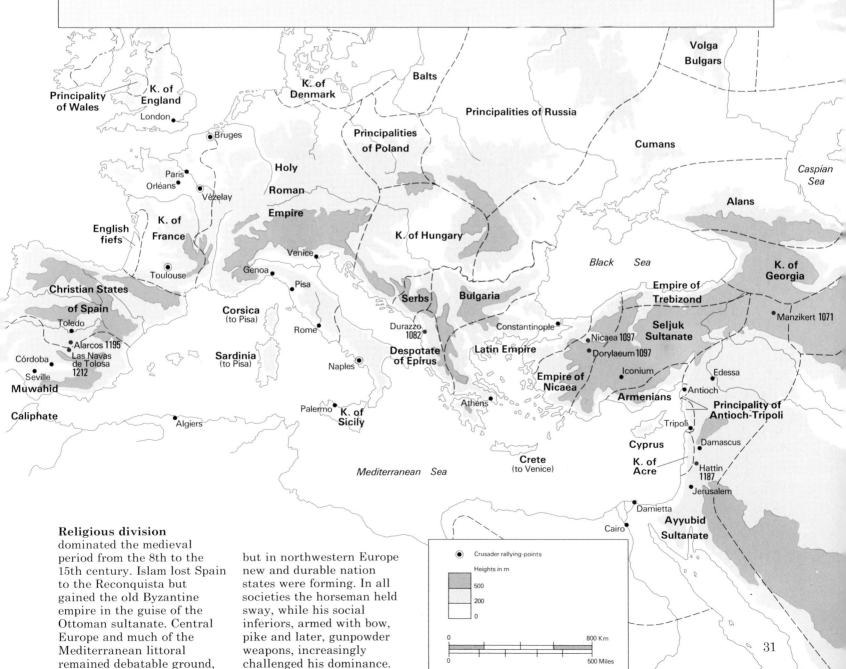

Religious division dominated the medieval period from the 8th to the 15th century. Islam lost Spain to the Reconquista but gained the old Byzantine empire in the guise of the Ottoman sultanate. Central Europe and much of the Mediterranean littoral remained debatable ground, but in northwestern Europe new and durable nation states were forming. In all societies the horseman held sway, while his social inferiors, armed with bow, pike and later, gunpowder weapons, increasingly challenged his dominance.

● Crusader rallying-points

Heights in m
500
200
0

0   800 Km
0   500 Miles

31

the feudal host, so called because these knights served in return for their *feudum* or fief. This fragmentation of authority, and its closer linking to military power, was a gradual but significant process that continued throughout the 10th and into the 11th century.

Viking invasion and settlement was just one example of the political disruption of the 10th century. In 911, Rolf, a Dane, established a state in northern France that was to become Normandy, the leading feudal principality of the following century. The counts of Anjou, Aquitaine, Flanders and elsewhere also established their effective independence from the French crown, maintaining their authority with knights and castles.

In England, the Viking impact resulted in unification rather than division. By 800 the kingdom of Mercia was the dominant force of the old Heptarchy of Anglo-Saxon Kingdoms, and its king, Offa, treated on equal terms with Charlemagne. In the 9th century Wessex rose to supremacy. Then came the Viking onslaught. This was slow at first, the crucial year being 865 when the Great Army came over from France for rich pickings. East Anglia was overrun and Northumbria collapsed two years later. Mercia succumbed in 874, and only the kingdom of Wessex stood against the invader.

King Alfred had inflicted defeats on the Vikings in 870–71, but was taken by surprise at Christmas 878. Attacked and driven into the marshes of the southwest, Alfred engineered a remarkable recovery after some hard fighting. By arranging a partition which gave the northern two-thirds of England over to the invaders, he won a breathing space. This resulted in a truce of 14 years, during which he reorganized the army so that parts of the host were on call all year round, and built an effective fleet to challenge the "sea-wolves" in their own element. He also arranged the construction of *burhs* – communal fortifications – which restricted the Vikings' lightning raids and denied them bases. These measures proved effective – when the Great Army came again in the 890s it was signally defeated. In the 10th century the kings of Wessex went onto the offensive, and England up to the Humber was gradually brought under their sway. The last Viking king of York was killed in 955. The Danes were to attack again, but the first wave had been defeated and absorbed.

Turning the tide

The kingdom of the East Franks proved similarly successful during the invasion period, and victory over the Vikings at the Dyle in 891 considerably reduced the Scandinavian threat. Meanwhile, the nomadic Magyars were also tackled and beaten. The method of defence was similar to that in England: development of fortified towns, defended by their citizenry (who also provided infantry on campaigns) and a strike force of heavily armoured men. But, as elsewhere on the continent, this meant cavalry, whereas England still retained the ancient Germanic and Scandinavian custom of fighting on foot. In East as in West Francia, the means of raising such troops became increasingly feudal. A document of 981 lists more than 2,000 mailed cavalrymen available from the empire's greatest lay and ecclesiastical magnates. For Otto I had been crowned at Rome in 955, and the Holy Roman Empire, based on a German and Italian axis, was to be a major

The strength of the **Vikings** lay in their mobility. Their longships sailed across oceans and rowed up rivers into the heart of the territories they desired to sack or settle. Once established on land, they often took to horse to enable them to campaign, or merely raid, more swiftly. The kingdoms of early medieval Europe were ill-equipped to deal with such a threat.

Two factors combined to change the situation: first, as the Vikings settled to create their own kingdoms, they stopped rampaging widely; secondly, a coordinated response on land and sea could be organized, as in England. Alfred of Wessex and his successors gradually conquered the Danes in the 10th century and established an English kingdom. In France, by contrast, Viking Normandy became a power for the future.

Map legend:
- Boundary of Christian west on eve of invasions
- Viking invasion routes
- Areas of Viking settlement

Map labels: Iceland 870–930; to Greenland (c.983) and Newfoundland (c.1000); Atlantic Ocean; Faroe Is; Shetland Is; Hebrides; Orkney; Iona; Norwegians; Swedes; Staraya Ladoga; Novgorod c.825; Britain; Lindisfarne 793; North Sea; Inishmurray; Bangor; Danes; Baltic Sea; Ireland; Limerick; Dublin 841; York; Lincoln; Cork; Wexford; Derby; Russia; Kiev c.825; Waterford; Leicester; Hamburg 845; Winchester; 834; Bremen; Hamwic; London; Dorestad; Ghent; Cologne; Quentovic; Arras; Aachen; 841 Rouen; Paris; Chartres; Nantes 799; Germany; Noirmutier; Poitiers; 844; France; 830; 859; Santiago de Compostela; 968; Nîmes; Arles; Toulouse; Pisa; Black Sea; Umayyad Caliphate; 859; Corsica; Córdoba; Sardinia; Seville 844; Balearic Is; 859; Constantinople 807,944; 859; Mediterranean Sea; Byzantine Empire; Sicily

military power throughout the medieval period.

Christendom on the offensive

Following the invasions of the 9th and 10th centuries, Christian Europe turned from desperate defence to attack. There was a new spirit of aggression and a new military means of carrying it out. The knight and the castle began to epitomize Western warfare: the former to win battles and the latter to secure territory. The Normans were in the forefront of this expansion. They were to be found in Spain, Italy, the Byzantine empire and, following the First Crusade, in the Holy Land. Increasingly, warfare with a religious motive justified territorial aggression. Crusades were fought by the Danes and Germans in areas of the Baltic and eastern Europe to defeat the heathen inhabitants, and to place Christian settlers on new, rich lands.

The Norman conquest of England

From 1016–42 England was under Danish rule. Viking attacks had resumed in the late 10th century, and became irresistible under the Danish king and his son Cnut. Poorly led by King Aethelred, the English paid tribute – Danegeld – but later fought stoutly under Edward Ironside. However, when he died in 1016 there was no one left to continue the fight. Cnut and his sons made little change to the kingdom's military structure, apart from introducing the professional household warrior (the *huscarl*), armed with the great two-handed Danish axe.

On the death of Edward the Confessor in 1066, Duke William of Normandy made ready to assert his claim to the throne. The new English king, Harold Godwinson, had already proved himself a formidable warrior against the Welsh. He now showed his qualities in a lightning march north to defeat the invasion of his brother Tostig and the Viking, Harald Hardrada, king of Norway. This enabled William to land his forces on the south coast, a real achievement involving the transportation of thousands of men and horses across the Channel.

Dashing south once more, Harold's forces took up a defensive position a few miles north of Hastings. The ensuing day-long battle was an indication of the way warfare was developing. The English still fought on foot, now wielding the Danish axe; the Normans assailed them with combined arms: infantry spearmen and bowmen, and knightly cavalry. The contest was equal until late in the day, when high-trajectory shooting by William's archers thinned the exhausted English ranks, wounding Harold in the eye. Deprived of leadership and inspiration, the defenders were finally routed by a determined cavalry charge. The battle won William the crown, but not the kingdom; this took another six years of hard campaigning.

The scenario just described – of good infantry being defeated by cavalry and missile fire – was to be replayed throughout the medieval world. Already, at Civitate in 1053, the Norman knights of Robert Guiscard had beaten the Pope's Swabian foot soldiers. This same adventurer attacked the Byzantine empire in 1081. At Durazzo the axemen of the Varangian Guard (now largely composed of exiled Englishmen) drove back Robert's cavalry, but were beaten by an infantry counterattack. The Norman influence on the East was to be profound. The Byzantine princess Anna Comnena claimed that "a mounted Frank

Norman success in warfare is indicated by an early 12th century carving of a knight (*left*), from Monreale Cathedral in Sicily. The mounted warrior is charging with his lance couched, crouching behind a large, "teardrop" shield. The flower and border decoration are not heraldic. The new, individual and hereditary designs of heraldry came into use about a generation later.

The Bayeux Tapestry (*above*) contrasts the fighting styles of France and Scandinavia. While the former used the mobility of the knight and the firepower of archers and crossbowmen, the latter stood rooted to the ground, swinging the great two-handed axe. But if infantry were well-armoured and determined, horsemen could make no headway. At Hastings the Normans and

their allies used feigned flight to draw the English from their hilltop position. Once on level ground and disordered by their pursuit, footmen became easy meat for the knights.

could bore his way through the walls of Babylon", so impetuous was his charge. The new knightly tactic of charging with the lance "couched" – tucked firmly under the arm to unite the impact of man and horse – proved a battle-winner.

In Spain, the Umayyad caliphate had reached a high point around 1000, but soon dissolved into a series of factional *taifa* kingdoms vulnerable to Christian attack. Here the Normans played a part in the early stages of the Reconquista, appearing at the siege of Barbastro in 1064. However, the capture of Toledo in 1085 sparked off a

---

"The Franks gave vent to a piercing war cry. I scorned death, thinking that everyone there was exposed as much as I. At the head of the Franks appeared a knight who had thrown down his coat of mail, unburdening himself in order to be able to overtake us. I hurled myself on him and struck him full in the chest. His body fell a good way from his saddle. Then I rushed at their knights, who were coming up in single file. They retreated. And yet I had no experience of fighting, for this was my first battle. I was mounted on a horse as swift as a bird; I dashed in pursuit of them to strike a blow within their ranks, without feeling the least fear of them."

Usamah ibn Munkid, a Muslim "knight", on the exploits of his youth

---

Muslim reaction, and the fanatical Berber Almoravide tribesmen invaded from Africa, defeating King Alfonso of Aragon at Zallacca in 1088. It took all the efforts of El Cid, through fierce border warfare, to restore the situation by

the end of the century. Norman adventurers also established states in southern Italy: Robert Guiscard in Calabria and Apulia, and Roger Great Count in Sicily, which he seized from the Muslims.

Holy warfare

The Byzantine empire was resurgent in the 10th century; Crete (961) and Cyprus (965) were recovered, the Russians and Bulgarians repelled (by 972), and southern Italy secured. John Tzimiskes (969–76) reintroduced a strike force of cataphracts – truly men of iron in their all-encompassing armour for man and horse. However, such troops were expensive to maintain, and fell out of use in the next century. In 1028, on the death of Basil II, the empire seemed secure, but weak leadership allowed the army to decline. Even so, when Emperor Romanus Diogenes led a force east of Lake Van in 1071 to repel the Turkish nomads, his defeat at Manzikert was totally unexpected.

Desperately short of troops, his successor Alexius I appealed to the Pope for mercenaries – and got crusaders instead. The conflict between Muslim and Christian was formalized by Pope Urban II in 1095, when he called for an expedition to rescue the Holy Land. This First Crusade was provoked by early Turkish invaders – the Seljuks – but such crusades were declared at regular intervals throughout the Middle Ages, until they finally withered under the assault of the Ottoman Turks in the 15th century.

The First Crusade was outstandingly successful, with the defeat of the Turks in battles at Nicaea and Dorylaeum in 1097. Then followed the siege of the great city of Antioch, which endured for eight months. In the course of

"It is indispensable for anyone who wants to give a blow with a lance to press his hand and his forearm against his side on his lance, and let his horse guide itself as best it can at the moment of impact. For if a man moves his hand, or attempts to guide his lance, the blow does no damage." (From a contemporary description by Usamah.)

Impregnable Antioch (*right*) fell in 1098, when Bohemond bribed a tower commander. A crusader describes the capture: "At dawn, our men who were outside in the tents heard an overpowering din break out in the city, so they hurried out and saw Bohemond's banner aloft on the hill. They all came running as fast as they could and entered the city gates . . ."

the siege, the Franks proved their superiority in battle on several occasions, usually under the command of the Norman Bohemond. No easterner could stand up to the knightly charge if it were properly delivered. The crusaders just got into Antioch – through treachery – before the arrival of a huge relief force. Then, besieged in their own turn, they were inspired by a miraculous relic – the Holy Lance – to defeat the Muslims. This enabled the crusaders to march on Jerusalem, which they captured after a short but hard siege, slaughtering the inhabitants.

Defence of the Holy Land

The new kingdom of Jerusalem was now faced with the problem of expanding and defending its territories. The first decades of the 12th century saw the gradual capture of the crucial ports, lifelines to the West, while the need to hold the interior resulted in the construction of the finest examples of castle architecture to be seen anywhere. From Margat in the north to Kerak in the south, these mighty fortifications were like "a bone in the throat of the Muslims". Because of the immense cost of building and maintaining these castles, their defence was soon entrusted to the Military Orders. The warrior monks of the Templars, Hospitallars, Teutonic Knights and others provided the cutting edge of militant Christianity. In addition to their fanaticism, they could draw on enormous financial resources from the lands donated in the West by pious laymen. The most powerful bankers of their day, their money bought military power.

However, the Holy Land could only be held while the Muslim world remained divided. Western knights were

also vulnerable to eastern tactics if carelessly led. The emir Zangi picked off Edessa, the most easterly crusader state in 1144, and in 1147 the French king's (Second) Crusade fell apart in the waterless deserts of Anatolia, harassed by archery attack. Under Saladin, who became lord of Egypt and Syria, the pressure intensified. Eventually, in 1187, the Latin king Guy made a mistake which delivered his army into the hands of the Muslims. The entire force was killed or captured at Hattin, leaving no garrisons to defend the kingdom's castles. As a result, these fell swiftly, provoking the Third Crusade. This failed to recover Jerusalem, but the efforts of Richard the Lionheart assured the crusader kingdom another 100 years of increasingly tenuous existence. The crusades of the 13th century were aimed at Egypt, but after initial successes the Fifth Crusade (1217–21) and Sixth Crusade (1248–52) foundered there. Jerusalem was recovered by treaty (1228–44), but the Christians had no answer to the rise of the Mamluk state. Gradually the fortified coastal ports fell. Finally, the heart of the kingdom, the great entrepôt of Acre, was lost in 1291, and the Military Orders retired to Cyprus to continue the struggle.

The wars that created Europe

William the Conqueror's Anglo-Norman realm was a powerful military organization which his sons fought over after his death. The youngest, Henry, lasted longest; coming to the throne on the death of William Rufus in 1099, he repelled the attack of Robert Curthose, duke of Normandy, in 1100–01, and at the battle of Tinchebrai (1106) he defeated and imprisoned Robert for the rest of his life.

Richard I, Coeur de Lion
(1157–99). A formidable jouster and a great general, Richard fully deserves his reputation as a man of almost legendary prowess. Whether in his native land of Aquitaine, or on the borders of Normandy, or in the Holy Land, he consistently proved himself the foremost soldier of his age.

His achievement on crusade was truly remarkable. First, he had to transport his army across the

Mediterranean – conquering Byzantine Cyprus on the way. Then, in the Holy Land, he besieged and captured the port of Acre while combating Saladin's superior forces, and shortly afterwards inflicted a signal defeat on the Muslim general at Arsuf (1191).

Philip of France's successes against Richard's brother John brought Richard back to Europe, where he was made captive (1193). However, his castle strategy in the Vexin, the cockpit of warfare between the Capets and the Angevins, enabled him to recover all lost possessions by 1198. Chateau Gaillard, his "saucy castle" at Les Andelys, was a constant reproach to French territorial aggression.

Richard's death during a siege in 1199 was swiftly followed by the British loss of Normandy, which probably would not have happened had he lived. Certainly, he was the only man who could have reconquered Jerusalem.

Overland travel to Jerusalem presented crusaders with problems of enormous distances and difficult terrain, so that after the mid-12th century large numbers went by sea. This reduced travel time from many months to a few weeks, and spared the valuable warhorses an exhausting journey. An eyewitness records the process of embarkation:

"We went aboard our ship at the port of Marseilles in the month of August [1248]. On the day we embarked, the door on the port side of the ship was opened, so that all the horses we wanted to take with us overseas could be put in the hold. As soon as we were inside, the door was closed and carefully caulked, as is done with a cask before plunging it into water, because, once a ship is on the high seas, that door is completely submerged." (Jean de Joinville, *Life of St Louis*)

However, travel by sea was still a risky business, as King Louis of France found on setting sail for Egypt from Cyprus in 1249:

"It was indeed a lovely sight to see, for it seemed as if all the sea, as far as the eye could reach, was covered with a canvas of the ships' sails. The total number of vessels, great and small, amounted to about 1,800. . . . [But] a violent raging wind, coming across the sea from the direction of Egypt, began to blow so strongly that out of 2,800 knights which the king had taken with him, there remained no more than 700 whom the wind had not separated from his company and carried on to Acre and other foreign parts; nor did they manage to rejoin the king for a very long time after." (*Ibid*)

Henry also beat his overlord, the King of France, at Bremule (1119), and his faithful Norman barons put down the revolt of Waleran of Meulan at Bourg Theroulde in 1126. Following Henry's death in 1136, the conflict between his daughter Matilda and nephew Stephen (who had acquired the crown) threw England into civil war. In 1141 Stephen was captured at Lincoln but then released, and the war dragged on until 1153, a year before his death. By that time, Matilda's husband Geoffrey of Anjou had overrun Normandy, and the Angevin "empire" was united under their son, Henry II.

An important factor in all these battles was the use of dismounted knights. At Tinchebrai, King Henry himself stood in the second line. Thus we should beware of thinking of the medieval period as dominated solely by the knightly cavalry charge. At the battle of the Standard in 1138, an Anglo-Norman force stood to repel the impetuous charge of the Scots. Dismounted knights stood alternately with archers, protecting them with lance and shield, while the missile men shot their arrows. In consideration of this it is apparent that the English tactics of the Hundred Years War, two centuries later, were not such an innovation.

The effective use of infantry was not restricted to northern Europe. In Italy the city communes produced good, solid footsoldiers, armed with pole arms and pikes and supported by powerful crossbows. At Legnano in 1176, such troops put up so stout a resistance against Frederick Barbarossa's German chivalry that the Emperor's forces were completely routed by a mounted counterpunch. The Italians rallied around the *caroccio*, or cart, which symbolized the city's authority just as the English raised a ship's mast at the battle of the Standard, thus giving the battle its name. Eastern spearmen were also capable of repelling a knightly charge if they stood firm. The household knights of Tancred, Prince of Antioch, could not be persuaded by threat or entreaty to risk their horses against the men of Shaizar in 1110. Infantry also played an important role in the most common form of warfare – the siege.

Castle strategy

Wars between the Angevin kings of England and the Capetian kings of France, which dominated the 12th century, centred around castle strategy. The borders of Normandy, a disputed land, became crammed with such fortifications, most notably Richard the Lionheart's great construction at Chateau Gaillard. All the Angevins were great castle-builders, as can be seen in many parts of England, most noticeably perhaps in Dover, the "Key to the Kingdom", besieged without success by the French prince Louis (1216–17).

The Angevins had the upper hand under Henry II and Richard, but John was outwitted by Philip Augustus, the French king. Normandy fell in 1204, largely because John could not trust his barons to hold their castles against the French. His attempt to recover the situation, involving a daring pincer movement, was defeated at Bouvines in 1214. His ally in the battle, the emperor Otto, almost achieved victory through the stalwart efforts of the Flemish pikemen, showing once again the significance of infantry; but it was not enough, and the French triumphed. Royal authority was further strengthened by the Albigensian Crusade, waged against the Cathar heretics of Toulouse

Castles were very varied in size and function. Gerald of Wales (c.1200) gives an affectionate portrait of his family seat of Manorbier, Dyfed (*above*):

"The fortified mansion known as Manorbier . . . stands, visible from afar because of its turrets and crenellations, on the top of a hill. Just beneath the walls there is an excellent fishpond . . . and a most attractive orchard. At the east end of the fortified promontory, between the castle and the church, a . . . stream winds its way along a valley."

At the other end of the scale, Caernarvon (*opposite*) is a massive site with an attached walled town. Its own tall walls imitated the imperial might of Constantinople; the only ingress two massive gateways. The King's Gate originally comprised six iron portcullises and five thick doors. Each of the huge towers formed self-contained fortresses, while still allowing rapid passage of troops around the wall-walk. A garrison of but a few men made it impregnable.

A remarkable spate of castle-building secured the Welsh conquests of Edward I. During his reign and that of his son, eight royal fortifications went up, five including walled towns. Beaumaris's concentric symmetry was begun in response to the Welsh revolt of 1294–95, but never completed. Caernarvon, built in 1283, is still the most impressive structure and symbol of Edward's imperial ambitions.

and the Languedoc. Here crusader ideology was used for secular ends, and for two decades (1208–26) the war was prosecuted with great cruelty, chiefly under the leadership of Simon de Montfort. The decisive battle was fought at Muret in 1213, when Pedro of Aragon came to the aid of Raymond, count of Toulouse. His forces were routed and Pedro was killed.

The wars in England now became internal. Henry III was forced to fight the Barons' War, in which Simon de Montfort (the crusader's son) triumphed at Lewes in 1264, only to be surrounded and crushed at Evesham in the following year. The architect of this victory was Prince Edward, who as Edward I was to conquer Wales and subdue Scotland. The first objective was largely achieved in the campaign of 1282, but the permanence of English rule had to be assured, and this was achieved in a way that shows the importance of castles. Great fortifications in the modern, "concentric" style were constructed from Flint in the north and around the coast to Caerphilly in the south. Beaumaris, Harlech and Caernarvon stand out as examples of military architecture. The last named resembles the walls of Constantinople – a reminder that Edward had served a military apprenticeship on crusade in 1270.

It was also Edward who, by employing large numbers of

Welsh archers, developed the tactics which were to stand the English in such good stead during the Hundred Years War. Englishmen were gradually becoming identified with the bow, which proved a formidable battlefield weapon. During the Scottish wars (after some initial setbacks) longbowmen shot the blocks of Scottish pikemen to pieces at Falkirk in 1298. It was Edward II's inability to deploy his

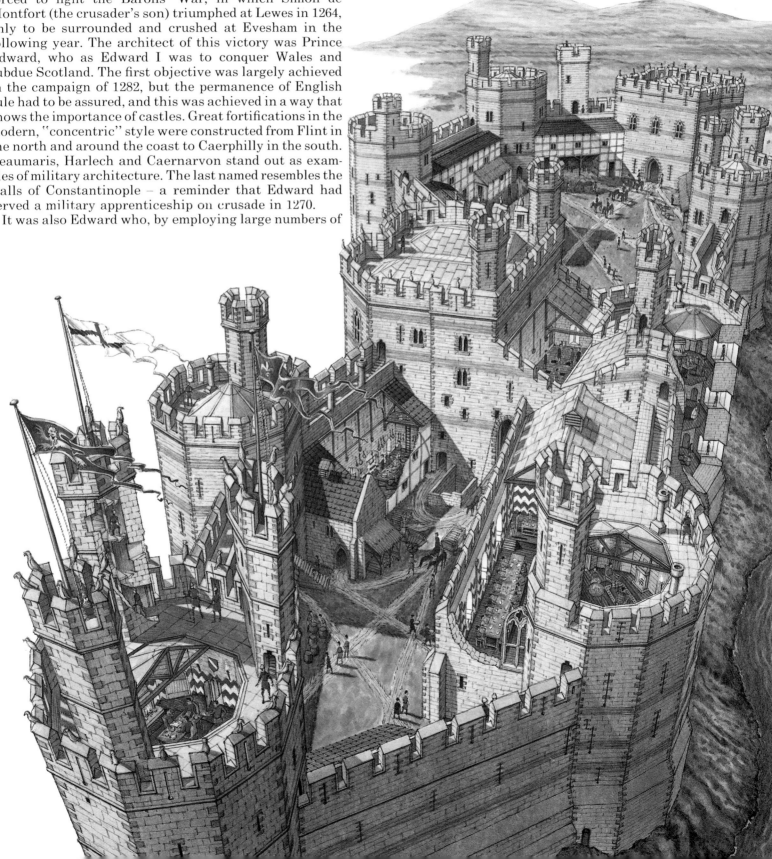

archers effectively which lost the battle of Bannockburn in 1314 (although this defeat was later avenged at Dupplin Moor [1332] and Halidon Hill [1333]).

Spain and Italy

The Spanish Reconquista continued throughout the 12th century, to be checked by the arrival of another group of Berbers, the Almohads, in 1145. These defeated Alfonso VIII of Castile at Alarcos in 1195, and it took a further two decades of campaigning before he was able to turn the tables on them at Las Navas de Tolosa (1212). As a result, his successor, Ferdinand, was able to capture Cordova and Seville. Naval power was employed to subdue the Balearic Islands in the 1220s, so that only the emirate of Granada remained of the Muslim power in Spain (although this was to survive until 1492).

The German emperors' preoccupation with their valuable Italian territories dominated central Mediterranean warfare in the 13th century. Frederick II won victory over the Lombard League at Cortenuova in 1237, but his bastard son Manfred, attempting to hold the kingdom of Sicily, was defeated at Benevento in 1266 by Charles of Anjou (the French king's brother). Victory at Tagliacozzo two years later secured this new Angevin state, but it was soon to be challenged by another expanding Iberian kingdom: Aragon. The Aragonese gained the upper hand in the war of the Sicilian Vespers, with naval victories at Messina in 1283 and Naples in the following year, in which Charles's son was captured. By 1302 they had conquered the region.

The Byzantine empire was going through a period of eclipse in the 13th century. The Fourth Crusade had been diverted by the Venetians to Constantinople, to instate a pretender to the throne, and after some confused negotiating the city was stormed by the crusaders, who elected one of their leaders as emperor. This was a remarkable achievement, only possible because they could use the Venetian fleet as ready-made siege towers. In many ways this capture of the city was more impressive than the conquest of 1453, when Byzantium was on its last legs. However, the Westerners were ill-suited to rule as the Byzantines had, and the first emperor disappeared in battle against the skirmishing Cumans in 1205. More significantly, the Latin rulers found that they could not manage the political problems posed by Constantinople's geographical position. Byzantine successor states in Europe and Asia Minor began to pick away the empire's territories. In 1261 Michael Paleologus, the founder of the last dynasty of Byzantine rulers, entered Constantinople in triumph. The Franks were confined to the Peloponnese, where they had already suffered a defeat at Pelagonia in 1259, but the Byzantine tendency to employ mercenaries soon betrayed them. The tough Catalan infantry of Jacques de Flor rebelled on their leader's death and established a state at Athens in 1311, which lasted until 1389. By then the shadow of the Ottoman Turks was already looming menacingly over the empire.

A feature of the 13th century was the "up-armouring" of the heavy cavalry. As plate began to be added to mail, knightly armour became far more expensive; horses too were caparisoned with mail, covered by cloth bearing heraldic devices. All over Europe the cost of maintaining a knight's equipment began to escalate. The kings of both

The knight, armed *cap-a-pie* and charging into battle, is the popular image of medieval warfare, fostered by modern movies. In fact, battles were rare events, for they were risky encounters which a general avoided if at all possible. In an age when society's leaders formed its military strike force, the consequences of defeat could be dire. The English victory at Poitiers, for example, saw the French king and the bulk of his nobility fall into enemy hands.

Ransom served to alleviate some of the dangers, but cost the prisoner's subjects dear, and negotiations could be long-winded. Strategy was based on the building, taking and holding of castles. Siege warfare techniques were sophisticated and varied, and, with the development of cannon, fortress design was adapted to counter the threat.

England and France vainly decreed that those able to afford the burden of knighthood should take it up. Now the term for a fully armoured warrior was a man-at-arms, a term which implied no particular social status.

Chivalry brought low?

As we have seen, infantry did not disappear in the early medieval period, and in fact retained its importance. But in the 14th and 15th centuries various types of foot soldier began to challenge the knightly cavalry, notably the English archers, Hussite handgunners and Swiss pikemen. These could, if determined enough, readily repel mounted charges while standing on the defensive. Thus at Coutrai in 1302, the communal militias of Flanders defeated the flower of French chivalry as a result of the rash charges of the knights. At Crécy in 1346, Edward III's small English army was able to shoot down the ill-coordinated assaults of the poorly led French.

Crécy was the first significant land battle of what became known as the Hundred Years War. In fact, this long-running contest between England and France fell into three phases. The first action took place in 1340 at sea, outside Sluys. Fought by both sides as a type of land action, there was little manoeuvre apart from grappling for boarding. The English deployed exactly as they would have done on land, with archers flanking the men-at-arms' ships. The inexperienced French were completely overwhelmed by a barrage of archery, quickly followed by hand-to-hand fighting. This effective tactic was to be the secret of English victories everywhere.

In fact, Edward and the Black Prince did not seek battle.

Their strategy was one of conducting destructive raids or *chevauchées*, combined with judicious sieges to break the enemy's will to resist. At Crécy, and Poitiers ten years later, they only accepted battle when trapped by a pursuing French force. At Poitiers, King John of France showed he had learnt the lesson that a cavalry charge could be shot to pieces, and dismounted most of his fully armoured men. However, he could not direct the battle from the front line and was eventually captured, along with most of his greatest lords, when the Black Prince organized a combined flank and counterattack. Essentially, the English held all the cards in battle. They proved this again at Navarette in 1367, when the Black Prince and the Constable of France, Bertrand du Guesclin, found themselves on opposing sides supporting claimants to the throne of Castile. But du Guesclin learnt from his defeat, and this experience, combined with the deaths of Edward and his son within a year, helped him to eventually turn the tables on the English.

The battle of Agincourt

The second phase of the Hundred Years War was one of French dominance. The fact that there were no great battles favoured the French, who raided and ravaged the south coast of England. Under the weak rule of Richard II, English possessions in France were recovered piecemeal. The third phase began in 1415 with the invasion of Henry V. At Agincourt he was forced to give battle, although outnumbered by at least three to one, but French numbers proved their downfall. Overconfidence led them to neglect the use of their own missile men. Initial cavalry charges

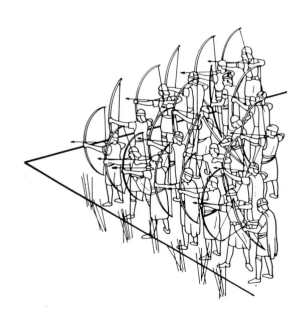

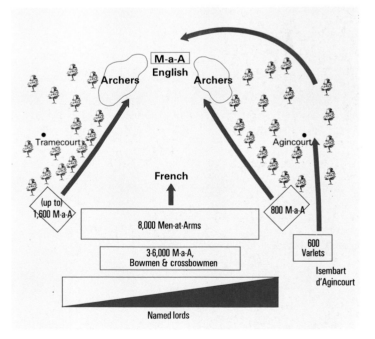

English longbowmen were capable of putting up to a dozen arrows in the air in one minute. This kind of rapid shooting was reserved for ranges under 100yds (91m) and could stop a cavalry charge in its tracks. Pits or stakes protected the archers if an enemy did reach them. It was not so much the size of the bow that mattered, as training from boyhood and deployment in large numbers. By perfecting these skills and tactics, the English won many victories.

English victory at Agincourt arose from discord in the French command, together with Henry V's bold initiative in advancing to occupy a strong position. After Crécy, the French had learned how to attack on foot, heads bowed, into the arrow storm, and in 1415 Marshal Boucicault had devised a plan to disrupt the English archers by attacking the flank and rear. But this came to nothing as the French lords jostled for glory in the front rank, only to be killed or captured.

foundered under the archers' storm of missiles and against their defensive stakes. In their flight the horsemen disordered the ranks of dismounted knights, who ended up so close-packed as to be unable to fight. The French were either slaughtered or captured.

"Sir Guillaume de Saveuse [leading the vanguard] and his two companions came in to attack, but their horses fell upon the stakes and they were themselves soon killed by the archers, which was a great shame; most of the remainder out of fear turned and moved back . . . It was by these, chiefly, that the French vanguard was put into disorder: countless persons began to fall and their horses, feeling the arrows falling upon them, started to bolt behind the enemy, many of the French following their example and turning to flight. Then, the English archers, seeing the advance guard in such disarray, emerged from behind their stakes, threw their bows and arrows aside and, taking up swords, hatchets, hammers, axes, falchions and other weapons, rushed into places where they saw gaps, killing and laying low the French without mercy, not ceasing to kill until the advance guard, which had fought but little, or not at all, was completely wiped out."

Jean de Waurin, French eyewitness at Agincourt, 1415

The key to English supremacy was the alliance with Burgundy. Henry's early death, resulting in minority rule in England, and the collapse of the Burgundian alliance in 1435, shifted the balance to the French. Joan of Arc had already boosted their morale, but her career spanned only three years (1429–31). More significant in the long term were Charles VII's reorganization of the French army, and the increased use of artillery. Normandy and Gascony were recovered in the 1440s. The attempt of John Talbot, Earl of Shrewsbury, at reconquest in the next decade ended with his death at Castillon, where cannon played a part in the English defeat.

Profit and Loss
It seemed to some that the greedy footsoldier was upsetting the balance of things. Complaints against the depredations of mercenary bands were very common. "Ecorcheurs" ("Flayers") was the notorious nickname for the numerous "free companies" who flourished in the environment of the Hundred Years War. In response there was a "Jacquerie" of outraged peasants who rose up against their tormentors in war-torn France.

In fact there was nothing exceptional in such behaviour. All classes fought for gain, through ransoming prisoners, stripping the enemy dead, and pillaging peasant and burgher alike.

The Hussites
Gunpowder weapons had been in use in Europe since the 1320s. They were chiefly used in sieges, owing to their slow rate of fire. By the end of the 14th century, every ruler had a siege train, and this in turn speeded the development of the low, thick-walled bastion fortifications. There were also a few crude handguns in use, as German chivalry were to discover in their crusades against the Hussites of Bohemia,

The success of the Hussite battle formation had a profound effect on warfare. In this print of 1480, showing an army on the march, the handgunners, halberdiers and defensive wagons have a new prominence. The armoured knight was not discarded; he now had a role in a force of combined arms.

followers of the proto-Protestant heretic Jan Hus, condemned and burnt at Constance in 1415. When Bohemia revolted against the Emperor Sigismund, he raised large armies to assert his authority.

However, the Hussites managed to combine religious fervour, iron discipline and battle-winning tactics. They proved unbeatable under the leadership of the remarkable Jan Zizka. The crusaders were routed at Prague in 1419, although this was a defensive encounter. Zizka's genius lay in the fact that he was able to make infantry more mobile, developing the ancient *wagenburg* tactic by packing the armoured wagons with masses of infantry wielding crude polearms, and supporting them with handguns and small field pieces. Finding his forces trapped outside Kutna Hora in the winter of 1422, he even risked a night attack against the surrounding crusaders. He broke through the lines and equally broke the German morale. Hard-bitten soldiers fled even at the sound of Hussite hymns following such victories.

Zizka died in 1424, but the Hussite momentum continued. They repelled renewed attacks in 1426–27, and then went on to the offensive, carrying out a series of raids into Germany over the next five years. However, tensions between competing religious groups eventually destroyed the Hussite state: the extreme Taborites, the driving force of the revolution, were defeated at Lipani in 1434 by the Moderates, and the monarchy was eventually reestablished. Nevertheless, the *wagenburg* became an established fixture in Eastern European warfare. It was used successfully by the Hungarians against the Ottoman Turks, and adopted by them in their later campaigns of

---

THE PLIGHT OF CIVILIANS

Throughout this period, the hapless peasantry suffered severely during the passage of an army, especially if it was engaged on a destructive cavalry raid – a *chevauchée*. These raids could be prolonged and far-reaching; in 1373 John of Gaunt led a *chevauchée* across France from Calais to Bordeaux. Noble knights and common soldiery alike acquired a taste for these destructive but profitable forays, unmoved by the misery they caused. One eyewitness account comes from the *Chansons des Lorrains*, a French epic poem of the 13th century, but similar episodes continued for generations. As military technology improved, the situation for defenceless civilians and peasants deteriorated, culminating in the horrors of the Thirty Years War.

"The march begins. Out in front are the scouts and incendiaries. After them come the foragers whose job it is to collect the spoils and carry them in the great baggage train. Soon all is tumult. The peasants, having just come out to the fields, turn back uttering loud cries. The shepherds gather their flocks and drive them towards the neighbouring woods in the hope of saving them. The incendiaries set the villages on fire and the foragers visit and sack them. The terrified inhabitants are either burned or led away with their hands tied to be held for ransom. Everywhere bells ring the alarm; a surge of fear sweeps over the countryside. Wherever you look you can see helmets glinting in the sun, pennons waving in the breeze, the whole plain covered in horsemen. Money, cattle, mules and sheep are all seized. The smoke billows and spreads, flames crackle. Peasants and shepherds scatter in all directions." (Translation: J Gillingham.)

The pike and shot era was a result of developments before 1500. Light field artillery (*above*) was employed widely in Burgundy, France and the Holy Roman Empire, where Emperor Maximilian I keenly supervised its production for his armies.

Infantry polearms were widely used in imitation of the Swiss (*right*) whose chief weapon was the halberd. This later print of Sempach (1386), illustrates well the violence of their attack.

conquest. Handguns became widespread only late in the 15th century, and this weapon alone could not challenge the mounted knight.

The Swiss

It was to be in conjunction with another weapon – the pike – that a new age of warfare was to dawn in the 16th century. The Swiss played a great part in making this weapon widely used. Scottish and Flemish pikemen had already proved that their spears could repulse a badly organized cavalry attack, although such bodies of often lightly armed footmen were vulnerable if the charges were supported by missile fire. The Swiss developed tactics of good battlefield mobility with their close-packed blocks of men armed with poles, well supported by crossbowmen, and also, wherever possible, cavalry.

The Swiss Cantons' path to military greatness began in the wars of independence from Austria. At Mortgarten (1315) the Swiss combined missile and halberd attacks with ambush in terrain that prevented the Austrians from employing their superior numbers. At Laupen (1339) pikemen and archers defeated the uncoordinated attacks of Burgundian cavalry and infantry. The Austrians copied the French and dismounted to assault at Sempach (1386), where initially they drove back the more lightly armed Swiss, but they soon tired and fell prey to a well-timed and disciplined counterattack. The result was Swiss independence by 1394.

The defeat of Burgundy

What made the Swiss the most sought-after troops in Europe at the end of the medieval period was their defeat of Burgundy in the 1470s. This was all the more astounding because the Burgundian duke, Charles the Bold, had developed a modern and well-equipped army, with regular ordnance companies on the French model. These combined heavy cavalry, infantry, English longbowmen (or those trained like them), and the most up-to-date mobile field artillery. However, his attempt to forge a European empire was met with defeat after defeat, and eventually death, at the hands of the Swiss. At Grandson in 1476, his carefully balanced force was routed by their skill and bravado. Now the Swiss were using true pike-blocks supported by halberdiers, and they manoeuvred with the élan of the Macedonian phalanx which they so closely resembled. At Morat in the same year, these pike columns marched into a killing ground devised by Charles, where they were surrounded on three sides by artillery, archers and cavalry. But they were so intelligently handled that, by utilizing speed and determination, they turned the tables completely, routing the opposition. Finally, in a hard slogging match at Nancy in 1477, with heavy casualties on both sides, the Burgundians broke first and Charles met his death in flight. The Swiss were also able to defeat the Emperor Maximilian at Dornach in 1499, establishing themselves as pre-eminent in Europe.

The pike also reached England, at that time a relative backwater. A body of pikemen were deployed on the losing side at Bosworth in 1485, where Richard III went down to defeat before Henry Tudor, but during the Wars of the Roses (1453–85), England had turned in upon herself. Her bowmen continued to be important in the early part of the

A Burgundian military camp (*left*) serves to represent a typical layout. Although lacking the ditches and palisades of Roman camps, wagons and tents still provided an obstacle to attackers. It was quite common to use such constructions to cover the rear of an army in battle, even before the development of fortified *wagenburgs*. In the foreground a crossbowman uses a windlass to load his weapon.

The battle of Pavia, 1525 (*right*). In the foreground, the knights still ride, armed with lance and sword and wearing full plate armour. Behind them stand their supplanters as rulers of the battlefield – footsoldiers armed with arquebus and pike. The best armour and the greatest élan were useless against them. By the end of the 16th century, cavalry were reduced to trotting up to the enemy and firing pistols.

16th century, chiefly against the Scots, and the victory of Flodden (1513) echoed that of the Standard almost four centuries earlier, as English bowmen shot down the over-enthusiastic Northerners.

On the continent, however, developments pointed in a different direction. Large bodies of pikemen were beginning to dominate the battlefield, supported by halberdiers or sword-and-buckler-men, and provided with missile power by troops wielding the improved handgun, or arquebus. The Emperor Maximilian had seen the way things were going and raised *landsknechts*, lightly armed pikemen, to counter the Swiss. Francis II of France, on the other hand, may have thought he had the measure of the Swiss infantry, following his victory at Marignano in 1515; but this was chiefly achieved through the use of artillery against some rashly deployed pike columns – always a vulnerable target to such weapons. At Pavia (1525) Imperial arquebusiers threw back the cream of his cavalry.

As we have seen, there was nothing especially new in this, but the emphasis was now increasingly laid on cheap, easily trained infantry, supported by lightly armoured cavalry. The man of iron was becoming redundant. When Sir Arthur Haselrigge raised a troop of horse for Parliament in the English Civil War, they were derided as "Lobsters". By this time, armour was no longer missile-proof, and it was discarded in favour of greater mobility. Some cavalry – cuirassiers – retained the breastplate into the 19th century, but the full *cap-a-pie* suit of armour had been purely decorative for 300 years. But then, as we have seen, the amount of armour worn was no guarantee of victory, even in the medieval period.

RULES OF WAR

Henry V issued ordinances laying down standards of behaviour to be enforced in the host by the Marshal and the Constable:

Obedience: "First, that all manner of men, of whatsoever nation, estate or condition, be obedient to our sovereign lord the King, and to his Constable and Marshal, upon pain of as much as he may forfeit in body and in goods".

Guard duty: "Also that every man be obedient to his captain to keep his watch and ward . . . and to do everything that a soldier should, upon pain of having his horse and harness taken into the charge of the Marshal, until such time that the offender has agreed with his captain following the decision of the court."

On the robbing of merchants bringing provisions: "Also that no one be so bold as to pillage or rob another of victuals or provisions which may be bought, upon pain of death . . . Nor may anyone rob another of forage or food, nor anything taken from the enemy upon pain of arrest at the King's will."

Attacking without the King's permission: "Also that no assault is to made upon a castle or fortification, by archers or other common men without the presence of a man of estate. If any assault is made, and the King, Constable or Marshal or any lord of the host orders it to be stopped, no one should be so bold as to renew it. And if any man should do so he shall be imprisoned and lose any profit he may have gained, and his horse and harness shall be taken into the charge of the Constable and Marshal."

— chapter four —
Storm from the East

Matthew Bennett

The Turkish repulse from Vienna in 1683 marked a turning point in history. For more than 1,000 years Christian Europe had been subject to constant attacks from horse-archer armies that swept all before them. They came from the rolling grasslands of the Central Asian steppe and their nomadic lifestyle made them expert in the art of war. The harsh environment taught them the importance of discipline, organization and planning, while the need to manage their flocks and herds gave them a mastery of logistics. They made grimly efficient rulers, since they applied their skills to the human populations they dominated. They were herders of men.

To the societies they attacked there seemed to be no rhyme or reason in the appearance, or equally sudden disappearance, of a formidable enemy on their borders. In fact, nomadic tribal movements came about as a result of what Arnold Toynbee has called the "conductivity" of the steppe. Political and military events far beyond the knowledge of Europeans, in China or Central Asia, produced a chain reaction. One tribe shunted another across the steppe until the weakest or most westerly ended up driven into Europe. Here they appeared as conquerors, although in reality they were often fleeing more ferocious foes.

The Huns

The name "Hun" has become a byword for barbarity, and their greatest leader, Attila, still represents cruel tyranny. This shows the tremendous psychological impact the Huns had on Christian Europe. It was their pressure, beginning around AD370, that forced the German tribes into the Roman empire and led to its destruction.

The people whom Attila led were known as the Black Huns – just one group of a great confederation that poured out of Central Asia in the 4th and 5th centuries. In origin these were the Hsiung-nu tribes, defeated by the Chinese Han emperors 200 years before. Attila's hordes terrorized Europe for two decades (AD434–53), operating from their base on the lower Danube.

"They enter battle in wedge-shaped masses, raising savage cries. Lightly equipped for swift motion, they suddenly divide into scattered bands and rush about in disorder dealing terrific slaughter, and because of their extraordinary rapidity they are never observed before they attack a rampart or pillage a camp."

Ammianus' description of the Huns, late 4th century

Although Hunnic armies were largely composed of light horse-archers, their aristocracy wore heavier equipment. Grave finds show the sort of cataphract armour for man and horse popular amongst the Sarmatians, whom the Huns had replaced on the Eurasian steppe. Also, Hun armies probably contained only a small proportion of ethnic Huns. Like other nomads, they relied for their expansion on large numbers of subject troops.

This gave the Huns an inherent weakness. Although they ravaged Greece and the Balkans in the 440s, the campaign of 451 met with defeat in central Gaul. Attila's army consisted of many German allies, including Gepids, Ostrogoths and Burgundians. Opposing them was an equally heterogeneous "Roman" force under Aetius, also largely composed of Germanic Franks, Saxons and Visigoths, in addition to Gallo-Roman forces. They met in battle at Châlons-sur-Marne near Troyes in June 451.

Although Attila's small Hunnic contingent fought well, the poor showing of their Sarmatian allies in the centre,

and the impetus of Aetius' Visigoths, drove them back into their wagon laager (circular defensive formation). This final defence was another feature of nomadic tribes, and, together with their premier weapon – the composite bow – arrows and wagons kept the enemy at bay.

Often heralded as the victory that saved Gaul, the outcome was more like a draw. Next year Attila campaigned successfully in northern Italy. But on his death in 453, the subject Germans rose in revolt, and the first, brief nomad domination was over.

The Avars

New Asiatic conquerors swept onto the Russian steppe.

Aggression, expansion and imperialism has characterized the last half-millennium of European history. It is easy to forget that before the development of gunpowder, nomadic societies, or states employing nomadic warriors, had the upper hand over agrarian peoples. Swiftness of movement, mastery of logistics and ruthlessness in pursuing their aims, made these raiders, from the Huns to the Ottomans, the terror of the West.

Only fortress walls could stand against them. Constantinople denied Huns, Avars, Arabs and Turks. Border regions like Hungary offered defence in depth. But when nomadic forces developed expertise in siege warfare – like the Mongols – countries lay helpless before them. Ottoman artillery finally breached Constantinople's thousand-year-old walls. For a century their sultans attacked by land and sea, even attempting a last, desperate assault on Vienna in 1693. But by then the balance of power had changed forever.

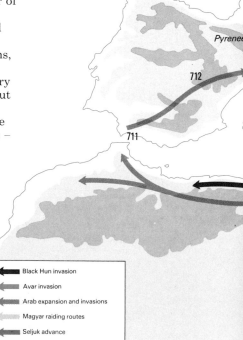

Pyrenees

712

711

	Black Hun invasion
	Avar invasion
	Arab expansion and invasions
	Magyar raiding routes
	Seljuk advance
	Mongol invasions
	Ottoman expansion and invasions

—TIME CHART—

451	Huns defeated at Châlons
c.555	Avars driven into Europe
by c.600	Cataphract, armoured horse-archer, mainstay of Roman/Byzantine armies
by 600	Slavs inhabit much of Eastern Europe
622	Flight of Mohammed from Mecca to Medina (*Hejira*)
from c.640	Muslim Arabs advance westwards
642	Umayyads capture Iran and Egypt
673	Greek Fire used by Byzantines
673	Failure of Muslim blockade of Constantinople
from 732	Rise of Carolingian empire
732	Poitiers: Franks stem Islamic advance into Europe
749	Abbasid Caliphate established
774	Charlemagne conquers Italy
800	Avar power broken
800	Charlemagne crowned Holy Roman Emperor
955	Lech: mobile German heavy cavalry defeats Magyar light horse-archers
c.1000	Magyar state becomes Christian Hungary
1040–1150	Seljuks ascendant in Persia and Asia Minor
1071	Manzikert: Byzantine cataphracts overwhelmed by more mobile Seljuk cavalry
1099	Crusaders capture Jerusalem
1100	Turcoman nomads form part of Arab armies
	Frankish kingdom of Edessa falls to Seljuks
1187	Hattin: horse-archer tactics end Latin kingdom of Jerusalem
1221–24	Mongol reconnaissance of southern Europe
1237–40	Mongol invasion of Eastern Europe
1241	Mongols defeat Poles and Teutonic Knights at Liegnitz, and Hungarians at Mohi
1244	Jerusalem finally lost to Turks
1260	Ain Jalut: Mongols defeated by superior Mamluk cavalry
c.1300	First Ottoman state formed in Asia Minor
from c.1350	Ottomans expand into Europe
1453	Siege artillery facilitates Ottoman capture of Constantinople
1529	Mohács: coordination of artillery, infantry and *spahis* gives Ottomans victory over Hungarians

This people had been known to the Chinese as the Juan Juan, and had recently been overthrown by Turkish nomads. The name Avars means "exiles". They joined with the remnants of the White Huns c.560 to establish a supremacy over the myriad Slav chiefdoms of eastern Europe. The Gepids were annihilated by their assault, and the Lombards were driven into Italy where they created a kingdom in 568. This accorded ill with Byzantine interests, for the emperors sought to use the nomads to police their enemies on the frontiers.

The Byzantine empire had overstretched its resources as a result of Justinian's grandiose attempts to reconquer the West. After his death in 565 there followed a long and exhausting war with Persia, continuing until 591. The Avar khan was quick to seize the opportunities offered and raided extensively in the Balkans. In 602, Avars even attacked Constantinople itself, although without success.

In 626, the Avars returned as allies of the revived Sassanid Persians, who had conquered all the eastern parts of the empire outside Asia Minor. This siege was equally fruitless, and resulted in the rebellion of their Slav subjects. There followed a great folk movement of Slavs into the Balkans and Greece, while the Avars remained on the steppeland. Thus, like the Huns before them, the Avars were responsible for a profound change in the human geography of Europe.

Charlemagne's campaigns of 791 and 796 wiped them from the map, prompting the Slav saying: "They perished like Avars." Nevertheless, the Avar era in Europe did bring some significant military developments. Higher saddles and better horse furniture, including stirrups, gave the mounted man a better seat. These, in time, enabled Westerners to develop the fearsome knightly charge.

The Arabs

When Mohammed died in AD632 he had welded the disparate tribes of Arabia into unity. More than that, Islam provided them with an aggressive ideology that made them very formidable. Contrary to popular legend, Arab armies were not entirely mounted. Many rode camels, although these were abandoned for much preferred horses as they became available, and the majority fought on foot. Good, close order and discipline gave them their victories. The best battle array was likened to ranks of the faithful at prayer; religious inspiration and military discipline went hand in hand.

Their opponents, the Byzantines and Sassanids, were ill-equipped to cope with this new threat. By AD628 Emperor Heraclius had recovered the eastern provinces through a bold, counter-punching strategy, but they were disaffected for religious reasons. A much larger Byzantine force was defeated by the Arabs at Yarmuk in 636, and within two years the whole of Syria was in Muslim hands. Egypt followed in 639–42, while Mesopotamia was overrun by 646. Sassanid Persia collapsed in equally dramatic style, losing its capital and western provinces by 649. Incredibly, in just two decades, under its first three caliphs, Islam had sprung to empire.

The third caliph founded the first Islamic dynasty, the Umayyads, but was then assassinated in 656 by Ali, the Prophet's son-in-law. The latter only ruled for five years, the Umayyads regaining power for another century, but

A heavily-equipped Avar horseman is shown in this 6th-century rock-carving (*above*). His style of equipment and the use of stirrups were gradually adopted by the Franks, giving them a strong cavalry arm.

The map (*left*) shows the balance of forces in Europe at the end of the 8th century, when the Arabs were assaulting Christian states in both East and West. They were checked by the Walls of Constantinople and the forces of the Franks. The latter counterattacked under their talented Carolingian dynasty. By 800 they had destroyed the Avars and created their own empire.

the dispute had far-reaching consequences. The result was a schism between Ali's followers, the Shiah, and the orthodox Sunni, which was to divide Islam politically.

However, Umayyad conquest continued apace, from Egypt as a naval base. Converted Berbers overran North Africa by the end of the century, and in 711 erupted into Spain. Here the Visigothic kingdom collapsed, leaving only vestigial Christian resistance in the northern mountains of Galicia. The Muslim wave swept on into southern France, and was only stopped at Poitiers in 732. This may only have been a raid, but it was severely defeated by Charles Martel, the first Carolingian. Despite the legend, there is no evidence that the Muslims were mounted, and were repelled by the solid Frankish foot; nor that the Franks were then inspired to take to horses. It is not clear even that the Berbers involved were actually Muslims. The point is that great religious and military movements were capable of hurling whole nomadic peoples across Europe, transforming the political map.

Meanwhile, Islam was also advancing in the East. Its armies swept east of the Black Sea to conquer Armenia by 717. Attacks then began on the Khazar nomads who at that time ruled the Russian steppe. Faced with a Christian emperor in Byzantium and Islam in Baghdad (founded 763), the khan elected for Judaism – and Khazar authority survived for another two centuries.

Constantinople came under attack in 673–78 and 717–18, but even in the first enthusiasm of conquest the Arabs could not take the city. The Byzantines also found a way to defend Asia Minor, by dividing it into strategic commands or *themes*. These provided mixed armies of horse and foot that matched those of the Islamic raiders. It was the cavalry's job to dog such expeditions, and restrain their ravages, while the infantry controlled the passes, prepared to pounce on marauders retreating with their booty.

Muslim expansion in the western Mediterranean

As borders crystallized in East and West, Dar-al-Islam began to fragment politically. The Abbasid dynasty, which came to power in 749, lost the impetus of expansion and even proved incapable of maintaining its authority. It was left to the Umayyads of Spain and the Aghlabids of Tunisia to conquer by sea. The Balearics (798) and Crete (823) fell to such attacks, while the Aghlabids began the reduction of Sicily and southern Italy. They held these possessions in the mid–9th century (c.840–80), and Muslim expansion continued in the western Mediterranean. Both dynasties profited from, and contributed toward, the disintegration of the Carolingian empire. Sardinia fell in 827 and Corsica in 850. Fraxinetum in Provence served as a base for raids until 915, while other Saracen pirates operated from the mouth of the Garigliano until 972. The Aghlabids, advancing from southern Italy, launched a full-scale attack on Rome in 846. The force was said to contain 11,000 men with 500 horses and 73 ships, and was only repelled with difficulty.

In the 10th century much of the Muslim West was reunited by the Fatimids of Tunisia, beginning in 909. They took over Sicily in 965 and, after their conquest of Egypt (967–72), had the most powerful navy in the Mediterranean. Despite this, Byzantium made a remarkable recovery under a series of gifted emperors, recapturing the strategic

The Walls of Constantinople (*right*) were to save the Byzantine empire on many occasions. They repelled Arab attacks twice in the first flush of the Muslim conquests. Constructed in the reign of Theodosius II (408–450) they ran all around the triangular peninsula which the city occupied. The landward side, across its base, was defended by double walls and a moat. Arab fleets were unable to storm the sea walls in the face of a new development: Greek Fire. This medieval napalm gave the Byzantines naval superiority. The Arabs learnt swiftly though, and in the 9th century were capable of mounting amphibious and seige operations such as the one that took Messina in 842–3 (*above*).

47

islands of Crete (961) and Cyprus (965). The Buywayhid emirates on the eastern frontier were also unable to make any headway. Meanwhile in Spain, while the Umayyads reconquered Morocco, the realm fell apart into *taifa* kingdoms, ripe for Christian reconquest, after the death of their great vizier al-Mansur in 1002.

By 1000 the Arabs had lost the religious impetus of conquest. The Fatimids were more interested in competing with their co-religionists than in the *jihad* or Holy War. To renew the war against Christendom, it would take more recent converts and those still imbued with the traditional steppe mentality. This was something the Seljuk Turks were soon to provide.

The Magyars
All had not been quiet on the Eurasian steppe during the Arab explosion. Following the collapse of the Avars, their place had been taken by the Bulgar khanate in the 9th century, which expanded into northern Greece and Thrace. Bulgaria was a problem for the Byzantines, and one which they gradually dealt with. By 1028, Basil II Bulgaroktonos (slayer of Bulgars) had eliminated the state entirely. In the last decade of the 9th century there appeared on the Hungarian plain a group of nomads who were to have a far wider impact – the Magyars.

These had been pushed into Europe by the Patzinaks, and, according to their own histories, formed a confederation of 180 tribes. However, they were divided for war into only seven corps, and then operated in hundreds and dozens. Equipped in the normal steppe manner with the powerful composite bow, their tactics were to skirmish and feign retreat. They created a swathe of devastated land around their kingdom, for they were not interested in further conquest.

The Magyars specialized in long deep raids into lands already left weak by Viking or Saracen attack. Between 899 and 937 they conducted devastating attacks of this kind. Their range and speed of movement were formidable, as they penetrated as far as northern Germany, southern France and Italy. Even after they were checked at Riade in 936 by Henry, Duke of Saxony, they were not done with. In the following year a Magyar force swept through places as far apart as Bremen, Orléans and the heel of Italy.

However, the German response meant the beginning of the end for their free-wheeling ways. As a direct result of Magyar incursions the Saxon dukes built up a cavalry force with which to defeat them. On the River Lech in 955, Henry's son, the emperor Otto I, inflicted a signal defeat on the nomads, capturing many of their nobles. The Magyars never recovered, but accepted Christianity at the end of the 10th century. As the Christian kingdom of Hungary, their state was to form a bulwark against Oriental aggression for the rest of the Middle Ages.

The Seljuk Turks
Just as the Magyar and Saracen threat was coming to an end, a new storm appeared in the East. The Turkic Ghuzz (an Arabic corruption of Orghuz) split into two parts to assault the West. The northern group, known to the Byzantines as the Cumans, drove the Patzinaks before them into the Danubian lands c.1060. Both tribes caused the empire considerable trouble for a century, but they

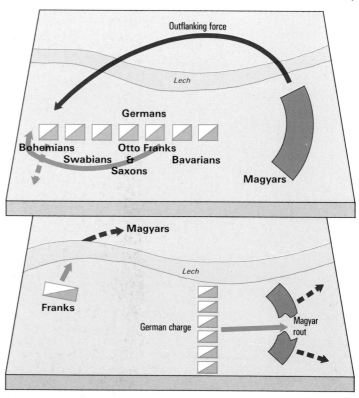

This 10th century silver dish (*above*) symbolizes the impact of the Magyars on Europe. A horseman leads a captive taken in one of the raids which terrified Europe. But only their nobility were so well-equipped; most Magyar cavalry were light horse-archers.

Battle of the Lech (*right*), August 955. King Otto's column of German heavy cavalry was moving south to attack the nomads. A Magyar flanking movement fell upon and routed the final division (*top*). But Otto had his men well in hand and sent the Franks back to counter the threat. Deploying his main battleline and instructing his men to charge closed-up, protecting themselves with their shields against Magyar archery, Otto won a signal victory (*above*).

were successfully contained by the Byzantine emperors of the Comnenian dynasty.

Not so the southern branch – the Seljuks. Their fierce bowmen overran the small Christian kingdom of Armenia, causing the inhabitants to migrate to Cilicia. A contemporary chronicler, Matthew of Edessa, commented particularly on how the Turkish arrows put the Armenians to rout. Then, under their great sultan Alp Arslan, the Seljuks burst into Asia Minor. Like all nomads they began with raids, penetrating deeper and deeper into Byzantium's rich agricultural heartland. Counterthrusts by the Greeks proved unable to dislodge them. Then, in 1071, the emperor Romanus Diogenes led a large army east for a showdown.

At Manzikert the two armies met in a battle that was to decide the fate of Asia Minor. The Byzantine force was a large one, allegedly consisting of 60,000 men. Advancing in two lines a bowshot apart, as the military manuals decreed, Romanus was disconcerted when the Turks failed to stand and fight. Using the classic steppe tactic, Alp Arslan withdrew his host until the Byzantines halted, believing there would be no engagement. Then, as Romanus attempted to return to his camp, the Turks pursued, harassing the Greeks with arrows. But the second line moved faster than the first, leaving it encircled. The finest soldiers of the Byzantine cavalry, including the "Schools" of the Imperial Guard, were shot down and the emperor was captured. The result was a disaster for Byzantium, and a classic nomad victory. At their leisure, bands of marauding Turks were now able to travel unopposed to the Bosphorus.

Other Seljuk bands moved into Syria, even capturing Jerusalem, and, in part, provoking the First Crusade. To the Byzantines, the crusaders seemed like another folk movement and they hurried the Westerners through their lands, seeking to recover territories in their wake. At Nicaea and Dorylaeum in 1097, the Seljuk Danishmend tribe were defeated by these new arrivals, but the crusaders swiftly moved south to their real goal.

The Seljuks of Rūm (as they called Anatolia) ruled until

"After we had set ourselves in order the Turks came upon us from all sides, skirmishing, throwing darts and javelins and shooting arrows from an astonishing range. Although we had no chance of withstanding them, or of taking the weight of the charge of so many foes, we went forward as one man. The women in our camp were a great help to us that day, for they brought up water for the fighting men to drink, and gallantly encouraged those who were fighting and defending them."

Anonymous First Crusader, Anatolia, 1097

---

they were subjugated by the Mongols in 1243. Their army consisted of two main parts: nomadic Turcoman tribesmen, the mainspring of the original conquests, and *ghulams*, or slave-soldiers of the sultan, who wore heavier equipment but lacked the nomads' tactical flexibility. In the 12th and 13th centuries the Seljuks developed sophisticated fortifications and fleets – as the Red Tower and boatyards at Alanya still show today.

The Berbers

Following the breakdown of Umayyad power in Spain, the

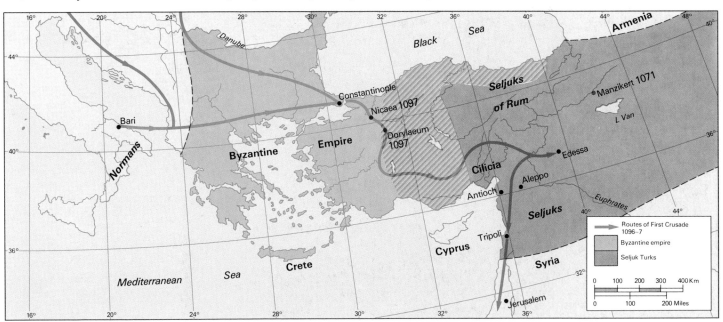

Asia Minor and the Levant saw profound changes in the 11th century. The Seljuk Turks irrupted into the area destroying the Christian kingdom of Armenia. Then the Byzantine empire, which had seemed so strong on the death of Basil II (1026), suffered a shattering defeat by the Seljuks at Manzikert in 1071. Emperor Alexius Comnenus (1081–1118) managed to recover some territory from the new Turkish emirates in Anatolia, but at a cost. This was because his main aid came from the West in the shape of a crusade. Its leaders were eager to establish their own states – which they did at Edessa, Antioch, Tripoli and Jerusalem. To the Byzantines, the arrival of the Franks was just as much an unpredictable folk movement as that of the Turks. The situation stabilized in the 12th century, with a Seljuk sultanate of Rūm, while the Syrian and Egyptian Muslims gradually recovered lost ground.

Christian kingdoms began to recover territory in the Reconquista. The capture of Toledo in 1085 provoked a reaction from the Almoravides of North Africa. Inspired by Islam, these Berbers were a formidable force. Clad in uniform black robes and capable of quick, well-disciplined manoeuvre, their combination of fast-moving cavalry and steady infantry threw back the Christians. Alfonso of Aragon was severely defeated at Zallaca in 1088. Despite the brilliance of Rodrigo Diaz (El Cid), his great warlord, the Muslims managed to regain some ground in the early 12th century.

When the Christian reconquest was once more moving forward, another Berber invasion occurred in 1145. The Almohades won a great victory over Castile at Alarcos in 1195. It took the Christians another two decades before they could in turn defeat the Muslims at Las Navas de Tolosa in 1212. Despite the fanaticism of the Berbers, the success of the Reconquista was now assured.

The Ayyubids

Meanwhile, in the Holy Land the crusaders had carved out a kingdom in Syria. Its very existence was an affront to Saracen pride, but fragmented Islam could do nothing about it. The 12th century, then, was a period in which notable leaders attempted to create a united state with which to recover the Holy Land.

Zangī, founder of the *atabeg* (regent) dynasty of the Zangids, was the first to make headway, recapturing Edessa in 1144. His son Nureddin saw off the Second Crusade (1146–7) and absorbed its goal, Damascus, in 1154. In 1169 Zangī took over Egypt, the powerhouse for reconquest, but it was left to Saladin to outshine his predecessors and found the Ayyubid dynasty. Born a Kurd, Saladin came to power through his own considerable abilities, and alone had the strength of character to hold the Syrian emirs to their task.

Saladin kept an army constantly in the field and wore the Christians down. It was his determination and guile which led to the destruction of the entire host of the Latin kingdom at Hattin in 1183. Stripped of their garrisons to provide this force, the crusaders' mighty castles fell easy prey to siege. Only Tyre was left to them. However, the arrival of King Richard on the Third Crusade gave Saladin his sternest test. Acre was recaptured for the Christians, and Saladin was defeated in battle at Arsuf in 1191. Richard had the generalship to be able to restrain his knights' charge until the crucial moment, routing the Saracen force. But he was compelled to leave the Holy Land before recapturing Jerusalem, as Saladin's war of attrition ground the crusaders down. Although surviving until 1291, the Latin kingdom never recovered, chiefly due to Saladin's achievements.

The Mongols

If nomad attacks on Europe may be likened to storms, the Mongol assault was of hurricane force. The Mongol leader Genghis Khan had created a truly irresistible army, although he did not live to lead it against the West – his concern was rather with the richer spoils of China. However, the Mongols developed an ideology of world empire, and nowhere was safe from their attentions. In 1220–21 the first "reconnaissance" swept around the Caspian Sea, encompassing the destruction of two eastern empires in its progress. The next assault was directed by

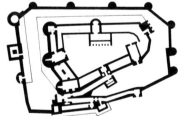

Krak des Chevaliers (*left* and *above*). The crusader states made stone do the work of men in defence, and their castles could only be reduced after long drawn-out sieges. Most horse-archer armies were ill-equipped to tackle fortifications (although this was not true of the Avars and Mongols with their experience of the Chinese art of war). Muslim leaders often had problems in keeping their armies together long enough to force a surrender. Saladin's success in capturing castles was due to the crusader defeat at Hattin (1187), which left no men to defend them. The Mamluks who conquered the Latins a century later had a strong siege train.

Batu, one of Genghis's four sons.

The Mongol army was formidable in its organization, discipline and loyalty, all carefully forged by Genghis. The simple decimal system, from troops of 10 to *tumans* (divisions) of 10,000, facilitated the transmission of orders. Western observers commented on the chilling silence with which the Mongols manoeuvred on the battlefield, following the signals of their horsetail standards. Political loyalty was assured by ruthlessness and the creation of a large imperial guard of 10,000 men, drawn from all the tribes of the confederation. There was no comparable force in Europe to challenge this military machine.

In the winter of 1237–38 the Mongols struck in Russia. Their hardy men and horses were untroubled by the bitter cold, and indeed the frozen rivers gave them extra mobility. Kiev and the other Russian principalities were totally overwhelmed. In the following year the Cumans and other nomads were mopped up. For the campaign of 1241–42 the invaders split into two. The northern force stormed into Poland. Duke Henry of Silesia, aided by the Teutonic knights, met it in battle. Charging impetuously, the Westerners fell into the Mongol trap, to be surrounded and annihilated at Liegnitz.

A few days later, on the River Sajo, Hungarian forces were just as utterly destroyed. Europe lay open to attack, but the death of Khan Ogedai caused the Mongols to return for a *kuriltai* (a meeting of all the tribal leaders) to elect another. Certainly this saved the Christian West, although it may be doubted that the Mongols would have made such rapid progress in the less suitable terrain west of the Alps. Their horse herds required extensive grazing, which was scarce once the Hungarian plain was left behind.

The Seljuks of Rūm were conquered at the same time. However, the main assault on the Arab world did not take place until the mid-1250s. In 1256, the Mongols won praise for destroying the terrorist Assassins at Alamut. But only two years later they sacked Baghdad and murdered the caliph, an act which rocked Islam. Once again, however, dynastic concerns caused them to withdraw, and it was only a small force that returned to conquer Egypt in 1260. Here, for the first time, they met an enemy worthy of them.

The Mamluks

As a result of the confusion brought about in Egypt by the crusade of King Louis IX of France (1248–54), the sultan was murdered in a coup (1250). The new dynasty was called Mamluk, a name that actually meant slave-soldier, and was used to describe the retainers of great Muslim lords. The mamluks of Egypt were drawn from warlike eastern tribes to form elite regiments. Under their own sultans they became the standard troop type.

Each mamluk was a cavalryman trained to use bow and lance to good effect. Like the *ghulam* soldiers, they fought in close ranks, and used archery for shock effect rather than the skirmishing tactics of the steppe nomads. Man (and often horse) wore armour covered by decorative robes, and the mamluks were equally adept at fighting on foot. They certainly learnt much from the contest with the Mongols in the late 13th century, and their sultans ensured a high level of training. There were even manuals to instruct them in the arts of war.

The Mamluk military organization was far from perfected when the Mongols first attacked in 1260. Sultan Qutuz was fortunate that political problems reduced the

Genghis Khan (c.1167–1227). The Mongol empire was created by the political and military genius of one man, Genghis Khan. After spending the first 25 years of his adult life uniting the Mongol and Tatar tribes of Central Asia, Genghis devoted his final two decades to the conquest of vast territories. In 1211 he invaded China. With the fall of Peking (1215), he turned his attention westwards towards the rival empires of the Kara Khitai and Kwarismians,

whom he conquered in only two campaigning seasons (1219–21).

Earlier nomads had often been halted by fortifications, but Genghis was quick to adopt useful military skills, and the Mongols soon developed an expertise in siege warfare, learned from the Chinese. He conducted his campaigns with utter ruthlessness, using subject troops to take the brunt of the fighting and bear the cost of siege assaults.

At the centre of Genghis's power was a relatively small force of 10,000 picked men, the fanatically loyal Guards. The size of Mongol armies was certainly exaggerated, but in warfare the ability to manoeuvre at speed and to achieve local superiority is more important than mere numbers. This was achieved by Genghis and his immediate successors, making the Mongol name universally feared.

> **MONGOLS AT PEACE AND WAR**
>
> "Each man has on average 18 horses and mares, and when their mount is tired they change it for another. They are provided with small tents of felt under which they shelter themselves against rain. They chiefly live on milk, thickened and dried to a hard paste. If speed is necessary, they can march for 10 days at a time without cooking food, during which time they subsist upon the blood drawn from their horses, each man opening a vein and drinking from his own little herd.
>
> "The men are used to remaining on horseback for two days and nights without dismounting; sleeping in their saddles whilst their horses graze. No people on earth can surpass them in fortitude under difficulties, nor show greater patience when suffering deprivations. They are perfectly obedient to their chiefs, and are maintained at small expense.
>
> ". . . When they come to engage in battle they never mix with the enemy, but keep hovering about him, discharging their arrows first from one side and then from the other, occasionally pretending to fly, and during their flight shooting backwards at their pursuers, killing men and horses as if they were fighting face to face. In this sort of warfare the adversary imagines he has gained a victory when he has in fact lost the battle; for, observing the mischief they have done him, they wheel about and, renewing the fight, overpower his remaining troops and make them prisoners, whatever the resistance. The horse are so well broken-in to quick changes of movement that, upon a signal given, they instantly turn in every direction; and by these rapid manoeuvres many victories have been obtained." (From the Journal of Marco Polo, c.1300.)

invading force to only two *tumans* (notionally 20,000 men). This meant that the Mamluks outnumbered the Mongols, whose forces also contained a high proportion of subject troops. However, few would have predicted the success that Qutuz achieved at the aptly named Spring of Goliath. He gave the Mongols a taste of their own medicine by feigning flight with his vanguard in order to lure the enemy into an ambush. Even then the battle was hotly contested, and only won by a final charge led by the sultan himself. The Mamluks' flexibility then came into play as they dismounted to pursue fleeing Mongols up steep hillsides.

Mongol attacks of 1281 and 1303 were also repelled, as the Mamluks became the chief power in the eastern Mediterranean. Sultan Baibars snuffed out the Latin kingdom by 1291, when Acre fell. In the 14th century Mamluk power expanded into Asia Minor, when they conquered the Christian kingdom of Armenia (1375), and only withdrew in the face of Ottoman pressure.

The Ottoman Turks

After the Seljuks of Rūm had been destroyed by the Mongols, Asia Minor split up into many small emirates. From one of these, in northeastern Anatolia, grew the Ottoman power. Victory over a Byzantine army in 1301 marked the first step of the Ottomans toward empire. Gradually their chiefs developed an aggressive and expanding state, based on *ghazi* warfare. This was a form of border raiding with a religious motivation; the Ottomans employed fanatical Dervishes from early in their history. Not that their army was restricted to Muslims alone. Flexibility, and the use of any tools that came to hand – human or technological – were a feature of the Ottoman rise to empire. So their armies consisted of dissident Greeks, and, once they were in Europe, Wallachians, Serbs and other "voynik" warriors.

Expansion in the face of declining Byzantine power was rapid. Bursa was captured, and became the capital in 1326, and raids across the Dardanelles into Thrace and northern Greece were already carving out territory there. The attacks took the form of *razzias* (raids) whereby the populations were penned in towns while the agricultural land was devastated, destroying the defenders' means of support. Political authority swiftly passed to the Ottomans, Edirne becoming the new capital after its capture in 1361. The Ottomans' objective was Constantinople itself, and they slowly tightened the noose around the city.

The cutting edge of Ottoman expansion lay in the Turcoman nomads and their style of warfare; but the sultans realized early on that their army needed a greater variety of troops if their state was to be maintained. They never neglected infantry, whether the irregular Azaps or the more famous Janissaries. This elite corps was formed in the mid–14th century, and originally depended on prisoners of war to fill its ranks. In 1438 the *devshirme*, a levy on Christian boys, was instituted. They were then brought up to be fanatically loyal to the sultan. A professional force, adept with bow, crossbow, and later, handguns, they wore distinctive, tall white caps.

However, the cavalry were not neglected. Heavier armour for man and horse marked the Spahi warrior, almost the counterpart of the western knight. Spahis also made up part of the *Kapikulu*, or Guard troops, and were usually reserved to deal the final blow in a battle.

This army was used to great effect, to crush the Serbs at Kossovo in 1389, and repel crusaders. The crusade of Nicopolis (1396) was an attempt to rescue Byzantium, hard-

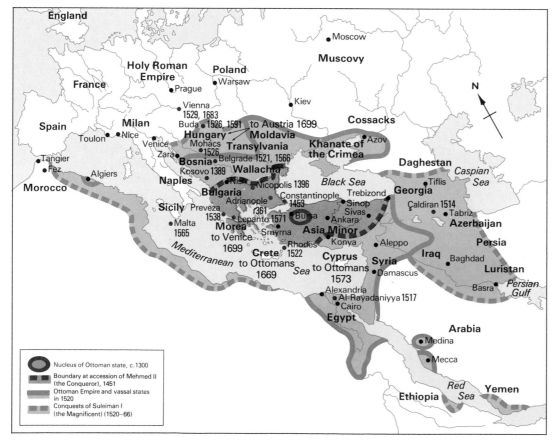

The growth of the Ottoman empire c.1300–1700. From the very humble beginnings of a small emirate in northwestern Asia Minor, the Ottomans rose to dominate the Islamic world and threaten the West. Cavalrymen by preference and tradition, the Turks learnt to incorporate infantry forces (notably the famous Janissaries) engineers and artillerymen into their armies. Their run of success as conquerors was only broken (briefly) by Bayezit's defeat at the hands of Tamburlane near Ankara in 1402. When Constantinople fell in 1453 it became the heart of the empire, from whence flowed victorious fleets and armies. Under Suleiman the Magnificent the Ottomans rose to the height of their power. Then, checked on land at Vienna (1529) and Malta (1565), and at sea at Lepanto (1571), they lost their impetus. Thereafter, a long, slow decline set in.

Legend (map):
- Nucleus of Ottoman state, c.1300
- Boundary at accession of Mehmed II (the Conqueror), 1451
- Ottoman Empire and vassal states in 1520
- Conquests of Suleiman I (the Magnificent) (1520–66)

pressed by Sultan Bayezit. Unfortunately, the impetuous French crusaders allowed themselves to be drawn into a classic Ottoman trap. Ignoring the pleas of their Hungarian allies for caution, the knights plunged forward to attack the swarms of Turkish light horse. These parted to reveal a field of stakes, behind which stood infantry archers, who took their toll on the crusaders' horses. Dismounting and struggling through the obstacles the well-armoured knights crushed the Ottoman foot including the Janissaries. But then the exhausted Westerners found themselves faced by the Spahis. Unable to fight any longer the entire force was ignominiously captured.

This battle summed up the Ottoman tactical system: to disrupt the enemy with horse-archers, weaken them with missiles and then counterattack with the heavy Spahis. They employed this method successfully for another two centuries, replacing the archers with chained-together cannon, as the new technology became available.

Although Bayezit was himself defeated by a resurgent Mongol empire under Tamburlane in 1402, the Ottomans were so well-established that they did not take long to recover. Nomad tactics were thus proved to be still supreme, but the development of gunpowder weapons was shortly to render them obsolete: the Ottomans used artillery to smash their way into Constantinople in 1453.

Under Mehmet the Conqueror they advanced on all fronts, even reaching Italy in 1480, the year of his death. Moreover, the Ottomans had also become a seapower, taking on the Venetians and driving them out of the Aegean. Under Ottoman tutelage, Barbary pirates began to ravage the coasts of the western Mediterranean. It was as if the bad days of the 10th century had come again.

The Ottomans continued to throw up great generals.

Selim I defeated Saffavid Persia at Çaldiran in 1514; the Mamluks of Egypt were overrun in the campaign of 1516–17. In both cases, old-style cavalry warfare was defeated by the use of handguns, cannon and wagons. This was a tactic that the Turks had learned from the Hungarians, who had resisted conquest in the later 15th century. Under Suleiman the Magnificent, Hungary was crushed at Mohács in 1526, however. The Sultan went on to besiege Vienna – and all Europe trembled. But the siege of 1529 failed; Ottoman resources had become overstretched.

The Turks were to make no further progress. They were repelled at Malta in 1565, and their fleet was destroyed at Lepanto in 1571. The vast empire began to stagnate. No further advances in military or naval technology were made, and the conservative troops refused to change previously successful tactics.

The quality of leadership offered by the sultans also declined significantly around 1600. In the 17th century it was a dynasty of viziers, the Köprülu, who instigated a brief revival. In fact the defeat at Vienna in 1683 proved to be a final fling. The new armies of Western Europe, equipped with rapid-firing muskets and artillery, now proved more than a match for Eastern horsemen.

European voyages of exploration and conquest had begun in the 15th century. Technological development now gave the Westerners a significant advantage. As Russia began to conquer Central Asia in the 16th century it only took a few shiploads of men with gunpowder weapons to rout the descendants of the terrifying Mongols. The Ottomans survived longer than other oriental powers, because they had in part adopted the new techniques of warfare; other nomad societies merely crumbled. The long reign of the steppe nomad was over.

The relief of Vienna by the army of John Sobieski, King of Poland, September 12 1683. Following the Austrian victory of St Gotthard in 1664, it was clear that Turkish armies were inferior in discipline, training and weaponry, especially artillery; but they still possessed overwhelming numbers. In 1683, the Austrians could raise only 30,000 to oppose 150,000. The emperor Leopold fled the city, which was invested by the Turks on July 17. But they lacked heavy artillery and mining operations proved insufficient. Vigorous sallies by Vienna's small garrison inflicted heavy casualties and Turkish morale was already low when Sobieski's army, 70,000 strong, approached the city. According to an eyewitness, "The Turkish forces covered the foot of the mountains, their white turbans presenting the effect of a great white carpet."

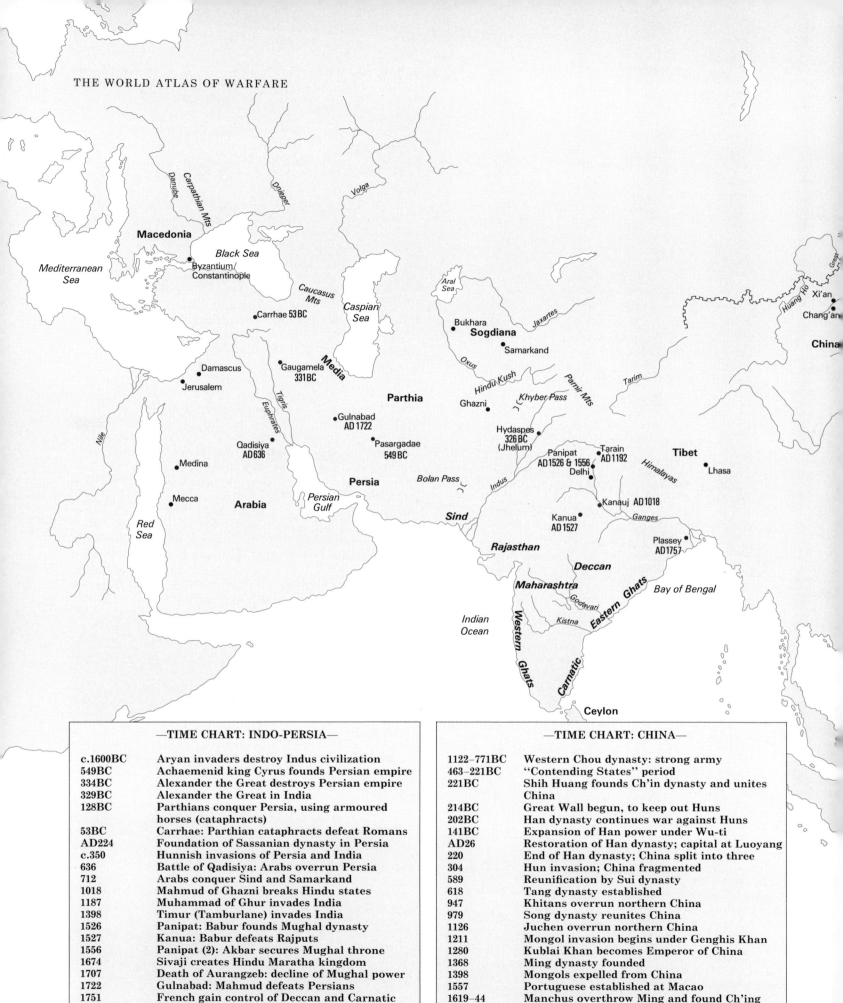

—TIME CHART: INDO-PERSIA—

c.1600BC	Aryan invaders destroy Indus civilization
549BC	Achaemenid king Cyrus founds Persian empire
334BC	Alexander the Great destroys Persian empire
329BC	Alexander the Great in India
128BC	Parthians conquer Persia, using armoured horses (cataphracts)
53BC	Carrhae: Parthian cataphracts defeat Romans
AD224	Foundation of Sassanian dynasty in Persia
c.350	Hunnish invasions of Persia and India
636	Battle of Qadisiya: Arabs overrun Persia
712	Arabs conquer Sind and Samarkand
1018	Mahmud of Ghazni breaks Hindu states
1187	Muhammad of Ghur invades India
1398	Timur (Tamburlane) invades India
1526	Panipat: Babur founds Mughal dynasty
1527	Kanua: Babur defeats Rajputs
1556	Panipat (2): Akbar secures Mughal throne
1674	Sivaji creates Hindu Maratha kingdom
1707	Death of Aurangzeb: decline of Mughal power
1722	Gulnabad: Mahmud defeats Persians
1751	French gain control of Deccan and Carnatic
1757	Plassey: British defeat Nawab of Bengal
1761	British destroy French power in India
1818	British defeat Marathas

—TIME CHART: CHINA—

1122–771BC	Western Chou dynasty: strong army
463–221BC	"Contending States" period
221BC	Shih Huang founds Ch'in dynasty and unites China
214BC	Great Wall begun, to keep out Huns
202BC	Han dynasty continues war against Huns
141BC	Expansion of Han power under Wu-ti
AD26	Restoration of Han dynasty; capital at Luoyang
220	End of Han dynasty; China split into three
304	Hun invasion; China fragmented
589	Reunification by Sui dynasty
618	Tang dynasty established
947	Khitans overrun northern China
979	Song dynasty reunites China
1126	Juchen overrun northern China
1211	Mongol invasion begins under Genghis Khan
1280	Kublai Khan becomes Emperor of China
1368	Ming dynasty founded
1398	Mongols expelled from China
1557	Portuguese established at Macao
1619–44	Manchus overthrow Ming and found Ch'ing
1840–42	Opium War: Britain annexes Hong Kong
1842	Five "Treaty Ports" opened to foreign trade
1850–64	Tai-p'ing rebellion

— chapter five —
War in the Orient

T A Heathcote

The military history of Asia covers an area as diverse as the ice deserts of Siberia and the jungles of Siam, or the sands of Palestine and the islands of the Pacific. This chapter concentrates on India, Iran, China and Japan before the dominance of the West. India and Iran shared a common military heritage, and, with China and Eastern Europe, a common military threat – the horse-riding barbarians of the Asiatic steppes.

The Orient is to most Europeans and Americans synonymous with the continent of Asia. Warfare in so vast an area covers almost every extreme of climate and terrain, and virtually every kind of society and culture. There have been wars among jungle tribesmen, hardy mountaineers, and riders of the endless steppes, as well as between highly sophisticated empires, with rich and ancient civilizations. Long before the first organized states had emerged in the Occident, there were, in the Orient, great empires and mighty cities with powerful armies, well-made weapons, and skilful generals. In classical and medieval times the successors of these empires were militarily far more significant than any European kingdom. But the military technology of the 19th century enabled European powers, particularly Russia, Britain, and France, to achieve the conquest of Asia.

—TIME CHART: JAPAN—

AD794–1185	Heian Period: emergence of warrior elite
1156–85	Gempei Wars between Taira and Minamoto
1185	Dan-no-Ura: Taira clan annihilated
1192	Minamoto Yorimoto establishes Shogunate
1192–1333	Kamakura period marks rise of warrior class
1274	First Mongol invasion
1281	Second Mongol invasion
1338–1573	Murimachi Shogunate
1467–77	The Onin Civil War
c.1542	Portuguese arrive in Kyushu
1560	Oda Nobunaga reduces power of feudal lords
1575	Nagashino: success of Nobunaga's arquebusiers
1590	Hideyoshi unifies Japan
1597	Hideyoshi's Korean expedition defeated
1600	Sekigahara: supremacy of Ieyasu
1603	Ieyasu forms Tokugawa Shogunate
1612	Christian missionaries banned
1615	Osaka: Tokugawa forces defeat Hideyori
1624	Isolation policy: Spanish traders expelled
1804	Russo-American trade mission rebuffed
1853–54	US naval mission under Perry negotiates trade treaty
1868	Meiji Restoration of the Emperor's power

India and Iran

Throughout most of its history the Indian subcontinent has been affected by invasions and internal wars. The earliest known Indian civilization, that of the Indus valley, was overthrown around 1500BC by invading Aryans from the Eurasian steppes, and these, in course of time, synthesized with existing inhabitants to form the distinctive religion and culture of Hindu society. To the usual three classes into which the early Indo-Europeans divided their society (priests and scribes; rulers and warriors; and peasants and traders) they added a fourth, the serf or menial, descendants of the original conquered inhabitants or the offspring of renegade Aryans.

It became an essential element of the culture of the Hindus, "the people of the Indus" as first the Persians and then the Europeans called them, that all civilized men belonged to one of these four great classes or *varna* (literally, colours) by a divinely preordained, immutable arrangement. Any social or physical contact between the different classes was forbidden. Barbarian invaders, jungle tribesmen, those who performed unclean tasks or otherwise ignored the Hindu code, were deemed "untouchable" and outside society altogether.

Hindu culture spread far beyond the regions of northern India settled by the Aryans. Its teaching was accepted in Bengal and the whole of the Indian peninsula, and spread to Ceylon, Burma, Indochina and Indonesia. The fourfold division of society (its original purpose long vanished) spread with them, with the effect of excluding the majority of the population from bearing arms. Only members of the *Ksatriya*, or warrior class, were expected to be soldiers. This, together with the system of castes (*jati*) or mutually exclusive occupational groups which evolved at a later date, (though based on economic rather than religious factors) paradoxically increased the size of armies in the field. Each separate logistic function, such as those of driver, grass-cutter, water-carrier, groom, armourer, bearer or sweeper was performed by noncombatants, hereditary specialists in one skill only. No Indian army, even in the 20th century AD, could take the field without these followers. In the medieval period, armies of up to a million people were recorded, but these figures would have included followers, together with merchants, sutlers, contractors, and other parasites, rather than disciplined soldiers. The ease with which bands of invaders could overthrow great Indian kingdoms owes much to the fact that only a relatively small proportion not merely of the population, but of the armies themselves, were conditioned to fighting in their own defence.

The four major arms of the classical Indian army were infantry, cavalry, chariots, and war elephants. Personal weapons varied according to the region and period but included the usual weapons known to early military history: swords and daggers of different types, thrusting and throwing spears, maces, battle-axes, and so on. Body armour and helmets were not generally worn until the medieval period. As in most armies of antiquity, missile action, whether by javelin, sling-stone, or arrow, all requiring personal agility, was left to the light infantry, while infantry of the line carried a stout spear and shield. Bows, made of bamboo, were 5–6ft long (needing to be steadied on the ground by the archer's foot) and shot long cane arrows. Shorter, composite bows were used by chariot and elephant troops.

Indian cavalrymen, though dashing riders and individually well-skilled at arms, always had difficulty in maintaining the size and strength of their horses, and were commonly less well-mounted than invaders from the Central Asian regions on which the Indian herd depended for replenishment. The light chariot of the Aryans was replaced by a heavier vehicle drawn by four horses yoked abreast and carrying a crew of four. Chariots ceased to be used in actual combat early in the Christian era, though long continuing in service, rather like horsed cavalry in the 20th century, to provide mounted infantry and for ceremonial parades.

Elephants-at-arms

The truly distinctive weapons system of Indian armies was the war elephant. Elephants had been both tamed and trained for war by the time of the first Indian ruler whose name is known to historians, King Bimbisara of Magadha (d.490BC), who owned a large and efficient corps of these sagacious pachyderms. In addition to his *mahut* or driver, who sat on the animal's neck equipped only with the *ankus* or goad, the war elephant carried another two or three men who fought from a wickerwork howdah, armed with bows, darts and quoits for missile action, and a specially long lance (*tomara*) for shock action. The elephant's tusks might be sharpened or lengthened with sword blades, and he might pick up enemy soldiers with his trunk, as well as trampling them underfoot.

The standard battlefield role of war elephants was in the assault, breaking up the enemy ranks, but they were also used in sieges, pushing over gates and palisades, or even as living bridges. The fighting elephants were usually bulls, selected for their size and ferocity. They were considerably outnumbered in most Indian armies by baggage elephants, mostly cows and smaller bulls, who combined a heavy load-carrying capacity with good cross-country performance.

Long before the rise of the first Hindu empire, the Indo-European tribes of Iran had coalesced into a kingdom which would be the greatest that the world had so far known. In 549BC, the Achaemenid King Cyrus (Kurus), ruler of the Persians, the people of southern Iran, conquered the Medes at the battle of Pasargadae. In 547BC at Sardis, he led his combined subjects to victory over Croesus, King of Lydia. He died in 530BC, in battle against the Scythian nomads of the Jaxartes Basin, having founded an empire that would stretch from the Nile to the Indus, and from the Aegean to the Arabian Sea.

A century and a half later, this empire was conquered by one of the most famous captains in military history, Alexander the Great of Macedon. In 331BC, he defeated Darius III at Gaugamela, and made himself master of the

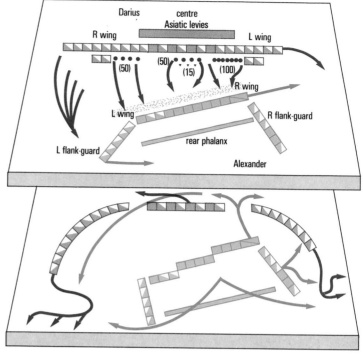

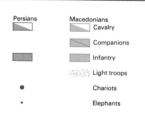

At Gaugamela, the Persian left-wing cavalry (*top*) raced ahead to intercept Alexander's oblique advance. The Macedonian phalanx charged the gap opened in the Persian line (*bottom*), rebuffed their right, and routed Darius.

Persians	Macedonians
	Cavalry
	Companions
	Infantry
	Light troops
Chariots	
Elephants	

Persian empire. After several campaigns to secure his position in Central Asia, he moved south to attempt the conquest of India. The major battle of Alexander's Indian campaign occurred in July 326BC, against a king known to the Greeks as Porus of the Paurava dynasty, ruling in the Punjab.

The two armies met on the banks of the Hydaspes (Jhelum). Alexander had successfully made opposed river crossings before, but appreciated that his horses could not be made to swim over in the face of elephants. Moreover, the monsoon had started and the river was in full spate. After several feints, he lulled Porus into a sense of false security and, under cover of a violent storm, made a night crossing several miles from his main camp. The Indian king sent cavalry and chariots to contain the bridgehead, but the cavalry were checked by Alexander's horse-archers (recruited during his campaigns in Central Asia) and thrown back by a charge led by Alexander himself. The chariots, attempting to drive to the rescue of their comrades, stuck fast in the mud and had to be abandoned. Porus, having located his enemy's main force, deployed in the conventional manner. On the wings were his remaining chariots, supported by cavalry. In the centre were his 200 war elephants, with infantry close supports between them, and the main infantry force in the rear. Alexander opened the attack with his horse-archers, to whom the Indians, struggling to get a grip on their long bows in the muddy ground, could make no effective reply. There followed a cavalry fight, in which the greater weight and cohesion of the Macedonians proved decisive, and the

The battle of Gaugamela (*opposite*), scene of Alexander's final defeat of the Persian king Darius, was fought on ground chosen by Darius: a levelled plain that would give advantage to his strength in cavalry. The time, however, was chosen by Alexander, who forced the Persians to stand to arms all night while his men slept. Then, threatening to turn Darius' right flank, he held off an attack on his own right, and destroyed the Persian centre with his heavy cavalry and infantry phalanx. Elephants were present but inconspicuous at Gaugamela, whereas later, during the Indian battle of the Hydaspes, they disrupted the Macedonian phalanx.

The Muslim rulers of India, like the Hindus before them, placed a great, though frequently unjustified, reliance upon the fighting qualities of elephants. Elephants were royal beasts, reserved for rulers. This principle was maintained by the British, who used elephants in their armies, pulling guns and carrying loads.

The war elephant, India's unique contribution to the battlefield, formed part of Indian armies from the 6th century BC to the 18th century AD. They took part in the campaigns of the Persian empire against the Greeks and Alexander the Great, and served in the campaigns of his successors around the Mediterranean. With Hannibal the Carthaginian they even crossed the Alps and reached the gates of Rome itself. The supply of elephants from India to the west was cut off by the rise of the Parthian empire, whose own new speciality, the cataphract, offered a passable substitute means of rapid shock action. War elephants were briefly re-introduced into Central Asian warfare by Mahmud of Ghazni. But the reliance of Indian tacticians on war elephants as a battle-winning system was often misplaced. Although they might, at first sight, strike terror into troops unfamiliar with them, even the best-trained elephants could be panicked, and turn into a common enemy trampling friend and foe alike. From the Romans to the British, experience proved time and again that well-drilled infantry had little to fear from war elephants, although on an unwieldy and ill-organized host their impact could well be decisive.

57

Indian horsemen retired through the elephant line. With a great beating of war drums, answered by the Greeks' trumpets, the elephants advanced, looking, so the Greeks said, "like a city on the move", with their turreted howdahs the towers, and their supporting infantry the curtain wall. The close-packed Macedonian phalanx made an easy target, but with their 16ft-long pikes they were able to hold off the Indian infantry and engage the elephants and their riders. Alexander's light troops harried the beasts as they wearied, though suffering casualties if approaching too close. Eventually, reminding the Greeks of ships backing water, the elephants retreated and were driven from the field, often badly hurt, or with their men killed or wounded. The phalanx charged home on the Indian infantry, and Alexander's cavalry closed in on both flanks. However, the prospect of a never-ending march to the east, through trackless jungles to a point where it was reported that the king of Magadha waited with elephants numbered in thousands rather than hundreds, was too much for Alexander's homesick men. Reluctantly he led them home through Sind and along the Arabian sea coast, fighting men or the climate throughout the journey, and suffering a serious wound which contributed to his death at Babylon in June 323BC.

A new weapons system emerged with the rise of the Parthians, originally nomads from the east Caspian regions, who by 128BC had replaced Alexander's Greek successors as the rulers of Iran. As the Parthians settled into their new conquests they no longer needed to leave their best horses out to pasture all the year but could grain-feed them, stalled in the imperial stables. The result was a new breed of horse, bigger and stronger than the hardy steppe ponies, and able not merely to carry an armoured rider, but to wear armour themselves. Thus evolved the cataphracts, or "bulwarked ones", the horsemen that would be the most powerful piece on the board of battle for over a thousand years. From the beginning, in conjunction with the more traditional horse-archers of the steppes, who could deliver the famous "Parthian shot" while riding away from an enemy, they proved a formidable combination. Their first great victory was over the Romans under Crassus at Carrhae (53BC), and a second was won against Mark Antony in Azerbaijan (36BC).

The coming of the Muslims

The centuries of warfare between first Rome and Parthia, and then Byzantium and the Sassanian empire of Iran ended with the emergence in the 7th century AD of a new power and a new faith, the Arabs and Islam. The Sassanian military system, which, since its reorganization under Khusrau I (531–579), had kept at bay the Byzantines in the west and the steppe nomads in the east, crumbled before this new onslaught. Its western armies were overwhelmed at Qadisiyya (636) and Niharand (642), and its eastern forces, with their homeland occupied, came to terms with the new regime, which offered salvation and lighter taxes. In theory the head of the whole Muslim world was the Caliph (Khalifa), God's shadow upon earth, governing (after 750) from Baghdad. In practice the outlying provinces were often independent states, such as that of Ghazni in Afghanistan, which in the 11th century AD became a powerful kingdom extending from the Oxus to the Punjab.

Cataphract · Horse-archer

c	Crassus		Heavy infantry		Baggage train		Parthian horse-archers
	Cavalry		Light troops		Parthian cataphracts (lancers)		

At the battle of Carrhae, (*left*) 53BC, the Romans, under Crassus, comprised 28,000 legionaries, 4,000 auxiliaries and 4,000 cavalry against 9,000 horse-archers and 1,000 cataphracts led by the Parthian Suren. Forming a square against the cataphracts, they became an easy target for the archers. When Roman light troops attacked the archers, they were destroyed by the cataphracts. The survivors of the main force escaped only after nightfall.

Horse-riding nomads (*above*) from the steppes of the Eurasian heartland presented a military threat to the settled populations of the civilized rimlands from the earliest recorded times. Their characteristic weapon was the short but powerful composite bow. Parthian steppe-riders produced a breed of horse that could carry armour, and so began the cataphracts, the forerunners of the mailed knights of medieval Christendom.

Mahmud (AD971–1030) deposed and succeeded his younger half-brother Isma'il as ruler of Ghazni in 997, after a brief war of succession following the death of their father Sabuktigin. In 1001, he turned his attention to India. Impressed by the wealth of the country, and its vulnerability to determined attack, Mahmud annexed most of the Punjab, and commenced a series of great raids deep into India.

With the proceeds of these raids he built up an empire covering all of modern Afghanistan and Pakistan, eastern Iran, and much of Soviet Central Asia. He maintained a standing army of over 54,000 horsemen and 1,300 armoured elephants (an arm which he re-introduced into Central Asian warfare) together with a camel-borne corps of mounted infantry. He recruited from a cross-section of his subjects, including Hindus, to encourage martial rivalry in battle and prevent combinations against himself. Troops without local ties were the readier to exact the high taxes needed to pay their wages.

The grasping nature of revenue officials led to Sultan Mahmud, as their employer, gaining a reputation for avarice which has outweighed his many estimable qualities. In war he was a bold and skilful general. In peace he was a wise and able governor.

Under their great sultan, Mahmud the Iconoclast, the troops of Ghazni, supported by religious warriors and adventurers from all Central Asia, raided deep into India. In the next century Ghazni itself fell to the state of Ghur, its former tributary, and in 1187 a prince of that state, Muhammad, led his own armies down into the rich plains of Hindustan.

The people on whom the attack fell were the Rajputs, descendants of earlier invaders who had come to be regarded as the embodiment of Hindu chivalry. All claimed to be the sons of kings (*rajaputra*), and their clan chieftains formed a haughty aristocracy, supported by armed retainers who scorned useful work but were ready to fight and die in defence of their lord's honour. Wars were fought over land, property, girls, or mere points of etiquette.

Conflict at Tarain

The Rajput kings set aside their quarrels to combine against the invasion of Muhammad of Ghur, just as their forefathers had against the raids of Mahmud of Ghazni. Still clinging to outmoded concepts of warfare that had failed to stop every invader since Alexander the Great, the Hindu armies, led by Prthvi Raja, the Lochinvar of India, met the Muslims at Tarain in 1191. Muhammad was wounded and retired with his army vowing revenge. The next year he came again in greater strength. At the second battle of Tarain, the Muslims' reliance on the traditional weapons system of the steppes, the mounted archer, proved justified. The vast Hindu host collapsed like a great ruined building. Their commander, the valiant Prthvi Raja was captured and slain.

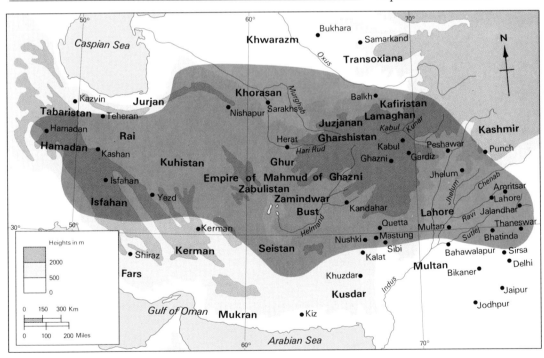

The Rajputs or "sons of kings" were mostly descendants of Huns and other invaders from Central Asia who conquered northwestern India in the 5th and 6th centuries AD. They took the place of the original *Ksatriya* caste to form a warrior aristocracy, with a code of chivalry derived from the ancient Hindu epics. Fiercely loyal to their clan chieftains, the Rajputs regarded warfare as the only truly honorable occupation, and regularly practised warlike pursuits.

The Ghaznavid empire was one of several in the course of history which united the Iranian plateau and the Indus valley, with the intervening mountains, under a common ruler. Sultan Mahmud himself ruled from Kashmir to the Zagros.

In consequence of Mahmud's campaigns, large areas of Iran were reduced to deserts, and great cities of Hindustan suffered irreparable damage. Nevertheless, he spent large sums beautifying Ghazni, and summoned many learned men to his court.

Mahmud's successors lost their northern provinces to Turkish nomads, and Ghazni itself to the Sultan of Ghur. It was destroyed in 1150.

In the campaigns that followed, all the great cities of Hindustan were taken. Muhammad then turned his attention to Khorasan, leaving his general Qutb-ud-din Aibak to complete the conquest of northern India. After his master's assassination by Muslim zealots in 1206, Qutb-ud-din became Sultan of Delhi, founding a state which by 1340 held sway over almost the whole subcontinent.

North India was spared the descent of the Mongols that devastated Iran, but shared the misfortune of an invasion by the Sultan Timur (Tamburlane) of Samarkand. Never-

"My principal object in coming to Hindustan was to accomplish two things. The first was to war with infidels . . . and by this religious warfare to acquire some claim to reward in the life to come. The other was a worldly object: that the army of the faithful might gain something by plundering the wealth of unbelievers."

Autobiography of Sultan Timur of Samarkand

theless, even his bloodthirsty troopers were apprehensive at the sight of 120 armoured elephants among the army waiting for them at Delhi, and the historians and scholars in his train were so alarmed by these anachronisms that, when asked by Timur where they wished to be placed, they replied "Among the ladies". Timur resorted to the old nomad tactic of defending his position with a lien of wagons and tethered buffaloes, which, with swathes of caltraps, broke the enemy charge and left the way open for

a decisive counterattack.

In 1525, Zahir-ud-din Muhammad, known as Babur (Tiger), prince of the house of Timur, invaded the plains of Hindustan. He had lost his Central Asian dominions to the Uzbergs, and was "tired of wandering about like a king on a chessboard" as he put it. He therefore decided to "place my foot in the stirrup of resolution, and my hand on the reins of confidence in God".

At Panipat, near Delhi, he encountered Sultan Ibrahim Lodi, who had 1,000 elephants and 100,000 men, but, in Babur's words "lacked experience of war, marched without order, halted or retired without judgement, and engaged without foresight". Ibrahim was provoked into attacking on a narrow front, against a line of stockades and wagons roped together, from the shelter of which Babur's cannon and matchlocks poured out deadly fire. Once more, the mounted archers of the steppes reduced an unwieldy Indian army to a helpless mob. Afterwards, they counted 20,000 enemy dead, among them Sultan Ibrahim, in total more than three times the number of Babur's entire army. A Rajput army of 80,000 horse and 500 elephants, under the veteran Rana Sangram Singh, now moved against Babur, hoping to restore Hindu rule. The battle of Kanua (March 16 1527) was a repeat of Panipat, with the decisive factor being Babur's matchlocks, on mobile tripods of his own invention. The Rajputs were "scattered like teased wool and broken like bubbles on wine".

Under Babur's successors, the Mughal empire extended over all of India. The second battle of Panipat (November 5

Babur's empire (*left*) was lost and then regained, except for its Central Asian territories, by his son Humayun, whose son Akbar completed the conquest of northern India, and pushed into the Deccan. The reigns of Akbar's son, Jahangir (1607–27), and grandson Shahjahan (1628–59) saw further conquests in the Deccan. Shahjahan's son Aurangzeb became ruler of the entire subcontinent.

Timur (Tamburlane) a Barlas Turk (*above*) who had already conquered most of Central Asia, campaigned in northern India in 1398–99. His 90,000 horsemen left a trail of slaughter and pillage even more terrible than that of Mahmud the Iconoclast. It was recorded by an eyewitness that Delhi was left so ruined that "for two whole months, not a bird moved a wing in the city".

"I felt that for me to be in a warm dwelling and comfort while my men were in the midst of snow and drift, for me to be within, enjoying sleep and ease, while my followers were in trouble and distress, would be inconsistent with what I owed them, and a deviation from that society of suffering that was their due. So I remained sitting in the snow and wind in the hole that I had dug out, with snow four hands thick on my head, back, and ears".

Autobiography of Babur, first Mughal Emperor

1556) brought his young grandson Akbar, with 20,000 horsemen and a well-served artillery park, victory over a combined force of Hindus and rival Muslims amounting to 100,000 men and 1,500 elephants. In his later years Akbar secured Kandahar and Baluchistan, and shortly before his death in 1605 commenced the conquest of the Deccan.

The campaigns of the Emperor Aurangzeb (ruled 1659–1707) saw the zenith of Mughal power. A puritanical defender of the orthodox Sunni branch of Islam, he conquered the Shiah sultanates of southern India, and by his intolerance drove the Rajputs into open rebellion. His merciless persecution of the Sikhs, or "disciples", (originally a peaceful sect, combining Islamic and Hindu ideals) turned them into a new martial class. The last 20 years of his reign were spent in a war against Marathas, a new Hindu power arising in the Deccan. The Maratha kingdom had been formed in the hills of the Western Ghats under Sivaji, a great guerrilla leader. The war that continued

In the siege of Fort Chitor, 1567, Akbar used exploding mines to bring down the walls of the historic Rajput stronghold. Akbar's policy towards the Rajputs combined conciliation and conquest, but he could be merciless towards those who refused to accept his supremacy. The capture of Chitor was followed by a massacre of its inhabitants, prompting other Rajput chiefs to accept Akbar as emperor in 1570.

under his successors drained the empire's resources. The Mughal armies, hampered by the entire imperial household, with a baggage train of 30,000 elephants and 50,000 camels, marched ponderously around the Deccan seeking a decisive battle. Swarms of Maratha light horsemen hung upon its flanks and harassed its supply lines. They destroyed the crops and pre-empted the revenue, leaving neither grain to feed Mughal men and animals, nor money to maintain armies. Eventually the Marathas marched openly, equipped and organized as conventional regular troops, and in 1737 raided outside the walls of Delhi itself.

Later developments in Iran

In Iran, new military groups emerged during the 16th

century. Its founder, Ismail, made himself Shah at the head of an army of his Turkish fellow tribesmen, militants in the Shiah cause. The very name, *qizilbash* (the red heads) derived from the 12 red stripes on their turbans, one for each of the early Shiah martyrs. Between 1501 and 1510, Ismail established the first kingdom to rule all Iran since the descent of the Mongol hordes.

The greatest of his successors was Abbas I (ruled 1571–1629). To reduce the power of the *qizilbash* tribal leaders, he formed a new military class, that of the *ghulam* or martial slave, descended from Circassians and Georgians captured in earlier northern campaigns. From these men he raised a regular army, paid from central revenues, with no local roots or allegiance to any but himself.

In 1722, an Afghan chieftain, Mahmud, ruler of the Ghalzai clan, marched on the Iranian capital Isfahan. At Gulnabad (March 8 1722) at the head of some 15,000 tribal warriors and other adventurers, he defeated a splendidly caparisoned but poorly organized Iranian army of over twice his own strength, and went on to take Isfahan where he ruled briefly before going insane.

He was eventually succeeded by another warlord, Nadir, who made himself Shah of Iran, drove out the Ghalzais, threw back the Turks and the Russians encroaching from the west and north, and invaded India. He sacked Delhi and gave his name there to a new word *nadirshahi*, a holocaust. In the end, his atrocities and paranoia grew too much even for his own generals, who, just in time to save their own heads, murdered him in 1742. The commander of his Afghan household troops, Ahmad Durrani, escaped with his men to found a kingdom in his own country. In the 1750s, raiding into the Punjab, he came into conflict with the Marathas. The last great Indian battle before the dominance of the Europeans was fought at Panipat on January 13 1761. The Maratha army, by this time far different from Sivaji's original "mountain-rats", took the field with all the traditional panoply of a Hindu host, accompanied by their families, followers, and an enormous baggage train. They allowed themselves to be blockaded within their camp by the Afghans, who were shrewd enough not to repeat the mistake made by Sultan Ibrahim against Babur on the same battleground. Instead they waited until disease and starvation made the Marathas attack. The Marathas' only alternative, for the cavalry to cut its way out and escape, was rejected partly because it was unthinkable to leave the dependents, and partly because Ibrahim Khan, leader of a mercenary infantry brigade trained in the European style, threatened to turn his guns on any of his employers who abandoned him. Once more the issue was whether Hindu or Muslim should be supreme in India. Once more it was decided in favour of Islam, and Panipat had much the same effect on the Maratha Confederacy as did Gettysburg on the Southern Confederacy a century later. Scarcely a family did not lose a relative, from the highest to the lowest. The enigmatic casualty report summed it up; "Two pearls have been dissolved, twenty-seven gold mohurs have been lost, and of the silver and copper, the total cannot be cast up".

Three and a half years previously, at Plassey (Palasi), a handful of British redcoats, supported by Indian auxiliaries trained and equipped as they were, had put to flight the whole army of the Nawab of Bengal. A new power, and a new way of fighting, had entered Indian military history.

China

The dynasty of the Chou, in northern China, lasted longer than any other in recorded history. In their first period, that of the "Western Chou" (1122–771BC), they maintained a strong government and army, organized in a quasi-feudal system, by which a thousand local states paid fealty to the monarch as the Son of Heaven. In the "Spring and Autumn" period (770–464BC) the fighting was less against barbarians and more among the great Chinese nobles, as each sought to gain control over the monarchy. In the final period, that of the "Contending States" (463–223BC) the great fief-holders, who had taken over their smaller neighbours, fought among themselves in a series of unstable alliances, each hoping to become himself the Son of Heaven by the defeat of all his rivals.

Throughout this period the most prestigious weapons system was the war chariot, a two-wheeled vehicle with a crew of three, consisting of a commander, who fought with the gentleman's weapon, the bow; his bodyguard, a spearman; and a driver. A code of honour was observed between charioteers of opposing armies. There were also well-drilled regular infantry, known as "Shields" or "Guards", but these lacked the glamour of the more dashing, or technologically impressive, branches. Cavalry did not form a part of any Chinese army until the 3rd century BC.

The period of Contending States ended with the victory of the Ch'in between 256 and 221BC. Like the early Chou, the Ch'in came from Shensi and were thought of as semi-Turkish barbarians by other Chinese. They had become militarily and economically strong by abolishing the quasi-feudal system in their own domains, and mobilizing all sectors of the population directly under the ruler. On setting up his new dynasty, Shih Huang Ti, (the first emperor), extended this principle to all China. The old noble families were relocated at the emperor's court, and their private armies disbanded.

It was Shih Huang Ti who ordered the building of the longest defence line on earth, the Great Wall of China. This

---

"Our grieved hearts are pained,
The grooms are distressed, exhausted,
The King has ordered Nan-chung
To go and build a fort on the frontier.

Long ago, when we set out,
The millets were in flower,
Now, as we return,
Snow falls upon the mire.
The King's service brings many hardships."

Chinese soldier's poem of the Western Chou period

---

massive feature, 1,600 miles of blockhouses, guard-towers, and curtain walls along which horsemen can ride six abreast, stands as a testament to the will of the emperor and the endurance of his subjects. Paradoxically, a measure designed to benefit the Chinese people brought them immense suffering. Frontier walls were not new concepts in China. Several such had been constructed by the Contending States. None had been built, however, on such a scale, nor under such relentless pressure to complete the work. Three hundred thousand veterans who had conquered China for their emperor were sent north to build the Wall.

Chariot detachment (*left*) from the terracotta army buried with the first emperor of China Shih Huang Ti. Chariots (*above*) were used as mobile command posts and as platforms for archers, who were of high social status. They moved at slow speed to ensure accuracy of shooting, with the close support of ten infantrymen to each three-man chariot. They were richly decorated with plumes, ribbons, and medallions awarded to their owners, and were retained for prestige and ceremonial purposes long after their combat usefulness was over. In the Spring and Autumn period of the Western Chou, a code of honour was observed between charioteers.

—THE GREAT WALL OF CHINA—

Walls to keep out a potential invader were not uncommon features of military architecture in the ancient world. The Median Wall of Babylonia in the 7th century BC, Hadrian's Wall and the Rhine-Danube *limes* of the Roman empire in the 2nd and 3rd centuries AD, and even "Offa's Dyke" on the 12th century Anglo-Welsh border, are all examples of this type of work. However, the 1600-mile extent of the Great Wall of China so far exceeds these structures as to be almost different in kind.

Frontier walls against the northern nomads began to be built about 300BC. These defensive walls were formed into a single system in 214BC by Shih Huang Ti, the first emperor of a united China. The result was a complex of garrison stations, fortified walls and signal towers (communicating through smoke signals by day and fire signals by night) making up the largest engineering and building construction project ever carried out. Several hundred thousand workers were conscripted for this massive task. Running east to west from the Chihli Gulf in the Yellow Sea to the edge of Tibet, the Great Wall marked the line where steppe nomads and Chinese farmers working the fertile loess soil confronted each other. Thus its significance was from the beginning as much geopolitical as military.

The Hsiung-nu, known to the West as the Huns (and equally dreaded), represented the chief threat that the Great Wall was built to counter. And yet, as was often the case with such walls, its military value came to take second place to its symbolic and economic use. Visible evidence of the power and prosperity of the states that had built them, such walls attracted the barbarians both to marvel at and to trade with the civilized societies within. Thus great markets grew up along the Wall at which Chinese peasants exchanged their produce for that of the non-Chinese nomads. Both sides became dependent on this trade for their material prosperity, and any disruption of it by one side or the other was liable to result in the very type of cross-border incursions it was designed to prevent.

The Great Wall was never in itself enough to defend China against a sufficiently large and determined invasion force, although it inhibited casual raiders, and at least made it difficult to return home with their loot. The Wall's most useful function was as a means of controlling the movement of civilian populations, in either direction, and as a customs barrier - tolls could be collected from all the merchants and caravans from Central Asia and the Silk Road that passed through its gates.

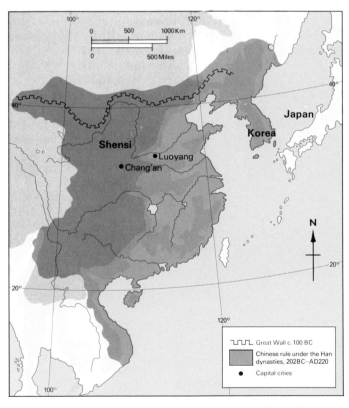

China was unified, and her border defences were strengthened by the first emperor. The primitive peoples to the west posed little threat. It was against the northern tribes that he built the Great Wall.

The Great Wall of China is not merely the most extensive construction in the whole history of warfare. It is the only item of man's handiwork that can be seen from beyond the atmosphere of his own planet.

They were joined by hundreds of thousands more, including impressed peasants taken from their fields, with the farms left to waste and their families to fend for themselves, convicts, civil servants who had failed in their duties, and scholars who had been slow to burn their books as ordered by the new regime. The Wall, on the edge of the empire, was on the edge of the Chinese world. Men died in their thousands from exposure, exhaustion, disease, and the brutality of overseers. Chinese poetry is full of references to the homesickness of the men, and the unhappiness of their loved ones far away.

The enemies against whom the Great Wall was built were the Hsiung-nu, mounted nomads, who would become known to India as the Huna, and to Europe as the Huns. Accustomed from boyhood to ride, and to shoot from the saddle against canine or human wolves who threatened his flocks and herds, the Hsiung-nu tribesman well merited not only the Wall to keep him out, but the large garrisons required to man it. Hostilities against the Hsiung-nu continued throughout the period of the Han dynasty (206BC–AD220).

Yet for all the efforts spent on the construction and periodic repair of the Great Wall, it was, from a military point of view, no stronger than the empire it was meant to defend. When China was powerful, her armies operated far beyond it. When she was weak, the barbarians could cross the Wall when they wished, or even be invited across as mercenaries (as were no less than 19 tribes of the Hsiung-nu between AD180 and 200).

The invasion of the Mongols

A thousand years later there was a double line of walls, neither of which could prevent the descent of the Mongol hordes under Genghis Khan, the self-proclaimed "Scourge

of God". The rulers of China, like those of Persia and Europe, reacted to the Mongol invasion first with arrogant and unjustified self-confidence, then with incredulity, and finally with horror. For two generations the Mongols swept all before them, and in 1280 Kublai Khan, Great Khan of the Mongols, and grandson of Genghis Khan, made himself Emperor of China. Eventually the exactions of the new dynasty grew too great for the people to bear. In 1368, the leader of a peasant revolt entered Peking, and founded the Ming dynasty.

Gunpowder, the one material which separates modern from ancient warfare, was a Chinese invention. Some kind of explosives were in use by Mongol armies, and by the 1330s Chinese artillerymen had weapons which discharged gunpowder-propelled missiles. Nevertheless the nature of Chinese society, dominated by a conservative official class, was less conducive than that of western Europe to technological innovation. It was therefore from the *Fu-lang-chi*, the feringhi, or Franks, in the persons of the Portuguese, that the Ming emperors first obtained mobile cannon and infantry firearms in the early 16th century.

The last dynasty of Imperial China, the Ch'ing, was founded by invaders from Manchuria in the mid-17th century. The Manchus threw off Ming overlordship in 1609, taking advantage of Chinese exhaustion following the war which drove out Japanese invaders from Korea (1592–99). In 1637, with Mongol allies, the Manchus conquered Korea for themselves. By 1681, they had extended their control over the whole Chinese empire. By 1760 all Mongolia, Eastern Turkestan and Zungharia were firmly under Ch'ing control with the ancient menace of the nomad horse-archer at last removed by the advent of firearms and field artillery.

In the 19th century the Ch'ing military system proved

The Hsiung-Nu or Huns were a nomadic people of Turkic stock. In the 3rd century BC they coalesced into an empire of the steppes that would, for several centuries, threaten all the civilizations with which it made contact. Its military strength consisted of horse-archers supported by Chinese technicians.

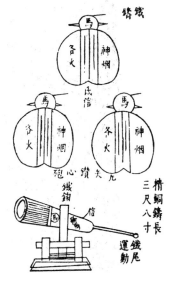

The Chinese invention of gunpowder led to the development of a substantial armoury. A 17th-century work (*above*) shows a range of bombs, grenades and various types of cannon.

In the Opium War, 1840–42 (*opposite*) paddle-steamers like HMS *Nemesis* of 700 tons, were ideal for inshore and river operations. In the action of January 7 1841 off Canton, she destroyed 11 Chinese war-junks using exploding rockets.

title *Shogun*, "Supreme Commander", completed the takeover of the imperial court by the provincial samurai. The title had been granted as early as the 8th century, but for the duration of particular campaigns only. Under the new regime it was that of a permanent, often hereditary, head of both civil and military government, while the emperors were left to carry out only ritual functions. In the 13th century its extensive powers were at times exercised by the Hojo family, acting as regents for puppet Shoguns, and it was to the Hojo that fell the task of defending Japan against the Mongols.

There were two Mongol invasions. The first, in 1274, brought 40,000 men, Mongols with their Chinese and Korean auxiliaries, to the shores of northern Kyushu. The local samurai, keen to emulate the feats of their ancestors, turned out to meet the invaders, supported by peasants fighting on foot with pikes and primitive swords.

The traditional individualist tactics of both groups made little impression on the well-drilled Mongols, whose complete victory was prevented only by a violent gale which drove their ships back to Korea. In 1281 they came again, with 4,400 ships and 42,000 soldiers. Meanwhile, the Hojo had trained men to fight, like the Mongols, in organized units. Nevertheless the samurai still vied with each other to be foremost in the attack, and eagerly manned naval cutters to harass the invaders as they landed. Weight of numbers, supported by ships' catapults, eventually allowed a beachhead to be established, and at the end of two months the Japanese position was critical. The emperor led national prayers for deliverance. The typhoon that followed wrecked the ships in which most of the invading

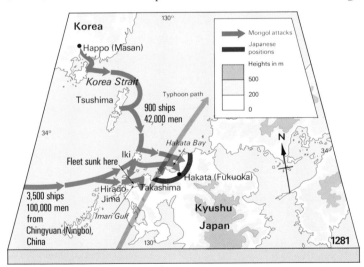

The Mongol invasion of 1274 (*opposite, above*) brought samurai warriors into the field with the same enthusiasm as that of the doomed youth of England in 1914. Like them, they found a new kind of warfare in which personal courage was outweighed by technology, as Mongol archers, and the fire bombs of their Chinese engineers (*opposite, below*) mowed them down.

Kublai Khan's second attempt to invade Japan in 1281 (*above*) was no more successful than the first. But despite the new coastal defences and revised tactics of the Japanese, Mongol superiority in numbers and armaments began to tell. National prayers for deliverance were followed by a great storm, the "divine wind" or *kamikaze*, which destroyed the invaders' fleet.

SAMURAI WARRIORS

Beginning in the early 17th century, the Tokugawa (Edo) period saw both the zenith of the samurai, in a rigidly stratified society, and their end, with the Meiji Restoration of 1868 and the abolition of the samurai caste. It is estimated that of the Japanese population of some 30 million during the Tokugawa period, fewer than 2 million were samurai.

The illustration shows two *daimyo* ("great names"; clan chiefs, or warlords) who led their followers to the great siege and battle at Osaka in 1614–15, where the defeat of Toyotomi Hideyori confirmed Tokugawa rule with its capital at Edo (Tokyo). Toyotomi is said to have deployed 90,000 fighting men at Osaka; Tokugawa numbers will have been greater.

Date Masamune (1565–1636) is seated on a lacquered camp stool and holds his *katana* (long sword) across his knees. He wears black-lacquered armour of what was then the latest type, *yukinoshita-do*, with a solid body-piece to give protection against arquebus balls. His *zunari-kabuto* (round helmet), surmounted by a gilded crescent moon, is fastened with stout cords: "After victory, tighten your helmet strings", ran a samurai proverb. His *sashimono* (identifying banner), on a staff fitted to a socket at the back of his armour, bears a sun disc: in battle, the banner would be carried by an aide.

Ii Naotaka (1590–1659) wears red-lacquered armour of *dangaiye-do* type, with laced plates rather than a solid body-piece. His helmet, ornamented with gilded metal "horns", is covered with a horsehair "wig", the intimidatory effect of which is magnified by his *mempo* (face mask).

army was embarked, leaving only unsupported shore parties to be mopped up by the defenders. Just as the English gave thanks to God for the storms that scattered the Spanish Armada, so the Japanese attributed their national survival to the Sun Goddess sending a "divine wind" or *kamikaze*, a name that would return to Japanese military history.

The 15th and 16th centuries in Japan were times of increasingly bitter civil war. The country was divided among great feudal lords or *daimyo*, who fought among themselves, enlarging their own domains at the expense of their weaker neighbours. Behaviour in warfare became less chivalrous. Even by the end of the Gempei Wars, older samurai had been deploring the way in which younger men ignored the conventions by deliberately shooting at horses, or attacking an enemy unready for battle. In this later period almost any form of treachery was considered acceptable. Great fortresses or castle-towns were built where the samurai, abandoning their own estates, took up residence as stipendary garrisons ready for immediate service. This period saw the emergence of a new kind of soldier, the *ashigaru* or "lightfoots". These were yeomanry infantry, who first appeared during the Onin War 1467–77, during which the daimyo, fighting each other in the streets of Kyoto from their great city palaces, called up not only their samurai, but retainers of humbler rank. As the demand for manpower increased, ashigaru were organized into trained units under samurai officers, and took their place in the line of battle as formed infantrymen. It was to them that was entrusted the new European weapon, the arquebus, imported into Japan from 1560 onwards.

At the battle of Nagashino in 1575, the power of the new weapon was convincingly demonstrated. The great daimyo general, Takeda Shingen, had fallen to a sniper's bullet in 1573. His son, Katsuyori, continued his wars, fighting under his famous banner inscribed with the old military text "Fast as the wind, firm as the mountain, free from sound as the forest, fierce as the fire". At Nagashino, Takeda Katsuyori, with 15,000 men, was opposed by an army of 38,000 led by Oda Nobunaga, a ruthless warlord and the first of Japan's three "great unifiers". Against the advice of his generals, and relying upon the superiority of the famed Takeda cavalry, Katsuyori decided to attack. Nobunaga deployed his 3,000 best arquebusiers behind staggered lines of palisades, and ordered each rank to fire in turn while the others reloaded. The charging horsemen and their followers were shot down before they could strike a blow. The mounted samurai had ceased to be sole masters of the battlefield.

The war against Korea

The second "great unifier" was Toyomoti Hideyoshi. By a combination of skilful alliances and brilliant victories in the field, he had made himself master of Japan by 1590. In 1592 he sent an expedition of 200,000 veterans against Korea. They swept the ill-prepared defenders before them and reached the Yalu river, leaving a trail of massacre and arson. The war dragged on. A Korean admiral, Yin Sunsin, playing Nelson to Hideyoshi's Napoleon, used armoured ships to defeat the Japanese navy and so leave the army stranded. The Koreans displayed an unexpected talent for guerrilla warfare, while conventional operations were undertaken by a large Chinese army sent to their aid. The last battle, So-chon, in 1598 was a major Japanese victory but, with all parties on the verge of ruin, Hideyoshi himself accepted that the war could not be won. He told his commander in Korea, "Don't let my soldiers become spirits in a foreign land." On his death in 1599 the expedition was abandoned and the army returned home.

Most of Hideyoshi's vassals then transferred their allegiance to Tokugawa Ieyasu (the last "great unifier"), an experienced general and shrewd politician, who had commanded during the Korean War. A league of western daimyo challenged his supremacy, only to be defeated at the battle of Seki-ga-hara, October 21 1600, with the loss of 40,000 men, many of them Korean veterans. The final

The sword has acquired a mystique in all societies in which it has been used, quite outside its primary function as a simple edged and pointed combat weapon. Making a sword that would not bend or break in battle, so leaving its user defenceless, always required great craftsmanship, raised at times to the level of ritual, involving a cost so high as to justify additional adornment. The cult of the sword everywhere long outlived its supremacy on the battlefield. In the West, the sword became the symbol of an officer's honour. In Japan it was regarded as the soul of the samurai, and special conventions governed its wearing and use.

Toyomoti Hideyoshi
("Monkey") (1536–1598) was
the son of a poor wood-cutter
in an obscure provincial
village. He was nicknamed
"Monkey" because of his
diminutive size, awkward
gait, and dark wizened
countenance. He attracted
the notice of Oda Nobunaga
(the first of Japan's three
"great unifiers") and rose to
power with him, eventually
becoming the ablest and
closest of his generals.

On his master's death in
1582, Hideyoshi fought to
take his place against
another of Nobunaga's
generals, Tokugawa Ieyasu.
Both men had fought at
Nagashino, and had seen the
effect of fieldworks and

disciplined fire against
traditional samurai dash.
After some cautious fighting,
the two finally came to an
agreement in 1584 which left
Hideyoshi as the most
powerful man in the empire.
After several successful
campaigns, all the main
islands were under the firm
control of a central
government for the first time
in Japanese history.

Hideyoshi was denied by
the Emperor the title of
Shogun, because of his
humble birth, but granted
that of Kwampaku or
Dictator. As the effective
ruler of the empire he proved
as great a statesman as he
was a soldier. He built great
palaces, encouraged artists,
reformed the currency and
the taxation system,
introduced laws to protect
cultivators, and encouraged
trade. He was capable of acts
of personal courage, moral
and physical, which made
allies out of enemies and
supporters out of rivals.
Despite his fatal decision to
order troops into Korea in
1592, leading to a disastrous
war which almost ruined all
parties, he may fairly be
considered one of the greatest
rulers, statesmen, and
generals of his century.

triumph of the Tokugawa at the battle of Osaka, June 1
1615, brought to Japan a peace that would last for almost
300 years.

Yet although the fighting was over, militarism was not.
Hideyoshi, who had begun his career as a horse-boy, was
determined to safeguard the social position of the samurai
class he had joined and ensure none followed his route into
it. Under the Tokugawa Shoguns who succeeded him, his
laws, rigidly segregating society into the classes of samurai, farmer, artisan and merchant, remained strictly in
force. The samurai, though required to live frugally upon
their stipends, were given a position of privilege as the
leaders of society, and the monopoly of military and civil
appointments. They alone were allowed to wear swords.
Moreover, they were allowed to use them at their own
discretion on any of the lower classes who offended them.
The disarmed commoners responded by adopting a wide
range of self-defence martial arts, using hands, feet,
quarterstaffs or rice-flails.

Later developments
Japan, isolating herself from the rest of the world after the
seclusion edict of 1638–41, failed to keep up with the
changing nature of warfare. The anger and shock when
first the United States Navy in 1853, and then the British
Royal Navy in 1863, steamed at will into Inland Sea, led to a
profound reorganization of Japanese social and military
institutions. The shogunate was abolished, the emperor
restored, and the privileges of the samurai ended. Many of
them suffered great hardship thereby, and their code,
Bushido, was rejected by the reformers as a relic of the
past. The national conscript army which took their place
as the Japanese defence force proved itself able to fight
well, in accord with western techniques and conventions of
warfare, though the revival of *Bushido* from the 1920s
onwards resulted in the reintroduction of attitudes towards a defeated enemy no longer acceptable in the
modern world.

**The Japanese adopted
Western military
techniques** so successfully
(in marked contrast to the
Chinese) that within 50 years
of doing so they were a major
world power. The new army
of conscripted peasants
defeated, to their mutual
surprise, a samurai rebel
force in 1877, and in 1882
landed in Korea to protect
Japanese interests there.
Rivalry between Japan and
Russia over control of Korea
(legally under Chinese
dominion) led to the outbreak
of war on February 8 1904
with a surprise attack on the
Russian naval base at Port
Arthur. The Russian Baltic
Fleet was destroyed at
Tsushima on May 27 1905
(*left*) in the greatest and most
decisive sea battle since
Trafalgar.

— chapter six —
"*Villainous saltpetre*"

Richard Holmes

We cannot be certain who invented gunpowder. But we do know that primitive cannon were used by the Chinese in the 11th century, and that a French chronicler attributed the defeat of his countrymen at Crécy in 1346 to English use of artillery. Yet whatever the effect of these early guns upon the morale of feudal horsemen or peasant levies, their real value was decidedly limited. They were cumbersome constructions, difficult to move and notoriously liable to burst, whose low power and slow rate of fire made them a questionable asset to field armies.

Early cannon were most useful in siege warfare. Medieval fortress builders displayed infinite cunning in confronting attackers with walls and towers, pierced with loopholes and machicolations, topped by battlements and protected by moats. Siege engines either hurled stones over walls or battered ponderously at gates and stonework: cannon brought wall and tower crashing down. When the Venetians attacked the fortress of Brondolo in 1380, their cannon felled part of the campanile on to the Genoese commander. In 1453, in what Sir Charles Oman calls "the first event of supreme importance whose result was determined by the power of artillery", the huge bronze cannon of Mohammed II breached the walls of Constantinople, ending the power of Byzantium which had dammed the torrent of Islam for nearly one thousand years.

The French King Charles VIII's invasion of Italy in 1494 firmly established the cannon in Western Europe and, in a

Pot-de-feu hand-cannon of the 14th century.

broader sense, marked the transition from feudal to modern warfare. Charles's bronze guns were sufficiently light to travel on field carriages but also strong enough to withstand the explosion of powerful propellants. Their fire levelled castle walls with unprecedented speed, and the French advance through northern Italy was, as Christopher Duffy has pointed out, the blitzkrieg of its day. However, the new artillery had many of the limitations of the old: in particular, its poor mobility restricted its use on the battlefield. At Pavia in 1525, a battle which ended the French domination of northern Italy begun by Charles VIII, Francis I was defeated by an Imperialist army which outflanked his main positions and brought him to battle where his heavy guns could not intervene.

Pavia did more than illustrate the continuing limitations of cannon. Although many of the infantry who fought there plied pike or halberd, many carried the arquebus, the infantry firearm which enjoyed increasing importance throughout the 16th century. In 1503 the Spanish general Gonzalo de Córdoba ("*el Gran Capitán*") beat the French at

Cerignola, first repulsing the determined attack of their heavy cavalry and Swiss mercenary pikemen with the fire of his arquebusiers, then counterattacking to rout the French and kill their leader, the Duc de Nemours. The arquebus was superseded by the matchlock musket, and the proportion of musketeers to other infantrymen grew. As Michael Roberts has acutely observed, the "military revolution" of the period 1560–1660 was in essence the result of an attempt to solve "the perennial problem of tactics: the problem of how to combine missile weapons with close action; how to unite hitting-power, mobility and defensive strength."

The cultural background
The revolution took place in the context of political and cultural events which gave it content and momentum. The intellectual currents of the Renaissance – from Niccolò Machiavelli's *The Prince*, with its exposition of a utilitarian morality which impelled many a war machine, to Leonardo's fantastic designs for weapons – helped shape military developments. The religious upheavals of the Reformation were the cause of, and as often the pretext for, conflict which lent ample scope to theorists and practical soldiers alike. The age was marked by repeated clashes between France, riven by domestic disputes, and the Holy Roman Empire, contending with internal unrest stemming from its disparate nature and tensions within its German heartland. Finally, the Turkish menace overhung Europe throughout the period, from the crushing Hungarian

Mohammed IV's army is defeated at Vienna, 1683.

defeat at Mohács in 1526 to the repulse of Mohammed IV's armies from Vienna in 1683.

Pikeman and musketeer
The armoured horseman clattered supreme across medieval Europe. Specialist infantry, most notably English longbowmen and Swiss pikemen, played their part in ending his ascendancy, but neither was a complete answer

—TIME CHART—

1337–1457	Hundred Years' War
1346	Crécy: longbow versus armoured cavalry
1419–36	Hussite Wars: Hussite wagon-fort tactics defeat armoured cavalry
from c.1430	Ottoman empire expanding in Europe
by 1450	Arrowhead bastion in use in Italy
1453	Constantinople falls to Turks after cannon breach its walls
1455	Wars of the Roses between houses of Lancaster and York begin in England
1492	Spain captures Granada, last Moorish outpost
1494	Italian Wars: French campaign characterized by use of mobile siege artillery
1503	Arquebusiers repulse pikemen at Cerignola
1525	Poor manoeuvrability of artillery in the field leads to French defeat at Pavia
1526	Turks crush Hungarians at Mohács
c.1535	Partition of Hungary
1562–98	French Wars of Religion
1566–1609	Dutch revolt in the Netherlands
1571	Defeat of Ottoman fleet at Lepanto
1590	Ivry: arquebusiers defeat caracoling *reiters*
by c.1595	Cohort-sized battalions devised by Maurice of Nassau
1598	Edict of Nantes
1609	Spanish ejected from Netherlands
1618–48	Thirty Years' War
1630	Gustav Adolf of Sweden invades Holy Roman Empire
by 1630	Swedish squadron and brigade formations perfected; 3-pounder mobile field guns attached to squadrons
1631	Swedish tactics crush *tercios* at Breitenfeld
1642–51	English Civil Wars
1643	The Duc d'Enghien (Condé) defeats Spanish *tercios* at Rocroi
1645	Naseby: disciplined New Model cavalry secures Parliamentary victory

Europe between the mid-14th and mid-17th centuries. It is necessarily a simplified view, since it deals with an era of almost constant warfare resulting in frequent territorial changes. For example, although it appears on the map as a well-defined entity, the Holy Roman Empire was a collection of states that were frequently at odds over political or religious issues. During the period c.1350–c.1650, the Holy Roman Empire experienced 38 major conflicts, culminating in the cataclysmic Thirty Years' War of 1618–48. To the West, one major conflict was the so-called Hundred Years' War of 1337–1457, which resulted in the loss to England of most of her territories in France.

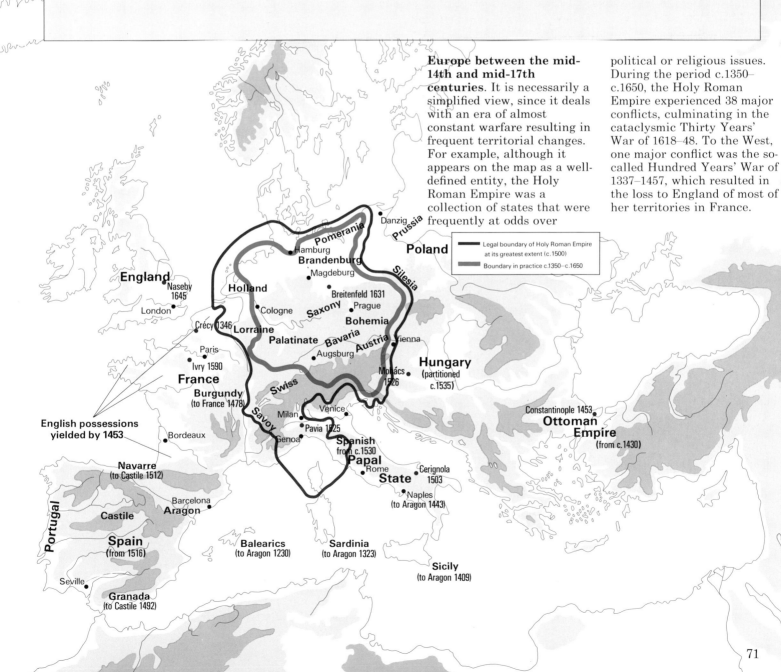

to the threat posed by the mounted knight. The bowman could not be improvised: the Swiss pikeman, with his pig-headed valour and arrant conservatism, was scarcely more flexible than the horsemen he scorned. The development of the infantry firearm, however, swung the balance in favour of the foot soldier. It did so slowly. For most of the period

---

"Would to heaven that this accursed engine [the arquebus] had never been invented, I had not then received those wounds which I now languish under, neither had so many valiant men been slain for the most part by the most pitiful fellows and the greatest cowards . . ."

Blaise de Montluc (1502–77)

---

firearms were inferior in range, accuracy and rate of fire to the longbow in skilled hands. Yet a musketeer could be drilled into passable efficiency in a matter of weeks, while an archer took years to train: in short, the new infantry of the 17th century could be mass-produced.

The Spanish *tercio*, so-called because it originally comprised one-third of an army, developed from a mixed force of arquebusiers, halberdiers, pikemen and muske-teers, subdivided into three *colunelas*. The musketeers gave the *tercio* its firepower: the pikemen protected them while they reloaded, and advanced "at push of pike" to capitalize on damage done by musketry. But the *tercio* in battle, its huge square of pikemen with "sleeves" of shot at the corners, was essentially a medieval formation, waste-ful of manpower and firepower alike, and had more in

common with a Macedonian phalanx or a column of Swiss pikes than the linear deployments of the 18th century. Nevertheless, reforms during the 1580s, which reduced its size to 1,600 men, and the tendency of Spanish commanders to deploy *tercios* in wedges of three or diamonds of four promoted flexibility, while the dogged spirit of its soldiers helped make the *tercio* a formidable instrument until the second quarter of the 17th century.

Maurice of Nassau, son of William the Silent and his successor as *stadhouder* of the Netherlands in 1584, spent most of his adult life attempting to eject the Spaniards from the United Provinces, an endeavour which was eventually crowned by success in 1609. Both his experience of the strengths and weaknesses of the *tercio* at close quarters, and his study of classical military writers, persuaded him that infantry units should be smaller and their formations shallower. His "battalion", modelled on the Roman cohort, numbered 550 men, and drew up 10 men deep on a front of 49. In most other respects, however, Maurice was a decidedly conventional soldier.

The reforms of Gustav Adolf

The same cannot be said of Gustav Adolf, who succeeded to the throne of Sweden in 1611 and swiftly gained military experience on the shores of the Baltic. Like Maurice, he was steeped in classical theory, and he went on to study the Dutch battalion and the reformed *tercio*. Gustav's "squad-ron" actually contained more pikemen than musketeers – 216 to 192. But it formed only six deep, and instead of the confusingly elaborate "countermarch", in which each rank

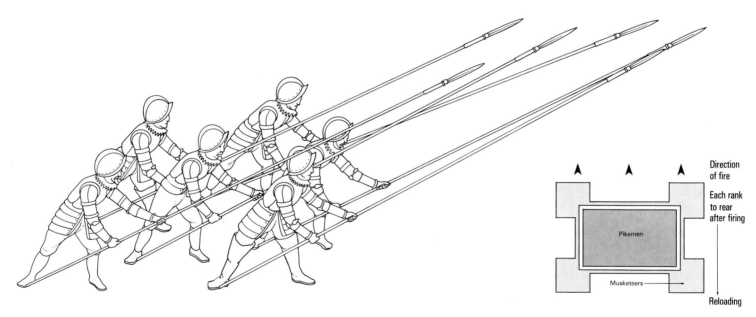

The battle formation known as the *tercio*, sometimes called "the Spanish square" from its country of origin, is shown in the diagram (*right*) in its mid-16th century configuration. It consists of some 3,000 men: a square of pikemen, with "sleeves" of musketeers at

each corner. After discharging their pieces, the musketeers in the front ranks "countermarch" to the rear, where the pikemen protect them during the lengthy process of reloading. When the musketeers' fire has done sufficient damage, the pikemen advance with their

14–18ft weapons, (*above*), to drive back the enemy. The *tercio* was a somewhat clumsy unit, although the deployment of three or four *tercios* in wedge- or diamond-shaped formations facilitated manouevre by allowing for some degree of mutual support.

Gustav Adolf (1594–1632), King Gustav II of Sweden, spent almost all of his short life at war, proving himself to be the most enlightened and progressive military leader of his time. In his earliest campaigns, against the Danes and Russians in 1612–17, Gustav showed both warlike and statesmanlike skills. His long struggle against Poland for pre-eminence on the Baltic coast was productive of military reforms that were to have a long-lasting effect

all over Europe.

Gustav's reforms embraced every aspect of military life, from recruitment and economic administration to the tactics of all arms. His innovations in infantry and cavalry tactics, and in particular his introduction of light, mobile field artillery for the support of both infantry and cavalry, made the Swedish army the most effective of its time.

In 1630 Gustav intervened on the hard-pressed Protestant side in the Thirty Years' War. Twice he defeated Tilly: at Breitenfeld, where fast-firing artillery and the firepower and manouevrability of musketeers and pikemen massed in "brigades" shattered the Imperialists; and at the River Lech, April 1632, where his artillery covered an audacious river crossing. On November 16 1632, Gustav was killed while attacking Wallenstein's Imperial army at Lützen.

of musketeers fired and was replaced by another, it practised a simpler form, with two ranks firing simultaneously. By 1630 the squadron doubled its ranks to fire a salvo, the heaviest concentration of musketry yet produced, and its pikemen were trained to attack vigorously to take full advantage of the salvo's effect. By 1630, too, Gustav had perfected his "brigade", the combination of two or three squadrons into a battle group of 1,500 to 2,000 men. The Swedish brigade gave proof of its prowess on the battlefields of the Thirty Years' War, and Gustav's victory over Tilly at Breitenfeld in 1631 established it as the supreme tactical instrument of its age.

Cuirassier and dragoon

Gustav's impact upon the mounted arm was no less marked. There had been several types of cavalry in the late 16th century. The armoured man-at-arms, descendant of the knight, sought to employ shock action against horse or foot, while a more lightly armoured horseman, often armed with a lance, scouted ahead of the army and pursued a broken enemy. "Shot on horseback" were armed with a wheel-lock "petronel" (a carbine or short musket) or pistol. The most successful pistol-armed horsemen were the German *reiters*, who perfected the *caracole*, trotting up to their enemy to fire pistols before wheeling off while the next rank repeated the manoeuvre.

By the early 17th century the cavalry was experiencing acute difficulties. The man-at-arms was already obsolete. His charge was often repulsed by steady pikemen, and making his armour bullet-proof imposed a weight penalty

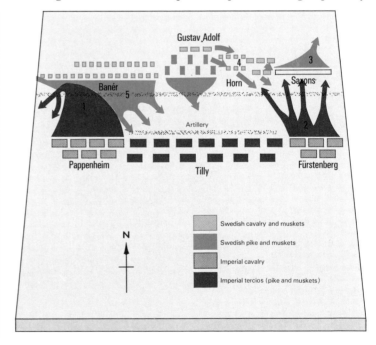

The battle of Breitenfeld (*above* and *right*) opened with Pappenheim's cavalry attack on the Swedish right (1) which was stopped by Banér's skilful deployment of alternate squares of horsemen and musketeers. On the Catholic right, however, Fürstenberg's horsemen

charged (2) and routed the Saxons who fled (3) abandoning their guns. Gustav turned his left (4) to seal off this breach in his flank, and musket volleys stemmed the repeated charges of the Imperialist cavalry. Meanwhile the Swedish right had completely

routed Pappenheim's cuirassiers by combined artillery and musket fire, which enabled Gustav to wheel his right and centre (5), capture the Imperialist artillery, and mow down the *tercios*. The Imperialist army fled as Gustav pressed home his attack.

which made it hard to find suitable horses in sufficient numbers. When Sir Edmund Verney was summoned to serve in full armour against the Scots in 1638 he wrote feelingly that: "It will kill a man to serve in a whole cuirass." By 1670 the man-at-arms had disappeared altogether, although the breast- and back-plate remained popular among heavy cavalry, who often retained the name *cuirassiers*.

If mounted shock was becoming more hazardous, mounted fire was certainly no substitute. Whatever the drill-book elegance of the *caracole*, its practical disadvantages were alarming. The horse was an uncertain weapon-platform, and the pistol-armed *reiter* courted disaster if he faced pikemen whose weapons kept him out of their ranks and musketeers whose weapons outshot his own. There were undoubtedly times when the *caracole* worked, for a body of horse, five ranks deep in the Dutch manner, with each trooper carrying two pistols, could generate substantial firepower. Yet all too often horsemen squibbed off their pistols well short of the infantry and made little effective contribution to the battle.

Gustav Adolf turned his cavalry into an army of decision. Initially six ranks deep, by 1632 only three, they were trained to charge at the trot, and instead of halting to fire they pressed home to use sword and pistol in the melee. In the early stages of the English Civil War there was disagreement between those officers who favoured the Dutch system, and those who preferred the Swedish one. The Royalists, under the impulsion of Prince Rupert, a veteran of the Thirty Years' War, quickly adopted the

"As I went up the hill, which was very steep and hollow, I met several dead and wounded officers brought off; besides several running away, that I had much ado to get by them. When I came to the top of the hill, I saw Sir Bevill Grinvill's stand of pikes, which certainly preserved our army from a total rout, with the loss of his most precious life; they stood as upon the eaves of an house for steepness, but as unmovable as a rock; on which side of this stand of pikes our horse were I could not discover; for the air was so darkened by the smoke of the powder, that for a quarter of an hour together (I dare say) there was no light seen, but what the fire of the volleys of shot gave . . ."

Captain Richard Atkyns, Lansdowne, June 5 1643

Swedish system: the Parliamentarians were slower to do so, but by 1644 the horse on both sides formed three deep and were trained to charge home. Gustav made some concession to providing cavalry with firepower by attaching platoons of musketeers to cavalry units, a practice sometimes followed in the English Civil War.

By 1660 most European cavalry were, like the "horse" of the Civil War, descendants of Gustav's troopers, armed with sword and pistol and often protected by a cuirass worn over the leather buff-coat which turned a sword's edge. They were supported by dragoons, who had originated as mounted infantry in the French service in the mid-16th century, taking their name from the short "dragon" musket which they first carried. Dragoons, who could be

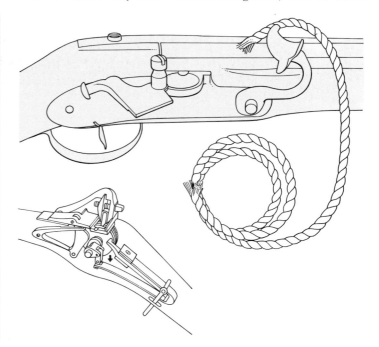

A troop of German *reiters* (*above*) during the Thirty Years' War, 1618–48. These armoured cavalrymen, equipped with the then newly-developed wheel-lock arms, suitable for use on horseback, first appeared on the battlefield around the mid-16th century. Their

tactics – notably the *caracole*, in which they advanced in successive ranks to discharge volleys – at first proved effective; but by the 17th century they had become less so in the face of steady ranks of pikemen supported by musketeers whose weapons outranged their own.

In the matchlock mechanism (*top*), a slow-burning match (porous cord steeped in a flammable compound) is held in the jaws of the cock. Pressure on the trigger pivots the cock so that the match touches off the powder in the pan, igniting the main charge in

the breech via the touch-hole.

Pulling the wheel-lock's trigger (*bottom*) causes a previously-wound clockwork mechanism to spin the abrasive wheel so that it strikes sparks from a piece of pyrite held in the cock. A pan cover protects the priming from damp.

—NAVAL WARFARE IN THE 16TH CENTURY—

The galley, for around 3,000 years the dominant warship in the relatively calm Mediterranean, saw its last great battle at Lepanto (painting: *below right*) in 1571, when a Christian Allied fleet of some 215 galleys decisively defeated a Turkish force of about 260 galleys. Six of the Allied galleys were of the largest Venetian type known as galleasses (bow and side views: *below left*). The galleass was some 145ft long and 27ft in beam. Propelled by around 50 oars, with five men to each, it also shipped three masts with lateen sails, and could, perhaps, cover 240 nautical miles (276 miles) in 24 hours. The bow view shows its forecastle, with eight heavy guns on swivel mountings allowing ahead- or quarter-firing. Two more heavy guns were mounted on the aftercastle, with around 15 lighter guns along the rails. The iron-shod ram remained a major weapon: in close-quarter fighting, as seen in the painting, the guns fired a salvo to throw the enemy into confusion before his ship was rammed and boarded.

A far heavier armament was mounted by the sailing warship *Henry Grâce à Dieu*, built for King Henry VIII of England. Launched in 1514, she was (with Henry's *Mary Rose* of 1513) the largest and most powerful warship of her time, displacing about 1,400 tons. Following reconstruction in 1536–39 she mounted the formidable armament of 21 brass ("bronze") heavy muzzle-loaders, cast in one piece for greater strength – comprising four "cannon", throwing 50–60lb shot, three 32-pounder "demi-cannon", four 18-pounder "culverins", two 9-pounder "demi-culverins", six "curtals" and two "cannon-periers", large-calibre pieces throwing 24lb stone shot – 14 breech-loading 6-pounder "port-pieces"; six 4-pounder "slings"; eight "fowlers"; 102 4-pounder "serpentines"; and around 100 rail-mounted "hand-guns". In the painting (*bottom right*), the ship is shown as she appeared after reconstruction. Note the ports for "chase guns" in the squared-off stern. The introduction at the beginning of the 16th century of gun-ports cut into the hull allowed guns to be mounted lower in the ship and facilitated their arrangement "on the broadside": Henry's warships were possibly the first to incorporate this revolutionary feature.

In the battle of Lepanto of 1571 (*right*), Venetian galeasses (*below*, bow and side views) mounted guns, but the primary weapon remained the ram.

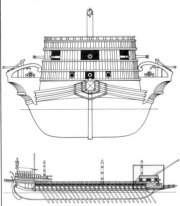

Henry VIII's *Henry Grâce à Dieu* (*far right*), from a contemporary painting, shows British development of naval firepower, including "chase guns" in the squared-off stern, and gun-ports cut into the hull. The drawing (*right*) shows the possible arrangement of the warship's gun decks aft of the mainmast, with the heaviest guns on carriages on the gun deck and main deck, and the lighter guns and rail-pieces above. The introduction of broadside armament meant that a warship could stand off and batter her opponent at "point blank" (i.e. effective) range of about a quarter-mile (for the bigger

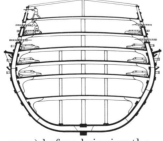

guns) before bringing the lighter weapons into action while closing to grapple and board.

cheaply mounted and needed no defensive armour, were widely used in the French Wars of Religion and the Thirty Years' War, so much so that the Duc de Rohan complained that "they ruined the infantry, every man desiring to have a nag so that he might be the fitter to rob and pillage." They provided mobile firepower to an army's van or rearguard, and in close country they had the special function of lining hedges. The gap between dragoons and cavalry proper narrowed steadily, and by the mid-19th century a dragoon would, no doubt, have been affronted if reminded of his origins as a cheap mounted infantryman. For light cavalry, European armies looked increasingly to "wild men" from the east: the 16th-century *stradiot* hailed from the Balkans, and in the 18th century the hussar, modelled on the Magyar horseman, became the *beau ideal* of the light cavalryman.

The nimble gunner

It was typical of Gustav Adolf's commitment to the military art that he was himself a skilled gunner. His interest was all the more surprising because, on the one hand, gunnery was – and continued to be – regarded with suspicion by many noblemen, and on the other, gunners themselves, with their guilds more reminiscent of medieval trade than modern war, were anxious to preserve the mysteries of their craft and remain independent of the formal military hierarchy. The Emperors Maximilian I and Charles V, and also King Henry II of France, had attempted to standardize the calibres of cannon, but their regulations were as honoured in the breach as the obser-

vance. In 1646 an English authority suggested that artillery ranged from the "cannon royal" firing a 63-pound shot, through cannon, demi-cannon, 15-pounder culverin, demi-culverin, 6-pounder saker, minion, falcon and falconet to the tiny robinet, but not all gunners, even in England, would have agreed with him.

Gustav standardized artillery in the Swedish service and, even more crucially, introduced a 3-pounder "regiment piece" in 1629. Unlike heavier guns, which might be moved on to the battlefield but were static once there, the regiment piece could be manhandled among the infantry, whence its cannister shot tore lanes through the enemy's ranks. By 1631 every one of Gustav's infantry squadrons had two or three light guns attached to it, and the salvoes of the infantry combined with gusts of cannister to produce a fire effect which perhaps at last equalled even that of the longbow in its heyday.

Bastion and trench

Important though Gustav's developments in field artillery were, the gunner made a more important contribution to sieges than to battles in open field. Siegecraft had long reflected the advantages in material and technique sought by attacker and defender, and improvements in artillery had a profound effect upon fortification, turning it into a branch of science akin to geometry. Because cannon fire directed at their bases brought high walls down, engineers began to build low. Conventional round towers made indifferent gun platforms, and high walls were all but useless for the purpose. Central to the new artillery

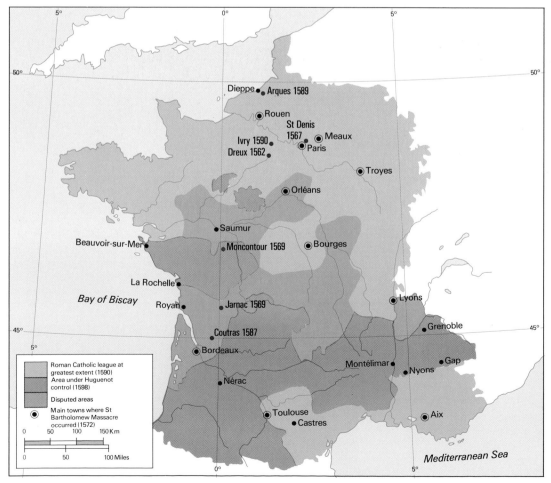

The French Wars of Religion, 1562–8. This series of Catholic versus Protestant conflicts was provoked by the massacre of Huguenots (French Protestants) at Vassy, March 1 1652, whereupon the Huguenots rose under the leadership of Condé and Coligny. There followed nine "Wars of Religion": 1652–63, with the indecisive battle of Dreux; 1567–68, the battle of St Denis, when the Huguenots threatened Paris; 1568–70, when the Huguenots were defeated at Jarnac and Moncontour; 1572–73, provoked by the slaughter of Huguenots in the Massacre of St Bartholomew's Day; 1575–76, which saw the foundation of the Catholic Holy League under Henry of Guise, supported by Philip II of Spain; 1576–77; 1580; 1585–89; and 1589–98, which saw Huguenot victories under Henry of Navarre (King Henry IV of France) at Arques and Ivry. The wars came to an end when Henry embraced Catholicism as the price of unity.

fortification was the bastion, projecting beyond the curtain wall and defending the wall and the face of the adjoining bastion with enfilade fire. Some bastions, like the Polish *basteja* or the gun towers on Henry VIII's castles at Walmer and Deal, were round. But although a low tower could sustain the discharge of cannon mounted on it, its circular form left "dead ground" at its front, where fire could not be brought to bear. The arrowhead bastion was first seen in mid-15th-century Italy: the bastion at Sarzanello, built by Genoese engineers in 1497, is the earliest example still standing.

The bastion became the quintessential ingredient of the "artillery front", in its classical form a star-shaped trace protected by a ditch, with an open slope, the glacis, rolling away before it. In place of high stone walls stood massive

---

"At this siege the Duke of Lorraine lost Mr Fouge a lieutenant general, and the best officer in his army; he was killed the night after they had taken the lower town. Being at supper with the Prince of Condé, in one of the next houses to the upper town, and making a debauch, he grew so drunk that he ran out in a fit of bravado at a back door with a napkin on his head to be discerned the better, and to provoke the enemy to shoot at him. The Chevalier de Guise and the Prince of Condé himself ran out after him to bring him back, but before they could hale him in, he received a shot which killed him.'

James Duke of York, Bar-le-Duc, 1652

---

banks of earth, faced with masonry or turf. From the enemy side there was little to see, for the parapet rose only high enough above the surrounding countryside for its guns to sweep the glacis. Outworks were thrown forward to defend the main front and keep a besieger at arm's length: the ravelin (*demi-lune*), a triangular fortification usually positioned between two bastions, was the most important. For the great age of fortress warfare we must look to the next chapter, but even by 1660 the artillery fortress exercised a growing influence on the conduct of war and, no less important, on the shape of the early modern state.

Gunpowder and government

The relationship between military developments and the rise of the European state was fundamental. As Sir George Clark wrote: "A state tightened its organization in order to be strong against its rivals, and the strength which it acquired in the contest for power strengthened its government at home." The declining military value of the feudal horseman mirrored, albeit imprecisely, the restriction of his political influence. The threat posed by overmighty subjects waned. When Franz von Sickingen was put to the ban of the Empire and sought refuge in his castle at Landstuhl in 1523 he might have expected to stand siege and emerge unscathed when the besiegers' patience wore thin. Not so in the dawn of the artillery age: the Emperor's gunners knocked his battlements to pieces about his ears and killed him. Private armies were outlawed: Henry VII's attack on "livery and maintenance" was designed to end them in England, and Peter the Great destroyed the *streltsi*,

Exacerbated by religious hatreds and by the employment of mercenaries who were often allowed to pillage in lieu of regular pay, the Thirty Years' War spread suffering among civilians throughout Europe. It is estimated that some 8 million people died in Germany alone. Jacques Callot (1598–1635) portrayed the horrors of the conflict in a series of

etchings, *Grandes Misères de la Guerre* (*above*). The murder of civilians by troops running amok is particularly associated with the sack of Magdeburg in May 1631: after the storming of the city by Imperial troops, Pappenheim's mercenaries engaged in an orgy of violence that ended in the burning of Magdeburg and 25,000 deaths.

A detail from the painting, *Embarkation for the Field of the Cloth of Gold*, portraying King Henry VIII's departure from England for his meeting with Francis I of France in 1520, serves well to illustrate an important feature of the coastal fortifications erected by Henry later in his reign, from 1539 onward (*above*). This detail shows a bastion, a fairly low-built tower acting

as a gun platform. This example takes the form of a round, detached blockhouse: bastions might also be of arrowhead configuration, often built out from keeps or curtain walls and sited so that they could support each other with enfilade fire.

part-soldiers, part-tradesmen, and wholly inimical to the spirit of his Russia, in 1698.

The new armies were not only subject to direct royal control. They were recruited, in growing measure, from the inhabitants of the state they defended and were uniformly trained and clad; their officers and NCOs took station in a formalized hierarchy; and they were more reliable than the mercenaries who had played such a large part in European history for centuries past (and who survived, often in the form of Royal – or Papal – guards, for years to come). They were also very much larger: as Michael Roberts has pointed out, Philip II of Spain dominated Europe with an army which probably did not exceed 40,000 men: a century later 400,000 men were required to maintain the ascendancy of Louis XIV. The battles of the Wars of Religion were fought by armies rarely larger than 20,000: at Ivry, Henry of Navarre's victory over Charles, Duc de Mayenne, in 1590, fewer than 40,000 men contended for the fate of France; at Breitenfeld in 1631, each side deployed some 40,000; at Malplaquet in 1709, 110,000 Allies fought against 80,000 Frenchmen.

New opportunities

The growth of armies increased opportunity as well as widening risk. The social distinction between horseman and footsoldier diminished, to the fury of many contemporaries, until neither birth nor wealth dictated a soldier's choice of arm. Although the nobility strove to retain its grip on command it was not universally successful, and war brought many a humble youth military rank, social status and, on occasion, hereditary nobility. It also gave employment to shoals of young men from the infertile or disputed peripheries of Europe: Scots and Irish adventurers left an enduring mark on the continent's armies.

But structural changes which began during the period were to end the career of one profiteer, the recruiting captain, a survival of the "free-lances" of the late Middle Ages, for whom service was a matter of contractual arrangement. In most European armies regiments were still owned by proprietor colonels who regarded them, at least in part, as a financial enterprise, but by 1660 these worthies were not the quasi-independent contractors of yesteryear. They held their commissions from the king, wore his uniform, not their own, and were subjected to the penetrating stares of a new generation of military bureaucrats. Paymasters and muster-clerks, intendants and commissaries, all materialized to further centralized control of armies and to ensure, within the limitations of the day, that soldiers were properly fed, clad and armed. The military bureaucracy offered its own prospects for advancement to men who found the pen more congenial than the sword, and it too might confer nobility: the great French war administrator Michel le Tellier was succeeded in 1668 by his son, the no less capable Marquis de Louvois.

Meeting the cost

Raimondo Montecuccoli (1608–81), an Italian general in the Imperial service, identified one of the problems of the military revolution when he said: "For war you need three things. 1. Money. 2. Money. 3. Money." War became more

The battle of Naseby began with Rupert's general advance, whereupon the entire Parliamentarian army moved forward to the edge of their high ground. Rupert's cavalry came under fire from Okey's dismounted dragoons, placed in ambush behind Sulby Hedges.

The second stage of the battle saw Astley's foot drive back the Parliamentarian centre (1); Ireton charged into Astley's right (2), but was repulsed in disorder (3). Meanwhile, however, Cromwell's front line had routed Langdale's men (4), and the Royalist reserve (5) wavered. Cromwell now led his second and third lines against Astley's left, while his first attacked from the rear (6); then Okey's remounted dragoons advanced (7), and the outnumbered Royalist infantry began to surrender. The King tried to rally his troops, but many had already fled.

Pre-eminent in battle in the age of the armoured knight, by the later 16th century the cavalry had undergone an eclipse, proving increasingly ineffective when confronted by determined pikemen with the support of musketeers. Gustav Adolf's reforms, however, had some success in transforming cavalry into a weapon of decision, and by the mid-17th century the typical cavalryman was modelled on Gustav's horseman, protected by a breastplate and felt hat or helmet, armed with pistol and sword and trained to charge home. The unarmoured dragoon, armed with a short musket, was initially little more than a mounted infantryman. In western armies the lance, once the predominant horseman's weapon, was almost totally abandoned. It was not until the early 19th century – from which period the illustration (*left*) of an Austrian lancer dates – that, inspired by Napoleon's successful use of Polish (and later French) cavalry armed with the lance, the lancer returned to the battlefield. Nevertheless, the development of firearms meant that after the 16th century, the cavalryman's major contribution often lay in reconnaissance and pursuit rather than the charge on the battlefield.

expensive as armies and navies grew larger and required more training and better armaments, and monarchs were hard put to make ends meet. Their search for money drove them to expedients like the sale of crown lands or the debasement of currency, and impelled them to negotiate with their Estates or to step beyond the constitution in their efforts to raise cash. Although they eventually succeeded, with the standing armies which became a feature of Europe as evidence of their triumph, it was often at the cost of civil war or insurrection, and always with the accompaniment of an increasingly efficient state bureaucracy.

Implications for the future

Artillery fortification played a special part in the process. Fortresses were not only expensive to build: they required continuous upkeep and occasional modification if they were not to become ruinous or outdated. As Christopher Duffy tells us: "You had to think in terms of decades or generations if you wished to form your construction engineers and gunners, assemble powerful siege trains, and win and consolidate state frontiers. The thing was fundamentally a matter of politics rather than technology." Indeed, great though the implications of military developments were for armies and the men who fought in them, they were even more important for society as a whole. "They were," as Michael Roberts wrote, "the agents and auxiliaries of constitutional and social change; and they bore a main share of responsibility for the coming of that new world which was to be so very unlike the old."

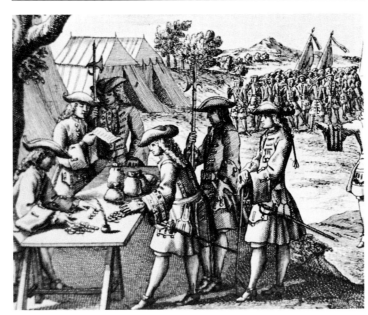

Paymasters check the muster roll (*above*) and hand out cash – a familiar sight by the later 17th century, when the increasing size of professional standing armies and the need for centralized control had given rise to a growing military bureaucracy. The paymaster's role was vital, for without regular payment discipline and morale must suffer. In Britain, determined attempts to minimize the need to "live off the country" by introducing a proper system of pay were made by Parliament during the Civil War: the Ordinance of 1645 stipulated that the "New Model" army should be both nationally controlled and regularly paid.

An aerial photograph of Deal Castle (*above*), the largest of three fortifications built along a three-mile stretch of the Kent coast in 1539–40, at the orders of Henry VIII, to protect the fleet anchorage known as the Downs. The parapets were originally more massive, rounded and with far fewer crenellations, but the compactness and basic simplicity that mark this as a fortress rather than a castle in the medieval sense are immediately apparent. Six semi-circular bastions project from the central keep into the dry moat, with six smaller bastions above. Thus, five tiers of guns are provided for: three firing from roofs and two through ports.

— chapter seven —
The age of the flintlock

Richard Holmes

The era of European history which began with Louis XIV's seizure of the reins of government on the death of Cardinal Mazarin in 1661 and ended with the restoration of the Bourbons in 1815 can be divided into three main periods. The first encompasses 1661 to 1715, and was characterized by conflict between France and her neighbours as Louis sought to expand French power. He was at first generally successful, pushing the borders of France towards her natural frontiers in the Low Countries and on the Rhine, but his dynastic ambitions for a Bourbon empire combining the thrones of France and Spain remained unfulfilled, and although the treaties which ended the War of the Spanish Succession (1701–14) deprived France of none of her territory, they dulled the lustre of the Sun King's prestige.

Between 1715 and 1789 the emphasis shifted to the east. The Treaty of Utrecht recognized the Elector of Brandenburg's title as King of Prussia, granted by the Emperor in 1701. Frederick William I (reigned 1713–40) continued the absolutist policy of his predecessors and laid the foundations for the military achievements of his son, Frederick the Great (reigned 1740–1786). The War of the Austrian Succession (1740–48) and the Seven Years' War (1756–63) saw Prussian fortunes ebb and flow until Prussia was left exhausted but in secure possession of Silesia and, no less to the point, with a formidable military reputation. The period also witnessed the increasing impact of Russia upon European affairs, thanks to the efforts of Peter the Great (1672–1725), whose victory over Charles XII of Sweden at Poltava in 1709 pointed the way to Russian domination of the Baltic.

For much of the final period, from the French Revolution in 1789 to Napoleon's defeat at Waterloo in 1815, France once more overshadowed Europe. The impact of revolutionary France shook old Europe to its foundations, and Napoleon's armies marked out an empire which eclipsed even the wildest ambitions of Louis XIV. But just as fear of Louis had conjured up an alliance which was too strong even for France, so Napoleonic France was unable to cope with the opposition of successive coalitions and the internal strain imposed by almost constant war.

18th-century limited war

The armies of the 18th century were large, impressively drilled and ferociously equipped, and the century was marked by frequent wars. Yet 18th-century war lacked both the destructive savagery of the Thirty Years' War and the sweeping manoeuvres of the Napoleonic era. Limited war was, in part, a reaction to the Thirty Years' War, which had depopulated parts of Europe. The barbarous character of that conflict drove reasonable men to search for alternatives, and both Grotius, whose *De jure belli et pacis* appeared in 1625 while the conflict was still in progress, and the Swiss jurist Emmerich de Vattel, author of *Droits des gens* (1758), argued that "natural law" should protect the individual against the horrors of war. The enlightened absolutism of monarchs like Frederick the Great and Emperor Joseph II of Austria also encouraged a rational approach to statecraft and a tendency to regard war as a political instrument with a finite aim.

The moderation of religious passions removed one source of hatred in war. There were moments of genuine nationalistic fervour on the odd occasion – as in Sweden during the Great Northern War (1700–21) – but most wars were dynastic struggles which did not engage popular sentiments. The officers of opposing armies were often more linked by common attitudes and interests than they were separated by the colour of their uniforms. Their soldiers saw in their opponents men much like themselves, alongside whom they might have fought in an earlier campaign. This is not to say that 18th-century war was a quiet and contemplative affair: Marlborough ravaged the Palatinate in 1704 in a manner reminiscent of a harsher age, while at Torgau in 1760, Frederick's army lost some 16,000 of the 44,000 Prussians who took the field. Nevertheless, men were not spurred on by ideology, and their rulers were more anxious to preserve the balance within Europe through alliances and compromise peaces than to batter an opponent into extinction.

Economic considerations

There were less lofty motives for the limitation of war. Mercantilism, the prevailing economic doctrine of the day, emphasized the importance of building up a trading surplus; this encouraged maritime conflict, like the Anglo-Dutch Wars, limited by scope and character, and prompted avoidance of the damage done to the domestic economy by both large-scale conscription and arms imports.

France played a central role in Europe between 1661 and 1815, and this map charts her fortunes during the period. Louis XIV sought natural frontiers on the Alps, the Pyrenees and the Rhine: though he met with considerable success in his early wars, the danger of a union between France and Spain inspired determined opposition, and the treaties of Utrecht and Rastatt, which ended the War of Spanish Succession, saw France retain her pre-war territory but ended the prospect of a dominant Bourbon state. The young French Republic found itself at odds with monarchical Europe, but, after acute military crises in 1792–93, its armies occupied the Austrian Netherlands and beat the Austrians in northern Italy: the Treaty of Campo Formio (1797) gave the Austrian Netherlands and Lombardy to France. Napoleon made further gains, establishing an empire which, at its height, spread French power from Spain to the Vistula and from Italy to the Baltic. The Congress of Vienna redrew the map of Europe following French defeat in the Napoleonic Wars, reducing France to the frontiers of 1789. The second Treaty of Paris (November 1815) embodied further changes, reflecting the "Hundred Days", and France lost a few frontier districts, most notably the Saar.

—TIME CHART—

1661	Death of Mazarin marks beginning of Louis XIV's personal rule
1662	Louis XIV invades Spanish Netherlands
1673	Investment of Maastricht: begins era of formal siegecraft
1687	Socket bayonet in use
1698	Iron ramrod introduced by Prussian army
1701–14	War of the Spanish Succession
1715	Death of Louis XIV
1740–48	War of the Austrian Succession
1740	Accession of Frederick the Great
1740	Prussia takes Silesia from Austria
1754–60	French and Indian Wars
1755	Utility of light troops demonstrated by British defeat at the Monongahela
1756–63	Seven Years' War
1757	Leuthen: artillery concentration and oblique-order tactics gain Prussian victory
1763	Treaty of Paris confirms British superiority in North America and India
1774	French artillery standardized
1775–83	American War of Independence: partisan forces play important role
1781	British surrender at Yorktown assures American Independence
1784	"Case shot" invented by Shrapnel
1789	French Revolution
by c.1790	Grenadiers and skirmishers attached to most European infantry battalions
1793	*Levée en masse*: French introduce universal conscription
1793–96	French introduce divisional system
1796	Napoleon's Italian campaign
from 1796	Napoleon develops corps system
by c.1800	Barracks in use for housing troops
1805	Trafalgar: Nelson's "cutting the line" tactic defeats French and Spanish
1805	Corps system facilitates Napoleon's victory at Ulm
1806	Operational ploy of *manoeuvre sur les derrières* gains Napoleon victory over Prussia at Jena
1806	Dissolution of Holy Roman Empire
1807–14	Peninsular War: Spanish guerrilla insurgency supported by British field army

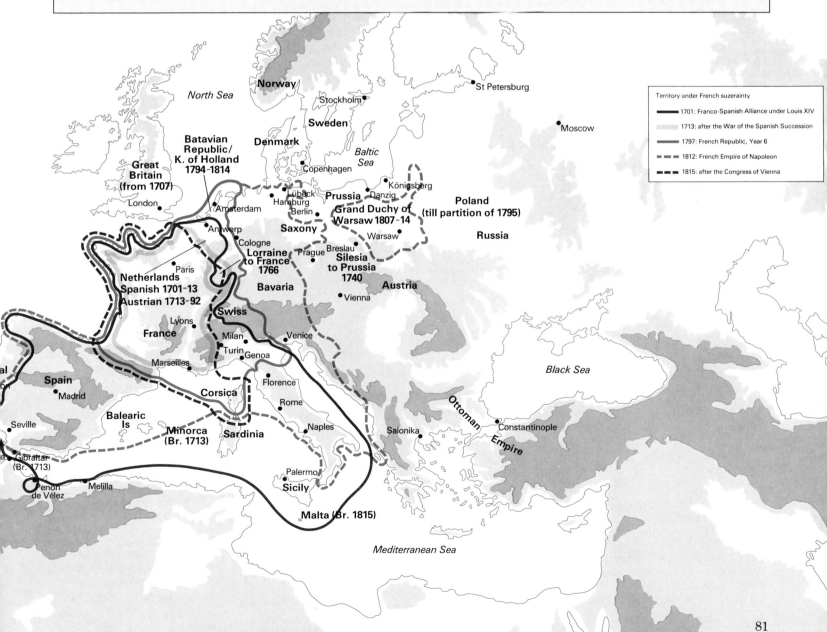

Territory under French suzerainty

1701: Franco-Spanish Alliance under Louis XIV
1713: after the War of the Spanish Succession
1797: French Republic, Year 6
1812: French Empire of Napoleon
1815: after the Congress of Vienna

Moreover, the ever-increasing cost of war was, as we shall see, a constraint in itself.

Historians differ on the extent to which logistics limited operations. Hew Strachan observes that armies had grown in size, but that there had been neither commensurate increase in the food supply nor marked improvement in means of transport. In consequence the armies of the period subsisted by requisition, moving into enemy territory as soon as they could, shackled to the roads by the need to move guns and ammunition, and compelled to march concentrated to protect themselves. Because this slow-moving column found it difficult to requisition enough food to keep itself supplied, it was forced to carry at least some food with it, and as the train grew in size movement was further impeded and the logistic bill lengthened. The magazine system became increasingly important, and 100 miles (160km) was believed to be the farthest an army could advance from its magazine without creating an intermediate one.

Martin van Creveld, however, discerns little evidence of trains acting as a brake on armies or "umbilical cords of supply" restricting their movement. He sees other reasons for limitation – suggesting, for instance, that the expense of standing armies and the effort put into their training made them too precious a commodity to expose to undue risk. There is a wider measure of agreement that fortresses impeded offensive operations, forcing an attacker either to reduce them by formal siege or to use part of his army to mask them while the remainder advanced. An army's battering train, without which it had no prospect of

mounting a siege, slowed it down: in 1708 the 18 guns and 20 mortars of Marlborough's siege train needed 16,000 horses and 3,000 wagons and took up 30 miles (48km) of road. The scarcity of good roads made matters worse, although Christopher Duffy, whose opinion cannot be dismissed lightly, is "impressed, almost disconcerted by the very high level of mobility of the monarchical armies".

Limited war had no single dominant cause. The philosophy of the Enlightenment probably had less impact on it than we might care to admit. Practical constraints of terrain and logistics, allied to a desire to minimize the damage inflicted on an expensive military instrument in an era when battle could usually be avoided if a commander wished to do so, were more significant. But at least as important was a strategic policy aimed not at the annihilation of the enemy or the deposition of his ruler, but at the attainment of limited objectives such as a province or a fortress-line. There is little evidence of a mismatch between the ends 18th-century commanders pursued and the means at their disposal.

The tools of war

Weapons changed relatively little throughout the era. The muzzle-loading flintlock musket remained the principal infantry firearm. Although it was improved as time went on, for instance by the introduction of the iron (and thus unbreakable) ramrod into the Prussian service by Leopold of Anhalt-Dessau in 1698, the weapon carried by the infantry at Waterloo was not strikingly different from that used at Blenheim more than a century before. Effective

The War of the Spanish Succession broke out after Louis declared that his grandson Philip V of Spain retained succession rights in France. England, Holland and the Empire joined the Grand Alliance against France, Spain and Bavaria. In 1703–4 Louis's attempt to take Vienna was thwarted at Blenheim by the combined forces of Eugène of Savoy and Marlborough. Villeroi's unsuccessful thrust into the Low Countries bled troops from Italy, enabling Eugène to defeat the besiegers of Turin. In 1708, renewed invasion of the Low Countries met with disaster yet again, although the Allied victory at Malplaquet was costly. The Archduke Charles, Allied claimant to the Spanish throne, took Madrid, and the British captured Gibraltar. But Allied success was tempered by defeats at Almanza and Brihuega, and, in 1711, when Charles became Holy Roman Emperor, peace negotiations at Utrecht and Rastatt ended the conflict.

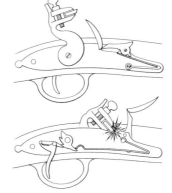

A grenadier loads and fires his musket. First he bites the end off the paper cartridge and trickles powder into the pan. Next, he tips the remaining powder down the muzzle, following it with the ball, and the paper case to act as wadding. The charge is tamped down with the ramrod before he cocks the musket and brings it to his shoulder. This sequence, simplified from the many movements taught on the drill-square, was often modified in battle.

A flintlock ready to fire (*left, above*) with its flint held in the jaws of the cock and the steel (frizzen) covering the pan filled with priming powder. When the trigger is pressed (*below*) the flint strikes the steel and sparks fly into the pan, igniting the priming powder and firing the weapon. Misfires were common: even if the priming ignited there was no guarantee that the main charge would follow suit.

range was short: at the end of the 18th century, a Prussian battalion fired at a target 100ft (30m) long and 6ft (2m) high, obtaining 25 percent hits at 225yds (205m), 40 percent hits at 150yds (137m) and 60 percent hits at 75yds (68m). A crack unit might manage six or even seven rounds a minute on the drill-ground, but it would be unlikely to exceed three on the battlefield. The musket was unreliable: perhaps one shot in five would misfire even in ideal circumstances, and wind or rain could worsen this. Some rifles were used, notably by German *jäger* units and North American irregulars, but they were slow to load and costly to produce.

The bayonet was a new addition to the infantryman's armament. The plug bayonet, jammed into the musket's muzzle, prevented him from firing once it was fixed, but the socket or ring bayonet, adopted by the French in 1687, fitted round the muzzle and offered no such impediment. The development of the bayonet made possible the abolition of the pike, which survived in an attenuated form as a badge of office for some officers and NCOs. Infantry tactics from Blenheim to Waterloo centred upon the relationship between the physical damage done by crashing volleys of musketry, and the shock – often more psychological than physical – of the advance with cold steel: we shall examine their evolution as this chapter proceeds.

The military revolution of the 17th century had established the importance of drill, and in the 18th century the care lavished upon it verged on the devotional. Not only did excellence in drill speed up weapon handling and tactical manoeuvre, but also the habits inculcated on the drill-ground helped condition men to carry out the mechanical processes of loading and firing in the smoke and terror of battle. We must not imagine that the battles of the age bore much relation to the neat figures in drill-books. G A Berenhorst's picture of the Prussian infantry at its business would have been familiar to many veterans:

"You began by firing by platoons, and perhaps two or three would get off orderly volleys, but then would follow a general blazing away – the usual rolling fire when everybody blasted off as soon as he had loaded, when the first rank was incapable of kneeling, even if it wanted to. The commanders, from subalterns to generals, would be incapable of getting the mass to perform anything else; they just had to wait until it finally set itself in motion forwards or backwards."

Improvements in artillery

Artillery changed far less between 1660 and 1815 than it had over the previous half-century or was to over the next. However, although the cannon that fired Napoleon's opening bombardment at Waterloo looked superficially similar to those which had accompanied Marlborough along the Danube, they had become far more effective weapons. In the 1740s Bélidor showed that powder charges could be reduced without loss of range, and Maritz improved the barrel-boring of cannon, developments which made for weaker charges, thinner barrels and hence lighter guns. In 1774 Jean Baptiste de Gribeauval reformed French artillery, standardizing field guns into 4-, 8- and 12-pounders, supplemented by 6-inch howitzers and mortars. The Gribeauval 12-pounder, far lighter than the cannon it replaced, could be drawn by only six horses, as opposed to the ten required for Frederick's "Austrian" 12-pounder, and its large wheels and iron axle-tree gave it a robust cross-country performance: the weapon and its derivatives became the mainstay of artillery parks across Europe.

Improvements in ammunition were equally significant. The solid roundshot was supplemented by canister – small balls enclosed in a tin container which burst as it left the muzzle – used against targets under about 500yds (457m). Other types of multiple projectile were used, such as grapeshot, similar to canister but using bigger balls, while both bar- and chain-shot, with their destructive effect on rigging, were favoured for sea service. Howitzers and mortars fired a cast-iron shell filled with powder, detonated by a fuze ignited by the flash of the discharge. In 1784 Lieutenant Henry Shrapnel of the British Royal Artillery produced his "spherical case shot", a thin iron ball containing musket balls and a burster charge ignited by a fuze, intended to explode above enemy infantry. However, the irregularities of fuzes, powder and metallurgy made both common shell and shrapnel uncertain in their effects.

It was perhaps the cavalry who looked most different but whose *modus operandi* changed least across the period. There were, broadly speaking, three types. First came the heavy cavalry, who had the task of demolishing enemy cavalry or infantry by the weight of their charge, delivered at full gallop in the Prussian service but elsewhere often at a fast trot or canter. But even the boldest cuirassier might find himself at a disadvantage when faced with resolute infantry who stood their ground. Infantry were trained to form square to repel cavalry, and providing they did so before the onslaught they were likely to survive: even if a cavalryman was determined to crash into the square, his horse often found the gaps between the squares a more

Frederick the Great (1713–86) suffered a harsh upbringing at the hands of his boorish father Frederick William I. In 1740 he inherited a useful army, a well-stocked war-chest and a centralized state, and promptly invaded the Austrian province of Silesia. He departed abruptly from the field of Mollwitz, his first battle, but went on to obtain Silesia by the Treaty of Breslau. In 1744 he embarked upon the Second Silesian War, recouping early failure by a brilliant victory at

Hohenfriedburg, and retaining Silesia by the Peace of Dresden (1745). The Seven Years' War witnessed two of his greatest triumphs, Rossbach and Leuthen, but also saw costly victories as well as outright defeats. The death of the Empress Elizabeth brought his admirer Peter III to the throne of Russia, enabling Frederick to emerge from a traumatic war with Silesia firmly in his grasp.

As a general, Frederick's ideas were not always sound, and he was sometimes a poor judge of men, but his resolution and audacity place him in the ranks of the great. As a man, he was both child of the Enlightenment, corresponding with Voltaire, and arch-cynic, breaking treaties and starting aggressive wars. He died leaving a state which had doubled in size and an army which enjoyed an unrivalled if overblown reputation.

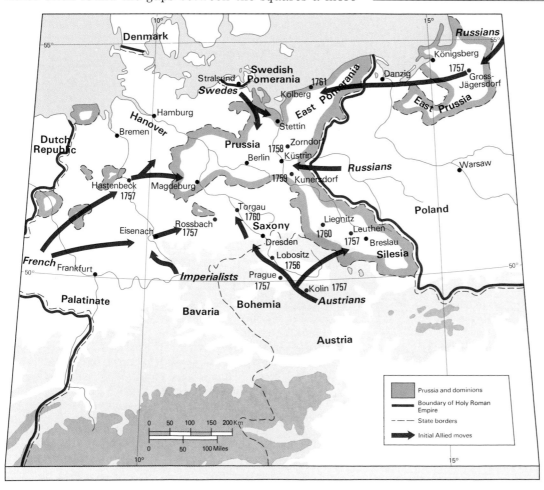

The Seven Years' War. In 1756 Frederick engulfed the Saxons after beating their Austrian allies at Lobositz. He then concentrated on the Austrians, beating them at Prague but losing at Kolin. In 1757 he crushed the French-Imperialist army at Rossbach before turning to beat the Austrians at Leuthen. The British, much encouraged, increased their subsidy to Frederick and fielded a new army under Ferdinand of Brunswick. After jabbing at the Austrian base at Olmütz, Frederick beat the Russians at Zorndorf, rushed back to face the Austrians and was routed at Hochkirch. Brunswick defeated the French at Minden, but Frederick lost to the Russians at Kunersdorf, and the Austrians occupied Dresden. Frederick beat the Austrians at Liegnitz and Torgau, but could not prevent the loss of Berlin. Though 1761 saw him under increasing pressure, he was saved by the death of the Empress Elizabeth, which took Russia out of the war.

attractive prospect. The overthrow of a French regiment by the dragoons of the King's German Legion at Garcia Hernandez, near Salamanca, in 1812 is a very rare example of solid infantry being broken by a charge, and it occurred because a wounded horse fell on the square, making a gap through which an enterprising dragoon urged his steed. If infantry were caught changing formation they risked destruction: at Albuera in 1811 three British battalions suffered 70 percent casualties when they were taken by surprise, although there a rainstorm both drenched their muskets and veiled the approach of French and Polish horsemen.

The role of the dragoon
The dragoon continued his rise to the ranks of cavalry proper, and in most armies made up the bulk of the mounted arm. He carried out most of his service on horseback and, like his comrades of the heavy cavalry, was expected to charge with the *arme blanche*. Yet he retained some of his old role, carrying a dragoon musket rather than the carbine used by other horsemen. In the French army, shortage of horses, a common wartime problem throughout the period, caused dragoon regiments to consist of three squadrons *à cheval* and two *à pied* after 1812.

The light cavalry displayed the widest variety of sartorial embellishment. The example of hussars, initially recruited in Hungary for the Austrians, spread across Europe, producing some spectacular extravagances as Hungarian dress was "improved" by military tailors. Polish cavalry had long been among the best in Europe,

and the successive partitions of Poland did not quell the martial ardour of the Poles. Napoleon equipped his two regiments of Polish *chevaux légers* with lances in 1809; he subsequently converted seven other regiments to lancers and formed two lancer regiments in his Guard. The lancer, dressed in the Polish style with his flat-topped *czapka* and plastron-fronted tunic, was widely imitated, but it was acknowledged that although the lance was deadly in the hands of an expert, it was far more difficult to master than the sabre. Light cavalry were vital for scouting, reconnaissance and screening, although they were also called upon to charge home on the battlefield when the situation demanded it.

"The troops marched on in a silence that could only have been produced by their reflections on having survived that great bloody day. But suddenly the quiet was broken by a grenadier who sounded the familiar hymn Nun danket alle Gott. *Every man was awakened as if from a deep sleep, and, transported with gratitude to Providence for their survival, more than twenty-five thousand troops sang the chorale with one voice through to the end. The darkness of the night, the voices of the troops, and the horror of the battlefield, where you stumbled on a corpse at almost every step, combined to give the episode a solemnity which was easier to feel than describe."*

Lieutenant J A von Retzow, Leuthen, December 5 1757

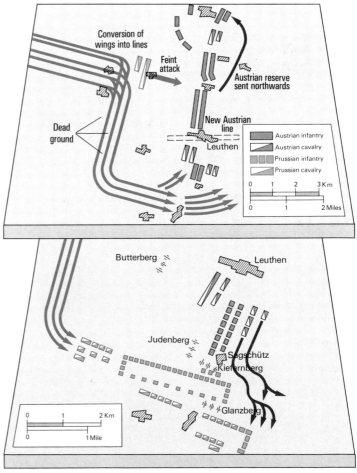

The infantryman could load and fire with his socket bayonet fixed. An infantry unit which remained steady confronted charging cavalry with musketry and cold steel. The square was the classic formation adopted against cavalry, although small bodies of horse were often beaten off by the simpler expedient of ordering the rear rank of a line to face about.

On December 5 1757, Frederick planned an oblique attack on the Austrian left, screening his advance behind low hills and feinting to make them strengthen their right (*top*). The Prussians took the Kiefenberg, repulsed a counterattack (*bottom*), and drove an improvised line from Leuthen before beating the cavalry on their left and routing the enemy infantry.

Siegecraft and fortification

Louis XIV's wars saw fortress warfare reach its height. The conduct of the siege of the Dutch fortress of Maastricht in 1673 by the French engineer Sébastien Le Prestre de Vauban (1633–1707) showed his technical mastery of siegecraft. Trenches were opened on June 17–18, and the governor, a brave and experienced officer, capitulated on July 1. During the intervening fortnight the siege had progressed with the formalism which typified such operations. The besieger opened his first parallel just out of cannonshot from the fortress. His engineers sapped forward, their zig-zag trenches offering poor targets to the defender's guns and protected on the enemy side by earth-filled wicker gabions. In good ground a sap might be driven forward 160yds (146m) in 24 hours, but it was an exhausting and dangerous process. A second parallel was dug some 300yds (274m) from the fortress; then the sappers pushed on again until a third parallel could be dug, ideally at the top of the ditch itself. Meanwhile, the besieger's gunners kept the fortress under bombardment from batteries in the second parallel, probably using 24-pounders, the workhorse of siege artillery, while mortars lobbed bombs into the body of the place. Once the third parallel was completed, the attacker could establish his breaching batteries and hammer away at the ravelin or bastion opposite. His engineers supported the attack with efforts of their own, digging beneath the fortress to explode mines.

The defender would not allow these measures to go unopposed. His gunners fired at the sap-head and took on enemy batteries as they were established. A determined·

governor would mount sorties at judicious moments, to discommode labourers at work on the parallels or to get into the besieger's batteries, where he might delay the siege by spiking guns – hammering files or nails into their touchholes – or do more permanent damage, bursting guns by firing them with a blocked muzzle. His own miners would be hard at work underground, their task often made easier by counter-mine galleries dug when the fortress was built, hoping to blow in the attackers' mines or to blow up a battery. Any one of these endeavours might be crowned with success: when the French besieged Turin in 1706 the garrison deftly exploded a mine beneath a battery "which scattered breaching cannon like straw", and at Fredriksten (also known as Fredrikshald and Halden) in 1718 Danish gunners, firing grapeshot in the dark towards likely spots for working parties, killed Charles XII of Sweden and thus ended the siege at a stroke.

Forced surrender

Such an outcome was unusual. It was more likely that the besieger would batter a practicable breach in the defences and summon the governor to surrender. If he refused, he risked the massacre of his garrison and the sack of the town when it was stormed. This convention was intended to prevent a useless effusion of blood, for it was argued that the defender had no hope of success at this juncture and his resistance was therefore unreasonable. It also reflected the fact that it was difficult to control troops committed to a storm: maddened by the fighting and spurred on by the prospects of rape, pillage and drink, they showed a

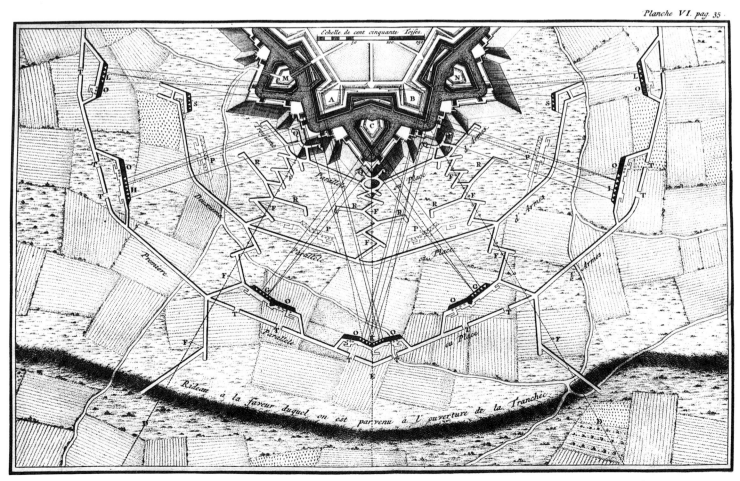

Marshal Sébastien Le Prestre de Vauban (1633–1707) came from a family of impoverished Burgundian gentry, and entered the army as a volunteer under the rebel Prince of Condé in 1651. Captured two years later, he was won over to the royalists by Cardinal Mazarin, and served his apprenticeship as an engineer under the Chevalier de Clerville. He succeeded his master as

Commissaire Général des Fortifications in 1678, was promoted lieutenant general in 1688, and became marshal of France, the first engineer ever to enjoy this dignity, in 1703. During a long and active career he directed nearly 50 sieges and produced plans for 160 fortresses. Vauban was a tough, practical man whose concern for the common man was unusual amongst the generals of his age. He worked prodigiously hard, spending his summers on campaign or on an endless round of fortress inspections, and his winters in Paris, poring over plans. At Maastricht in 1673 he pioneered the attack by parallels, and the principles he established survived as long as the classical fortress. Not only scores of fortresses, but also the imposing Maintenon Aqueduct, built in 1685 to carry water to the new palace at Versailles, commemorate this tireless and talented engineer.

distressing propensity for sacking a town even if their commanders wished otherwise. The British army was among the best-disciplined of the Napoleonic period, but the scenes which followed the stormings of Badajoz and Ciudad Rodrigo in 1812 were reminiscent of the savagery of the Thirty Years' War.

The essence of Vauban's siegecraft was to apply the defensive strengths of earthworks to the attack in a manner more prodigal with sweat than blood. His fortresses, revealing a similar desire to husband human resources, emphasized the role of artillery. He produced three systems, starting with a bastioned trace like that used by his countryman the Comte de Pagan, and ending, in the splendid fortifications of Neuf-Brisach on the German frontier, with massive bastion towers, covered redoubts in the ravelins, and recessed curtain walls with casemates which would remain immune from fire until the last moment. Distinguished though he was, Vauban was not unique. The great Dutch engineer Menno van Coehoorn (1641–1704) merits laurels of his own, while the Marquis de Montalambert (1714–1800) recognized that a siege was in essence an artillery duel, proposing that fortresses should be huge multistorey gun-towers which would dominate the firing capability of any attacker.

The decline of fortress warfare

Given the difficulty of charting the evolution of the military art as events actually unroll, it is not surprising that engineers haggled over the merits of rival systems of fortification long after fortress warfare had lost the

Vauban's own diagram of the "regular attack" (*left*). Note the three parallels connected by zig-zag saps. The first parallel is armed with batteries sited to take the defence in enfilade, their shot ricocheting along the terreplein on which the defender's guns would be sited.

The lofty citadel of Bitche in Lorraine was built by Vauban in 1683, making good use of commanding ground: it held British prisoners during the Napoleonic Wars. Fort St Sebastien (*top*) is connected to the town by defence works, forming an entrenched camp in which a field army could take refuge. Just as cannon had made medieval fortifications obsolete, so artillery developments in the late 19th century imperilled Vauban's finest creations.

A 16th century German cannon. These guns had a devastating effect on the high towers and walls of medieval fortification, and provided the impulsion behind the development of the geometrical artillery fortresses of the 17th and 18th centuries. The cost of procuring such weapons helped limit the military

power of great nobles; trains of artillery in royal hands threatened the refuges of overmighty subjects, and fortress-building increasingly became an element of state policy.

eminence it reached in the period 1660–1715. Christopher Duffy, the leading modern authority on fortification, suggests that fortress warfare declined neither solely because of the growing superiority of attack over defence, nor even because the ability of fortresses to block choke-points diminished as armies became more mobile. He points instead to changes in the structure of armies, and it is to these that we must now turn our attention.

The baroque armies

The baroque was the dominant artistic style of the 17th and 18th centuries. Characterized by vigour and exuberance, it expressed the sense of power and creativity of the age, and its armies revealed the same qualities. They were built on a grand scale: Louvois increased the strength of the French army from 30,000 to 120,000 for the Dutch War of 1672–78, and to 360,000 for the War of the League of Augsburg in 1688–97. The Prussian army increased from 18,000 in 1688 to 40,000 in 1713. Under Frederick it grew still larger, to 133,000 in 1751 and 190,000 on his death in 1786. Like the ornamented facades and sweeping staircases of baroque architecture, these armies were costly. Louis XIV's wars overloaded an economy which even the efforts of Jean-Baptiste Colbert, his capable Minister of Finance, could not equip to bear the strain. Louis spared no expense: Neuf-Brisach alone had cost nearly 3,000,000 *livres* by 1705. Frederick used a variety of methods, from taxation, loans and excise duties to a huge subsidy from the British and successive debasements which reduced the value of Prussian coinage by well over half, to meet the cost of his wars.

The zeal and inventiveness of his officials, coupled with the monarch's own parsimony and readiness to plunder occupied areas, enabled Frederick to survive financially, although in the mid-1770s approximately 13 million of the state's total income of nearly 22 million *thalers* went to the upkeep of the army.

The need to pay for war produced friction between monarchs and representative bodies. In 1763 the British government attempted to tax the North American colonies to meet the debts of the Seven Years' War and the charges of frontier defence, and the dispute flared into the War of Independence in 1775. The French, scenting an opportunity to deal with an isolated Britain, joined the war in 1778, and their regular troops and the Comte de Grasse's fleet paved the way for Cornwallis's surrender at Yorktown in October 1781. In helping to end monarchical rule in North America the *ancien régime* had unwittingly increased its own peril: by 1786 nearly half the crown's revenue was absorbed by the interest on the national debt. The privileged classes – nobility, church and magistrature alike – resisted the attempts of Louis XVI's ministers to achieve fundamental economic reform, and in 1789 the government took the desperate step of summoning the Estates-General, which had not met for 175 years. The financial burden of war must be ranked high among the long-term causes of the French Revolution.

The human cost of 18th-century war was no less striking. The Seven Years' War killed more than 500,000 men on the continent, and casualties and disruption reduced the population of Prussia by at least the same amount. The

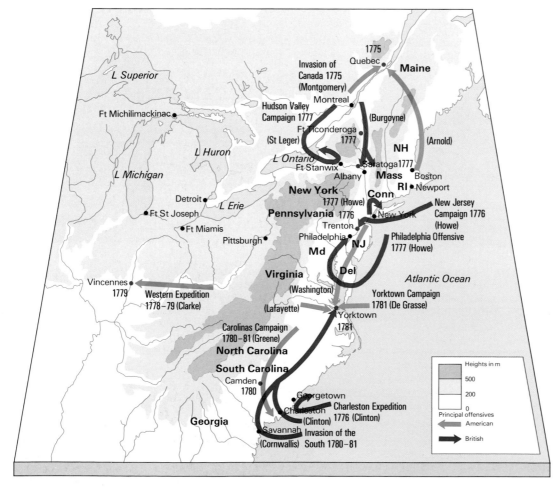

The American War of Independence. The first clash between British and Americans came in April 1775. In June Gage emerged from Boston to win the battle of Bunker Hill, but his successor, Howe, was later forced to evacuate the city. American thrusts into Canada failed in 1775, and in 1776 the British sent Howe into New York, forcing Washington back. In 1777, however, British advances from Canada ended in failure. Howe did better, but although he took Philadelphia he failed to destroy Washington's army, and Clinton, his successor, marched back to New York in 1778. The British took Savannah in December, and in 1780 Clinton moved south in strength, taking Charleston. Cornwallis won a hollow victory at Guilford Courthouse in 1781 and withdrew to Virginia. The Americans now enjoyed French assistance: deprived of naval support by de Grasse's fleet, Cornwallis surrendered at Yorktown.

nobility bore a heavy share of the burden: by 1702 three of Colbert's sons and one of his grandsons had perished; the Prussian noble house of Belling lost 20 of its 23 male members in the Seven Years' War. This ghastly butcher's bill bore witness to the fact that most penetrating wounds of the abdomen proved fatal, and the majority of amputations were followed by death in a foul and overcrowded field hospital. Moreover, if musketball and grapeshot killed thousands, disease felled tens of thousands. Service at sea was especially hazardous: when Commodore George Anson circumnavigated the globe in 1740–44 he lost 1,300 men from disease and only 4 by enemy action.

Nobility and rank

The armies of the 18th century bore more signs of the absolutism of the monarchs they served than of the prevailing concept of Enlightenment. The nobility were harnessed to the state, and hierarchical tables, from the French *Ordre de Tableau* of 1675 to Peter the Great's comprehensive Table of Ranks of 1722, defined the rungs of the ladder and laid down the rules by which it might be ascended. Prussia depended upon the nobility, mainly the tough *junker* squirearchy of East Prussia, to officer her army. In 1786 there were only 700 bourgeois officers in an officer corps of 7,000, and most of the non-nobles could be found in the least prestigious arms, such as the engineers and artillery. The Prussian officer enjoyed a lofty social status, leading a surprised Saxon to comment that "every ensign or cornet considers he is as important as a minister of state".

In contrast, the tone of French society remained definitively civilian. France retained the legal fiction of demanding military service under the old feudal terms of *ban et arrière-ban* until the end of the 17th century, but the reforms of Louvois formalized a system of commutation payments which made more money available to finance the standing army. If a French nobleman chose to serve he might purchase a commission, and would expect to reach higher rank than his bourgeois comrade: moreover, the period after the death of Louis XIV witnessed the "aristocratization" of the army at the expense of its bourgeois elements. After the Seven Years' War, the Duc de Choiseul carried out some much-needed reforms, reducing the influence of proprietor colonels and improving military education; Saint-Germain continued the process in the 1770s. The nobility fought back, and in 1781 a decree was passed, requiring the possession of four degrees of nobility for commissioning. This step was probably intended as much to appease the old nobility by barring the newly ennobled as to keep out commoners. Yet despite aristocratic reaction, the work done by Choiseul and Saint-Germain provided a useful foundation for the Revolutionary and Napoleonic armies.

The British army was much freer from aristocratic influence. About two-thirds of its officers purchased their commissions according to tariffs established in 1720 and subsequently updated; many of the remaining third had been "volunteers", who served in the ranks in the hope of attracting favourable notice, or NCOs, who were commissioned in large numbers in wartime. The officer of the

The fighting in North America during the Seven Years' War and War of Independence encouraged the formation of irregular units whose dress and tactics reflected the realities of campaigning in the backwoods. Rogers' Rangers were raised in 1756 by Robert Rogers, a farmer turned trapper who had had several brushes with the law. The Rangers responded to French and Indian harassment of American settlements by attacks on French outposts and Indian villages, paying their opponents back in the same hard coinage of scalping and torture. Their considerable success in irregular warfare attracted the attention of many British officers, and the Rangers were more influential than their small numbers (nine companies in 1758) might suggest. During the War of Independence, Rogers cultivated both sides, eventually raising a company which fought for the British.

In a hard-fought battle at Fontenoy on May 11 1745, the Duke of Cumberland was defeated by Marshal Saxe. A British Guards officer, Lord John Hay, raised his flask and toasted adversaries of the *Gardes Françaises*, begging them to fire first: *"Tirez les premiers, Messieurs les Français."* There was little elegant about the fighting that followed. The British pressed their attack in a narrow corridor between Barry Wood and Fontenoy village under heavy fire, and were eventually checked by the counterattack of the Irish Brigade, five regiments of "Wild Geese" in the French service. The linear formations of the 18th century infantry are well depicted in this painting by Felix Philippoteaux. Their close-range musketry caused frightful casualties: each side lost some 7,000 men at Fontenoy.

period was far from being the redcoated booby of popular mythology. There was certainly a sprinkling of incompetents, and the Duke of Marlborough complained of absenteeism, but it is hard to disagree with J A Houlding's verdict that "the British army was . . . led by an officer corps of the most considerable experience, made up of men who, by and large, entered the service for life and got by on steady, competent service."

The baroque armies presented a pleasing appearance in their wide-lapelled coats and tricorne hats, but there was little pleasure to be had from service in their ranks. In the 17th century there was nothing dishonourable in a gentleman serving as a private soldier, especially if he was able to "trail the puissant pike". But during the 18th century the status of private soldiers declined. The British army supplemented voluntary enlistment by drafting "all such able-bodied, idle and disorderly persons who cannot upon examination prove themselves to exercise some lawful trade or employment", and one of Frederick's regiments was filled with criminals obligingly sent to him by the Landgrave of Hesse-Kassel.

Soldiers' lives were dominated by the demands of the drill-ground and they were subjected to discipline which brought with it the constant risk of corporal punishment, whether from a casual blow of the sergeant's stick or the more deliberate attentions of the cat o' nine tails. However, once a man had finished his recruit training he might, at least in peacetime, expect to spend much of his time off duty, helping with the harvest or even taking up a civilian trade. For much of the period soldiers were billeted on private households or lodged at inns in garrison towns: billeting was a source of friction between monarchs and their subjects, the more so because it was sometimes used as a means of punishing refractory citizens. By the end of the 18th century, troops were often housed in purpose-built barracks, the long austere blocks which are a feature of so many fortresses.

The rise in foreign recruitment
The 17th century had seen the birth of national armies in which the mercenary played a declining part. The steady increase in the size of armies in the 18th century presented problems. Sweeping native males into their ranks denuded the workforce and reduced income from taxation, serious considerations at a time when war, famine and pestilence reduced population and money was scarce. Thus, the new national armies of the 18th century found themselves increasingly dependent upon foreign recruitment. Frederick's army usually contained less than half native-born Prussians: the remainder were foreign mercenaries or, as the army's strength was sapped by casualties, conscripted prisoners of war. In 1709, 81,000 of the 150,000 British soldiers in the field were actually foreign, either exiles, often French Huguenots, serving for pay, or regiments from German states hired out by their rulers. The battles of the period were filled with ironic confrontations: at Fontenoy in 1745 the Irish Brigade in the French service played a prominent part in defeating the British; at Trenton in 1776 King George's cause was defended by Colonel Johann Rall's Hessian brigade.

The high proportion of mercenaries in 18th-century armies was one of the reasons for their most prevalent disease: desertion. Christopher Duffy calls desertion "the bane of the Prussian army", and notes that the Regiment Garde in Potsdam lost 3 officers, 93 NCOs, 32 musicians and 1,525 men by desertion in the period 1740–1800, not to mention 29 executions and 130 suicides. During the Seven Years' War 80,000 men deserted from the Russian army, 70,000 from the French and 62,000 from the Austrian.

The 18th-century tactical debate
Before considering the Revolutionary and Napoleonic Wars we must first review 18th-century innovations in tactics, for most developments of 1789–1815 have earlier origins. The infantry, as we have seen, fought in line to make the maximum use of their firepower. But they marched in column, and deploying into line of battle was complex, time-consuming and difficult to conceal: an opponent who preferred not to offer battle could usually slip away. Nor was this the only difficulty. Practical soldiers knew that it was hard to control troops in line who, as Berenhorst has told us, tended to become caught up in an unproductive firefight. The Chevalier Folard, burrowing into classical sources – like Maurice of Nassau more than a century earlier – advocated the reintroduction of the phalanx and sketched out a massive pike and musket column. Marshal Maurice de Saxe also sought classical inspiration, and in 1732 recommended the "maniple" of two battalions formed eight ranks deep to combine fire and shock. Behind these theories, and others like them, lay a distrust of the effectiveness of the musket and a conviction that determined attack with cold steel was decisive. Frederick caught the prevailing mood, experimented with an advance without fire, and even issued a 13ft (4m) pike to some men in the rear rank. He discovered at Prague in May 1757 that the enemy's artillery and infantry did fearful damage to an advancing unit before it could close, and thereafter he reverted to his earlier preference for musketry.

In 1772 the Comte de Guibert produced the *Essai général de tactique*, one of the most influential works of the period. He favoured the three-deep line, and agreed that small columns moving rapidly, their front screened by skirmishers, were useful in the attack. The column came into its own for manoeuvre, and Guibert developed a method of deploying from column into line very rapidly. The Duc de Broglie experimented with various formations at the Camp of Instruction at Vassieux in 1778, and the French provisional drill-book of 1788 and the ordinance of 1791 were undogmatic, recommending the formation best suited to the circumstances and stressing the value of *l'ordre mixte*, a combination of line and column. Elsewhere the more formalized Prussian system triumphed, and in Britain David Dundas's *Principles of Military Movement* (1788), despite its Prussian origins, introduced a clear, simple system of drill to an army that urgently needed it.

Guibert, a child of the Enlightenment, wrote of patriotic citizen soldiers. The fighting in North America during the Seven Years' War, and especially Braddock's defeat by a smaller force of French troops and their Indian allies on the Monongahela river in 1755, encouraged the British to experiment with locally raised light troops. The 60th Royal Americans were trained for skirmishing in the backwoods and their approach to discipline smacked more of Guibert than of Frederick. In the American War of Independence both sides used light troops with less than formal discipline

—FIGHTING SAIL—

In the 1750s the British standardized the classification of their warships, with three rates of "line of battle ships". A third rate mounting 74 guns on two gundecks required the wood from 2,000 fully-grown oak trees, together with the elm and beech needed for her planking. Her 110ft (33m) mainmast would have come from the Baltic, though some were obtained from New Brunswick.

The fourth and fifth-rate men-of-war or frigates (32–44 guns) fell outside the classification of "ships of the line". The illustration (*below*) shows a frigate engaged in battle. The crew of the 12-pounder (*top*) are using ropes and pulleys to run their gun out after loading, while the 24-pounder below has been run out, and its captain is about to fire by pulling the lanyard attached to its flintlock. The surgeon carries out an amputation, without benefit of anaesthetic, on the orlop deck. Provisions are stored in the hold: biscuit and salt meat were the staple diet on British warships. The crew slept in hammocks on the dank gundeck, while the warrant officers and junior officers enjoyed marginally greater comfort.

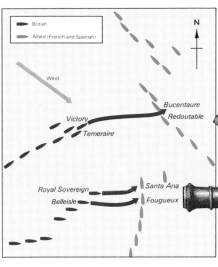

Trafalgar, October 21 1805.
Nelson's tactic of "cutting the line" enabled his ships to fire down the length of enemy ships, causing greater damage than a more conventional close-range broadside battering. The result was decisive.

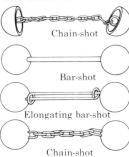

Chain-shot

Bar-shot

Elongating bar-shot

Chain-shot

The naval cannon fired a variety of projectiles. The roundshot was most common, and was used to damage hull, superstructure and masts. The cut-away barrel (*above*) shows a roundshot loaded onto a serge-covered gunpowder cartridge. Chain-shot, bar-shot and elongating bar-shot (*top*) were designed to cripple a ship by cutting rigging.

as an adjunct to their regular forces. However, the attractive image of the self-reliant militiaman should not make us lose sight of the crucial role played by Washington's Continental Army with its European instructors.

Light troops had been used in Europe well before they proved their merits in the New World. Frederick was confident of beating Austrian regulars, but he found their cunning Croatian light infantry elusive opponents. Again, the connection with Guibert is apt, for Prussian regulars could not be trusted to fight dispersed: "Open order," writes Hew Strachan, "was socially and politically unacceptable in Prussia." Frederick raised "free battalions" to deal with the Croats, but their performance was so discouraging that they were disbanded en masse. Other European armies were more successful, and by 1789 the concept of using skirmishers to cover the main body of a battalion had gained wide acceptance. Infantry battalions included a light company, composed of nimble men who shot well, to carry out this function, and a grenadier company of tall, stalwart fellows for use when brute force was required.

If light troops were one valuable legacy bequeathed to Napoleon by the 18th century, improved artillery was another. We have already noted Gribeauval's influence on the construction and organization of artillery in France. The Chevalier du Teil emphasized speed of movement and concentration of fire in *De l'usage de l'artillerie nouvelle . . .* (1778), a work read by Napoleon in 1788–89 when he was a subaltern at the artillery school at Auxonne, then commanded by du Teil's brother. No less worthy of note were the parallel efforts of Prussian and Austrian gunners. Frederick was at first unimpressed by artillery and in his penny-pinching way cavilled at the expense of the arm. But he recognized the importance of concentration of fire, using massed 12-pounders to deadly effect at Leuthen in 1757, and his experiments with *artillerie volante* looked forward to the use of artillery as a mobile striking force. Finally, the 18th century foreshadowed a key organizational change: in *Principes de la guerre des montagnes* (1764–71) Pierre Bourcet described the organization of an army into divisions, and Broglie briefly used the divisional system in 1760.

The Napoleonic Wars

On September 20 1792 an Austro-Prussian force under the Duke of Brunswick advanced on a French army at Valmy. The clash between the threadbare revolutionaries and the automata of old Europe never materialized, for the allies shied away after an inconclusive cannonade. Yet the episode is a telling one. Like all other organs of state, the French army had been transformed by the Revolution. Many noble officers had fled abroad and their places had been taken by ambitious NCOs to whom the newly formed National Guard also offered an avenue for advancement. Eightyfive percent of lieutenants in the French army in 1793 had been sergeants four years before and most of their men had enlisted since the Revolution. The declaration *La patrie en danger*, which followed the outbreak of war in 1792, produced a fresh surge of volunteers and there were more than 400,000 men under arms in 1792. The new army was of mixed quality. Its leaders included competent ex-NCOs and rapidly promoted junior officers, while enthusiastic young artisans rubbed shoulders with sullen

peasants in its ranks. But the events at Valmy showed that it would stand and fight and that its artillery, which had suffered less from officer emigration than other arms, remained formidable.

Valmy was a false dawn. The army was short of men by early 1793, and an attempt to conscript 300,000 met with only partial success and sparked off an uprising in the Vendée. That summer there was unrest across France, Toulon was occupied by the British and Valenciennes fell to the Prussians. The Convention decreed the *levée en masse*, declaring all Frenchmen "requisitioned for the needs of the armies" until the emergency passed. It may be that French strength attained 1,000,000 in mid-1794, but by 1797 casualties and desertions had reduced it to below 400,000. Nevertheless, the French army had done more than survive. It had become highly politicized, kept to its task by peoples' representatives with the armies who sent the disloyal or the unsuccessful to the guillotine. Between 1793 and 1796 it was reorganized into *demi-brigades* of one regular and two volunteer battalions, and all-arms divisions were established with two *demi-brigades* of infantry, one of light infantry and one of cavalry, together with artillery, staff and rudimentary services.

This force was largely the creation of Lazare Carnot (1753–1823), a member of the Committee of Public Safety and a former engineer officer. Carnot was not content to exercise central control over his armies, using the *Bureau Topographique* as an embryo general staff and communicating with the frontiers via Claude Chappe's semaphore system. He ordered commanders to eschew the geographical objectives of 18th-century war in favour of the enemy's

At its apogee, Napoleon's empire dwarfed the achievements of Louis XIV. In 1803 France obtained the Rhine frontier by a rationalization of Germany. Though Napoleon lost command of the sea at Trafalgar, 1805 saw him defeat the Austrians at Ulm and the Russians and Austrians at Austerlitz. Victory brought more territory under his control, and enabled him to create the Confederation of the Rhine. When Prussia intervened she was smashed at Jena, and the Russians were beaten at Eylau and Friedland: the Treaty of Tilsit recognized earlier changes and created the Grand Duchy of Warsaw. Strengthening of the Continental System led to extension of French influence in Sweden and Spain, though in the latter case it opened a "running sore" as the French fought Spanish guerrillas and a British army. In 1809 the Austrians resumed the war, only to be defeated at Essling and Wagram, and forced to

make a chastening peace. The empire began to crumble in 1812. Napoleon invaded Russia with a massive army, winning battles at Smolensk and Borodino and taking Moscow, but the Russians fought on, forcing him to make a costly retreat. Prussia joined Russia, and though Napoleon beat them at Lützen and Bautzen, when Austria intervened he was defeated at Leipzig. He was forced to abdicate in 1814. The Hundred Days saw his brief restoration and ended in defeat at Waterloo.

armies, which were to be brought to battle wherever possible. France, her ranks filled by the *levée en masse*, could stand heavy casualties: her opponents could not. Moreover, Carnot recognized that the real value of the divisional system lay in the ability of divisions to march divided, thus using road-space and requisitioning areas more efficiently, but to fight united. The revolutionary regime was less successful in its attempt to revitalize the navy, which had suffered badly from indiscipline and the loss of officers: nevertheless, it enabled several food convoys to reach France from America in the desperate days of 1794.

The quality of manpower produced by the *levée en masse*, allied to the pragmatic tone of the 1791 ordinance, enabled the French army to employ fluid tactics, with small columns breaking into loose skirmish lines to lap against an enemy formation, melting away if it tried to charge and pressing its withdrawal if it fell back. The *Armée du Nord* developed a characteristic tactical style which spread to other French armies, with a high proportion of skirmishers (in 1795 nearly a quarter of French infantry was light) moving ahead of shallow columns.

The advent of Napoleon

The political reaction of 1793–94 replaced the Jacobin extremists by more moderate men. The army found itself freed from paranoid invigilation, and at the same time took on a political role of its own. In October 1795 Napoleon Bonaparte, now a rising young general who had played a major part in expelling the British from Toulon in 1793, dispersed Parisian insurgents with "a whiff of grapeshot".

He gained fresh laurels from his campaign in northern Italy in 1796–97, although an expedition to Egypt ended badly because the British destroyed the French fleet in Aboukir Bay (battle of the Nile) in August 1798. On November 9 1799 Napoleon took a leading part in a perilous *coup d'état*: he was appointed First Consul and in 1804 became Emperor.

From 1799 to 1815 Napoleon's impact upon European affairs was nothing short of colossal. Although we cannot ignore his accomplishments in the legal and administrative spheres, it was upon his achievements as a soldier that his power depended, and these rested firmly on 18th-century foundations. Napoleon's background as a gunner and his familiarity with the military authors of the age helped mould his views, and he inherited a military machine tempered by war and infused with a fervour rarely seen in the armies of monarchical Europe. Until 1812 he was able to produce sufficient soldiers to fill the ranks of a huge army – there were 520,000 men in the field in 1808, with a further 180,000 in depots or garrisons – by drawing on an annual contingent of conscripts, using laws based on legislation of 1798. When Frenchmen ran short, he adopted the old expedient of recruiting foreigners: in 1812 half the *Grande Armée* was foreign. Finally, he was able to capitalize on the superb officer material released by the Revolution. He re-introduced the rank of marshal in 1804, and, though the men he appointed were of varying ability, their general youthfulness mirrored that of the remainder of the officer corps, and whatever their failings as independent commanders, most were men of unquestioned personal courage and tactical flair.

The corps system was fundamental to Napoleon's operational art. Each corps consisted of two to four infantry divisions, a brigade or division of cavalry, artillery, engineers and train. A corps could march and fight independently, but Napoleon's success rested on using his corps in unison, most classically in his favourite *manoeuvre sur les derrières*, employed at least 30 times between 1796 and 1815, with the battles of Ulm (1805) and Jena (1806) among its most dramatic examples. It involved pinning an enemy to his position by a feint attack, and then moving, with that astonishing rapidity which typified Napoleon at his best, to the enemy's flank or rear, where, ideally, Napoleon would occupy a natural barrier, forcing the enemy to fight or capitulate. Napoleon's opponents usually found themselves fighting off-balance on ground not of their choosing, and soon the awe he aroused contributed to their psychological dislocation. The Duke of Wellington, who never fell under his spell, observed that most of the Emperor's adversaries were half-beaten before battle even began.

The offensive was the essence of the Napoleonic battle as well as of the Napoleonic campaign. Napoleon favoured envelopment, launching what was often a surprise attack to unbalance an enemy army already preoccupied with a threat to its front. As the enemy focused on the unexpected danger, Napoleon produced his *masse de rupture*, which was committed after massed artillery had softened up the point of attack. "It is with artillery," wrote Napoleon, "that war is made," and its concentration featured prominently in his victories. The infantry would carry out the first part of their advance in column formation, deploying into *l'ordre mixte* as they neared the enemy line. Cavalry assisted in three different ways – by attacking to force the enemy to form square and thereby reduce his firepower, by

---

"At four o'clock our square was a perfect hospital, being full of dead, dying and mutilated soldiers. The charges of cavalry were in appearance very formidable, but in reality a great relief, as the artillery could no longer fire on us: the very earth shook under the enormous mass of men and horses. I shall never forget the strange noise our bullets made against the breastplates of Kellermann's and Milhaud's Cuirassiers, six or seven thousand in number, who attacked us with great fury. I can only compare it, with a somewhat homely simile, to the noise of a violent hail-storm beating upon panes of glass."

Ensign H R Gronow, Waterloo, June 18 1815

---

crashing through the breach when one was made, and by relentlessly pursuing the beaten foe. Whatever debts Napoleon owed to the 18th century, his approach to pursuit was definitively modern. He beat the Prussians at Jena on October 14 1806: his troops were on the Elbe by the 20th and reached Berlin on the 24th. November saw the fall of the remaining Prussian fortresses. Only 33 days after the battle of Jena not a man, horse or gun remained under Prussian command.

It is easy to paint too rosy a picture of the *Grande Armée* and its leader. Although Napoleon had an astonishingly powerful influence on his soldiers, reinforced by his understanding of the importance his men attached to

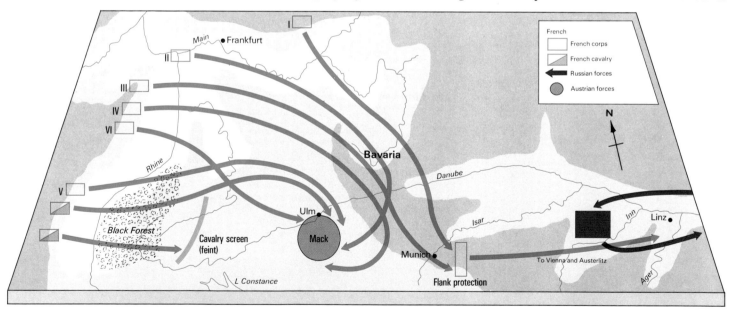

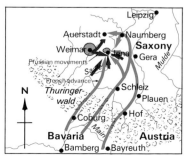

Napoleon's campaign at Jena, (*left*) shows his merit in dealing with an uncertain situation. His forces advanced in three columns, searching for their foe, and when a large Prussian force was finally identified on October 13, he concentrated to meet and defeat it.

In 1805 the Austro-Russian Allies sent a force into Bavaria to cover the approach of Russian armies. Napoleon proposed to destroy the Austrians in Bavaria before the Russians could intervene. He launched his army on August 26 with each corps moving along an independent line of march,

requisitioning as it went. Mack's army was swallowed up at Ulm (*above*), capitulating on October 20. Napoleon then pursued the Russians who had been marching to join Mack, taking Vienna on November 12, and going on to win, on December 2, a sweeping victory at Austerlitz.

public recognition, he was plagued by the age-old problem of desertion, especially as the war dragged on and the knife of conscription cut deep. Half the conscripts from the Haute-Loire could be expected to desert, and there were 8,000 deserters in hiding in the Lyons military district alone. He could be cruel and capricious and, certainly after 1812, made serious mistakes. His failure to recognize the importance of seapower and the impossibility of fully enforcing the Continental System in order to starve Britain of her European trade cost him dear, and his confidence in his own destiny prevented him from settling for a negotiated peace. Waterloo found him tired and ill, and both campaign and battle showed grave errors of judgment. Many of the difficulties he experienced in 1813–15 can be linked to his tendency to over-centralize. His headquarters was an instrument which responded only to its master's touch. Berthier, his chief of staff, was described as "a secretary of a certain rank", and few of his marshals showed real aptitude for independent command.

The threat and example of the *Grande Armée* brought changes among its opponents. The British army remained small by continental standards and, despite a thread of military enlightenment exemplified by Sir John Moore, was in most respects an 18th-century army in the best sense of the term, emphasizing robust discipline, voluntary recruitment and strong regimental identity. It fought well in the Peninsular War in Spain (1808–14), opening a "running sore" in the French flank, and Wellington established himself as the master of the defensive battle, using ground to protect his army from the worst effects of French fire. Britain's naval superiority, confirmed at Trafalgar in 1805, meant that the danger of French invasion was never great, and Britain as a whole neither had large-scale contact with the fervour of revolutionary France nor needed to conjure up large armies to face it.

The Austrian, Prussian and Russian armies, too, recognized the merits of skirmishing and adopted the divisional system. They also produced some remarkably capable leaders, among whom the Austrian Archduke Charles (1771–1847) deserves particular recognition. But the monarchies found it difficult to appeal to their citizen's sense of duty in the same way as the French, for patriotic duties implied political rights. Their conscription was partial, with the well-to-do usually escaping. Although they sometimes appealed to nationalistic sentiment, notably in the case of Russia in 1812, they remained cautious about arousing heady passions. Prussia was a notable exception. The earthquake of Jena toppled the conventions laid down by Frederick. Under Scharnhorst and Gneisenau the Prussian army was reformed, with the admission of bourgeois officers and, ultimately, a national call to arms and universal conscription.

The Congress of Vienna, which sought to put the world back to rights after the upheavals of 1789–1815, could redraw the map of Europe to restore "legitimate" governments, reward the victors and penalize France by a war indemnity rather than substantial territorial loss. It could not hide the fact that what had passed since the Revolution was utterly unlike what had gone before. The legacy of Napoleon could not be quantified, although that did not prevent men from attempting to do so: but it was to cast long shadows into the 19th century and beyond.

The young Napoleon in 1797 (painted by David, *above*), his Italian campaign behind him and the disastrous Egyptian adventure to come. Many would argue that "Napoleon the Emperor was in many respects less the soldier than had been General Bonaparte". It is certainly true that his operational flair ebbed as often as it flowed in his later career.

French cuirassiers (*left*) charge a British square at Waterloo.

— chapter eight —
The age of factory war

Richard Holmes

A century to the day separated Napoleon's defeat at Waterloo on June 18 1815 from the end of the French Artois offensive of World War I. The battlefields were 70 miles apart, and it is not too fanciful to suggest that many of the combatants of 1915 were great-grandsons of the men who fought at Waterloo. Yet a veteran of Waterloo would have found the soldiers of 1915, with their long-range weapons and drab clothing, almost like creatures from another planet.

In the century between the battles the face of war had been transformed. New weapons had revolutionized tactics; new currents of military thought and improvements in transport and communications had combined to bring about shifts in strategy; the shape and size of armies had altered dramatically. Behind all this was the teeming factory, product of the Industrial Revolution and fount of the machinery of war in an era of fundamental change.

The impact of the Industrial Revolution

Historians disagree over the precise relationship between military demands and technological innovation, but it is certain that the impact of industry on war was profound and complex. At one level, technology enhanced the range, accuracy and killing-power of weapons, made possible their mass production, and enabled the men who used them to move more efficiently. James Watt's development of the commercially viable steam engine in 1776 aided the exploitation of coal and powered the hammers and rolling mills of the iron industry. Iron output rocketed, rising in Germany from 85,000 tons in 1823 to more than 1,000,000 tons in 1867 and almost 15,000,000 tons on the eve of World War I. Even more significant was the Bessemer Converter's impact on steel production, leading to a fourfold increase in output in Britain and Germany between 1865 and 1879. Less dramatic, but no less important, were developments in fields as widely separated as road-building and textiles. Standardization of machine tools facilitated mass production of weapons and the stockpiling of spares, and contributed to the consistent performance of small arms and artillery alike. Although private industry remained indispensable, for all the *laissez-faire* of the period, state arsenals like Springfield and Harper's Ferry, St-Étienne and Châtellerault, Woolwich and Enfield, enjoyed a widening role.

All this was momentous enough. But at another level technology helped ensure that the population was better clad and fed. Advances in medical science, from Jenner's experiments with smallpox vaccine to Pasteur's work on bacteriology and vaccination and Lister's on antiseptics, increased life expectancy, as did great strides in public health administration. Concern for social welfare saw the increase of public education and a growing corpus of legislation regulating the employment of children. The population of Europe grew rapidly, from 187 million in 1800 to 266 million in 1850, 401 million in 1900 and 468 million by 1913. This occurred despite accelerating emigration to the New World and a decline in the birth rate which began in France and Ireland in the middle of the century and was general in Europe by its end.

The military significance of social concerns was not lost upon contemporaries: the Prussian decision of 1839 to limit child labour was prompted by fear that the people of a post-industrial civilization were less suited than countrymen to military service. Whatever the Prussians felt about the human products of industrial society – and recruit medical statistics from World War I do indeed suggest that twisted limbs and hollow chests were features of the factory age – the demographic explosion of the 19th century made the recruitment of a mass army possible, just as its industrial and technological developments meant that such an army could be armed, clad and fed. Finally, one need not share Karl Marx's views to recognize that the advent of an industrial proletariat had immense political implications.

The challenge of conscription

For most of the century there was no agreement among soldiers, nor among politicians, that a mass army was desirable. The principle of obligatory military service had been widely accepted in the 18th century, but universal conscription was shunned until the Convention's *levée en masse* decree of August 23 1793. This radical measure proved short-lived, and although subsequent French legislation retained the principle of universal service it was eroded by generous exceptions. Postwar legislation, in 1818, 1832 and 1855, maintained the theory of conscription and the practice of wide exemptions, and it was not until the Prussian victory over Austria in 1866 that the French seriously attempted to introduce universal conscription. The *Loi Niel* of 1866 went much further than previous measures. The size of the annual contingent was determined by the Legislature: its first portion would spend up to five years with the colours and four on the reserve, while the second portion trained for five months and spent the remainder of its nine years on the reserve. Young men who did not form part of the contingent were to serve in the *Garde Nationale Mobile,* although restrictions on its training were so great as to render it almost useless.

The struggle for the *Loi Niel* reflects a wider debate. Most French officers favoured a system which would keep the soldier with the colours long enough to acquire the true military spirit which short service could never produce. The political right, too, admired long service because it created an army set apart from society, which could be relied upon when revolution threatened. The left complained that a long period of conscription made the soldier a stranger in his own land: it argued also that long-service soldiers were likely to support repressive regimes.

This conflict struck chords across the Rhine. The Prussian Army Law of 1814 established the principle of universal conscription, although exemptions were allowed. Recruits were to serve for three years with the colours and two with the reserve before going on to the *Landwehr*; these terms were modified in 1833 to produce two years' regular and five years' reserve service, followed by eleven years in the *Landwehr*. The latter, a citizen militia with its own organization, was seen by its supporters as a link between army and nation, and enjoyed great popularity during the short-lived liberalism in Prussia at the end of the Napoleonic Wars. The *Landwehr* soon lost its independent status, coming under the regular army's control in 1819. The performance of the Prussian army in 1848–49, when it secured Berlin and helped to crush the

—TIME CHART—

1815	Napoleon defeated at Waterloo
1816–17	Prussian General Staff reorganized
1838	British army adopts percussion system
1840s	Prussian army re-equips with Dreyse breech-loading needle-gun
1851	British army adopts Minié rifle
1853–56	Crimean War
1856	Prussians form field telegraph units
1859	Franco-Austrian War in northern Italy
1860s	Prussian army re-equips with steel breech-loading artillery
1861–65	American Civil War
1861	Confederate victory at First Bull Run
1862	Battle of Antietam
1863	Battle of Gettysburg
1866	French army adopts Chassepot rifle
1866	Austro-Prussian War: battle of Königgrätz
1870–71	Franco–Prussian War
1880s	"Interrupted thread" breech-block developed by de Bagne
1880s	British adopt khaki uniform for field service
1884	Germans introduce Mauser bolt-action rifle
1885	French army introduces smokeless *Poudre B*
1888	British army adopts Maxim machine gun
1894–95	Sino-Japanese War
1897	French introduce quick-firing 75mm gun with hydraulic buffer
1898	Spanish-American War
1899–1902	Second Boer War
1903	US General Staff established
1904–05	Russo-Japanese War
1904	Japanese use howitzers at Port Arthur
1906	British General Staff established
1907	German army introduces field-grey uniform
1908	British Territorial Army formed
1912	US Quartermaster Dept formed
1914	World War I begins

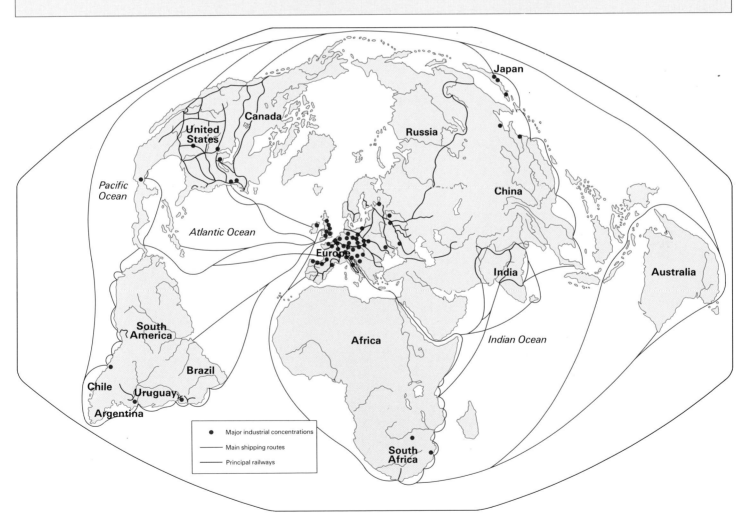

- ● Major industrial concentrations
- — Main shipping routes
- — Principal railways

Europe dominated the 19th century world. Its factories consumed imported raw materials, and its manufactured goods found growing markets abroad. Yet already the balance of power was shifting. In the United States, the North's output during the Civil War was a prime example of factory war. Russia was less buoyant. In mid-century Alexander II implemented much-needed reforms, but a reaction followed. Russian expansion in the Caucasus, Central Asia and the Far East led to conflict with Turkey and defeat in the Russo-Japanese War. European empires were a feature of the period. Although most of the states of South America achieved independence, the exploration of Africa was followed by the establishment of European colonies.

revolutionary movement in the rest of Germany, could not conceal the fact that it was too small, and the *Landwehr* too inefficient, to lend Prussia much weight in the council-chambers of Europe.

The reform of the Prussian army was the work of General Albrecht von Roon, appointed War Minister in 1859, with the patronage of Prince William – regent in 1858 and king in 1861 – and the political support of Otto von Bismarck. After a prolonged crisis, William and his supporters triumphed: conscripts were to serve for three years with the colours and four with the reserve before spending five years with the *Landwehr*, now in effect a second-line reserve. Roon's system showed its prowess in 1866, and in 1870 the Prussians and their allies mobilized nearly 1,200,000 men in a little over two weeks. In the same period France mobilized fewer than 570,000 men, many of them reservists whose discontent and lack of training bore witness to the failure of French military service legislation. During the later stages of the war the French raised a variety of troops who sometimes fought surprisingly well, but the contrast between French *francs-tireurs* (irregular troops) and *Gardes Mobiles* and the better-organized Prussians was starkly clear. The Franco-Prussian War effectively ended the debate on the composition of European armies. Some discordant voices on the left championed a citizen army like the Swiss, but universal conscription on the Prussian pattern, accepted by France in 1872, was henceforth the rule. Periods of service varied between states and fluctuated with political pressures and demographic trends.

Britain stood apart, her professional army backed by part-time militia and volunteers. This seemed acceptable, for her strategy accorded pride of place to the navy, leaving the army to garrison colonies and provide expeditionary forces, like that sent to the Crimea in 1854–56. The system creaked ominously even during the Crimean War, and all but collapsed during the Second Boer War of 1899–1902,

which took almost 450,000 British soldiers to South Africa: the 256,000 regulars and regular reservists were fleshed out with militia, volunteers, yeomanry and colonial contingents. The war led to a thorough examination of the army's structure, but the political case against conscription proved irresistible, and R B Haldane, Secretary of State for War, brought the Territorial Army into being in 1908. Its part-time volunteers provided for home defence while the regular army, reinforced by the Special Reserve, took part in operations overseas.

There were some similarities between the British and American experience, not least in the reluctance of demo-

---

US CIVIL WAR: MANPOWER

Estimated Military Population (White Males, 18–45)
Union: c.3,500,000 + Free Negroes
Confederacy: c.1,000,000

Total of Enlistments: 1861–1865
Union: 2,778,304 (includes multiple enlistments)
Confederacy: 750,000 (authoritative estimate)

Comparative Strength of Armies
US Regular Army (1860): 16,367 men

	Union	Confederacy
1861 (end)	575,917	326,768
1862 (end)	918,191	449,439
1863 (end)	860,737	464,646
1864 (end)	959,460	400,787
1865	1,000,516	358,692

(Figures represent total strengths: at times, absenteeism ran at up to 35 percent in the Union and more than 50 percent in the Confederacy.)

---

cratic politicians to permit conscription in peacetime. The Revolutionary War had initiated the tradition of a dual army, a citizen militia stiffened by a small regular force. During the Civil War there were too few volunteers to sustain the Union and Confederate armies, leading to the introduction of conscription by the Confederacy in April 1862 and by the Union in March 1863. The Union's

Prussia's deployment was complete by June 6, and on the 15th she attacked Austria's allies Hanover and Hesse-Cassel, subduing them despite a Hanoverian victory at Langensalza (June 27), and invaded Saxony, with the Elbe Army pursuing the Saxons into Austria. Moltke's ambitious "forward concentration" resulted in widely separated columns crossing the frontier, beating the Austrians at Nachod (June 27) but suffering a rebuff at Trautenau on the same day. The Austrians concentrated at Sadowa: on July 3 the First and Elbe Armies attacked them frontally while the Second moved in from the north.

Map legend:
Prussian advances
Austrian defensive moves
Railway
Heights in m
500
200
0

Soldiers on both sides posed for the photographer, often wearing finery which would prove inappropriate on the battlefield. Here Colonel J K Jordan of the 31st North Carolina, a Southern regiment, wears a pre-war blue uniform, rather than the "butternut" grey of the Confederate army, as well as large epaulettes which would have made him an obvious target. Most Confederate officers showed their rank on collar badges and "Austrian knots" (vulgarly known as "chicken guts") on their cuffs, while their adversaries favoured epaulette straps.

instead that the regular army should be enlarged and made easily expansible, with "National Volunteers" to reinforce it in wartime. His suggestions were too ambitious, but when the Spanish-American War broke out in 1898, federal-controlled volunteer units were raised alongside state volunteer regiments to produce a wartime army of 274,700. The army shrank when the war ended, but the evidence of unpreparedness encouraged reform and the Militia Acts of 1903 and 1908 did much to create an effective reserve. In some respects the United States had more in common with Britain than with the powers of continental Europe, for her geographical position enabled her to avoid conscription and encouraged her to devote a growing proportion of defence expenditure to the navy. In 1914 her navy was the third largest in the world: her army was scarcely bigger than that of Belgium.

Strategy in theory and practice

Like its technology, the military thought of the 19th century looked to the past as well as the future. Antoine Henri Jomini, a Swiss who served in the French and Russian armies, was heavily influenced by the formalism of 18th-century theorists like Lloyd and Bülow, and recalled the days when wars were concerned with the balance of power rather than national survival. While he acknowledged that war was "a terrible and impassioned drama", made complex by its physical and moral conditions, he argued that the application of governing principles brought victory. In his most influential work, *Précis de l'art de la guerre* (1838), Jomini analysed the Napoleonic period, emphasizing the importance of applying mass to the decisive point at the proper time. His concern for movement and concentration led to a preoccupation with an army's base of operations, its objectives, and the line of operations between the two. This was not merely sterile geometry, for it enabled him to deduce the importance of interior lines, which enabled an army to concentrate to

Conscription Act produced only 6 percent of the army's manpower, but threat of conscription combined with the attraction of bounties to provoke large-scale voluntary enlistment: more than 1,000,000 men signed up in the last two years of the war. The Confederacy drew 20 percent of its strength from conscription, but again the threat of the draft persuaded many men to enlist.

The Civil War over, the United States reverted to its familiar pattern of a tiny regular army – 27,442 in 1876 – backed by militia, soon to be styled the National Guard and numbering more than 100,000 by the early 1890s. The role and quality of the reserve provoked much discussion. Emory Upton, a general with a brilliant Civil War record, admired the German system, although he recognized that conscription in peacetime was unacceptable. He proposed

A Union poster, (*above*) issued in Chicago in December 1863, appeals for recruits for Mulligan's Irish Brigade. Patriotism is

reinforced with the offer of bounties and the warning that this is the last chance to avoid conscription.

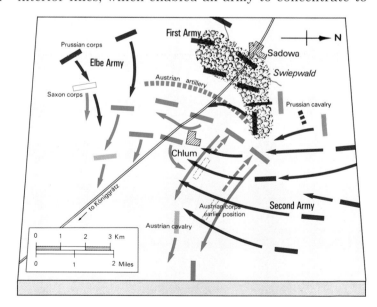

The battle of Königgrätz (Sadowa, *above*), 1866 ended the Austro-Prussian war at a stroke after a campaign of only seven weeks. Moltke's

troops were armed with breech-loading needle-guns and took advantage of Austrian mistakes to press home their advantage.

99

meet separate threats. He linked the logic of interior lines with his distaste for the nation in arms by suggesting that small, well-trained forces were better able to capitalize upon these principles than cumbersome mass armies.

Jomini had his shortcomings – his prejudices did indeed lead him to bend history to fit his analysis – but still his ideas gained wide currency. So pervasive was his authority that when the French were preparing the Italian expedition of 1859 they turned to him for assistance, while in Britain both MacDougall's *The Theory of War* (1856) and Hamley's *The Operations of War* (1866) relied heavily on Jomini. Hamley's work helped mould the ideas of a generation of senior British officers: in 1914 Sir John French, commander-in-chief of the British Expeditionary Force, was dissuaded from retiring on the fortress of Maubeuge by remembering a line of Hamley's. Jomini's long-lasting influence was not confined to Europe. The revival of the United States Military Academy at West Point under Sylvanus Thayer saw the extension of French principles to North America, and both Dennis Hart Mahan's *Outpost* (1847) and Halleck's *Elements of Military Art and Science* (1846), widely read by the officers who fought the Civil War, depended upon Jomini's analysis of the Napoleonic period.

Clausewitz and Moltke

If Jomini looked back to the 18th century, his Prussian counterpart Karl von Clausewitz looked forward to the 20th century. Clausewitz fought at Jena in 1806 and served briefly with the Russians before rejoining the Prussian army. He was appointed director of the *Kriegsakademie* in 1818, and died while Gneisenau's chief of staff in the Polish campaign of 1831. His book *Vom Kriege*, published posthumously, has suffered from inadequate translators and from commentators who have blamed Clausewitz for the events his writings foreshadowed.

Drawing on his own experience and diverse intellectual strains, Clausewitz described a war far distant from the measured conflict envisaged by Jomini. War, he declared, was the realm of uncertainty and chance, filled with physical exertion and danger. It was characterized by the element of friction. "Everything in war is very simple," he wrote, "but the most simple thing is very difficult. The difficulties accumulate and end by producing a kind of friction that is inconceivable unless one has experienced war." Good training, and combat experience, helped overcome this friction, and the historical study of war provided invaluable lessons. Morale was of fundamental importance: moral factors were "the precious metal, the real weapon, the finely-honed blade". Clausewitz placed war in its political context by stressing that it was the continuation of state policy by other means. Its object was the destruction of the enemy's means of resistance: battle was the means by which this was achieved. But he warned that war remained the expression of a political purpose, which had to be "the supreme consideration in conducting it". Unusually amongst his contemporaries, he saw the defensive as the stronger form of war, although he wrote of this in the context of the strategic defensive, in which the defender fell back on his own communications while the attacker's own strength diminished until "the culminating point of victory" was reached and the defender could take the initiative and counterattack.

For years Clausewitz remained a prophet without honour. *Vom Kriege* was not widely read even in Germany and lacked French and English translations until 1852 and 1873 respectively. But among the handful of men influenced by it was Helmuth von Moltke, Chief of the Prussian General Staff from 1857 to 1888. Moltke was a pragmatist, arguing that the precepts of strategy hardly went beyond common sense: their value lay in their application to a particular case. Like Clausewitz, he recognized the value of the converging attack in which forces united on the field of battle against the enemy's front and flanks. His concepts were in accord with the technology of the mid-19th century: the railway made strategic deployment more rapid, the telegraph permitted communication between divided forces, and breech-loading weapons made frontal attacks costly.

Moltke's campaigns against Austria in 1866 and France in 1870–71 exemplified his concepts, and his general staff formed the nervous system of massive armies brought together by conscription and swept to the frontier by rail. In 1866 the Prussians flowed into Bohemia in five streams, in a "forward concentration" which astonished contemporaries – and which might have exposed Moltke to defeat if the Austrians had reacted more astutely. The Austrian commander, *Feldzugmeister* Ludwig von Benedek, sent inadequate forces forward to check the Prussians while he

General Karl Maria von Clausewitz (1780–1831) was born near Magdeburg and served in the Prussian army against the French in 1793–94 before attending the Berlin military academy, where he attracted the interest of Scharnhorst. He served on the staff, but was captured at Auerstadt in 1806. After his release Clausewitz assisted Scharnhorst with the reform of the Prussian army, and was appointed military tutor to the Crown Prince. His intense dislike of Prussia's enforced military collaboration with France in 1812 drove him, like many of his comrades, to enter the Russian service. A staff officer in the 1812 campaign, he helped negotiate the Convention of Tauroggen which brough Yorck von Wartenburg's Prussian corps out of the war. He was readmitted to the Prussian army in 1814, and was chief of staff of Thielmann's 3rd Corps in the Waterloo campaign.

In 1818 Clausewitz was promoted major general and appointed commandant of the *Kriegsakademie*. The triumph of the reactionaries in the period following the Napoleonic Wars helped limit his influence, and he had little impact upon the academy's syllabus or teaching. He had begun work on a comprehensive analysis of war as early as 1816, but he died in the great cholera epidemic of 1831 while chief of staff of Gneisenau's force dispatched to put down the Polish insurrection, with his work unfinished. His widow edited his writings, which appeared as *Vom Kriege* in 1832–34, and through them Clausewitz was destined to enjoy a far greater influence after his death than he ever had during his life.

concentrated around Josefstadt, and in a series of border battles in late June the Prussians burst through the cordon. Benedek planned to join his ally, the Saxon Crown Prince Albrecht, at Jitschin, but was too slow to do so, and Albrecht extracted himself from an attempted envelopment with considerable skill.

On June 30 Benedek realized the seriousness of his plight and warned the Emperor Francis Joseph that catastrophe was imminent, begging him to make peace. The Emperor refused, and Benedek's morale improved as he took up a defensive position in the hills northwest of Königgrätz. It was not until the evening of July 2 that Moltke, hampered by inadequate cavalry reconnaissance, discovered where Benedek was. On July 3 Prince Frederick Charles's Second Army attacked Benedek's front while the Crown Prince's First Army groped towards his flank. The Austrians were hotly engaged against Frederick Charles when the Crown Prince bit into their right flank. By nightfall, despite the bravery of Benedek's infantry and cavalry and the disciplined fire of his gunners, the Austrians were routed with a loss of nearly 43,000 men.

Moltke's pragmatism was confirmed by the events of 1870–71. Three German armies under General von Steinmetz, Prince Frederick Charles and the Crown Prince of Prussia crossed the French frontier after efficient mobilization and concentration. Moltke hoped to encircle the

"There in front of the church – I can never forget the sight – stood the sous-intendant militaire*, hemmed round on all sides with the angry multitude, clamouring and howling for food. He was crying like a child, clasping his hands above his head: 'I have nothing – I can give nothing; they have left me in the lurch'."*

Pastor Klein, Froeschwiller, August 1870

French north of the Saar, but his plan was thwarted by Steinmetz's obstinacy and the Third Army's slowness. On August 6 the army commanders became entangled in battles against his wishes: in the north, the First and Second Armies hammered the French 2 Corps at Spicheren; in the south, the Crown Prince beat Marshal MacMahon at Froeschwiller. The French Army of the Rhine was beset by numerous difficulties, not least the fact that Napoleon III attempted to command in person with the assistance of a cumbersome staff. As his five northern corps fell back on Metz, the Emperor left the army, entrusting command to Marshal Bazaine, who attempted to retreat on Verdun. Moltke's conduct of operations over the next days followed no preconceived plan – and sometimes went dangerously wrong – but shows him as a brilliant opportunist. As the French set off towards Verdun, Moltke ordered pursuit, only to discover that Bazaine

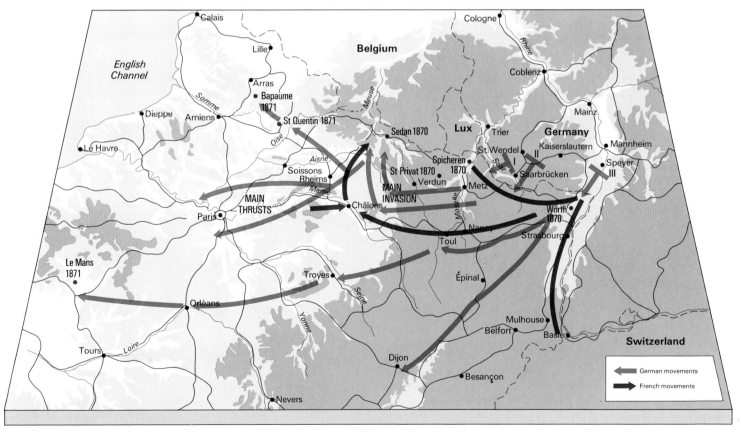

In August 1870 Moltke defeated the Imperial armies in Alsace and Lorraine and then invested Paris (September 20). The First Army reduced fortresses in the Prussian rear before marching north, fighting inconclusively around Bapaume in January 1871. The French Army of the Loire defeated the Bavarians near Orléans (November 9) but was soon swamped by forces released by the surrender of Metz, and plans to relieve Paris came to naught: Le Mans fell in mid-January. In the south, too, the French were beaten despite some local successes, and, after an armistice had been agreed elsewhere, they withdrew to Switzerland.

had barely left Metz, and that the leading elements of the Third Army, far from snapping at the French army's tail, had stumbled into its jaws. The battle of Vionville – Mars-la-Tour on August 16 saw the Germans check the French advance, persuading Bazaine to pull back onto the long ridge running through St-Privat to Gravelotte. On the 18th the Germans came up in strength, and although the French held their ground with their customary determination, their right eventually crumbled under artillery fire and flanking attack. Bazaine withdrew into Metz, which he was never to leave as a free man.

The end for the French

MacMahon assembled the three right wing corps of the Army of the Rhine at Châlons, forming the nucleus of the Army of Châlons, which set off on August 21 in the hope of joining Bazaine, who was expected to break out of Metz. MacMahon's march was beset by changes of direction resulting from contradictory information on Bazaine and the need to pick up supplies from the railway. Moltke formed the Army of the Meuse, under the Crown Prince of Saxony, by splitting the Second Army, leaving its rump with Frederick Charles to besiege Metz. The Third Army and the Army of the Meuse set off after MacMahon, though it was not until the 26th that German cavalry found the Army of Châlons. MacMahon might have saved himself by

retreat but, spurred on by an anxious government, he lurched onwards. On August 30 one of MacMahon's corps was surprised at Beaumont, and on the 31st MacMahon slid his army into position around Sedan, unaware that Moltke was moving in from the southeast while the Crown Prince marched up from the southwest.

The battle of Sedan, fought on August 31, was a foregone conclusion. The French were encircled, their thick lines of infantry deluged by fire to which their own gunners could make no effective response. MacMahon was wounded early in the day, and his successor failed to snatch the slender opportunity of a breakout. By late afternoon the situation was hopeless, and Napoleon, who had accompanied the army, offered King William his surrender. The French had suffered 17,000 casualties and lost 21,000 prisoners during the fighting: the surrender yielded another 83,000. German losses totalled only 9,000, a small price for a victory which ended a dynasty.

Sedan did not end the war. Metz surrendered on October 27, but the new Government of National Defence struggled on, raising troops in the provinces and defending Paris with a determination born of desperation. The second phase of the war tried Moltke as sorely as the first, for the defeat of the Imperial armies ushered in the nation in arms, the very spectre from which Jomini had averted his gaze. The colossal energy of Léon Gambetta, Minister of the

Helmuth von Moltke (1800–91) joined the Prussian army in 1822, and almost immediately attended the *Kriegsakademie*. Clausewitz was then the academy's director and his classic work *Vom Kriege* had a considerable impact upon the young man.

In 1835, Moltke was given leave to visit Turkey where he served with the Sultan's army. He published an account of his travels in 1839. Foreign service combined with literary skill made Moltke unusual amongst his Prussian contemporaries, and helped attract the patronage of the Hohenzollerns. In 1855, he became aide-de-camp to

Prince William, heir to the throne. When, in October 1857, his patron became regent, Moltke was appointed Chief of the Great General Staff. He strengthened his hold on royal favour while serving as chief of staff to the Austro-Prussian force during the Schleswig-Holstein War in 1864. In 1866, success in the Austro-Prussian War and the victory at Königgrätz made him a figure of European importance. In 1870 he masterminded the defeat of the armies of Imperial France making Prussia the dominant power on the continent. This brought Moltke the title of count, and later his field marshal's baton, awarded in June 1871. He was released from his post in August 1888, and died three years later.

Moltke was more than merely a slick practitioner of staff work. He was quick to grasp the potential of the railway, and his deployments in 1866 and 1870 were landmarks in military history. His operational command, notably during the Sedan campaign of 1870, marks him out as a general of the highest ability.

Bazaine held a strong position on high ground (*above*), though his right was exposed. German attacks were repulsed until the Saxons turned the northern flank, leading to partial French collapse.

Interior and virtual dictator in the provinces, and Charles de Freycinet, engineer turned war administrator, and the efforts of a host of civilian specialists whose talents were conscripted to meet the demands of war, kept the Armies of National Defence in the field long after military wisdom announced their collapse. Their energy could not stave off defeat: an armistice was concluded on February 28 1871, and France made peace on humiliating terms.

The Franco-Prussian War was to help shape the events of 1914. Two points should be noted here. One is the enthusiasm with which the French subsequently discovered Clausewitz, blending his emphasis on battle with their mystique of *furia francese*, a reaction against defensive-mindedness, and Ardant du Picq's insistence on the supremacy of morale, to produce a doctrine of the offensive. The other is the war's effect on an increasingly militaristic Germany. The rise of the Chief of the General Staff diminished not merely the Minister of War but the Chancellor himself. If the French discovered Clausewitz after the war, then the Germans neglected some of his wisest precepts. Moltke's successors forgot that the field marshal had warned them that no plan could be relied upon after first contact with the enemy: they forgot, too, that war was the handmaiden of politics, not vice versa.

For all his pragmatism, Moltke paid scant attention to the American Civil War: he was said to have observed peevishly that little could be learned from armed mobs chasing one another about the countryside. This would have been a grave misjudgment, for the Civil War was pregnant with lessons for the future. Military experience in North America was stretched thin to stiffen a Union army which reached 1,000,000 men at its peak and a Confederate army rather less than half as strong. In a sense these forces were – like the Armies of National Defence – a foretaste of what was to come. The major wars of the 20th century were to be characterized by a national mobilization which bit deep into the fabric of society, just as the demands of war, economic as well as military, bore down on the North and crushed the South.

The role of the railway

Moltke might also have observed the role played by railways. His own campaigns relied heavily on the railway, but as E A Pratt was to write, without railways the Civil War "could hardly have been fought at all". The North gained immense advantage from the fact that it contained all but 9,000 of the 31,000 miles of track in the country: its railroads were better planned, built and maintained than their southern counterparts. Railways in Europe were invaluable for concentration at the opening of a campaign, but during the Civil War they were used for the large-scale movement of troops between fronts, much as they were to

Shell and cartridge were still separate: fixed ammunition, with shell and cartridge combined, made possible the quick-firing field artillery of World War I. The development of recoil buffers in the 1890s was also crucial: until then guns rolled back on firing.

A steel breech-loading Krupp 70lb field gun. Krupp favoured the sliding breech-block rather than the threaded systems used by most other manufacturers.

103

be in world wars in Europe. The North's superiority in railways underlined its formidable industrial muscle which meant that, time and again, the achievements of Southern commanders were rendered nugatory by the North's ability to replace lost equipment. The North's arsenals, public and private, produced fewer than 50,000 small arms in 1860: during the war they turned out 2,500,000. The South imported 600,000, despite the blockade, manufactured some of its own, and picked up not a few by battlefield scavenging.

The economic disparity between the two sides had a decisive influence upon strategy. The Confederacy's only hope lay not in conquering the North, for that was manifestly beyond its resources, but in fighting on until Union weariness or European intervention ended the war. President Jefferson Davis faced the choice of whether to defend the borders of the Confederacy or to permit Robert E Lee to invade the North, as he did on three occasions, in the belief that victories there would accelerate the erosion of Union morale. Davis also faced a conflict of strategic priorities. The Eastern theatre of war was clearly important, for the rival capitals were close and capture of either would have disproportionate repercussions. The wider Western theatre was no less crucial, and the Chattanooga– Atlanta area was especially vital because the Confederacy's main lateral railroads ran through it. In the event, Davis preferred the defence of the borders of the Confederacy to Lee's "offensive defensive", and he compromised by dividing the Confederacy into departments whose commanders could defend their own areas and cooperate in meeting major thrusts by transferring reserves by railway. It was a strategy designed to buy time during which the Union, and perhaps England and France, would conclude that the South could not be crushed.

Strategy of the Union
Union leaders realized that they could win only by conquering the South. In the opening stages of the war Lieutenant General Winfield Scott produced his so-called Anaconda Plan, designed to throttle the South by a naval blockade combined with a thrust down the Mississippi to split the Confederacy. This solution, slow but certain, did not commend itself to politicians or populace, for whom "On to Richmond" was an irresistible cry. It also failed to appeal to Abraham Lincoln, who pressed his generals to "destroy the rebel army" in a decisive battle. His urgings sometimes persuaded them to embark upon projects about which they had misgivings, and it was less easy to destroy an army in the geostrategic circumstances of North America than Lincoln recognized.

The fact that Union strategy was less than single-minded was not a matter of simple incompetence, though there was enough of that in an army which had promoted many regulars beyond their capability and had also to accommodate generals whose political claims to rank were better than their military credentials. It took time for capable men to emerge, and it is a tribute to the flexibility of the North that Grant could rise to lieutenant general in three years and Upton could be a brevet major general at the age of 24. It also took time for the North's might to make its weight felt, and for its leaders to recognize that it was the mace, not the rapier, that would kill the South. Much credit is due to Grant, who took command of the Union

armies in March 1864 and at once announced his intention of putting simultaneous pressure on the sagging Confederacy, using "all parts of the army together, and somewhat towards a common centre". In Union strategy during the last 12 months of the war we see striking elements of modernity, most notably the realization that a

US CIVIL WAR: RESOURCES			
Transport (1860)	**Union**	**Confederacy**	**Total**
Railways:	22,085 miles	8,541 miles	30,626 miles
Horses:	4,417,130	1,698,328	6,115,458
Draught animals:	1,712,320	1,657,308	3,369,628
Agriculture (1860)			
Improved farmland:	106,171,756 acres	57,089,633 acres	
Unimproved farmland:	106,486,777 acres	140,021,467 acres	
Industry (1860)			
Industrial establishments:	110,274	18,026	128,300
Capital investment:	$949,335,000	$100,665,000	$1,050,000,000

belligerent's strength lies in his human and economic resources. The Civil War was, in an important sense, the shape of war to come.

The railway and the telegraph made possible the style of war waged by Grant and by Moltke. The railway transformed war, enabling mass armies to be transported to concentration areas and to be supplied from railheads. Its

US military railroads. Field Hospital at City Point, Virginia. The locomotive on the left is the General Dix, built by Baldwin in 1862. The other is the General McClellan, built by the New Jersey Loco Works in 1862.

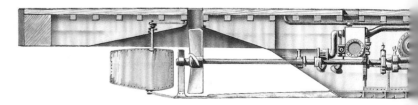

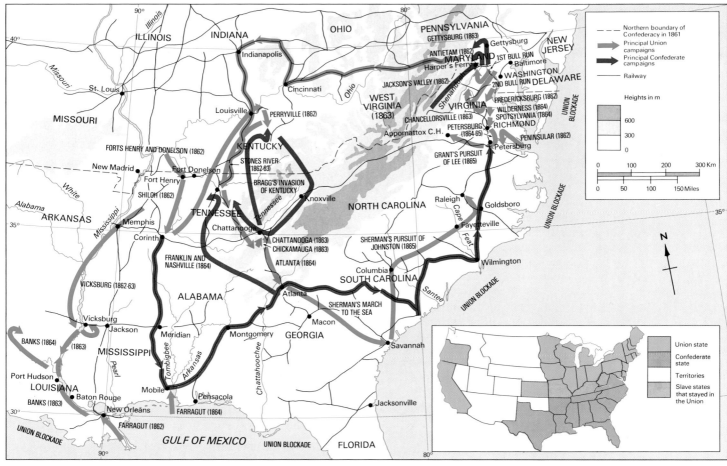

The map (above) shows the area affected by the American Civil War, 1861–65, and the course of the principal campaigns. (*Inset:* the respective areas of the states of the Confederacy – Alabama, Arkansas, Florida, Georgia, Louisiana, Mississippi, North Carolina, South Carolina, Tennessee, Virginia (from which pro-Union West Virginia seceded to become an independent state of the Union in 1863) and Texas – and those of the Union, which included both "free" and "slave" states and territories.) On the main map, note particularly how the railway network facilitated the movement of troops, much to the advantage of the Union, and how, in 1864–65, Sherman's march through Georgia "from Atlanta to the sea" and Grant's successful Petersburg campaign to the north split apart the Confederacy and made Lee's surrender at Appomattox inevitable.

CSS Merrimac

Navies experimented with ironclad warships and shell-firing guns from as early as the 1840s. Though most remained full-rigged, warships were increasingly steam-driven. The US Navy fell into Union hands at the outbreak of war, but the Confederates refloated the scuttled *Merrimac* and refitted her as an ironclad, her deck shrouded by timber faced with iron plates. She mounted three guns to each broadside, one fore and aft, and a ram projected from her bow. *Monitor* was built to fight her, with an armoured superstructure surmounted by a revolving turret with two 11-inch guns. On March 8 1862 *Merrimac* (renamed *Virginia*) destroyed two wooden vessels in Hampton Roads, and on the 9th duelled inconclusively with *Monitor*. *Monitor* later foundered in a gale and *Merrimac* was scuttled.

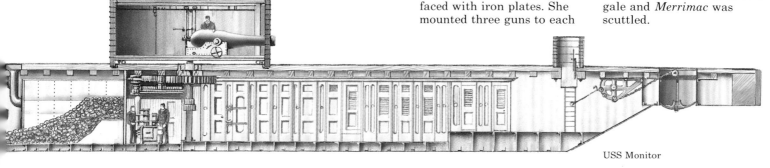

USS Monitor

military potential was recognized quickly. The British moved a battalion by rail in 1840, the Russians a corps of 14,500 men six years later, and the French in their Italian campaign against Austria in 1859, transported 604,381 men and 129,227 horses by rail. But it was Germany upon which the railway seemed to confer the most attractive benefits. It offered the enticing possibility of operating on interior lines against potential enemies on three sides, and permitted the great strategic *Aufmarschen* of 1866 and 1870.

Alongside the railway (quite literally, for their lines often accompanied one another) ran the telegraph. During the Napoleonic Wars semaphore telegraphs had been used to link fixed points, but their limitations were obvious. The electric telegraph, perfected in 1829 and made more useful by the introduction of Samuel Morse's code after 1850, was first used during the Crimean War, when its results were not altogether helpful. Problems with decoding were solved in time to permit Napoleon III to intervene in the conduct of operations, confusing an already temperamental French high command. The Prussians formed field telegraph units in 1856–57, and these played an important part in the campaigns of 1866 and 1870–71, when the telegraph's main use was in connecting general headquarters to the capital and in keeping it in communication with major formations: it was of little value forward of corps headquarters.

Further improvements in communication

The telephone, patented by Alexander Graham Bell in 1876, permitted voice conversation, but like the telegraph it depended upon wires. In 1885 the Germans experimented with linking artillery to forward observers by telephone, a development of far-reaching importance, for it would enable artillery to become more than a long-range direct fire weapon. Indirect fire, by which gunners engaged a target they could not see, depended not only upon improvements in artillery but also upon the link between observer and gun. During the American Civil War observers in static balloons had communicated with the ground by telegraph, and they were soon to do so by telephone.

Marconi's work on wireless was to have a profound effect on communications. In 1895 he transmitted a signal one mile; six years later he sent one across the Atlantic. But early equipment was bulky, and the dictates of security and the risk of jamming limited its value. Like the telegraph and telephone, the early wireless was more useful for keeping higher headquarters in touch than for tactical communication. Forward of corps headquarters, armies relied on dispatch riders on horseback or motorcycle, and on staff officers and commanders in motor cars.

The contrast between ancient and modern in communications was paralleled in transport. The railway provided high-speed strategic transport, and the motor lorry, available in small but growing numbers after the turn of the century, showed the way ahead. Forward of railheads, however, the horse took pride of place in 1914 as it had a century before. Armies' demands for horseflesh were scarcely less voracious than those for manpower. When Grant crossed the Rapidan in May 1864 his command included 125,000 men and more than 56,000 horses and mules, and even in October 1918 the American Expeditionary Force (for all its 30,000 trucks) had 163,000 animals to its 1,974,400 men. The supply of horses to armies was an industry in itself, and feeding them remained a major concern into and beyond World War I. In 1914 the 84,000 horses of a single German army ate almost 2,000,000lbs of fodder every day, and fodder formed the largest single item in British stores shipped to France in World War I, narrowly outstripping ammunition.

The burgeoning size of armies made the task of feeding their soldiers quantitatively greater, but improvements in transport and the development of preserved rations made logistics more efficient. Preserved food, from the boiled beef in glass jars of the Napoleonic Wars, through the pickled meat, hardtack biscuits and desiccated vegetables of the American Civil War to the tinned food of World War I, was never to replace fresh rations, but it gave logisticians increasing flexibility, especially where terrain and climate made the supply of fresh food difficult.

The professionalization of logistic services also helped. In the second half of the 19th century, transport services became uniformed branches of the army. The British Commissariat merged with the Military Train to form the Army Service Corps in 1888. The United States was slower in moving towards a consolidated logistic service, but in 1912 the Quartermaster Department at last came into being. It was not only drivers who found themselves part of the army's chain of command. Civilian officials, often uniformed members of the Commissariat or *Intendance*, had long been responsible for the purchase, control and issue of food and equipment. Their militarization sprang partly from the growing professionalism which characterized officer corps during the period, and partly from a

British shipping in Balaklava harbour, with tents and prefabricated huts on the inhospitable uplands. Troops lived in tents for the first terrible winter.

revulsion at the administrative chaos of many 19th-century campaigns.

For the British, the Crimean War was epitomized not merely by the heroic folly of the Light Brigade, but by the sufferings of the expeditionary force as it wintered outside Sevastopol. Men had no warm clothes, little shelter and no firewood. They died of exposure while bundles of clothing were flung into the harbour at Balaklava to form makeshift jetties; so ponderous was the Commissariat's procurement system that many of the stores ordered for the first winter of the war arrived only in time for the second. The plight of British troops was publicized by W H Russell of *The Times*, and staff and Commissariat were widely criticized. In a similar way the Franco-Prussian and Spanish-American

—DEVELOPMENTS IN UNIFORM—

Military uniform may be traced back at least as far as the armies of the ancient world, and in Europe its standardization was bound up with the growth of standing armies. Uniform tended to be founded on civilian dress, modified to facilitate and reflect its wearer's function. It served numerous other purposes. Lofty headdress gave the soldier added height, while epaulettes or padded shoulders increased breadth and fur on headdress or pack contributed to general ferocity. National colours encouraged feelings of community, aided recognition in battle, and, no less pertinently, facilitated bulk manufacture of cloth. Regimental distinctions – badges, braid, buttons, coloured facings – played their part in promoting *esprit de corps*.

Uniform reflected the imitation that was indeed the sincerest form of flattery. The dominant military nation in Europe found its dress aped across the world, and a specific expertise was often marked by distinctive dress which persisted after its real reason had disappeared. Grenadiers, for instance, could not sling muskets to throw grenades wearing the three-cornered hat. They took to a conical cap, which became their distinguishing mark long after the grenade was obsolete.

Much uniform was woefully impractical. The double-breasted coat was the basis of 18th century infantry dress. Its cloth shrank, and the demands of economy led to its being cut so skimpily that the lapels could not be buttoned across in cold weather. Dragoon helmets had practical origins, but peacetime fashion increased height and multiplied embellishments, making them unsteady on horseback and brain-boiling in hot weather.

The rigours of active service led to modifications. The British and French made local adjustments in North America in the 18th century: coats were shortened, lace removed and headdress simplified. The example of German *jäger* encouraged the formation of rifle units clad in dark green. Nevertheless, as long as infantry carried short-range weapons, there was little demand for more sombre dress.

In the 19th century, there was pressure for utilitarian dress as weapons made dispersion and concealment important. Some British units wore khaki (from the Urdu for dust-coloured) during the Indian Mutiny, and after the First Boer War, Britain adopted it for field service. German troops sent to China in 1900–01 also wore it but field-grey was eventually adopted for war and manoeuvres. The Russo-Japanese War persuaded its belligerents to introduce khaki, and the Austrians took to "pike" grey in 1909. The United States flirted with khaki in 1898 and introduced an olive drab service uniform in 1902. Yet it was not until April 1915 that French infantry adopted horizon blue to replace the blue and scarlet of yesteryear.

French
zouave
(1854)

British
11th Hussar
(1854–56)

French Imperial
Guard Grenadier
(1804–1815)

German
jäger
(1804–1815)

British
Army:
Boer
War
(1899–1902)

Wars pointed the way to reform.

The general staff was central to many of these developments. Most armies had a system centred upon two principal staff officers: the adjutant general, responsible for discipline, appointments and promotions, and the quartermaster general, with his wider concern for movements and quartering. The quartermaster general foreshadowed the modern chief of staff, but the French army, for all that it instituted the appointment of chief of staff in 1792, reduced him to the position of an adjutant writ large: even Berthier was only "a secretary of a certain rank". The long peace which followed the Napoleonic Wars discouraged the development of the staff: although staff schools were established in France in 1818 and in Russia in 1832, staffs remained pools of administrators. There was no capital staff to advise the government and formulate strategic doctrine, and without it the staffs of field armies remained mere scribblers, lacking unity of purpose and professional identity.

The General Staff in Prussia

For the genesis of the general staff we must look to Prussia, whose General Staff came into being in 1817 when Department II of the War Ministry, the old quartermaster general's department, received that title. Its officers were already carefully selected and trained. It grew in influence in the 1820s, but even after Moltke took over in 1857 it was overshadowed by the monarch's Military Cabinet and was mistrusted by senior officers. The staff's lack of status is demonstrated by the fact that Moltke was neither directly involved in military reorganization nor consulted on plans for war with Denmark in 1864. The mistakes of the Prussian commander and chief of staff of the Austro-Prussian force sent to Schleswig led to Moltke's appointment as chief of staff, and the storming of the fortress of Düppel in April reflected well upon him. In June 1866 the Chief of the General Staff was given authority to issue orders to field formations without passing through the Ministry of War, and with the 1866 campaign the General Staff came of age.

In 1870 Moltke's supremacy was virtually unquestioned. King William, presiding over Royal Headquarters, relied exclusively on his advice, and quartermaster general and intendant general were his disciples. Moltke's own staff formed three sections: movements, rail transport and supply, and intelligence. With the aid of his "demigods" (the 3 colonels heading the sections) 11 officers, 10 draughtsmen, 7 clerks and 57 other ranks, Moltke controlled armies which, by the end of the war, had 850,000 men in the field.

Prussia's example was widely copied. Austria-Hungary and Russia reorganized their staffs in the German image, and even the French followed suit, setting up the *École Supérieure de Guerre* in 1876, dissolving the closed Staff Corps in 1880, and appointing their first proper chief of the general staff in 1890. French military reform faltered after the short-lived "Golden Age" that followed the Franco-Prussian War, and the Dreyfus affair of the mid-1890s split army and nation. It was not until 1911 that steps were taken to create an effective high command. The chief of the general staff became France's senior military officer and commander of her armies on mobilization. The *conseil supérieure de guerre*, a forum for corps commanders, was

restored (its predecessor having perished in the aftermath of the Dreyfus affair), and the *comité supérieure de la defense nationale* was established to bring soldiers and politicians together at the highest level.

The United States owed its general staff to Secretary of War Elihu Root, appointed in 1899 while memories of the Spanish-American War were fresh. Well aware of the opposition to increased military preparedness outside the army, and of resistance to the general staff concept within it, Root moved deftly, setting up the Army War College in 1900 and creating a general staff in 1903. In Britain the general staff was established in 1906, among the reforms which followed the Boer War, and unleashed its infant energy on the task of planning for war with Germany.

By 1914 general staffs were universal among the armies of the civilized world. Their officers formed what Spencer Wilkinson called "the brain of an army", providing expert advice, refining war plans, and furnishing field formations with the technicians of operations, intelligence, movement and supply. Yet their sheer professional competence brought with it introspection, single-mindedness, and a tendency to subordinate the political to the military.

Weapons and tactics

We should not be surprised that tactics often lagged behind technology, for the innovations of the 19th century were most marked in the field of weapon development. Almost every war saw the appearance of some new weapon whose use and influence were conjectural, and even when a tactical point was made with clarity in battle, pressures of reaction, swirls of inter-arm rivalry, jostlings within the hierarchy and misreading of military history all helped obscure matters. And technology tapped the kaleidoscope constantly, changing the picture so swiftly that the cleverest of men found it hard to peer into the future.

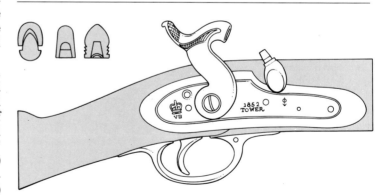

The principle of rifling was well known, but the need for a tight fit between bullet and barrel caused difficulties. In 1841 Captain Delvigne devised a "cylindroconical" bullet, and Captain Claude Minié of the Musketry School at Vincennes developed it further. The system had several variants, but its essence was that the explosion of the charge expanded the base of the bullet into the rifling.

The period is dominated by one deadly fact: the increase of firepower. In the second quarter of the century infantry weapons were improved first by the replacement of the flintlock by the more reliable percussion system, and by rifling, which increased range and accuracy. The Minié rifle, firing a conical bullet whose hollow base expanded with the explosion to grip the rifling, was the characteristic weapon of the 1850s and 1860s. Its performance was far better than that of the smoothbore: it could hit formed bodies of troops at 800 yds (750m) and its heavy bullet was still lethal at 1,000 yds (900m). In practice its effectiveness was reduced by the poor weapon-handling which was common in battle: in the most lethal encounters of the Civil War, like Bloody Lane at Antietam, the infantry fought it out at close range, not with the volleys of the drill-ground

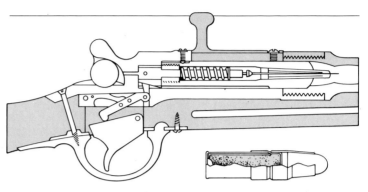

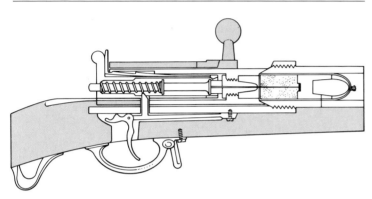

The problem of achieving a gas-tight seal at the breech long perplexed weapon designers. Antoine Chassepot partly solved it by using a rubber obturating ring. His rifle, strikingly superior to the needle-gun, was initially rejected by the French army: only the pressure of Napoleon III brought about its adoption.

The Dreyse needle-gun was the first breech-loader generally adopted for military service. Its "needle" firing pin entered the cartridge to strike the primer at the base of the bullet. It greatly increased the infantry soldier's firepower, and enabled him to load and fire lying down.

but with a constant ripple of fire as men loaded and fired as fast as they could.

The breech-loader was a notable improvement. Captain Patrick Ferguson patented a breech-loading rifle in 1776, but the Dreyse needle-gun, ordered on a large scale by the Prussians in 1840, was the first breech-loader to be generally issued. The Dreyse was far from perfect: its thin firing pin, which had to pass through the cartridge to strike the primer at the base of the bullet, soon broke, and as the weapon became worn it leaked gas at the breech. Its trajectory was not as flat as that of the Minié and it was sighted up to only 800 yds (750m). But it could be fired twice as fast, at about seven rounds a minute, and was simpler to manage in battle. Many mistrusted the breech-loader, but its performance in 1866 chastened the critics. The French introduced the Chassepot in that year: it was sighted up to 1,600 yds (1500m) and its designer partly solved the problem of obturation by a rubber seal at the breech. The Americans had experimented with breech-loaders before the Civil War, and during it the Sharps breech-loading rifle and the Spencer and Henry magazine carbines were used increasingly by Union troops.

A revolution in small arms

The first breech-loaders were single-shot weapons whose large-calibre lead bullets were propelled by black powder. In the 1880s the metallic cartridge was produced, filled with smokeless powder which propelled a smaller bullet to long ranges. This made possible magazine-fed bolt-action weapons like the German Mauser (1884), French Lebel (1886) and British Lee-Metford (1889).

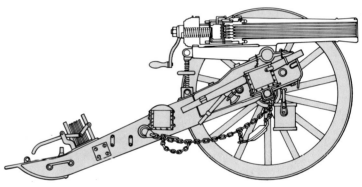

The Mitrailleuse superficially resembled a conventional field gun, but its bronze barrel enclosed 25 13mm rifled barrels. These were loaded simultaneously with a 25-round block, fired, by cranking the handle, in a *rafale* or burst. About three bursts could be fired in a minute.

Inventors had long experimented with multi-shot weapons. The first militarily useful machine guns, the American Gatling and the French *Mitrailleuse*, were multi-barrelled and consequently heavy, with the appearance and some of the limitations of a conventional artillery piece. The Gatling saw service in the Civil War, and the *Mitrailleuse* was used in 1870, when it suffered from fragility and from the French preference for keeping it back with the artillery rather than forward with the infantry. Between 1883 and 1885 Hiram Maxim developed a machine gun which used the force of the recoil to extract the empty case, cock the firing pin and load another round. The weapon needed only one barrel, cooled by a water jacket, and fired 600 rounds per minute. It caused a sensation when Maxim demonstrated it, and was adopted by Britain in 1888 and by Germany and Russia soon afterwards.

The developments which increased the effectiveness of small arms also influenced artillery. The French put rifled guns to good use in 1859, but during the Civil War roughly one-third of the Army of the Potomac's guns were smoothbore bronze "Napoleon" 12-pounders: General George B McClellan had decided that the wooded countryside over which the army would operate presented few opportunities for rifled guns. The Confederates, wrestling with serious procurement problems, used a greater variety of guns, including some steel breech-loading English Whitworths.

Breech-loading artillery shared with small arms the problem of achieving a gas-tight breech. Armstrong in England and Krupp in Germany developed sliding breech-blocks, but the most efficient system was the "interrupted thread" developed by de Bagne in the 1880s. No less serious was the problem of metallurgy. Guns for land service were traditionally made of brass or bronze, because iron burst if cast too thin or rendered the gun immobile if cast too heavy. Steel was the metal of the future. The Bessemer Converter and the Siemens-Martin open hearth process improved manufacture and brought down cost. The Prussians re-equipped with rifled steel breech-loaders in the 1860s, and after the Franco-Prussian War all major armies followed suit, although it took the British, who briefly adopted the Armstrong as early as 1859, until 1885 to decide that the day of the muzzle-loader was over.

Black powder provided both propellant and bursting charges for early rifled breech-loaders. Its smoke impeded concealment and observation and its bulk and irregular burning reduced efficiency. The perfection of smokeless powders, such as the French *Poudre B* of 1885 and British cordite of 1890, was a major advance: more powerful and controllable than previous explosives, they produced higher muzzle velocities and greater range. Shells filled with the new explosive and fitted with percussion or time fuses burst with greater effect and predictability, high explosive on the target and shrapnel above it.

For centuries guns had recoiled on discharge and had had to be run back into position after each shot. The hydraulic buffer, developed by Langlois in France and Erhardt in Germany, absorbed recoil, leaving the carriage stationary, while recuperator springs returned the barrel to its firing position. The detachment could now remain closed up around the gun, protected against rifle fire by an armoured shield. The French 75mm field gun of 1897 could fire up to 20 rounds a minute, the task of its gunners made easier by the development of "fixed" ammunition, with the charge and projectile united by a brass case.

Firepower and infantry

The new weapons transformed the battlefield. The relationship between fire and shock lay at the heart of the tactical debate. Although infantry had long produced their prime effect by fire, delivered from linear formations, the psychological shock of the advance with the bayonet was important. The conviction remained that unless troops

Hagerstown Pike, Antietam, 1862. The American Civil War was the first conflict in which photographs of the human cost of war were readily available. Many photographers were not above rearranging the debris of battle to make more dramatic shots, but Alexander Gardner's photograph of Confederate dead, one of a series that was put on display by Mathew Brady's New York City gallery in October 1862, seems genuine enough.

"Not anything that one reads can come up to the realities of a field of battle after an action; the horrible sights of the mangled bodies and wounded men praying for the love of God for a drop of water and begging one not to let them bleed to death, are far worse than the actual fighting."

Captain Hugh Hibbert, The Alma, September 20 1854

were prepared to press forward they would be unable to capitalize on damage done by their fire, and battle would become an inconclusive firefight. The instructions issued to the French troops sent to the Crimea stressed the primacy of offensive action, emphasizing that once enemy ranks had been thinned by the fire of skirmishers, moving in open order in front of battalion columns, the columns would charge. In practice bayonets rarely crossed: even at Inkerman, a close-range struggle in the fog on November 5 1854, only 6 percent of casualties were caused by the bayonet. French instructions of 1859 were similar, suggesting an *ordre mixte* of alternate deployed battalions and columns. The Austrians relied on the firepower of their Lorenz rifle, but in the close country of Lombardy fast-moving French columns crossed the fire-swept zone very quickly. Experience in North Africa and the Crimea made the French redoubtable opponents, and their officers had

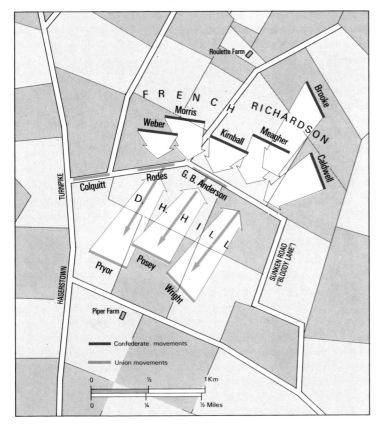

The centre battle, Antietam, September 17 1862. Some of the fiercest fighting on "America's bloodiest day" took place along the sunken road where D H Hill's Confederate division grappled with the Union 2 Corps. The fighting showed the destructive power of close-range infantry and artillery fire directed at frontal assaults.

few equals when it came to charismatic leadership.

The Franco-Austrian War was easily misinterpreted. The Austrians concluded that their sloth had lost them the war, and their 1862 regulations preached a *stosstaktik* of vigorous attacks. In 1866 they unleashed these against the Prussians, whose infantry, armed with the needle-gun, replied with an unprecedented weight of fire. Even when they worked, Austrian tactics were costly, and at Königgrätz, where they failed, they were the major cause of the grim disparity in casualty figures.

The Prussians developed tactics that looked remarkably similar to those used by the French in 1859: their 1847 regulations prescribed company columns which would attack after a preliminary fusillade. The 1861 regulations continued to emphasize close formations and, in the aftermath of 1859, envisaged company columns moving quickly to make good use of the ground. The company column was the basis of Prussian tactical organization in 1866, although in practice columns tended to coalesce into a dense firing line.

Fire and manoeuvre

The French strove hard to draw balanced conclusions from the Austro-Prussian War. While official policy championed firepower, there remained great affection for old-style tactics based on skirmishers, columns and assault with the bayonet. In the first battles of 1870 the Chassepot gave a good account of itself: Prussian infantry made little impression unless their gunners had battered the French into insensibility or their attacks had unravelled positions from the flanks. Frontal assault against steady infantry armed with breech-loaders was suicidal, as the experience of the Prussian Guard at St-Privat demonstrated. When the Guard successfully assaulted Le Bourget two months later, its casualties were far fewer, for this time it went forward in open order, with its men widely spaced, using cover and supporting one another by fire. Parallel develop-

"Away to our left and rear some of Bragg's people set up the 'rebel yell'. It was taken up successively and passed round our front, along our right and in behind us until it seemed almost to have got to the point where it started. It was the ugliest sound that any mortal ever heard – even a mortal exhausted and unnerved by two days of hard fighting without sleep, without rest, without food and without hope."

Ambrose Bierce, Chickamauga, September 1863

ments in the Civil War had already produced much the same result. Infantry on both sides used heavy skirmish lines to cover formed bodies. Major General Lew Wallace described an advance by Union troops: "Now on the ground, creeping when the fire was hottest, running when it slackened, they gained ground with astonishing rapidity, and at the same time maintained a fire that was like a sparkling of the earth."

After 1871 there was wide recognition that fire dominated the infantry battle, and discussion centred upon the relationship between firepower and offensive tactics. In France, belief in the offensive coupled with doubts about the ability of conscripts to cope with a fluid battle encouraged tactics which had a familiar ring. At about

440yds (400m) from the enemy the firing line would be reinforced and bayonets fixed. It would fight its way forward to 165 or 220yds (150–200m), and open rapid fire while reserves closed up: the entire line charged on the colonel's signal. British, German and United States manuals enshrined similar tactics: victory in the firefight followed by assault. There was reluctance to think through the practicalities of the last few hundred yards of combat, and a British officer observed that the infantry relied on "a superior volume of fire coupled with complete indifference to heavy losses".

". . . two fellows put at me. The first one fired at me and missed. Before he could again cock his revolver, I succeeded in closing with him. My saber took him in the neck, and must have cut the jugular. The blood gushed out in a black-looking stream: he gave a horrible yell . . ."
A Union trooper, Brandy Station, June 9 1863

The plight of the cavalryman was no more comfortable. The increase of firepower had reduced the utility of shock action. There were few occasions in the second half of the century when cavalry managed to charge home, and for every successful charge there were a dozen when horsemen were stopped dead by rifle fire. Traditionalists argued that unless the cavalryman was equipped with the *arme blanche* he would never dare to press his attack. Reformers maintained that sword and lance were so much useless baggage: cavalry should provide mobile firepower. The "mounted rifles" lobby gained strength, but generally the conservatives triumphed. In the British army, French's charge at Klip Drift in 1900 was used to justify shock

action, and, following a brief recession, sword and lance were restored to their eminence, although it was an eminence shared with the rifle. The cavalry had some reason for clinging to the charge. Its obsolescence then was less clear than it seems today, and the charge lay at the very kernel of the cavalryman's functional expertise and could not be easily jettisoned.

The debate over the charge distracted attention from the cavalry's other functions. Its performance in reconnaissance was unimpressive, and, although the cavalry raiders of the Civil War – J E B Stuart for the Confederacy and Grierson for the Union – attracted publicity disproportionate to their real effect, their forays could not conceal the fact that army commanders were often fighting blind. The British learned painful lessons in South Africa, and the cavalry on both sides in the Russo-Japanese War proved as bad at reconnaissance as their comrades had done elsewhere. The multiplication of firepower made its other tasks harder: increased ranges made it difficult for cavalry to exploit the fire effect of other arms, and pursuit was impeded by the ease with which a knot of determined infantry could hold up the most enterprising squadron of horse. By 1914 there was probably more cavalry in the world than ever before, but in the opening moves of World War I there was the usual sequence of lost contacts and missed opportunities.

Technology had placed artillery within measurable distance of revolutionizing the battlefield, yet in the process it had apparently lost some ground. Artillery had been the major casualty-inflicter in the wars of the first half of the 19th century. But as infantry firepower grew, so the proportion of casualties inflicted by infantry grew too: small arms caused about 40 percent of casualties in the Mexican War of 1846–48, but some 85–90 percent in the

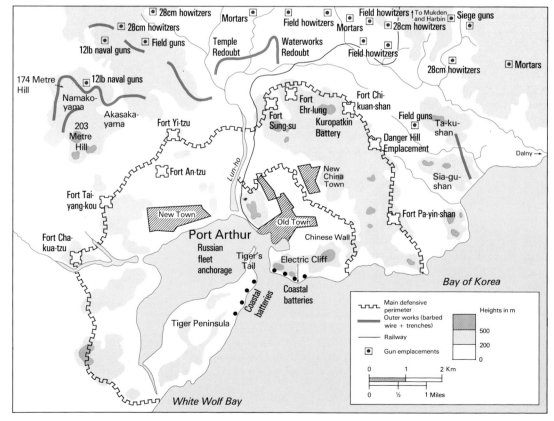

By the Treaty of Shimonoseki (1895) China ceded the Liaotung Peninsula to Japan. European pressure forced Japan to withdraw, and Russia leased the tip of the peninsula and its warm-water ports. The Japanese navy attacked Port Arthur without warning on February 8 1904, and the army occupied Korea and fought Kuropatkin's field army in Manchuria. Nogi besieged Port Arthur, whose incomplete fortifications failed to cover 203 Metre Hill, which gave observation of the fleet and harbour installations. Nogi took it in December, bringing the ships and defences under artillery fire: heavy howitzers (*right*) were particularly effective. Attacks on Russian positions were costly, foreshadowing events of 1914–18, but on January 2 1905 the demoralized Russian commander surrendered.

wars of the 1860s and 1870s. Until the development of the field telephone the range of artillery was effectively the limit of vision, and in many circumstances the rifle enjoyed the same range: in the American Civil, Franco-Prussian and Boer Wars gunners regularly fell victim to infantry fire. Even after the telephone made indirect fire possible,

The Federal observation balloon *Intrepid*, Virginia, May 1862.

artillery was seen primarily as a direct fire weapon. Senior commanders believed that direct fire was prompt and effective, while artillerymen were sensitive to comments about skulking behind cover.

Field fortifications, a feature of many wars of the period, also reduced the effect of artillery. The most striking example of their use was Osman Pasha's defence of Plevna against the Russians in 1877. Turkish riflemen inflicted horrific casualties on the attackers, whose losses of about 38,000 exceeded the total size of the Turkish garrison.

Japanese troops haul up an 11-inch howitzer, Russo-Japanese War, early 1905.

The battery in position

Russian artillery fire was, in contrast, ineffective: a Russian general complained that one of his batteries might fire all day to kill a single Turk.

Yet this was only half the story. For all its limitations, artillery was emerging as an arm of decisive importance. Initially, most armies saw it primarily as a means of augmenting the firepower of their infantry and only secondarily as a counter-battery weapon. Heavier guns were held back to be concentrated at the right moment: in 1870 the French kept their 12-pounders in the corps artillery reserve, leaving the divisional 4-pounders to sustain the battle until the reserves came into action. In 1866 the Prussians kept their guns to the rear on the line of march, deploying them slowly to answer Austrian gunners who were trained to fight forward to support their infantry, which trusted in shock rather than in its own firepower. The war persuaded the Prussians that early deployment was vital, and in 1870 their guns moved well up, came into action quickly, and, their percussion fuzes more effective than French time-fuzes, simply outshot their opponents.

Fall of the fortress

The duel between gunner and sapper was an old one, each modifying his techniques to match the other's advances. Improvements in artillery induced engineers to strengthen fortresses by lowering their silhouettes, reinforcing them with concrete and mounting guns in steel cupolas. In 1888 the French bombarded one of their older forts with shells filled with the new explosive melinite. The experiment persuaded them to add a second skin of concrete to their forts, with a burster layer of sand between the skins to absorb the force of exploding shells. In 1904 the Japanese siege of the Russian fortress of Port Arthur gave a clue of what might come. The defences of Port Arthur were incomplete, but the Japanese lost 15,000 men trying to take them by assault and settled down for a formal siege. The Japanese commander, General Nogi, had sent for some 280mm howitzers to bombard warships trapped in the harbour. When they arrived the Japanese were unable to observe the harbour properly, so the howitzers were set to work on the forts, where their 700lb shells pierced earth and concrete to burst inside. Such an impressive performance was of interest to the Germans: in Krupp's factory at Essen work began on the weapons which were to dash the hopes of a generation of fortress-builders.

The popular image of armies on the eve of World War I as conservative and unthinking institutions is something of a caricature. There was indeed failure to grasp the effect of new weapons, but I S Bloch, whose *Modern Weapons and Modern War* predicted much of what was to happen in 1914–18, was not unique. Many authorities recognized that a future European war would have devastating social, economic and political implications if it was allowed to continue. Some concluded that such a conflict would be unacceptable: it was widely said in 1914 that the war would be "over by Christmas", because the consequences of fighting on to mutual destruction were unthinkable. Others, like Britain's Lord Kitchener, prophesied a war of exhaustion. Few realized that the fusion of the technology and intellectual currents of the 19th century made possible a new sort of war, well-described by Raymond Aron as "the absolute form of war whose political stake the belligerents are incapable of specifying".

chapter nine

The Star of Empire

Anthony Clayton

A prestigious British imperial honour was that of the *Order of the Star of India*. The Order was well-named. Empire was a star, and this chapter concerns the West European nations and individuals that followed that star, by sea, to Africa, Asia, America and Australasia. There, the greatest maritime empires in the modern period, those of Spain, Britain, France and Holland enjoyed a splendour brief but glittering, and other countries – Germany, Portugal, Belgium and Italy – enhanced their roles upon the world stage. Three military dimensions of imperial rule are outlined; acquisition, rule (including the uses of indigenous manpower to serve wider imperial causes), and the end of empire either through withdrawal or ejection.

Few generalizations can be made about empire. The reasons for the acquisition of imperial possessions were as varied as the styles of imperial rule, and again in turn as varied as the colours of imperial sunsets. The complexities do not admit even of any universal assertion that possession of empire was negative and colonial emancipation positive. Nor was colonial rule invariably exploitative. At their best – when accountable to an observant legislature – latter-day colonial administrations produced records of solid achievement offset by only negligible authoritarianism.

The earliest imperialists, mainly Iberian, sought territory for its wealth, raw materials and in particular gold. These were quickly followed by others who saw the possibilities of estate agriculture founded upon slave or cheap labour, and later yet others concerned with the acquisition – and military protection – of investment and tied markets. But from the earliest years, wider purposes were seen by European governments: the wish of a small country like Portugal or Holland to remain independent in a period of European conflict; France's need to find indigenous soldiers to supplement her metropolitan army; the strategic value to Britain or France of a certain territory; a need to provide a country with a religious or secular *mission civilisatrice* (a role heightened by perceptions of barbarities held about indigenous polities); a Bismarckian need for a card to play in international politics; or Italy's need to find territory for an increasing home population. In the 19th and early 20th centuries, as governments became more democratic, pressures built up for the extension of formal empire as opposed to the "informal empire" of trading stations. Formal empire secured the territory against a foreign rival and brought responsibility to colonial administrations. In military terms, this meant a move away from informal company or locally recruited forces to a more extensive use of metropolitan forces, or metropolitan control of formal indigenous regiments.

Men and women as individuals were attracted to empire not only for reasons of commercial profit, but also for reasons of idealism and service, for the pull of the exotic, for the retention of a lifestyle fading in the metropole, or simply for personal emancipation. Soldiers were attracted to empire by a similar mix of motives, usually career-orientated – adventure, promotion and medals, an escape from the tedium of metropolitan garrison life or unpalatable advances in military technology and also, particularly in the case of France, a social idealism.

Further uses of military power

After the suppression of primary resistance, the next sets of military commitments, two sides of the same coin, were (1) campaigns mounted to repress protest or nationalism,

European colonial possessions in 1913. By the early 20th century the last vestige of Spanish colonial rule in the New World had disappeared with the loss of Cuba. The British and French empires were approaching their greatest territorial extents and the lesser German, Netherlands, Italian and Portuguese empires were established. The British empire was larger and wider than that of France, a consequence primarily of British sea power. The French were largely constrained by that sea power to North and West Africa, but these regions also reflected the long-lasting French geopolitical concept of certain essential trans-Mediterranean unities. The coming of the new land-mass empires, with US interest in Panama and Russian interest in the Pacific can also be seen.

—TIME CHART—

1520–60	Spain acquires empire in South America	1900–1902	British use "concentration camps" against Boer guerrillas
by 1600	Portugal acquires trading stations in Africa, India and the Far East	1919	Amritsar massacre
by 1750	Holland, Britain and France acquire possessions in North America, West Indies, India and Southeast Asia	1919–39	Two Non-Cooperation Campaigns and almost continuous North West Frontier operations in India
from c.1750	Britain using Indian troops	1920s	European empires in Asia and Africa at their greatest extent
1760	British defeat French in India	1936–39	Arab revolt in Palestine
1760	French surrender to British in Canada	from 1942	Emancipation of Africa
1783	United States gains independence	from 1945	Emancipation of Southeast Asia
by 1805	British gain supremacy in India	1946–54	Franco-Vietnamese war (Dien Bien Phu, 1954)
1810–1828	Emancipation wars in South America	1947	Madagascar Revolt
from c.1820	Holland using East Indian troops	from 1947	Independence of India, Pakistan, Burma and Ceylon
by 1830	Dutch ascendant in East Indies		
from c.1830	France using indigenous *Tirailleurs* and *Spahis*	1948–56 (60)	Malaya insurgency: helicopter first used
1857–58	Indian Sepoy Mutiny (Great Mutiny)	1952–59	Greek EOKA campaign in Cyprus
1880–1913	Western powers acquire possessions in Africa	1952–56 (60)	Kenya Mau Mau uprising
from 1881	Worldwide anticolonial reactions	c.1958–62	French deploy electronic defence lines against Algerian insurgents
1896	Battle of Adua	1961–74	Spreading insurgency in Portuguese Africa
1898	Machine guns secure British victory at Omdurman	from 1962	West Indian states gain independence
1899–1902	Second Boer War (South African War)	1982	British recapture Falklands from Argentina

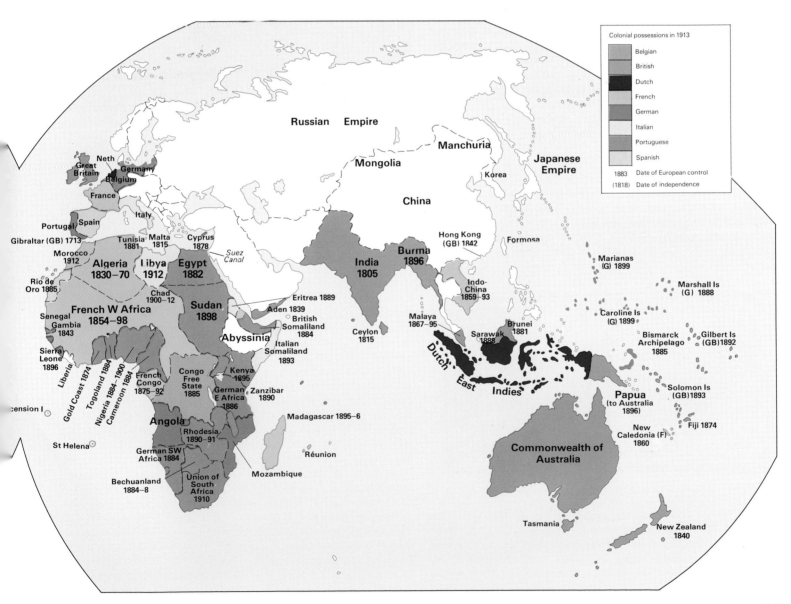

generally in groupings of secondary resistance, and (2) the training of indigenous soldiers for use in these campaigns, in peacetime displays of power, and in world wars. Vitally important workhorses of the British and French empires respectively were the Punjabi Muslim riflemen and the Algerian *Tirailleurs*.

The end of empire offers the widest variety in the uses of imperial military power. The early British American and Spanish American empires suffered military defeat. After World War II, declining metropolitan enthusiasm for empire, anti-colonial sentiment in the new great powers (the USA and the USSR), together with mounting criticism at the United Nations, and ever-growing nationalist protest in colonial possessions, all combined to bring about an unexpectedly quick end to the age of colonial empire. Britain's waning imperial military power was deployed to try to ensure that the conditions of withdrawal were acceptable. France could not mount this role and both lost her empire and suffered humiliation with strife in her own army. The Dutch grudgingly withdrew from Indonesia in the late 1940s and Portugal was defeated militarily in her African territories in the early 1970s.

Finally, in all three phases there was the widest variety in the numbers of soldiers, especially metropolitan troops employed. In some immensely significant campaigns the numbers were minuscule, in others prodigious. Most imperial campaigning was based on a technological superiority – from the Spanish use of the horse in the 16th century to the firearms of the 17th and 18th century, rifles, machine guns and pack artillery of the 19th and wireless, lorries and aircraft in the 20th century.

Acquisition

After earlier landings on Haiti, Cuba and Panama, Spain's major military conquests were of the Aztec confederacy in Mexico and the Inca empire in northwest South America in the years 1520 to 1540. Spain's *conquistadores* were adventurous and experienced soldiers fighting at their own expense with religious fervour and ferocity. Their assets

Spanish Conquistadores with their most potent weapon – the horse.

were light cavalry, cannon, and infantry skilled with the long two-edged sword but also using arquebus and crossbow. These enabled very small forces to defeat vast armies whose weapons were limited to slings, swords, clubs, axes, javelins, bows and darts. Cortés defeated a 40,000 Aztec army with 600 men, 15 horses and 14 guns – Pizarro's first

victories over the Incas were secured by 180 infantry and 62 horsemen, the latter terrifying the Incas. The most effective, guerrilla-style resistance was that of the Maya peoples of Yucatán, crushed only in 1697 but to rebel on occasions later. All South America (less Brazil, taken by Portugal) fell to Spain, as did the Philippines later.

Conquest profited the *conquistadores* little. The Spanish crown, needing gold and other wealth for its European wars, assumed control. Spain's great age was, however, short-lived; the increasing manpower bill and the drugs of gold and silver soon debilitated the entire economy.

After the 17th-century occupation of Tangiers, Britain's major colonial campaign was in Canada in 1754–60. The aim was to prevent a French encircling move from Quebec

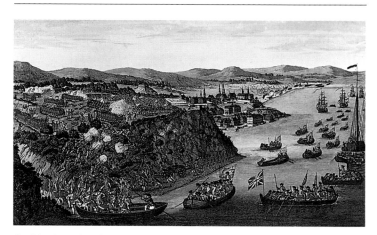

Combined naval and military power ensured British victory at Quebec.

to Louisiana. After initial reverses, seapower enabled Britain eventually to field a force totalling 25,000 regular and 25,000 provincial troops. In 1758, the British experienced success at Louisbourg and reverse at Ticonderoga; in 1759, victory was clinched with a spectacular cliff assault by British regulars upon Quebec. In the attack, Wolfe, the commander, was killed. French counterattacks failed – Canada was secured.

Seapower was not adequate for the retention of the British American colonies which, no longer needing protection against France, rose in revolt in 1773. In time, the British built up a force of some 25,000, many being German mercenaries; the colonists already had state militia of men trained as frontiersmen. The British secured an early pyrrhic victory at Bunker Hill in 1775 and in 1776 recaptured New York. The next year saw further minor successes, notably Howe's capture of Philadelphia. But these were totally offset by the disaster of Burgoyne's doughty but unsuccessful Hudson River campaign and surrender at Saratoga, and the 1778 French entry into the war, necessitating diversion of British troops to Europe and the West Indies. Two years later, encouraged by the recapture and apparent loyalty of Georgia and failing to appreciate the solidarity of Americans farther north, Cornwallis led a British force into the Carolinas, but he gained only limited success. In frustration he moved into Virginia. There, the revolt behind him of the Carolinas, a temporary loss of sea control to the French and the inability of Britain's New York force to move, forced Cornwallis to surrender at Yorktown in September 1781. Although never actually defeated in battle, the war and

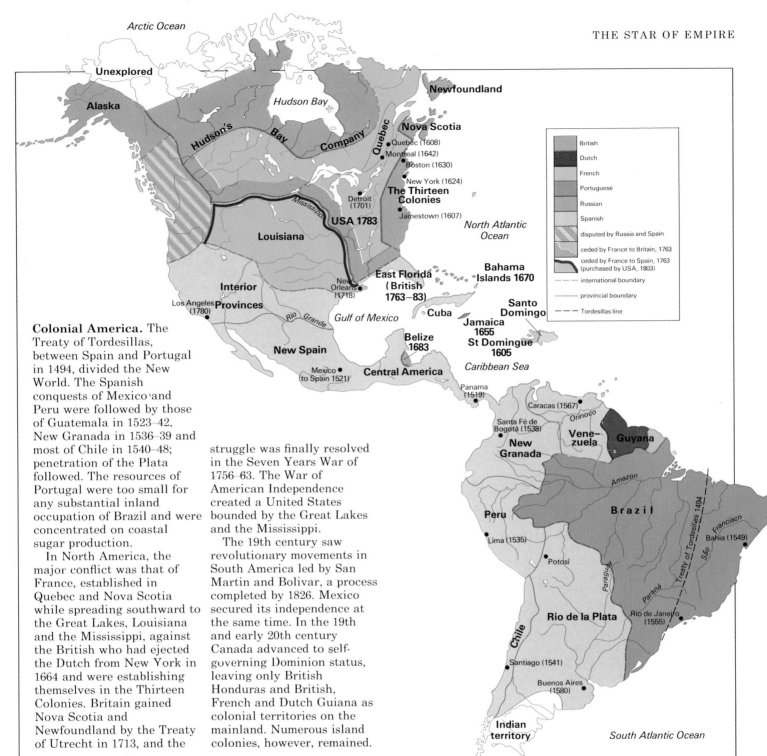

Map labels:

Arctic Ocean
Unexplored
Alaska
Hudson Bay
Newfoundland
Hudson's Bay Company
Nova Scotia
Quebec
Quebec (1608)
Montreal (1642)
Boston (1630)
New York (1624)
Detroit (1701)
The Thirteen Colonies
Mississippi
USA 1783
Jamestown (1607)
North Atlantic Ocean
Louisiana
Bahama Islands 1670
East Florida (British 1763–83)
New Orleans (1718)
Interior Provinces
Los Angeles (1780)
Rio Grande
Gulf of Mexico
Cuba
Santo Domingo
Jamaica 1655
Belize 1683
St Domingue 1605
New Spain
Mexico (to Spain 1521)
Central America
Caribbean Sea
Panama (1519)
Caracas (1567)
Orinoco
Santa Fé de Bogotá (1538)
New Granada
Vene-zuela
Guyana
Amazon
Peru
Brazil
Lima (1535)
Treaty of Tordesillas 1494
São Francisco
Bahia (1549)
Potosí
Paraguay
Paraná
Rio de Janeiro (1555)
Rio de la Plata
Chile
Santiago (1541)
Buenos Aires (1580)
Indian territory
South Atlantic Ocean
Falkland Is

Legend:
British
Dutch
French
Portuguese
Russian
Spanish
disputed by Russia and Spain
ceded by France to Britain, 1763
ceded by France to Spain, 1763 (purchased by USA, 1803)
international boundary
provincial boundary
Tordesillas line

Colonial America. The Treaty of Tordesillas, between Spain and Portugal in 1494, divided the New World. The Spanish conquests of Mexico and Peru were followed by those of Guatemala in 1523–42, New Granada in 1536–39 and most of Chile in 1540–48; penetration of the Plata followed. The resources of Portugal were too small for any substantial inland occupation of Brazil and were concentrated on coastal sugar production.

In North America, the major conflict was that of France, established in Quebec and Nova Scotia while spreading southward to the Great Lakes, Louisiana and the Mississippi, against the British who had ejected the Dutch from New York in 1664 and were establishing themselves in the Thirteen Colonies. Britain gained Nova Scotia and Newfoundland by the Treaty of Utrecht in 1713, and the struggle was finally resolved in the Seven Years War of 1756–63. The War of American Independence created a United States bounded by the Great Lakes and the Mississippi.

The 19th century saw revolutionary movements in South America led by San Martin and Bolivar, a process completed by 1826. Mexico secured its independence at the same time. In the 19th and early 20th century Canada advanced to self-governing Dominion status, leaving only British Honduras and British, French and Dutch Guiana as colonial territories on the mainland. Numerous island colonies, however, remained.

The battle of Yorktown, October 1781 (*left*), marked Britain's decisive defeat in the War of American Independence. Cornwallis's 8,000-strong force had been penned into a corner on the York river, Virginia, by Washington and, following de Grasse's earlier victory at Virginia Capes, could be neither reinforced nor evacuated. Driven from their advanced redoubts on October 14, and prevented from breaking out by a storm on the night of the 16th, the British surrender was concluded on October 19.

British control of America was lost.

The British in India

The British occupation of India was achieved in stages. Initially the East India Company's activities were limited to trading posts at Madras, Bombay and Calcutta, but French ambitions playing upon the stage of the disintegrating Mughal empire sparked conflict with the 1746 seizure of Madras by Dupleix. Madras was restored to the Company in 1748, but the years 1751–61 saw an indirect conflict in the Carnatic between Britain and France, both using local rulers as surrogates, while Britain also fought the ruler of Bengal from 1756. British success was largely due to two men; Stringer Lawrence, who organized small but effective Company forces, European and Indian, maximizing European military skills, and Robert Clive, who led the forces to success. One regular British regiment was committed from 1754. The final defeat of the French was clinched by Eyre Coote's victory at Wandewash in 1760, and Bengal was assured by victory at Buxar in 1764.

The next stage saw British control established over all southern India. The means was a mix of metropolitan British regiments, the Company's European regiments (mostly ex-regulars) and the growing number of Company Indian units with British cadres, all cemented by a professionalism that could vanquish the sizeable armies of Indian princes. Tipu's state of Mysore was reduced by Cornwallis's investing of Seringapatam in 1792, and destroyed by Harris's capture of the fortress in 1799. Two

Robert Clive (1725–1774), founder of British India, first attracted attention when, as a lowly Company clerk, he served as a volunteer in the fighting against the French near Madras. His first major command was a diversionary attack upon, and the successful defence of, Arcot in 1751, a decisive reverse for the French. In 1756 he was given command of the expedition to retake Calcutta, seized by the Nawab of Bengal, Siraj-ud-Daula. The city retaken, Clive proceeded first to destroy Siraj-ud-Daula at Plassey and then to retain his own nominee as Nawab against Mughal and Dutch challenges. His military success was founded upon skilled use of small contingents of disciplined troops. At Plassey, Clive commanded 3,200 men with 9 guns against 50,000 men and 53 guns.

Twice governor of Bengal (1758–60 and 1765–7), Clive unwisely accepted a cash reward for his services, tarnishing his first administration. His second, however, brought major achievements, notably the dual system of administration with the Nawab; he also sharply curbed corruption. But later, Clive had to rebut Parliamentary criticism, a humiliation that, with his moody temperament, led him to take his own life in 1774.

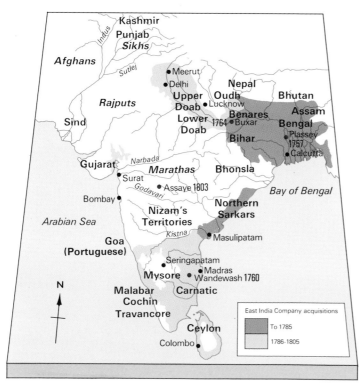

British territorial control in India was first asserted, at the expense of France, over the Carnatic coast in 1746–60. Clive's victory at Plassey and Munro's at Buxar extended British rule into Bihar and Bengal. Entry into Oudh followed. The defeat and fall of Tipu in 1792–99 assured British ascendancy in Mysore and Malabar. Wellington's victories, notably at Assaye (1803) extended Company authority into Rajputana.

Aliwal, last of the major Sikh War battles, cost the Sikhs all their guns and over 3,000 dead, after charges by the 16th Lancers had broken two Sikh squares. The Lancers are wearing Polish-style *czapka* headdresses and thick uniforms – the latter proving a cause of serious health problems in the Indian heat. Waterloo veterans of the Regiment claimed the Sikhs had fought "as bravely as the French". British losses were comparatively light.

wars against the Mahratta Confederacy, the first beginning in 1775 and the second ending in 1805, were in practice decided by Wellington's victory at Assaye in 1803.

The final stage, 1806–1853, saw several successful campaigns in the north of India and in Burma, together with a disastrous attempt to occupy Afghanistan. The actual events included the suppression of Sepoy mutinies in 1806

". . . a dreadful station, the barracks on the edge of a large open plain, nothing but sand and a wild jungle around us, the game came out of the jungle into the barracks looking for water, and the Bheels came at night to rob, digging through the sand under the buildings."

Sergeant Major Sowrey, 6th Foot, writing of Disa, India in 1833.

---

and 1824, a war against the Gurkhas in 1814–16, a final Mahratta campaign 1817–1819, an expedition to Rangoon to counter a Burmese threat to Bengal in 1824 and the occupation of Sind in 1843. In India, the Punjab and the Northwest alone remained, these involving two tough wars against the Sikhs. In the first, 1845–46, General Gough's British/Indian forces, 18,000-strong with artillery, achieved success in the ferocious battles of Mudki, Ferozeshah and Aliwal. The most notable actions in the second, 1848–49, were Gough's battles of Chilianwala and Gujarat. A brief second campaign in 1852–53 annexed lower Burma and a defeat of a Persian army in 1856–57 completed the military occupation.

The British had, however, overstretched themselves with an 1839 expedition into Afghanistan, ejected and destroyed by a winter uprising in 1841–42; British prestige but not authority was re-established by a brief occupation of Kabul at the end of 1842.

A rich prize secured

Conquest then changed to retention, the main challenge being that of the great Sepoy mutiny of 1857–58, an event combining both military and political protest. Beginning at Meerut, spreading to Delhi and Oudh, the mutiny involved the Company's Bengal army and a very few units of the Bombay army, but none of the Madras army: it was, however, accompanied by fearful massacre. Decisive events were the British retention of Lucknow, despite a grim siege, and the storming and recapture of Delhi. Thereafter India passed to the British Crown and the Company's forces became the Indian Army.

The remaining military operations acquiring empire were, with one exception, small-scale. Some operations undertaken to punish or suppress lawlessness developed into, or were soon followed by, campaigns of annexation. A series of campaigns in the late 19th century secured India's northwest and northeast frontiers. The most notable campaign was perhaps the 1878–1880 Second Afghan War, which included a temporary punitive occupation of Kabul, and Robert's famous march from Kabul to Kandahar. Another campaign in 1885–92 annexed Burma, one more in 1875–76 annexed Perak in Malaya. Three small wars against the Maoris, the last ending in 1866, secured New Zealand. Minor expeditions were mounted in 1866–67 and 1870 in Canada. British troops were used to suppress revolt in Jamaica in 1865 and Barbados in 1875–76, and to annex British Honduras in 1866–67 and 1872.

THE BRITISH IN INDIA

Immediately prior to the Indian Mutiny in 1856, British Indian forces had entered Persia to ensure the security of British rule in western India. In the course of the fighting at Bushire, Captain J A Wood led the Grenadier Company of the 20th Bombay Native Infantry into the attack on the fort (*below*). He was the first man on the parapet, and although struck by seven musket balls, continued to command the attack, which proved successful. For this he was awarded the Victoria Cross.

The British military achievement was far from being merely repressive. India's infrastructure gained greatly from the army. Engineer officers made massive contributions in the fields of road construction – for example the extension of the Grand Trunk Road after the end of the Sikh wars; architecture, including a number of public buildings in major Indian cities; advice and preparation for railway construction; marine works, in particular harbours, moles and dry docks; surveys of all kinds; telegraphs; mints; financial services; and engineering education. Army medical staffs were of particular value in public health, preventive medicine and research into tropical diseases.

Techniques and processes applied in such massive construction projects as the Curzon Bridge (*bottom*) owed not a little to the pioneering achievements of army engineers.

The need to secure routes to India led Britain to Africa with the permanent military occupation of Cape Colony in 1805. Defence of the colonists involved a reluctant Britain in a series of small frontier wars and finally in a war with the powerful Zulu kingdom in 1878–79. At the battle of Isandhlwana, mass envelopment by Zulu impis destroyed a British force before technological superiority could develop. Black power destroyed, Britain's next conflicts were with the doughty Boer settlers. In the First Boer War, General Colley's force was defeated by the Boers at Majuba in 1881. Rhodes's designs, both commercial and imperial, involved British forces in Matabeleland (Zimbabwe) in 1893 and 1896 and Bechuanaland in 1896–97.

Britain's second main area of concern was Egypt, occupied by Wolseley after a brief campaign and the battle of Tel-el-Kebir in 1882. Egypt was not, however, secure without the Sudan, where British forces were unsuccessful against the Mahdi in 1884–85, but returned victorious under Kitchener in 1895–96. Minor campaigns also occurred in Ashanti in 1873–74 and 1895–1900, in south Nigeria (Benin) in 1897, north Nigeria 1897–1900, Kenya in 1896, and Somaliland, where the British sustained sharp initial reverses.

The Second Boer War

Britain's last great imperial victory was the Second Boer War of 1899–1902, a full-scale war involving 200,000 men including Australians, Canadians and New Zealanders. The war fell into three phases. The first, to the end of 1899,

saw Boer sieges of Mafeking, Kimberley and Ladysmith with defeats over the British at Stormberg, Magersfontein and Colenso. Britain's foremost soldiers, Roberts and Kitchener, then assumed command, and a powerful march on Bloemfontein and Pretoria secured the formal surrender of the Boer commander, Cronje. Two years of guerrilla warfare followed, with the future imperial statesmen, Botha and Smuts, among the Boer guerrilla leaders. Just in time, the war began a process of transforming the British army from an Asian expeditionary force to a modern army fit to fight in Europe, though colonial skills of marksmanship, fieldcraft and improvisation remained invaluable.

Indian troops were used in a number of operations, notably in New Zealand, Egypt, Sudan and Kenya, in China (1859–60), and also in a punitive expedition into Abyssinia in 1867–68. Success was obtained by superior fire power, notably from the breech-loading Snider rifle, volley-firing, the development of quick-firing and later the Maxim machine gun, steadiness in squares and on occasions the use of cavalry. At Kitchener's 1898 battles of Atbara and Omdurman, for example, British dead totalled 48 against 10,000 Dervish corpses found on the field. Communication lines were crucial, campaigns involving vast numbers of porters and animals. Climate and terrain were often difficult, disease killing more than the opponent; frequently, campaigns had to be finished before a seasonal change. The occasional reverses were noted at home by critics of empire, and more awkwardly, by ill-wishers in foreign chancelleries.

The battle of Omdurman, September 1898, was the first major battle in which a fully automatic machine gun, the Maxim, was used. Although the terrain was unfavourable for defence, the Maxim gun was able to mow down the advancing soldiers of the Mahdist army, despite all their courage. The Maxim was supplemented by the Lee-Metford rifle, the British Army's first repeating infantry rifle.

The star of empire led France to Africa and the Far East. After Waterloo and again after 1870, France lacked self-confidence; many saw a colonial empire, offering both wealth and military manpower, as the answer. Apart from securing Senegal, the Restoration and July monarchies achieved little in Black Africa. But in 1830, one of the last acts of the Restoration monarchy – in a fruitless attempt to save itself – was to invade Algeria. An army of 37,000 metropolitan Frenchmen under Bourmont seized Algiers in 1830. Bourmont and his successors also began the raising of local units and the Foreign Legion, the future *Armée d'Afrique*, to replace the unwilling French conscripts. Only limited military progress was made until the arrival of General Bugeaud in 1841. Bugeaud, with an army progressively reinforced to a total of over 100,000 Frenchmen and locally recruited indigenous, defeated the forces of the Western Algerian resistance leader, Abd el Kader, in a series of actions, notably the 1844 battle of Isly (fought at 140°F), and began the pacification of the Kabylies. This work, and the initial moves southward to the Sahara oases was continued from 1853 onwards by Randon, and was culminated in 1857 by MacMahon's victory at Icheriden. There followed small-scale operations extending French authority farther southwards, Saharan camel companies being formed for some areas. The French forces embarked upon a draconian repression of all local resistance, the most serious revolts occurring in 1871 and 1916. In 1881, a French force of 38,000 occupied Tunisia in a brief campaign.

A battalion of the 1st Tirailleurs leaves Algiers for Morocco, 1911. France's 20,000-strong army in Morocco in 1914 included 19 Algerian and Senegalese *Tirailleurs* battalions, 4 Tunisian or Algerian *Spahis* regiments and 5 *Legion Étrangère* battalions. The 9 *Zouave* battalions contained many from the Algerian *colon* population. The French metropolitan army contributed only 3 corrective battalions of convicts and 7 battalions of regular Colonial infantry. French artillery and machine guns ensured victory over far superior numbers.

THE RELIEF OF MAFEKING

In general, British public opinion only rarely expressed itself vehemently on imperial issues, and British statesmen generally ensured that they did not pursue vote-losing imperial policies. A major exception was the public blaming of Gladstone for the death of Gordon and loss of Khartoum in 1885, one of the causes of Gladstone's fall.

However, one imperial military rescue in the Second Boer War, the relief of Mafeking's seven-month siege by Colonels Mahon and Plumer in May 1900, led to the greatest outbreak of mass popular exuberance ever recorded in Britian. As the news spread, theatres and music hall shows were interrupted, vast crowds appeared spontaneously in the streets of London and other cities, singing and dancing. In the north, brass bands played and factory sirens hooted, church bells pealed in town and country. Alcohol flowed, policemen were embraced. The next day houses were decorated, trains, carriages and ships were festooned and ordinary people wore red, white and blue buttonholes. Imperialism had climaxed, and a new word entered the English language.

Major General (later Lieutenant General Sir Robert) Baden-Powell, Mafeking's defender, was able to use his status as a national hero to launch the Boy Scout movement in 1908. The eyewitness illustration of Mafeking celebrations in Piccadilly (*below*) shows a placard of the hero prominently displayed.

During the siege, special money was issued (*bottom*), the notes being reproduced photographically by the ferro-prussiate process, according to a design prepared by Baden-Powell himself. "The wolf" (or, more accurately, "The wolf that never sleeps") was the name by which he was known to the natives, and this was applied by the whites to the 4½ inch howitzer, made by Major Panzera from a drainpipe, and firing round cannon balls, which appears in the centre of the design.

In contrast to the big Algerian campaigns, France's acquisition of West Africa was with a mix of small forces locally recruited (*Tirailleurs Sénégalais*) and ambitious French *Troupes de Marine* officers, operating against a political background of on-the-ground local rulers capable but oppressive, and of indifference or ignorance in Paris. A limited start up the Senegal river was made in the 1850s by Faidherbe, using his newly formed *Tirailleurs Sénégalais*. The 1870–71 defeat gave France a fresh incentive for expansion, and from 1877 a series of small campaigns were mounted by *Marine* majors and colonels who often deliberately exceeded their instructions, both in terms of the limits of advance and the means employed. First, the demi-states of Senegal and northern Guinea, then the Tucolor state of Ahmadu, the coastal kingdom of Dahomey and the Mande empire of Samory with its large regular army were all destroyed. Only the 1892 Dahomey expedition – 3,450-strong and including a Legion detachment – was of any size. Joffre's force that entered Timbuktu in 1893 totalled only 397 military and 700 porters. Success was achieved by firepower, the Chassepot and later Gras and Lebel rifles, and 80mm mountain guns. Columns would take sheep, cattle and sometimes camels on the march, and form squares when attacked, holding fire until the attackers were 50 paces distant. Continued resistance was crushed, often with severity.

The advance from coastal stations into Equatorial Africa was generally less violent until converging from different directions, French columns reached Chad where a series of small campaigns against local demi-states continued from 1900 to 1918. A brilliant effort was made in 1897–98 by a small column under Captain Marchand to lay a claim to the Sudan, but the effort was thwarted in a tense international atmosphere by the British. A major 18,000-man campaign, mismanaged by the metropolitan army, occupied Madagascar in 1895. Revolt and resistance, however, lasted from 1896 to 1905, to be suppressed by Gallieni again with severity, but also post-occupation reconciliation and reconstruction.

In all France's African campaigns, disease was a major problem. In 1830–34, 8,322 men died in Algeria, with 9,686 in 1840 alone. Quinine dosing, pioneered by Maillot, saved the French army. But the lesson was soon forgotten. In the 1895 Madagascar campaign, 4,613 soldiers died of disease, mostly malaria, against 25 killed in battle. These losses reinforced the quest for indigenous troops.

French initiative in the Far East

Napoleon III's attempt to establish European rule in Mexico, involving some 30,000 men between 1861 and 1867, failed disastrously, but he did acquire Cochin China, including Saigon, and a protectorate over Cambodia. Incidents and the death of the emperor of Annam in 1883 led the Jules Ferry government in Paris to embark on a military campaign to annex Annam and Tonkin. The campaign, a bloody one in which the Vietnamese were at one point supported by China, lasted four years. Exaggerated reports of a reverse at Long San occasioned the fall of

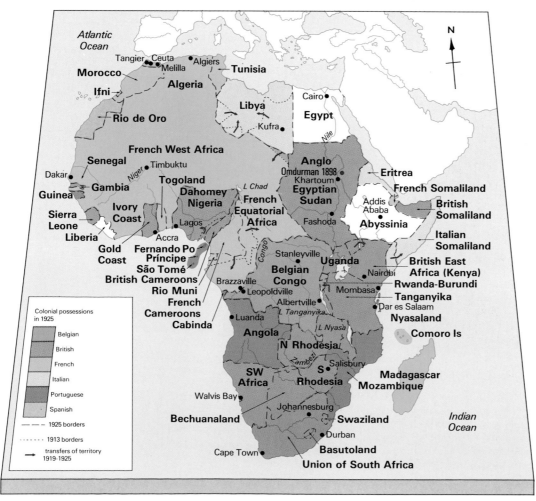

The map shows the allocation of the colonies of the former German empire in the form of League of Nations mandates: South Africa receiving southwest Africa, Britain receiving Tanganyika and small portions of Cameroun and Togo, France receiving the larger portions of Cameroun and Togo, and Belgium receiving Rwanda and Burundi. The map also shows the minor frontier changes made in favour of Italy's colonies of Libya and Somaliland. From 1922 Egypt enjoyed a flag independence as a British client state; Ethiopia, Liberia and – as a British Dominion – the Union of South Africa, were the only fully independent territories on the continent.

Ferry in 1885. The campaign had begun with a French force of but 3,500 *Marines*; by 1885, Generals Brière de l'Isle and his successor, de Courcy were commanding over 30,000, including North African *Tirailleurs*, *Spahis* and the Legion. The Annam empire collapsed, and six years later, threatened by an invasion from the sea, Siam ceded Laos to the French.

The military occupation of Morocco began cautiously, largely as a consequence of great power politics, in 1908, with France taking advantage of feeble sultans, and a little later of the power of major Berber traditional rulers. The arrival, at a moment of crisis in 1912 of General Lyautey, perhaps Europe's greatest proconsul, began a new, brilliant stage. Deploying a force reaching 70,000 predominantly North or Black African *Tirailleurs*, and using aircraft and wireless, Lyautey extended French authority in Morocco, installing military administrations concerned with reconciliation, economic development and medical services. Using second-line and reserve units, he maintained a French military assertion throughout World War I, his ablest general, Poeymirau, even marching across the Middle Atlas in 1916. After 1918, Lyautey resumed operations in the unoccupied mountain areas, but the major revolt, beginning in Spanish Morocco, of Abd el Krim produced a crisis in 1925–26. Pétain, dispatched from Paris to command, eventually and quite unnecessarily deployed 72 battalions, with cavalry, artillery and aircraft. After the rising a lull followed until 1931, when in three campaigns, Generals Huré and Giraud established the authority of

Rabat over Morocco's mountain fastnesses, using a force of some 40,000 troops and irregulars. France's other major colonial possessions, the League of Nations mandates for Syria and Lebanon, were part of her 1918 victor's reward.

Other European imperialist activity

Portugal extended her African empire inland in the 19th century by means of small military parties and armed settlers or traders. The commercial dimension to the acquisition of empire in Africa was highlighted by the ruthless penetration of the Congo basin by King Leopold of the Belgians, at first in a personal capacity. Besides France, Portugal, Belgium and Britain, Germany too joined in, occupying Togo, Cameroun, Southwest Africa and German East Africa by forceful penetration of small military detachments in the 1890s. Italy rashly attempted a major military expedition against the Ethiopian empire, General Baratieri's 14,000-strong army sustaining a bloody defeat at the battle of Adua in 1896; she was forced to be content with Eritrea and part of Somalia. Her attention then moved to Libya, snatched from Turkey in 1911–12 but to require continual anti-guerrilla campaigning until 1932, and later to Mussolini's renewed invasion of Ethiopia in 1935–36. The Ethiopians fought bravely against the Italian mechanized forces, artillery, bombing and poison gas, but could not survive. The Italians, under de Bono and later Badoglio, totalled about 250,000. Their forces included Italian regulars, Fascist Blackshirts, Eritreans, some unreliable Somalis

Louis Hubert Lyautey (1854–1934) combined to a unique degree the qualities of military commander, administrator and philosopher of colonialism. Before his greatest work as Resident-General in Morocco from 1912, Lyautey had served in Indochina, Madagascar and Algeria. He had also written on the colonial role of an army officer, a role he saw as professionally desirable and personally ennobling.

In Morocco, Lyautey's campaigns were conducted in accordance with his "oil stain" strategy. Occupation

of an area would be followed not only by an outward extension of the area pacified, but by the immediate introduction of economic and social development schemes: roads, crops, markets and medical services. This work was carried out by military officers, a development of Bugeaud's policies.

Lyautey ensured that respect was given to the Moroccan Sultan and Muslim institutions. He also strove to prevent large-scale French settlement. Where suitable or necessary, he retained traditional chieftains, but in his eagerness to avoid a static policy he also began the education of a new Moroccan middle class. He described himself as "le galvanisateur" but his successors lacked his vision. Lyautey's conservatism and belief in the importance of ceremonial however led to criticism and, blamed somewhat unjustly for lack of preparation for the 1925 Rif rising, he was recalled to France in 1926.

POLITICS AND PACIFICATION

France's final pacification campaign in Morocco (1931–4) was closely linked to domestic political considerations. It was important that the campaign was completed before the anticipated heavy fall in numbers of conscripts available for the metropolitan army – a birth-rate decline consequent on World War I. After pacification, North African units would be needed to garrison France. As a correspondent of the Paris magazine *Voilà* wrote, "It is with the flesh and blood of partisans, Moroccan riflemen and the Foreign Legion that France is completing her conquest of Morocco."

The four French columns in Morocco, each the size of a small division, were almost entirely composed of North and Black African *Tirailleur* units and irregulars stiffened by battalions of the Foreign Legion. Men of the Legion are seen (*below*) in a North African barracks in the early 20th century. They were used to garrison Sahara fringe forts and certain Moroccan hill forts, which they often built themselves, as well as the roads leading to them.

and a few Libyans. The best Italian and indigenous units went on to fight well in the later World War II East Africa campaign, but the majority were lacklustre.

Holland began the formal conquest of Indonesia in 1619, with the seizure of Jakarta where the Dutch survived three massive 17th century sieges. In the 18th and early 19th centuries they expanded their settlements and vassal states in Java, Celebes, the Moluccas and Borneo against continuing opposition and guerrilla insurrection. After a brief British occupation, an agreement in 1814 enabled the Dutch to look beyond Java again. An attempt to occupy Bali in 1846–49 failed, but the occupation of southern Borneo began in the mid-19th century, and a protracted war from 1873 to 1908 fastened Dutch control upon Sumatra. The years 1900 to 1910 saw the complete occupation of Bali, Celebes and the other islands. Although from the 17th century the Javanese manufactured cannon and firearms to supplement the *kris* of their infantry and long pikes of their cavalry, Dutch weaponry was generally superior. The Dutch East Indies Company soldiers were a mix of men recruited in Holland, together with Buginese and Amboinese. After formal Dutch rule from 1816, soldiers were largely drawn from Holland, and Madura, north Celebes and Surakarta. Dutch rule was established at the Cape in South Africa in 1652 as a victualling station for ships en route to the East Indies. It was extended inland by farmers informally organized into "commandos".

At the end of the 19th century, a New World maritime imperial power, the United States, appeared. In a 115-day-long war with Spain in 1898, Dewey destroyed the Spanish Pacific fleet in Manila Bay and Sampson destroyed the Spanish Atlantic fleet off Santiago de Cuba. After the latter victory, the Spanish forces in Cuba surrendered to the rather less professional US army. The USA took the Philippines and Puerto Rico as spoils of war. Five years later, a US cruiser supported a Panamanian revolutionary movement seeking secession from Colombia. The Panama canal was then completed and secured by the US army.

Rule

Britain's vast empire was retained through displays of military or naval power, British and Indian, with, when necessary, the applied force of new control technology – aircraft (notably the De Havilland DH9A bomber), wireless, armoured cars and lorries. In Iraq, for example, in 1920, 50 battalions, mostly Indian, were committed in suppressing a rebellion, but by 1929 the imperial garrison was reduced to 2 battalions, a few local units, and 6 RAF squadrons controlling Kurd insurrection. Aircraft also played the decisive role in the destruction of Mohammed Abdille Hassan's 1919 uprising in Somaliland; in small-scale operations from 1919 to 1934 in Sudan; and, operating with armoured cars, in controlling the Aden hinterland and the mandated territories of Transjordan and Iraq. India remained the major commitment, the main events being the Afghan war and Waziristan campaign of 1919–21; containment of the first Non-Cooperation campaign of 1919–22 and the Moplah rising of 1921; fierce conventional war scale fighting on the northwest frontier throughout the 1930s; and the second Non-Cooperation campaign of the early 1930s. The Amritsar massacre of 1919, when British-officered Indian troops opened fire on a nationalist rally, killing over 380 people, was of special significance. On the

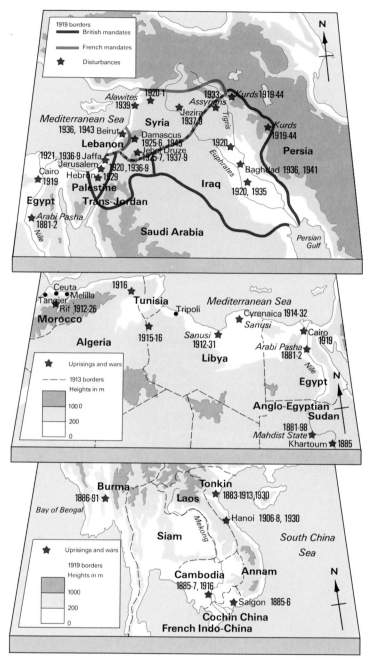

Colonial flashpoints.
Colonial rule met with both primary and secondary resistance and on occasions local revolt. Primary resistance opposed incoming imperial authority and often harassed that authority for some time; it was based on existing polities or ethnic groupings. Secondary resistance was formed from new groups that had emerged from the colonial occupation such as labour, or communities that felt especially threatened by colonial rule. This resistance could cross ethnic boundaries. Local revolts were the product of local misrule. Most of the flashpoints on these maps represent primary resistance. Examples of secondary resistance here include the Druze risings in Syria, Khartoum 1924 (Egyptian military), Hanoi 1930 (neo-Communist), and the Arab revolts of 1929 and 1936 in Palestine. The Batna (Algeria) 1916 uprising is an example of local revolt – against the French military draft.

one hand it gave an enormous boost to the Indian nationalist cause; on the other it led to such domestic criticism in Britain that never again were British military commanders able to abuse their power to such excess.

Palestine, important for its oil pipe line and overall strategic significance, saw an Arab revolt in 1929 with a much more serious rising in 1936, initially Arab but from 1939 Jewish. Over 20 battalions had to be committed at particular crises. Other territories that saw uprisings were Burma in 1930 and Cyprus in 1931. There was minor immediate post-world-war unrest in the Caribbean and in Malta, with more serious unrest following the depression in the Caribbean, notably Trinidad in 1937 and Jamaica in 1938. These were contained by local "Planters Rifles" colonist units, sometimes supported by warships or detached British battalions.

Military power in the 1919–39 era was normally distributed on the basis of approximately 60,000 British (45 battalions and 5 cavalry regiments) and 150,000 Indian troops (some 110 battalions and 21 cavalry regiments) in India, 6 British battalions and 3 cavalry regiments in Egypt, detached British battalions in a number of strategically important colonies, and a small number of locally recruited African battalions in African territories.

Despite unrest, imperial military unity was impressive. In World War I, Canadian, Australian, New Zealand, South African and Indian troops fought in France; Australians, New Zealanders and Indians at Gallipoli and in the Middle East; South, East and West Africans and Indians in East Africa with West Africans also in West Africa; and West Indians in Palestine. The World War II effort was even greater. The massive Indian Army contribution included East and North Africa, Syria, Italy, Malaya and Burma. Canadians fought in northwestern Europe, Italy and Hong Kong, South Africans in East and North Africa and Italy, Australians in Greece, North Africa, Malaya and the Pacific; New Zealanders in North Africa, Italy and the Pacific, East and West Africans in East Africa and Burma, Malays in Malaya and Maltese in Malta. But the emergence of anti-colonial sentiment was evidenced by Indians and Burmese who chose to fight for Asia's first modern maritime empire, Japan.

Soldiers of empire

After the occupation operations, France met only limited secondary resistance in her empire in the years prior to 1945, the 1925 Druze rising in Syria, and unrest in the late 1920s in Equatorial Africa and in 1930 in Indochina. Order was maintained by sizeable military formations. The normal establishments in the pre-1939 years were 4 divisions of troops in Morocco, 3 in Algeria, approximately 1 each in Indochina, Syria and Tunisia, 14 battalions in French West Africa, 5 in Equatorial Africa and 4 in Madagascar. Most of the soldiers in these garrisons were non-Frenchmen, indigenous *Tirailleurs* and *Spahis*, conscripted in many territories; the Legion was also an important contributor. In most territories, small numbers of professional regular *Coloniale* infantry or in the case of North Africa, *Zouaves* and *Chasseurs d'Afrique*, supplied the French contribution.

Soldiers from the empire were used extensively in France's national wars from the mid-19th century onwards. Small Algerian *Tirailleurs* units served in the

Crimea, Italy and Mexico; 9 battalions fought in the 1870–71 Franco-Prussian War. In World War I, 172,000 Algerians served, 158,000 in combat units, mostly on the Western Front. Tunisia supplied 54,000 men and Morocco 37,000. Some 160,000 Black African *Tirailleurs Sénégalais* served in France, as did over 40,000 Malgaches. Yet further *Sénégalais* fought in the Balkans and in Africa against Germans, Moroccan primary resistance or the Senussi. In 1920, 129 battalions of North African *Tirailleurs*, 4 *Spahis* regiments and over 50 battalions of *Tirailleurs Sénégalais* were serving France's interests in France, Germany, the Balkans, the Levant and North and Black Africa. By the late 1930s, 43 battalions of North African *Tirailleurs* and 18 of *Sénégalais* were in the permanent peacetime garrison of

---

"Yes, it's as hard as anything in the world, I suppose. But I don't regret going for it. I wanted to live a man's life. What should I have been doing if I had stayed in Paris? Cocktail Parties? Night Clubs? There will be plenty of time for those when I have finished with the Legion, and I shall enjoy them all the more for having seen something of this as a contrast. As for the university I have learnt more about men and things in the Legion than any professor could have taught me".
Corporal Ortiz, American serving in the Foreign Legion, Morocco 1933

France. By May 1940, the totals had reached some 93 and 24 respectively, together with further colonial personnel in *Spahis*, artillery and logistics units, and over 50 further North African battalions or other units deployed in North Africa and Syria.

France's post-1940 military renaissance was initiated by a handful of *Sénégalais* units in Equatorial Africa that rallied to de Gaulle. The *Armée d'Afrique*, carefully preserved by General Weygand, fielded three small divisions mostly of North African *Tirailleurs*, in the Tunisian campaign, and three excellent North African divisions plus a number of other units and several thousand Moroccan irregulars in the Italian and Liberation campaigns. Black African *Tirailleurs* also contributed significantly to operations in the Mediterranean and in France. Soldiers and politicians alike concluded that military capability and empire were synonymous, given this massive contribution from indigenous colonial populations.

German colonial rule was so severe that, even prior to

French Moroccan troops entering Tunis, May 1943.

1914, it had led to riots in Cameroun, Hottentot risings and the major 1901 Herero revolt in Southwest Africa, and the 1904–05 Maji-Maji uprising in German East Africa. These were all suppressed drastically, by General von Trotha in the Herero case almost to a point of genocide. German-recruited African soldiers fought well in the various World

---

"God has decreed that France represents Power, and that is why I shall fight with you because that is His Will. But I worry about you because you are a Christian and condemned to damnation. But because I like you well, if I see you near to death I shall recite the Muslim prayer for those about to die in battle."

Moroccan Tirailleur to his French officer, World War II

War I campaigns, notably so under General von Lettow-Vorbeck in a resolute four-year campaign in East Africa. After the war, Germany lost her colonies. The Netherlands experienced small uprisings in Java in 1926 and Sumatra in 1927 but no major revolt. They were, however, never able effectively to reimpose their authority on their Indonesian empire after World War II. Neither Belgium nor Portugal faced any significant military uprising in the core colonial era. Italy lost her empire following military defeats in World War II.

Departure
In addition to the pre-1945 reasons – the large gaps filled by

indigenous troops in French strategy and, on occasions, internal domestic situations – other factors in France forced the army to resist colonial nationalism. These factors included both a role in French reassertion against "les Anglo-Saxons", and the traditional pull-factors of colonial soldiering now seen as offering a haven against a rapacious consumer society as well as escape from new unpalatable military technology. The French army was never able to draw breath from the war before the major colonial uprisings, the instability of Fourth Republic governments often led to concession being portrayed as weakness, and the French centralist tradition saw nationalism as dissidence – heresy against French culture.

On the ground, as difficulties worsened, vast numbers of men were deployed and horrific excesses committed, often – but not always – in response to deliberate nationalist use of barbaric terror. The numbers of men, including thousands of indigenous, led French officers to fear for the safety of these men after any French withdrawal. The French excesses committed – a result of the total revolutionary warfare theories of elite regiment officers, mainly middle-ranking and many lower middle-class – led to anxiety, almost guilt feelings over the possibility of defeat and to an intensification of campaigning. Withdrawal appeared ever more unthinkable. In Algeria, parachute soldiers not greatly interested in the protection of *colon* estate owners, found an unexpected common identity with the blue-collar *colons* of Algiers city, with unfortunate and regrettable consequences.

A company of *askaris* in German East Africa prior to World War I (*above*). These excellent soldiers, mainly Lukuma and Nyamwezi, served the Germans loyally in the four-year German campaign in East Africa under General von Lettow-Vorbeck. The campaign

showed how a small well-led force could tie down a much larger force. It was also notable for the appalling suffering and casualties among military porter labour conscripted by the British – over 44,000 Kenyans dead.

Victors and vanquished at Dien Bien Phu. General Giap (*above*), points to the map with Ho Chi Minh and members of the Party Political Bureau. French elite troops (*right*) are photographed landing in a desperate attempt to secure the garrison. Giap achieved his victory with massive concentration of artillery fire power, which the French had failed to anticipate, and with successive attacks in waves by his peasant revolutionary army.

However, the French defeat was not without its epic

qualities: some of the Foreign Legion and Colonial Infantry parachute troops who volunteered for an already doomed mission had only made one or two previous parachute descents. They and the majority of the Algerian and Moroccan units in the garrison, supported by West African artillerymen, generally fought to the last before being overrun. Many of the Foreign Legion soldiers were Germans, and the suicidal nature of the battle accorded well with the cynical, despairing mood of the Germany they had left.

126

The campaigns in Indochina

French effort to retain Syria and Lebanon was insignificant, but that for Indochina, already destabilized by the Japanese occupation, was massive. The years 1945 to 1954 fall into three phases. The first of these was an old-fashioned colonial campaign built upon control of cities, roads and static rural posts. This collapsed with the emergence of a regular Vietminh guerrilla army, culminating in the 1950 frontier fort disasters of Dong Khe and Cao Bang with the French being forced to withdraw from most of north Tonkin. The next phase, the year 1951, saw the brilliant application of traditional North African tactics by de Lattre de Tassigny. Simultaneous attacks by converging powerful mobile columns, emerging from a chain of well-protected forts and now supported by napalm airstrikes, secured some spectacular reverses for the Vietminh at Vinh Yen, Mao Khe and Ninh Binh. De Lattre's final victory, at Hoa Binh, was however costly and de Lattre himself returned to France to die of cancer.

In the final catastrophic phase, de Lattre's successors, Salan and Navarre, tried to hold the line of forts, with diminishing success in face of the mounting Vietminh challenge. Seized with a Verdun style fatalism, Navarre decided to concentrate on a smaller number of larger strong points, the most important to be at Dien Bien Phu, where it was thought an impregnable fortress could also provide a base for operations in either Laos or Tonkin. After an epic struggle, Dien Bien Phu, badly sited, was overrun in 1954 by weight of numbers and Vietminh artillery domination; 15 battalions plus other units, many elite, being annihilated. The defeat was shattering; the French were left with no option but to withdraw, no political aims having been secured.

French all-services strength in Indochina rose from 100,000 in 1948–49 to 145,000 in 1950, 188,000 in 1952 and 200,000 in 1953–54. Of these, in July 1953 the ground force totals were 54,800 French (conscripts could not be sent to Indochina against their will), 19,000 Legion, 29,500 North Africans, 18,000 Black Africans and 53,000 Indochinese in French service. To this must be added a further 177,000 serving in the French-sponsored associated states.

Elsewhere in the French empire, a similar inflexible military response was meeting nationalistic challenge. Forces each of some 7,000, mostly North and Black Africans, had to be used to suppress nationalist uprisings in Algeria in 1945 and in Madagascar in 1947. Policies mostly unyielding, *la politique du sabre* of military Residents-General in Tunisia and in Morocco required troop levels that increased progressively to 40,000 and 106,000 respectively in 1955, by which time the Algerian uprising had also broken out. Paris again had no option but to grant both territories independence, with few remaining special advantages for France.

France's last and most disastrous campaign, in Algeria, ran from 1954 to 1962. After a hesitant start, the French army gained an ascendancy and then control – at a cost of the deployment by 1959 of over 400,000 men in 16 divisions, reinforced by several thousand more ancillary levies. The

400,000 included French regulars, conscripts, Legion, Black Africans and a large number of loyalist Algerians. The French tactics were based on military administration, zones held by static garrisons (*quadrillage*), heavily fortified lines of wire and mines on Algeria's borders, the effective if drastic clearance and control of the major cities, and the dispatch of rapid-pursuit elite regiment forces, helicopter-borne, to chase nationalist insurgent groups on the move. The strain of the bloody campaign,

---

"The 1st May 1962 – Marseille. They made us move from the boat to the train as if in shame. In the squares, on the jetties everywhere the green and white flag of the FLN and red flags. But where was our own flag? And where were the local military authorities . . .? That day, returning to France after many months [in Algeria], I no longer felt at home."

Sergeant B Laprugne, 3rd Hussars, French Army

---

however, broke the Fourth Republic in 1958. Despite the successes of his brilliant field commander, Challe, the president of the Fifth Republic, Charles de Gaulle, came increasingly to accept that military control was no political solution. His policies were challenged by a group of generals, including Challe, in an abortive coup attempt in Algiers in 1961. With an army now divided and embittered, the terms for Algerian independence that de Gaulle had finally to accept in 1962 contained even fewer concessions to France. Over 1,000,000 French *colons* fled to France to

Helicopter-borne Foreign Legionaries occupy a mountain area of Algeria, 1960. Operations mounted by General Challe were driving Algerian nationalist insurgents towards the Ligne Morice, a massive wire and mine fortification system on the Algerian-Tunisian border, thus giving France a military ascendancy. Tactics involved the traditional *Armée d'Afrique* concept of an attack launched simultaneously from several directions. But military control does not equate with political victory, a point not grasped by certain French generals who, in 1961, attempted a coup to prevent de Gaulle's more concessionary policies.

escape the inevitable violence, and several scores of thousands of Algerian loyalist irregulars were butchered by the victorious nationalists.

If French decolonization met with cognitive dissonance, that of Britain met with Common Law pragmatism. In the final months in India, amid communal strife, British battalions endeavoured and assisted Indian units to endeavour, to save life. In Palestine from 1945 to 1948, a force totalling two divisions of British troops in search of a just solution strove to stem Jewish terrorism until withdrawal was ordered. In a long campaign in Malaya from 1948 to 1960, a force at its peak of 10 British, 8 Gurkha, 5 (later 7) Malay, 3 African and 1 Fiji, 1 Australian and 1 New Zealand battalion contained and suppressed a largely Chinese Communist jungle insurrection using a doctrine brilliantly developed by General Templer. A friendly independent Malaysian government resulted.

Gerald Templer (1898–1979), Britain's outstanding post-1945 soldier, was appointed to full political and military command in Malaya early in 1952. He imposed a successful strategy that was to determine British army counterinsurgency doctrine elsewhere. This strategy centred upon control fusing military, police and civil administration, with the modernization of the police, intelligence services and a local home guard. Malaya's own regiment was also greatly expanded.

In the field, like Lyautey, Templer worked outwards from areas cleared. He used special forces, troop-carrying helicopters, bombing and air support.

Templer encouraged rapid political progress with elected local and state councils and clear promises for full independence. At village level, he urged on imaginative rural development. All these activities were explained by expanded information services. A final ingredient for success was Templer's own volcanic energy.

In Africa, the British garrison in Egypt, at its height during the Iran crisis of 1951–52 totalling 16 battalions, 2 armoured and 7 artillery regiments, contained Egyptian partisan activity until formal withdrawal in 1956, following the short-lived 1955 agreement. Out of keeping with the general British pragmatism was the abortive 1956 Anglo-French Suez expedition, supposedly tasked to secure the canal but in fact covertly seeking the fall of President Nasser and the reassertion of a British hegemony. One consequence of this humiliation was the need to dispatch small forces to protect friendly regimes in Muscat in 1957 and Kuwait in 1961.

From 1952 to 1956 a force, at its maximum 5 British and 6 African battalions, was deployed with police and Kikuyu home guards in suppressing the Mau Mau uprising in central Kenya, all in the context of a radical political and economic restructuring of Kenya acceptable locally and to Britain. In several other African colonies, one or two local battalions underpinned the colonial withdrawal. In the case of Southern Rhodesia's unilateral declaration of independence in 1965, however, British action was limited to naval patrols attempting to enforce sanctions by closing Mozambique ports to Rhodesian trade, in particular oil. The Greek EOKA uprising in Cyprus from 1954 to 1957 engaged at its height 18 British units of battalion size in a mix of village and urban riot control and massive sweeps in the Troodos Mountains; the independence agreement retained two sovereign bases for Britain. The final major

Middle East decolonization campaign was that of Aden from 1955 to 1967. A border campaign from 1955 to 1958 was followed by a lull until 1963 when fighting flared in the Radfan. This spread to Aden itself from which, despite the defection of locally recruited Arab units, 7 battalions and supporting arms conducted a skilled operational withdrawal covered by the Royal Navy in 1967. In this campaign, no political aims were secured. The 1960s also saw the British army's most professional decolonization campaign, the containment of Indonesian inspired revolt and incursions into British, and later Malaysian, territories on Borneo. The campaign, involving at its height British, Gurkha, Australian, New Zealand and Malaysian battalions with British armour and artillery, was a skilled mix of jungle patrolling, intelligence and special force operations, and naval helicopter usage in a campaign of notable professional skill. In minor operations in the Caribbean, one or two battalions had to be deployed operationally at intervals from 1948, one to protect British Honduras against Guatemala, one to control political excitement in British Guyana in 1953 and again in 1962–66, and one in Jamaica in 1960. A naval mounted assault recapture of the Falkland Islands, occupied briefly by Argentina in 1982, involved approximately two brigades of troops, one brigade including a Gurkha battalion.

Further decolonization in Africa

Portugal believed, like France, that her "overseas provinces" were essential for national survival. An uprising in Angola in 1961 was drastically suppressed, but guerrilla warfare opened in Guinea-Bissau and Mozambique in 1963. Ten years later, General Spinola, using air to ground strikes, was containing a limited conventional assault of insurgent infantry and artillery in Guinea, but the defeat by a massive FRELIMO insurgent enveloping movement of the forces of General Kaulza de Arriaga, concentrated in northern Mozambique, acted as catalyst for the 1974 collapse of the Lisbon authoritarian government. Independence then followed. The Belgian failure to make any satisfactory preparation for a post-independence government in the Congo led to a collapse of authority in the territory at independence in 1960. A large United Nations force comprising *inter alia* units from Sweden, Ireland, Malaya, Indonesia, India, Guinea, Ethiopia, Senegal, Ghana and Nigeria was assembled, initially to save life and restore order. Later the same force was effectively used to end a secessionist regime in the mineral-rich Katanga (Shaba) province.

Britain intervened in her three former East African territories at their request in January 1964 following army mutinies, but British operations after that date have been small-scale, such as the provision of personnel in the Commonwealth-monitoring force in Zimbabwe in 1979–80. France has intervened in a number of her former African territories since 1960 and still retains garrisons in both West Africa and Djibouti. The post-independence institutions, the (British) Commonwealth and France's less formal relationships contain no military obligations binding either the former imperial power or the successor states to fight for each other's causes.

Although problems of European settlement territories remained in southern Africa, the star of maritime colonial empire had finally set. The new titans upon the world stage are those of land masses, the USSR, China and the USA.

RETREAT FROM EMPIRE: DATES OF INDEPENDENCE			
1941 Ethiopia	**1960** Cameroun, Central African Republic, Chad, Congo, Cyprus, Dahomey, Gabon, Ivory Coast, Madagascar, Mali, Mauritania, Niger, Nigeria, Senegal, Somalia, Swaziland, Togo, Zaire	**1967** Democratic Republic of Yemen	**1977** Djibouti
1945 Lebanon		**1968** Equatorial Guinea, Nauru	**1978** Dominica, Solomon Islands, Tuvalu
1946 Jordan, Philippines, Syria		**1969** Ifni (to Morocco)	**1979** Kiribati, St Lucia
1947 India, Pakistan	**1961** Kuwait, Lesotho, Sierra Leone, Tanganyika	**1970** Fiji, Tonga	**1980** St Vincent, Vanuatu, Zimbabwe
1948 Burma, Ceylon, Israel	**1962** Algeria, Jamaica, Trinidad & Tobago, Western Samoa	**1971** Bahrain, Bangladesh, Qatar, United Arab Emirates	**1981** Antigua-Barbuda, Belize
1949 Indonesia	**1963** Dutch New Guinea (to Indonesia), Rhodesia, Zanzibar	**1973** Bahamas	**1982** Burundi, Rwanda
1951 Libya		**1974** Grenada, Guinea-Bissau	**1983** St Kitts-Nevis
1954 Laos, North Vietnam, South Vietnam	**1964** Malta, Malawi, Tanzania (Tanganyika and Zanzibar), Zambia	**1975** Angola, Cape Verde, Comoros, Fernando Po (to Equatorial Guinea), Mozambique, Papua New Guinea, São Tomé & Príncipe, Western Sahara	**1984** Brunei, Malaysia (Malaya plus Sarawak and Sabah)
1956 Egypt, Morocco, Sudan, Tunisia	**1965** The Gambia, Maldives, Singapore		
1957 Ghana, Malaya			
1958 Cambodia, Guinea	**1966** Barbados, Botswana	**1976** Seychelles, Timor (to Indonesia), Vietnam (united)	

— chapter ten —
World War I: Advance to deadlock

Richard Holmes

Shortly before eleven o'clock on the morning of June 28 1914 Gavrilo Princip, a Bosnian student, shot the Archduke Francis Ferdinand, heir apparent to the throne of Austria-Hungary, and his wife Sophie in Sarajevo, capital of the Austrian province of Bosnia. The assassination was not the work of accomplished killers: it was so clumsily executed that only an error by the Archduke's chauffeur enabled Princip to fire the fatal shots.

Princip and his accomplices were idealistic young men swept along by revolutionary undercurrents which were the product of hatred of Austria, desire for Serbo-Croat national unity, and the influence of the Serbian *Narodna Odbrana* and "Black Hand" secret societies. It is difficult to allocate blame for the assassination, but the involvement of Serbian officers in the secret societies points at the very least to culpable negligence on Serbia's part.

Mainsprings of conflict: the political background

The events which followed are comprehensible only in the context of the military developments of the preceding 40 years, and of the alliances which had marched the states of Europe into two rival camps. One conflict provided the springboard for another. The Franco-Prussian War of 1870–71 left the French embittered: not only had they suffered humiliating defeat, but they had also been forced to pay a large indemnity and to relinquish the frontier provinces of Alsace and Lorraine. Moltke, the architect of Germany's victory, had prophesied that "what our sword has won in half a year our sword must guard for half a century", and France's rapid recovery from the war bore out his sagacity. The indemnity was speedily paid and the Third Republic grew in strength and confidence. Its army was extensively modernized, with the acceptance of the principle of universal military service in 1872, the establishment of the *École Supérieure de Guerre* in 1875, and the adoption of new weapons.

Alongside military reform went concerted efforts to ensure that France would not stand alone in a war with Germany. In August 1891, France and Russia agreed to consult if peace was threatened, and in 1893 this hardened into a formal military agreement. Long-standing rivalry, popular hostility during the Boer War and colonial friction impeded a similar understanding with Britain. The visit of Edward VII to Paris in May 1903 and President Loubet's return visit in July did much to improve Franco-British relations and to pave the way for the Entente Cordiale. This had no overt military implications, but several influential British officers, most notably Henry Wilson, Director of Military Operations at the War Office 1910–1914, believed implicitly in a German challenge to British power. "Unofficial" staff talks – which, the British government emphasized, were not politically binding – resulted in a war plan which would take an expeditionary force to northern France on mobilization.

The transformation of British foreign and military policy in the first decade of the century had profound implications. British strategy was redefined: it had hitherto emphasized naval power rather than a continental commitment, and the decision to send the British army to France in the event of a Franco-German war showed the new assertiveness of the army under its recently formed general staff. When war came, Britain's participation was to ensure two things: supply of fresh manpower when the massive conscript armies of her allies and opponents were beginning to run short, and a naval supremacy which would grip Germany in a relentless blockade.

A bitter and truncated France was one of the products of the Franco-Prussian War: the Second German Reich, founded when William I of Prussia was proclaimed emperor (Kaiser) on January 18 1871, was another. Prussia, pre-eminent among the states of the new Reich, was dominated by the military and the bureaucracy: her middle classes were denied political power commensurate with their economic importance. Her impact was most marked in military affairs. Thirteen of the *Kaiserheer's* army corps came from Prussia, two from Bavaria, one from Württemberg and one from Saxony: the remaining one was based in Alsace-Lorraine. The war ministries and general staffs of the German states, theoretically independent, were subordinate to the Prussian War Ministry and General Staff, and the Kaiser topped the pyramid in his capacity as "Supreme War Lord".

Preliminaries to war

There is no unanimity among historians over the motives for Germany's policy before World War I and the extent to which her leaders actively plotted war. Fritz Fischer suggests greater deliberation than most others would accept, but it is beyond dispute that the General Staff, perturbed by the growing strength of Germany's rivals, saw merit in a preventive war. Yet if there is doubt about Germany's motives, the pattern of events is clear enough. In the years before 1914 Germany embarked upon a linked policy of *Weltpolitik* and navalism, building up her fleet and pressing for territory and influence overseas. This, coming at a time of increasing economic competition, alienated much British opinion, reinforcing the ties between Britain and France. Germany, for her part, was drawn closer to Austria-Hungary and offered support for her policy in the Balkans, even if the Russians intervened to protect their fellow-Slavs in Serbia. In January 1909, the younger Moltke, chief of the German General Staff like his uncle before him, told Franz Conrad von Hötzendorf, his Austrian counterpart, that "the moment Russia mobilizes, Germany will also mobilize. . . ."

It was not just against Serbia that Conrad's hawkish zeal was directed. Although Italy had concluded the Triple Alliance with Germany and Austria-Hungary in 1882, she subsequently also reached agreement with Britain, France and Russia, leading an exasperated Kaiser to wonder who she actually intended to support. Conrad saw Italy as a probable enemy and in 1911 suggested a pre-emptive strike when she was fighting the Turks in Libya. There were grounds for conflict between Italy and Austria: many Italians lived under Austrian rule and there were demands for the "rectification" of the Austro-Italian frontier. So, despite her membership of the Triple Alliance, Italy's allegiance was far from certain.

Turkey's position was clearer. The Balkan crisis of the early 1900s was largely a consequence of the decline of

—TIME CHART—

1879–1918	Austro-German alliance
1881–95	Austro-Serbian alliance
1894–1917	Franco-Russian alliance
1904	Franco-British *Entente Cordiale*
1908	German army commissions first Zeppelin
1911	Italians use aircraft against Turks in Tripolitania
1912–13	First Balkan War: Turkey loses European territories
1913	Second Balkan War: Bulgaria loses newly gained territories
1913	German military mission to Turkey
1914	
June 28	Archduke Francis Ferdinand assassinated
July 23	Austria delivers ultimatum to Serbia
July 28	Austria declares war on Serbia
Aug 1	Germany declares war on Russia
Aug 3	Germany declares war on France
Aug 4	Germany invades Belgium
Aug 4	Britain declares war on Germany
Aug 7	German warships *Goeben* and *Breslau* reach Dardanelles
Aug 12	Austrians invade Serbia

Aug 12–21	Serbs push back Austrians on the Jadar
Aug 14–22	French repulsed in Lorraine
Aug 15	Russians enter East Prussia
Aug 16	German 420mm siege howitzers destroy last of Liège forts
Aug 20	Russian advance checked at Gumbinnen
Aug 23	Battle of Mons: BEF withdraws
Aug 20–25	French defeated in the battle of the Frontiers
Aug 23	Austrians invade Poland
Aug 26–31	Russians defeated at Tannenberg
Sept 5–10	Germans fall back from the Marne
Sept 11	Austrians retreat from Galicia
Sept 13–14	Allies rebuffed on the Aisne
Oct 9	Antwerp surrenders
Oct	German warships bombard Odessa
Oct 30–Nov 24	German advance checked at first battle of Ypres
Nov 2–5	Britain, France and Russia declare war on Turkey
Nov 7	Japanese take Tsingtao on Chinese coast from Germans
Nov 25	Germans take Lodz
Dec 10–24	Allied Western Front offensive

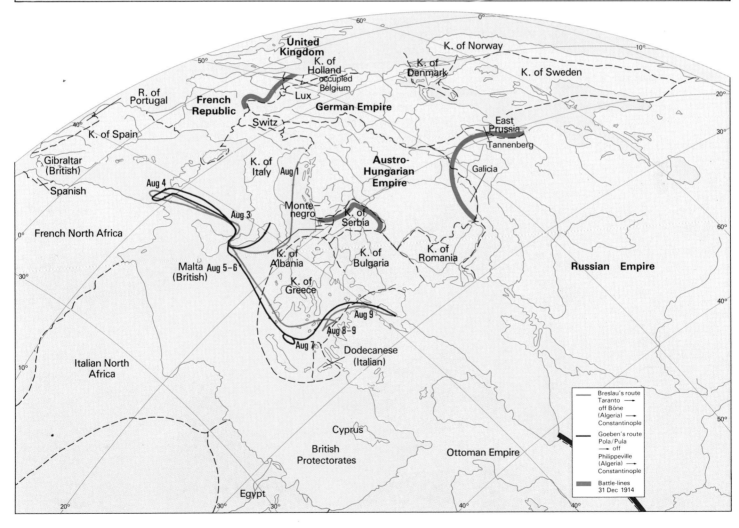

The map shows the area affected by World War I and the battlelines that had been established up to Dec 31 1914.

Turkish power, a process accelerated by defeat in the Russo-Turkish War of 1877–78. The Balkan states of Romania, Bulgaria and Serbia were anxious to expand and, lying as they did at the fringes of Austrian, Russian and Turkish areas of interest, were drawn into the web of great power politics. In 1912 Bulgaria, Greece, Montenegro and Serbia attacked Turkey, then preoccupied by the fighting in Libya, and obtained territory by the terms of the Treaty of London. In 1913, however, the Second Balkan War saw Serbia and Greece, latterly with Romanian and Turkish support, quickly defeat Bulgaria. Serbia almost doubled in size as a result of these two wars and her ambitions were thoroughly aroused. Turkey's defeat in the First Balkan War was a blow to German prestige, for the Turkish army was German-trained: in 1913 the Germans sent General Liman von Sanders to reorganize it, an act that at once affronted the Russians and strengthened the links between Turkey and Germany. These were economic as well as military. The Berlin-Baghdad railway, built with German capital by German engineers, was not part of a grand political design, but it naturally increased Germany's interest in Turkey and aroused both British and Russian suspicions.

The crisis which followed the assassination of Francis Ferdinand was the last in a series which had begun with the Moroccan affair of 1905. Almost any one of these might have led to war, but in 1914 war came with a suddenness which astonished even the participants. Count Leopold von Berchtold, the Austrian Foreign Minister, obtained assurance of German support in his task of punishing the Serbs, and on July 23 delivered an ultimatum to Belgrade.

Although the Serbs were prepared to compromise, they could not accept the ultimatum as it stood, and on July 28 Austria declared war. The Russian government attempted to mobilize against Austria alone, but, discovering that plans for partial mobilization did not exist, ordered general mobilization. Germany warned Russia that this would bring a German response, and, well aware of the Franco-Russian alliance, asked France for a promise of neutrality. This was not forthcoming, and Germany declared war on Russia on August 1 and on France on August 3.

Britain was in a difficult and uncertain position, with her Liberal government reluctant to go to war despite the military agreement with France. German violation of Belgian neutrality in breach of the 1839 treaty to which Britain was signatory pushed the Asquith cabinet over the brink, and on August 4 Britain declared war on Germany. This declaration triggered yet another. The German warships *Goeben* and *Breslau*, pursued by British vessels, were allowed to take refuge off Constantinople in August, and in late October they steamed into the Black Sea to bombard the Russian port of Odessa, leading the Entente Powers (Britain, Russia and France) to declare war on Turkey in the first week of November. Italy initially resisted the slide, but eventually declared war on Austria in May 1915. She did not join the war against Germany for another 14 months, but by that stage, with operations under way on three continents, it was already a world war.

Technology and timetables: the military background
It is impossible to study the events of 1914 without being struck by the fact that mobilization and war had become

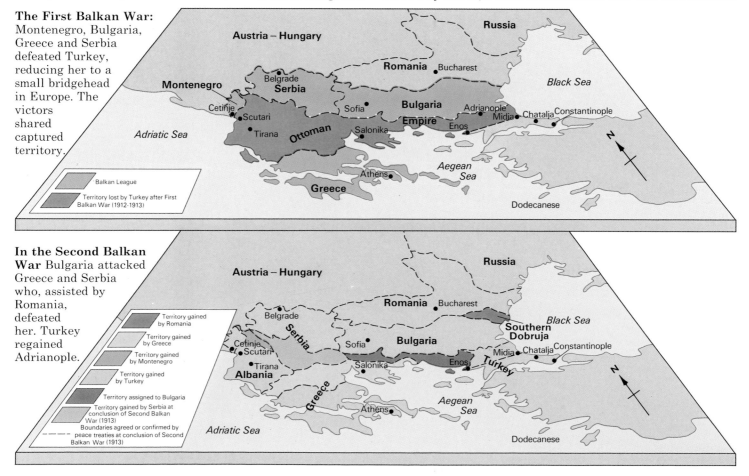

The First Balkan War: Montenegro, Bulgaria, Greece and Serbia defeated Turkey, reducing her to a small bridgehead in Europe. The victors shared captured territory.

Balkan League

Territory lost by Turkey after First Balkan War (1912-1913)

In the Second Balkan War Bulgaria attacked Greece and Serbia who, assisted by Romania, defeated her. Turkey regained Adrianople.

Territory gained by Romania

Territory gained by Greece

Territory gained by Montenegro

Territory gained by Turkey

Territory assigned to Bulgaria

Territory gained by Serbia at conclusion of Second Balkan War (1913)

Boundaries agreed or confirmed by peace treaties at conclusion of Second Balkan War (1913)

synonymous and simultaneous. This was the tragedy of 1914, and it had its roots not in the malevolence or frailty of man, but in his inventiveness and industry. The development of the railway in the second half of the 19th century had, as we have seen, immense military consequences. The web of European railways had continued to thicken, and the development of strategic railways in Russia had special significance. In the early years of the century Russia could dispatch a mere 200 trains daily to her western borders, while Germany could send 650 across the Rhine at Cologne alone. By 1910, however, the Russians could mobilize at the rate of 250 trains per day; by 1914, 360. The refinement of the bureaucracy of mobilization and plans for concentration by railway meant that even a day's delay could have fatal results. As Norman Stone has written: "The railway, chief production of nineteenth-century civilization, had proved to be the chief agent in its destruction."

Men in their millions

Although the railway was the quintessential ingredient of the lethal brew, another was no less important. European war was very much a matter of manpower, and all the major

German conscripts are given an enthusiastic send-off, 1914

belligerents except Britain had armies raised by conscription. Systems differed in detail but had the same aim: to channel the manhood of the nation into its army. In Germany every man who was fit for military service was enlisted into the *Landsturm* at the age of 17. Three years later he joined the regular army for two years before going onto the reserve for five. Recruits to the cavalry and horse artillery spent three years with the colours but only four with the reserve. All reservists were recalled for two weeks' annual training in September. A man joined the *Landwehr* at 27 and served in it till he was 39, when he joined the *Landsturm*, where he remained until the age of 45. The country was divided into military districts, housing the corps, divisions, brigades and regiments of the active army. On mobilization, reservists were recalled to the colours: some brought regular formations up to strength, while others went into reserve regiments and divisions which shadowed their regular equivalents and were ready for war scarcely less quickly. The older men of the *Landwehr* and *Landsturm* were available for lines of communication duties and home defence. This process took the army from its peacetime strength of 700,000 men to a staggering 3,840,000 in six days.

Just as the growing strength of her rivals was used to

MOBILIZATION AUGUST 1914

For many, the outbreak of war caused grief encapsulated in Sir Edward Grey's remark: "The lamps are going out all over Europe; we shall not see them lit again in our lifetime." Captain James Jack shared Grey's gloom, writing: "One can scarcely believe that five Great Powers – also styled 'civilized' – are at war, and that the original spark came from the murder of one man and his wife. It is quite mad, as well as being quite dreadful." Herbert Sulzbach lacked Jack's experience of combat but felt uncertain at the prospect. "You feel that a stone has begun to roll downhill," he wrote, "and that dreadful things may be in store for Europe."

Young men, regulars or reservists, often saw things differently. "The spirit of patriotism of that age had to be experienced to be realized," recalled Richard Gale, shortly to enter Sandhurst: "The whole nation rallied behind the colours." Lieutenant Alan Hanbury-Sparrow declared: "Now we are to find out if our training has been right; now we are to find out what war is really like, and now to find out if we are – men." Lieutenant Erwin Rommel observed: "All the young faces radiated joy, animation and anticipation. Is there anything finer than marching against an enemy at the head of such soldiers?" The populace of London, Paris and Berlin sent their men off amidst scenes of delirious enthusiasm, singing patriotic songs, and pressing flowers on soldiers marching to the great railway termini from whose maws they clattered away to war.

Comparative strengths of the belligerents: July 1914
Regular armies:

Allies	**Central Powers**
Britain: 254,000 men	Germany: 860,000 men
France: 790,000 men	Austria–Hungary: 1,300,000 men
Belgium: 117,000 men	
Russia: 1,300,000 men	
Serbia: 190,000 men	

(Estimated peacetime strengths: war strengths, including reservists, were considerably higher. Germany, for example, was estimated to have some 1,885,000 men under arms by the end of August 1914.)

Total number of men mobilized in 1914–18 (estimates):

Britain: 5,704,416	Greece: 200,000
British empire: 2,950,000	Romania: 1,000,000
France: 7,800,000	Japan: 1,500,000
French empire: 394,500	Turkey: 2,850,000
Belgium 380,000	
Russia: 19,000,000	Germany: 13,250,000
Serbia: 450,000	Austria–Hungary: 9,000,000
Portugal: 60,000	Bulgaria: 950,000
Italy: 5,615,000	USA: 3,800,000

Naval strengths in July 1914:

Britain: 22 *Dreadnought* battleships, 40 pre-*Dreadnought* battleships, 9 battlecruisers, 48 armoured cruisers, 71 light cruisers, 225 destroyers, 75 submarines

France: 4 *Dreadnought* battleships, 21 pre-*Dreadnought* battleships, 25 armoured cruisers, 3 light cruisers, 81 destroyers, 67 submarines

Germany: 16 *Dreadnought* battleships, 30 pre-*Dreadnought* battleships, 6 battlecruisers, 15 armoured cruisers, 33 light cruisers, 152 destroyers, 30 submarines

justify Germany's military expenditure, so the burgeoning size of the German army provoked a response. France had reduced military service to two years in 1905, but in August 1913, despite fierce resistance from the Left, this was raised to three years with the colours and fourteen on the reserve. In the summer of 1914 France put 4,000,000 men into the field. In Russia, the "Great Programme" adopted in June 1914 decreed three-year service and a total peacetime army of almost 2,000,000 men. This was passed too late to affect mobilization the following month, but in the event Russia managed to mobilize nearly 6,000,000 men.

An irreversible situation

The recall of hundreds of thousands of reservists and their dispatch to the frontiers was the fruit of detailed planning. Such plans could not be changed at the last moment, and their authors knew that there was nothing unique about their own deliberations: allies and opponents would be doing the same thing, and doing it as fast as they could. Once mobilization had begun, it was impossible to stop and difficult to change without great risk. When Moltke was summoned to see the Kaiser on August 1, the monarch suggested that the political situation demanded only war with Russia: plans for sending the bulk of the army to the west could be shelved: "I answered his majesty," wrote Moltke, "that this was impossible". The deployment of an army a million strong was not to be achieved by a mighty battle of encirclement. Most of the divisions allocated to the western theatre would deploy north of Metz and would swing through Holland – the "Maastricht Appendix" – Belgium and northern France, with the last man on the right "brushing the Channel with

his sleeve". The outer German army would swing around Paris and the wheeling armies would snap back against the French left rear, winning a decisive battle which would end the war in the west at a stroke.

Moltke modified the plan in several important ways. First, he decided not to violate Dutch territory, and in consequence the wheeling armies would be canalized through Belgium. He also decided to allocate stronger forces to the Eastern Front, leaving only 55 of Schlieffen's 71 divisions north of Metz. Moltke is often accused of tinkering, like the sorcerer's apprentice, with something he did not understand. This criticism overlooks the fact that Schlieffen's plan was flawed. It did not take into account the delay likely to be imposed upon the advance by the fortresses in its path; it underestimated the effect of the Clausewitzian element of "friction" upon the troops and commanders carrying out the wheeling march, and upon the logisticians who supplied them; and it did not recognize that the original plan was – as Schlieffen himself admitted – one for which even the mighty German army was not strong enough. Finally, by subordinating political judgements to military imperatives, it resulted in a violation of Belgian neutrality which brought Britain, with her navy and empire, into the war.

Military reform in Russia

It was not simply that Schlieffen's plan was imperfect: the conditions which had prevailed when he last amended it no longer applied. Schlieffen had based his calculations on the state of Russia in 1905, when she had just lost the Russo-Japanese War and was riven by disorder. Between 1908 and 1914 the Russian army underwent far-reaching reform,

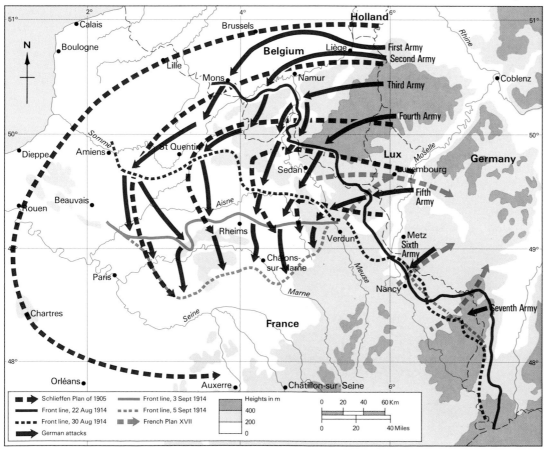

The Schlieffen Plan was the product of Schlieffen's determination to win a war on two fronts. Recognizing that the depth of Russia meant that he could win only "ordinary victories" there, he aimed at achieving decisive victory over France. He intended to shun the heavily fortified Franco-German border, and to advance instead through the Dutch "Maastricht Appendix" and neutral Belgium, his outer army wheeling west of Paris. Even Schlieffen recognized that the plan was impractically ambitious; Moltke modified it, and the friction of war produced further changes in the first weeks of fighting. Vigorous French attacks into Alsace and Lorraine made Moltke's task easier, but he unwisely permitted his Sixth and Seventh Armies to counterattack.

largely at the hands of General V A Sukhomlinov, Chief of the General Staff and later War Minister. Sukhomlinov has had a worse press than he deserves for, working within a system which was bureaucratic, conservative and faction-ridden, he and his associates increased the army's overall strength and quality and speeded up mobilization, although they were able neither to create an efficient high command nor to ensure that there was coherent war-planning. Much of Russia's increased defence expenditure was spent on fortresses or on the swarms of cavalry in which Russian commanders placed excessive confidence. Nevertheless, when Russia mobilized in 1914 she did so more rapidly than Schlieffen had expected and fielded a much better army than he would ever have faced. The improvement of the Russian army, coupled with the news that Britain and Russia had commenced naval talks, convinced the General Staff that the opportunity for a preventive war was passing, and in May 1914 Moltke told Conrad that "any adjournment will have the effect of diminishing our chances of success."

Germany's tactical viewpoint

It was upon the French army that the General Staff's attention was most narrowly focused. French and German soldiers might have agreed that the war of 1870–71 made another one inevitable. They would, however, have disagreed over the tactical lessons of that war – and of the subsequent South African and Russo-Japanese conflicts – as well as over the effect that recent technological developments would have upon the battlefield. This disagreement not only coloured operational and tactical doctrine, but also influenced training and procurement policies. The Germans thought that the massive increase in firepower (examined in a previous chapter) placed the tactical advantage in the hands of the defender. Yet, because the geopolitical logic of Germany's position argued in favour of offensive operations, German soldiers must either find their opponents' flanks – hence Schlieffen's fascination with Cannae – or, as Schlieffen put it, "try, as in siege warfare, to come to grips with the enemy from position to position, day and night, advancing, digging in, advancing again, digging in again, using every means of modern science to dislodge the enemy behind his cover." This interpretation of military developments not only gave birth to the Schlieffen plan itself, but also lent fresh impetus to German interest in artillery, especially the heavy howitzers which could deal with French and Belgian fortresses and the field howitzers whose high-angle fire could drop shells on troops in field fortifications. In 1914 the Germans enjoyed a clear lead in heavy artillery, with 575 heavy guns and howitzers in their field armies compared to the Russian 240 and the French 180.

A German 21cm heavy howitzer in Champagne. Note the use of brushwood to conceal ammunition from air observation.

Colonel General Helmuth von Moltke (1848–1916) was nephew of the architect of Prussian victory in 1866 and 1870, and succeeded Schlieffen as Chief of the General Staff in 1906. He lacked his uncle's iron will: nevertheless, he identified some of the defects in the Schlieffen plan, and paid particular attention to its logistic problems.

Whatever Moltke's virtues as a staff officer, he proved an inadequate commander.

Bad news from the Russian front weighed heavily on him, and eventually induced him to divert troops from the west. He was reluctant to impose his will on his army commanders, permitting Crown Prince Rupprecht of Bavaria to counterattack the French advance in Lorraine, and failing to direct effectively the movements of the three right-wing armies. In part his difficulties stemmed from being remote from the battle, and in part from his lack of mental robustness. Sick and nervous, he was relieved of his duties on September 14, but stayed on at headquarters to preserve appearances until November 3. Moltke's tender conscience was part of his tragedy. On September 7 he wrote of the misery the war was causing: "Terror overcomes me when I think about this, and the feeling I have is as if I must answer for this horror . . ."

The French position

France's lack of heavy guns was no accident. In 1909 a staff officer questioned about heavy field artillery had replied: "Thank God we don't have any! What gives the French army its force is the lightness of its cannon." A trail of twisted logic had put French military thought into a straitjacket. Theorists found it easy to conclude that the defensive-mindedness of the Imperial armies had brought about their downfall in 1870, and Ardant du Picq's views on the supremacy of morale gained wide currency. As the French army recovered from defeat its planners began to consider offensive operations, but there was a fundamental failure to align demands for offensive strategy with the tactical consequences of increased firepower. The doctrine of the offensive was propounded at the *École Supérieure de Guerre*, most actively by its director, General Ferdinand Foch, whose belief in the primacy of morale was enshrined in the dictum *"victoire, c'est la volonté"* ("the will to win is the first necessity for victory"). Foch did not, though, neglect *sûreté* (protection) and firepower. His disciples were less discriminating, and one of them, Colonel Grandmaison, director of the Bureau of Military Operations, was especially influential in infecting the army with belief in *offensive à l'outrance* ("attack to the limit").

The 1913 Regulations embodied this belief. "The French army, returning to its traditions", they proclaimed, "henceforth knows no law but the offensive," and Plan XVII adopted in May 1913, was the offensive writ large. On mobilization the overwhelming weight of the French armies would thrust forward in a massive attack either side of Metz, with the First and Second Armies to the south and the Third, Fourth and Fifth Armies to the northwest. The British Expeditionary Force was to concentrate north of Hirson on the French left, with almost nothing between it and the coast.

The authors of the rival plans were not blind to their opponents' intentions. Schlieffen had described the sort of offensive embodied in Plan 17 as "a kindly favour", which could be comfortably contained by the German forces opposite it. It would result in the French left being weak and would make the battle of encirclement easier. Moltke was less single-minded, and suggested that such an attack might justify modification of the plan by bringing forces down from Belgium to take the attacking French in the flank. The French, for their part, had some evidence that the Germans were planning a wide outflanking movement, but did not see how they could provide troops for it without using reserve formations in the front line, something which the French, with their mistrust of reservists, would not countenance. Intelligence reports which told otherwise, and officers who warned of a German offensive through Belgium, were ignored. If the Germans did manage to extend their flank into Flanders, it could only be by thinning their centre, where the French blow was to fall. "If they come as far as Lille,"; declared General de Castelnau, Deputy Chief of the General Staff, "so much the better for us!"

The mobilization of 1914 has passed into history as the last flare of popular euphoria in a world which was soon

It had been argued that the world's economic interdependence would prevent major war. Economies and financial institutions soon felt its impact, which lasted long after the fighting had ended.

to pass out of sight for ever. In the great capitals of Europe war was greeted with wild enthusiasm. "Whoever failed to see Paris, this morning and yesterday," wrote the poet Charles Péguy on August 3 (he would be dead before the month was out) "has seen nothing." Factionalism was swamped by what Paul-Marie de la Gorce has called "the

The German ironmaster Hermann Gruson began to experiment with cast-iron armour for forts in 1868, and went on to develop the armoured cupola. It was especially suitable for the subterranean forts built after 1870, and the Belgian engineer Brialmont fortified Namur, Antwerp and Liège with underground works protected by ditches and mounting guns in armoured cupolas. The disappearing cupola was widely used: it lay almost flush with the ground until raised to firing position.

Liège blocked the path into Belgium, and the Germans devoted much attention to the problem of dealing with its forts. Moltke thought them so badly sited that it might be possible to sieze Liège by a *coup de main*. His gunners were less sanguine, and asked Krupp for heavy guns to deal with them.

extraordinary national discipline" that existed in every European country. In France, President Poincaré's call for a *union sacré* was almost universally heeded; in Berlin, the Kaiser could say: "I do not see parties any more; I see only Germans." Not all actors in the drama were gripped by the same euphoria. A French sergeant noted that the young peasants in his barrack-room were "sick at heart", and in a Breton village the mayor's reading of the order for general mobilization produced only "a petrified dumbness". These dissenting murmurs were drowned by the overwhelming clamour for war.

Opening moves: the Western Front

The invasion of Belgium began early on August 4 when troopers of General von der Marwitz's cavalry corps trotted across the frontier 70 miles east of Brussels. Behind them three German armies, 34 divisions in all, prepared to make their mighty swing. Kluck's First Army was to take the outer flank, scything down through Brussels, on into Flanders and west of Paris. On its left, Bülow's Second Army and Hausen's Third were also to march through Belgium and into northern France. In the centre, forming up between Trier and Metz, were the 20 divisions of the Fourth and Fifth Armies, while the defence of Alsace-Lorraine was entrusted to the 16 divisions of the Sixth and Seventh Armies, from August 8 under the unified command of the Sixth Army's Crown Prince Rupprecht of Bavaria.

The fortress of Liège blocked the advance into Belgium, its forts covering the Meuse crossings and the gap between the Ardennes and Holland. A detachment from the Second Army, under General von Emmich, was given the task of

dealing with it. Emmich's first attempt ended in bloody repulse, but on August 6 Major General Erich Ludendorff, fortuitously on the scene, took over a brigade whose commander had been killed and pushed his men through the outer ring of forts, persuading the Belgian commander, General Leman, to abandon the citadel and take refuge in Fort Loncin. Ludendorff's achievement was to make him the first of the war's popular heroes and bring him to the notice of a high command already in need of reassurance. But it did not clear the way through Liège, whose forts still blocked the advance. The Germans now recognized that the forts were immune to the guns with their field armies. Two mighty 420mm siege howitzers were hastily entrained at Krupp's factory at Essen, though it was not until 12 August that they were in a position to fire. The 420s and the Austrian-made Skoda 305mm howitzers smashed the forts to fragments. Loncin was the last to fall, on August 16. Leman was pulled unconscious from its ruins; Emmich, with a chivalrous gesture, returned his sword to him.

There was little enough German chivalry in Belgium that summer. The Germans had a low opinion of the tiny Belgian army and hoped that diplomatic pressure would make the Belgians yield free passage. The popular and energetic King Albert had already declared that Belgium would defend herself whatever the cost, although he had little success in modernizing his army or imposing unity on his squabbling generals. Nevertheless, Belgium resisted German blandishments and fought harder than the Germans had ever expected. The result of the struggle could never be in doubt: Belgium was swiftly overrun. Antwerp held out, with British help, until October 9, and much of its

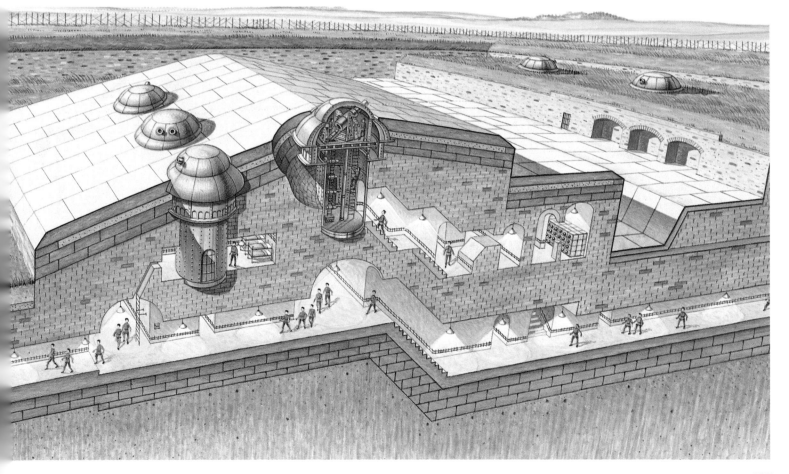

garrison broke clear. But Belgian resistance was not fruitless. It cost the Germans time, shook Moltke's confidence and provoked his soldiers into acts which redounded to their discredit.

Tales of German atrocities in Belgium were exaggerated, but they were founded on fact. Most Germans had not been in action before. They were surprised by Belgian resistance, and the shock and confusion of war fanned their worst instincts. Suspected snipers were shot out of hand and hostages were executed. Even Moltke admitted that: "Our advance in Belgium is certainly brutal." The small town of Andenne was burned and more than 200 of its inhabitants were shot; Tamines was sacked and nearly 400 of its citizens were killed. On August 25–26 part of the city of Louvain, including the library with its fine collection of medieval manuscripts, was burned. In part these acts reflected overreaction by excited young men; in part, though, they were evidence of deliberate use of terror to break an enemy's spirit. Their effects were far-reaching. Before the war was a month old, real bitterness had entered the spirits of many combatants, bitterness which helped fuel a determination to see the business through to the end and to shun a compromise peace.

As the Germans elbowed their way into Belgium, the French commander-in-chief, General Joseph Joffre, studied reports of their movements in *Grand Quartier Général* (GQG) at Vitry-le-François on the Marne, roughly equidis-

tant between his army headquarters. Joffre was not the exuberant and excitable Frenchman of Anglo-Saxon imagination, and his refusal to panic was to save France over the weeks that followed. But in early August his imperturbability was no real asset, for he was disinclined to pay much attention to events in Belgium. He sent General Sordet's cavalry corps lumbering off to find out what was happening, and got on with the preparations for his offensive.

The accelerated development of the popular media in preceding years led, in World War I, to the first great propaganda campaigns. This poster exhorts the public to "Put Strength in the Final Blow" (and vicariously bayonet a hated Hun) by purchasing War Bonds, one of the British government's several war loans schemes. Other posters of the period showed the enemy's alleged "atrocities", often in near-pornographic detail.

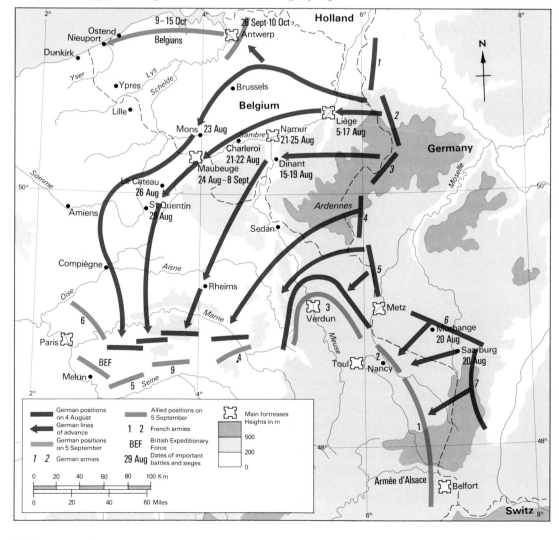

France's preoccupation with the offensive ensured that the main strength of her armies lay along the Franco-German border, with only the BEF and three French territorial divisions between the French 5th Army and the Channel. The wheeling German armies forced back the BEF and the 5th Army, but the Germans edged eastwards, and on August 31 Kluck's 1st Army swung in front of Paris rather than enveloping it to the west. This enabled the 6th Army to threaten the German flank, and paved the way for the battle of the Marne.

138

Its preliminary moves went well enough. On August 8 French troops entered Mulhouse in Alsace, only to be forced out two days later. Joffre duly reorganized his southern forces to create an Army of Alsace, which resumed the attack while, to the north, the French First and Second Armies set off against their first objectives, Morhange and Sarrebourg. Crown Prince Rupprecht's soldiers, obedient to the instructions of Moltke's *Oberste Heeresleitung* (OHL), gave ground before them, and Sarrebourg fell on the 18th. Things looked promising for the French, who could already scent the prospect of recovering the lost provinces and reaching the Rhine.

Trouble in the north

Not all Joffre's commanders shared his enthusiasm. The Fifth Army, on his left, was commanded by General Charles Lanrezac, a brilliant but caustic officer, who viewed German deployment in Belgium with alarm and begged to be allowed to swing his army north to meet it instead of attacking northeastwards into the Ardennes. Joffre let him move a corps to Dinant, but told him that his fears were groundless. On the evening of the 15th, however, even GQG could not ignore the reports of Germans crossing the Meuse at Huy and news that 1 Corps was hotly engaged at Dinant. Lanrezac was ordered to turn his main body to meet the German threat and to collaborate with the British Expeditionary Force under Sir John French,

coming up into line on his left, in doing so. His right flank was to support the Fourth Army, which was now to spearhead the attack into the Ardennes.

The picture presented to Moltke, as OHL moved up from Berlin to Coblenz on the Rhine on the 16th, was no clearer than that at GQG. The thrust into Lorraine tempted him to abandon the northern swing in favour of a shorter jab into the French flank, but he decided to persist with the original plan. He was less resolute when Rupprecht, tired of fighting a defensive battle against an enemy who seemed to be faltering, demanded permission to counterattack. Such a move was inimical to the spirit of the plan, but Moltke weakly let Rupprecht have his head.

The storm broke on the 20th and raged until the 24th in what became known as the battle of the Frontiers. In Lorraine, Rupprecht sent the First and Second Armies reeling back, although Foch, now commanding a corps in front of Nancy, fought with a tenacity which marked him out for greater things. In the centre, the Third and Fourth Armies lurched to destruction in the splendid and absurd fashion decreed by the 1913 Regulations. Lines of infantry, conspicuous in their red trousers and long blue greatcoats, went forward against machine guns, while the German artillery battered them mercilessly, its fire directed by spotter aircraft. Both French armies, far from gaining ground, were forced to retreat. On the French left, Lanrezac fought the battle of Charleroi against the Ger-

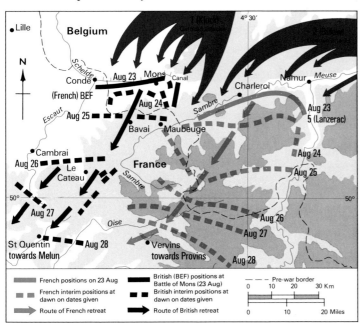

Sir John French's BEF was badly exposed on the flank of Lanrezac's 5th Army, but before his peril was fully apparent French gave battle at Mons. The ensuing retreat took the BEF through Le Cateau, where Sir Horace Smith-Dorrien of 2 Corps checked his pursuers, to Melun on the Seine.

"We dug little pits for ourselves – rabbit-scrapes really – lit a fag, and wondered what the Germans would look like when they arrived. There were a lot of fir trees about 400 yards away across the canal and then some higher ground beyond. Were they watching us? And then I suddenly remembered it was my birthday. And no one had sent me a birthday cake . . .

It was like sitting in a sack and not knowing what was going on outside except that, whatever it was, it was noisy and dangerous. The place was swarming with Germans and by mid-morning they had even got round behind us but not in any great numbers. At times you didn't know if you were firing at friend or foe. But in spite of our casualties, we managed to hang on. By noon we had lost all but one of our company officers. I remember feeling very hungry, and wished I was back home."

Private Tom Bradley, The Middlesex Regiment, Mons, August 23 1914

man Second and Third Armies. He was, as he had predicted, badly mauled, and fell back on the evening of the 23rd. On the same day the BEF fought the battle of Mons, giving Kluck's infantrymen a taste of what their comrades were meting out to the French elsewhere. Then, with the French Fifth Army falling back on its right and its left flank open, the BEF began a long retreat which would take it through Le Cateau, where General Sir Horace Smith-Dorrien's 2 Corps paused to administer a sharp rebuff to its pursuers, almost to the gates of Paris.

Joffre showed his stature in the days that followed. He issued a note to rectify the worst tactical errors of the previous week and remorselessly dismissed the hesitant or incompetent. He created the new Sixth Army, made up of troops from Lorraine, to take up position on the French left, and ordered his retreating armies to check the

Germans by rearguard actions and counterattacks. He did his best to encourage Sir John French, who got on badly with Lanrezac and whose mercurial personality and lack of confidence in the French inclined him to take the BEF out of the line altogether. Yet even Joffre could not stop the retreat, and on September 1 he ordered a withdrawal to the line Verdun – Bar-le-Duc – Vitry-le-François – Arcis-sur-Aube and the Seine through Nogent, with the BEF at Melun. The government left Paris on the 2nd, leaving the city's newly appointed military governor, General Joseph Galliéni, with instructions to defend it to the last.

A decisive counterattack

Moltke, who had moved OHL up to Luxemburg on August 29, wavered as Joffre remained firm. The crisis on the Eastern Front had persuaded him to send off two corps from his right wing, and Rupprecht's success in Lorraine dissuaded him from replacing them with troops from the Sixth and Seventh Armies. The three wheeling armies, their soldiers exhausted by marching and fighting, were depleted by casualties and the need to garrison Belgium and mask fortresses. Shortage of troops, a desire to pinch in the French flank and, above all, the absence of a strong directing will, all contributed to a gradual edging to the east, away from the coast – and in front of Paris.

The German turn gave Joffre his opportunity for a counterattack. The battle of the Marne, which began on September 6, was the decisive clash of the campaign. It was not a crisply planned or neatly executed manoeuvre, but a confused struggle between weary men. On the French left, Maunoury's Sixth Army – reinforced by troops from Paris, sent out in taxi-cabs by Galliéni – struck Kluck's flank.

Marshal Joseph Jacques Césaire Joffre (1852–1931) was born near Perpignan to a *petit bourgeois* family. A temporary engineer officer in the Siege of Paris he was commissioned into the regular army in 1871. After spending the first years of his career building fortresses in France, he asked for an overseas posting, and commanded an engineer company in Indochina before becoming chief engineer at Hanoi. Other colonial appointments followed: in 1893 he established French power at Timbuktu, and later served as Galliéni's chief engineer in Madagascar. He became Director of Engineers at the War Ministry in 1905, commanded a division in 1906, a corps in 1908, and became Director of Services of the Rear in 1910.

In 1911 he was the compromise choice for the newly merged posts of Chief of the War Ministry Staff and vice-president of the *conseil supérieure de la guerre – de facto* commander-in-chief on mobilization. Although he was slow to admit the bankruptcy of Plan XVII, his moral strength infused his army at its hour of need. In 1915 his authority was increased to include armies in all theatres, but he came under increasing criticism, and in December was replaced by Nivelle. He was not a great general, but in 1914 held a shaken army together when his "imperturbable calm and . . . rough good sense" were invaluable.

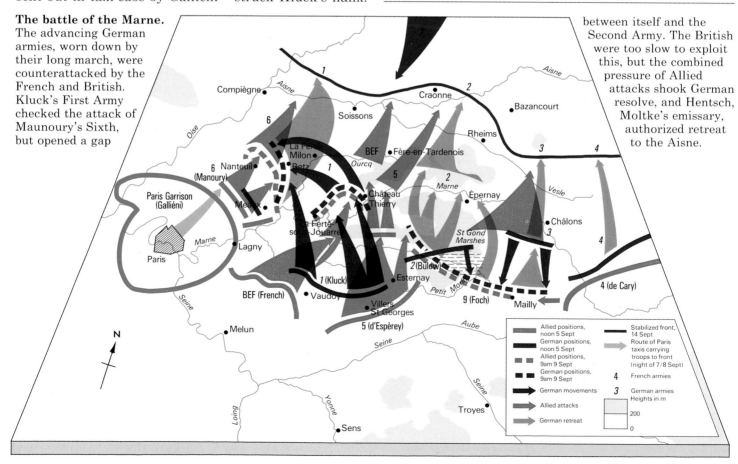

The battle of the Marne. The advancing German armies, worn down by their long march, were counterattacked by the French and British. Kluck's First Army checked the attack of Maunoury's Sixth, but opened a gap between itself and the Second Army. The British were too slow to exploit this, but the combined pressure of Allied attacks shook German resolve, and Hentsch, Moltke's emissary, authorized retreat to the Aisne.

	Allied positions, noon 5 Sept	Stabilized front, 14 Sept
	German positions, noon 5 Sept	Route of Paris taxis carrying troops to front (night of 7/8 Sept)
	Allied positions, 9am 9 Sept	
	German positions, 9am 9 Sept	**4** French armies
	German movements	**3** German armies
	Allied attacks	Heights in m
	German retreat	200 / 0

Kluck turned to meet the attack and checked it, but a gap opened between the German First and Second Armies. The BEF, induced to attack by Joffre's passionate entreaty, was too slow to exploit this, but Bülow was alarmed by the threat to his flank, and told Moltke so. The French Fifth Army, now under the energetic Franchet d'Esperey, advanced across the Petit-Morin, and on its right Foch's new Ninth Army attacked vigorously.

The Germans retreat

On September 8 Moltke, staggering under the weight of his responsibility, sent Lieutenant Colonel Hentsch to visit the army headquarters on his behalf. There is some doubt as to how wide Hentsch's powers were, but it is certain that he was already pessimistic and that on an exhausting drive along roads in the rear areas of fighting armies he saw sights which made him gloomier still. He concluded that a general retreat was the only solution and ordered a withdrawal. It was the ultimate irony of the first month's fighting that the German failure on the Marne was essentially moral rather than material. The battle was still in doubt: either the reinforcement of the right wing or the encouragement of Rupprecht, tantalizingly close to a breakthrough, might still have swung its balance in Germany's favour. But Moltke was a sick and demoralized man and his defeatism pervaded OHL. Joffre, whatever his intellectual limitations, remained robustly confident, and something of his determination was transmitted, through commanders whose will he dominated, to the troops who achieved what the French were not unjustly to call "the miracle of the Marne".

The Germans fell back to the Aisne on September 9–13 in

SPY MANIA

During the war, espionage was common, although it was probably less prevalent than contemporaries imagined. Colonel Nicolai, head of German intelligence, complained: "Strong scepticism prevailed in the army commands regarding the possibility and the usefulness of espionage in mobile operations." Nevertheless, in 1914 fear of espionage produced a spy mania. Lieutenant Spears, a liaison officer with the French, wrote that "at least twenty people" were shot at Rheims in mid-September. A French officer excused this, saying: "Justice has little to do with it. Our duty is to see that no spy escapes, whatever the cost may be." Some civilians were shot out of hand. Corporal Ashurst saw an NCO shoot a civilian, explaining "that he had caught the civilian telephoning to the enemy from the cellar . . ." Security became obsessive: the French were ordered "not . . . to allow their movements or operations to be watched by civilians," and a British poster (*below*) warned against careless talk.

LEAKAGE of INFORMATION

EXTRACT FROM ARMY ROUTINE ORDER.

It is forbidden, except in the course of duty, to **DISCUSS OR REFER TO ANY MOVEMENT OF TROOPS,** or to the situation of any body of troops, or to operations of any kind whatsoever.

Where evidence is forthcoming that any Officer, Warrant Officer, Non-Commissioned Officer or Private Soldier has disobeyed this order, he will be tried by Court-Martial.

The first trenches of the war were primitive affairs, but by late 1914 they had features which were to become standard. The garrison of this German trench are standing on the firestep, looking over a parapet reinforced with sandbags and revetted with hurdles. Duckboards cover a sump in the bottom of the trench behind them.

British "walking wounded" (*right*) make their way back from the line after receiving attention at an advanced dressing station. Medical services in the field had improved considerably by World War I, but increased firepower led to a high casualty rate: overall, an estimated 8 million men were killed in action and more than 21 million wounded.

Field Marshal Sir John French (1st Earl of Ypres) (1852–1925), served in the navy before being commissioned into the 19th Hussars in 1874. He first saw action with the Gordon relief expedition in 1884, and in 1888 he took command of his regiment. His career faltered after a scandal caused by his vigorous love life, but in 1897 he was given command of a cavalry brigade. During the Boer war, victories at Elandslaagte and Klip Drift made his reputation. In 1912, although he had not been to Staff College, he became Chief of the Imperial General Staff. During this period he was well to the fore in the tactical debate, arguing that cavalry should be trained and equipped for shock action.

Promoted field marshal in 1913, he resigned as CIGS following the Curragh incident in March 1914, but was employed to command the BEF in August. The failure of the French plan and the exposed position of the BEF made his task difficult, and his mercurial personality and awkward relationship with Kitchener and the French did not help. In 1915 his offensives proved costly failures, and working up a political row over ammunition shortages worsened his plight. He was succeeded by Haig in December, and served first as commander-in-chief Home Forces and later as Lord Lieutenant of Ireland. French was temperamentally unsuited for high command and never really grasped modern war, but has been well described as the most distinguished English cavalry leader since Cromwell.

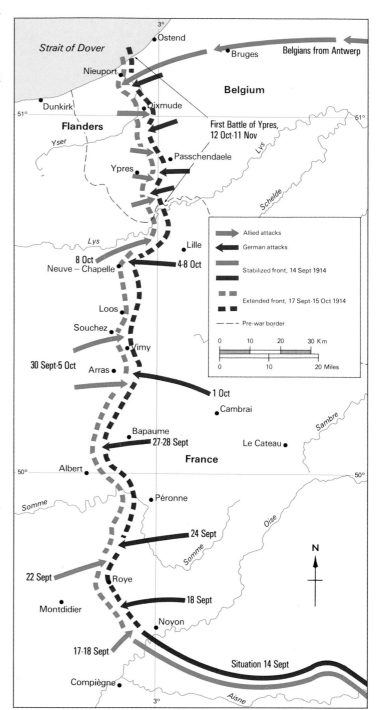

weather whose grimness matched their mood, and dug in on the slopes overlooking the river. When the Allies came up to attack them on the 13–14th the fighting gave an early demonstration of what field fortifications and machine guns could achieve. Sir John French recognized that the character of the war was changing fast. "I think the battle of the Aisne is very typical of what battles in the future are most likely to resemble," he told the King.

"Siege operations will enter largely into the tactical problems, the *spade* will be as great a necessity as the rifle, and the heaviest calibre and types of artillery will be brought up in support of either side."

"The Race to the Sea"

French found the prospect unappealing and, with his cavalryman's instinct, yearned for the open flank on the Allied left. Joffre agreed to the move. Both sides were already edging towards the sea, like two wrestlers struggling for a foothold, and when the British arrived in Flanders in early October, hoping to hook around the open flank, they collided with German troops bent on exactly the same task.

The result was a bitter battle around the little Belgian town of Ypres. It lasted until late November, and there were moments, the most desperate on October 31, when the line between the Germans and the Channel ports was wafer-thin. It held, albeit at frightful cost to the BEF, and as winter set in the Western Front froze into immobility.

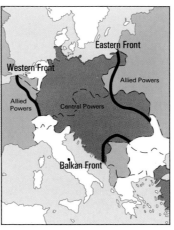

After the battle of the Marne and the German withdrawal to the Aisne, the rival armies extended northwards in an abortive search for an open flank. The manoeuvre became known as "The Race to the Sea", and it resulted, not in the outflanking battle that each side hoped for, but in the gradual establishment of a trench line running to the North Sea.

The Belgians held the low-lying ground along the Yser on the extreme left, with French units linking them to the BEF in the Ypres salient. From the British right, around Armentières, the French line ran due south towards Soissons, whence it swung away between the Marne and Aisne to Verdun, running back to Saint-Mihiel, to the pre-war border in southern Lorraine and on to the Swiss border south of Mulhouse.

Five months' fighting had cost the adversaries dear. The French army had lost 955,000 men, the Germans rather fewer than 900,000. The BEF's casualties – 90,000 by the end of November – had fallen heavily on the old regular army: in the battalions which had fought on the Marne and at Ypres, there remained, on average, only one officer and thirty men who had landed in August. Reputations, too, had perished. Moltke was replaced by General Erich von Falkenhayn on September 14, although he stayed on at OHL until November 3. Joffre had disposed of many French generals: among the coming men were Foch, now commanding the northern group of armies, and Pétain, a passed-over colonel when war broke out, now commanding a corps in front of Arras. French remained at the head of the BEF, although he had already attracted criticism and there had been a stormy interview between him and Lord Kitchener, Secretary of State for War, in Paris on September 1. The most notable casualty was the hope that it would all be over by Christmas.

Opening moves: the Eastern Front

Germany had long expected to deal with France before Russia could intervene and her deployment in August 1914 reflected this conviction. Lieutenant General Max von Prittwitz und Gaffron's 135,000-man Eighth Army held the Eastern frontier. The speed of Russia's mobilization, and her determination to honour her agreement with the French and launch an offensive as quickly as possible, threw German calculations out. The Grand Duke Nicholas, appointed commander-in-chief on August 2, lost no time in moving to his field headquarters at Baranovichi on the Moscow-Warsaw railway line, and General Jilinsky's North-Western Army Group, 650,000-men strong and made up of General Rennenkampf's First Army and General Samsonov's Second Army, prepared to advance. Rennenkampf was to make for the 30-mile wide Insterburg Gap between the defences of Königsberg and the Masurian Lakes; he would engage the Germans there while Samsanov moved up south of the lakes against the German right flank and rear.

The Russian sledgehammer was less awesome than it seemed. The Grand Duke's *Stavka* was hastily cobbled together and the friction between its members bore testimony to the Russian army's factionalism. Senior commanders mirrored this discord: Rennenkampf and Samsonov got on especially badly. Russian forces on the East Prussian front were weakened by a last-minute attempt to mount another offensive, through Poland; by insistence on fully garrisoning fortresses; and by detaching divisions to cover the flanks of the attacking armies. These errors whittled down Russian strength: the Second Army's 304 battalions, 111 cavalry squadrons and 1,160 guns were reduced to 188 battalions, 72 squadrons and 738 guns. This gave it a bare superiority over the German Eighth Army's 158 battalions, 78 squadrons and 774 guns,

which was inadequate for a full-blooded offensive, especially in view of the fact that for several days the two attacking armies would have a 60-mile gap between them and thus could not count on mutual support. Russian logistics were overstretched and the change of railway gauge at the frontier did not help matters. The supply of artillery ammunition was to cause particular problems. The Russians, like all combatants in 1914, had underestimated the amount of ammunition their guns would fire, and although the ammunition available to Jilinski's gunners seemed lavish by peacetime standards, it was to prove inadequate, often because it could not be got forward to the batteries that needed it.

On August 15 the First Army crossed the border into

Captured Russian troops drag their machine guns into captivity under the eye of a German guard.

East Prussia. Prittwitz decided to deal with Rennenkampf before Samsonov's thrust had developed, and Samsonov was not to reach the border until the 20th. Three and a half German corps moved up to Gumbinnen to face Rennenkampf, while the fourth went south to watch Samsonov. Prittwitz was blessed with independently-minded subordinates, and one of them, General von François of 1 Corps, went beyond Gumbinnen: on August 17 he mauled a Russian division at Stallupönen, despite orders not to fight. He fell back to Gumbinnen, whence he pressed Prittwitz for permission to counterattack. Prittwitz was undecided. The bulk of his army was in a solid defensive position behind Gumbinnen, on the line of the Angerapp River. But the plight of the civilian population, exposed to Russian invasion, weighed heavily on him, and he knew from intercepted radio messages – Russian communications were woefully insecure – that Rennenkampf had been ordered to halt while Samsonov caught up. News that Samsonov's men had crossed the frontier made up Prittwitz's mind: the opportunity of dealing with Rennenkampf would never be better.

A temporary setback

The battle of Gumbinnen, fought on August 20, went badly wrong. On the left, François cut up yet another Russian division, but General August von Mackensen's 17 Corps, just up from the Angerapp, lost 8,000 men and dragged back in its retreat 1 Reserve corps on its right. Prittwitz was shocked by the battle, and the report from 20

Corps in the south that Samsonov was on the move depressed him further. He decided to break off the action and retire behind the Vistula, a move which meant giving up the whole of East Prussia.

Prittwitz telephoned Moltke at Coblenz to tell him the news, adding that, as the waters of the Vistula were low, he might not be able to hold the river-line. Moltke, his mind bent to the problems of the Western Front, was horrified and, after his staff had managed to talk to the Fifth Army's corps commanders individually, he determined to replace Prittwitz and his chief of staff by men he could trust. As chief of staff Moltke selected Ludendorff, who left by train for the east on the evening of August 22. He had already studied the situation and issued preliminary orders. These were very much in accordance with the solution already devised by Colonel Max Hoffmann, the Eighth Army's deputy chief of operations and the sharpest brain on that army's staff. François was to break off the battle at once and move south to support 20 Corps, while Mackensen and Below regrouped and followed. The move was a complex one, but precise staff work and the good frontier railway system made it possible. The choice of commander for the Eighth Army was more difficult: Moltke eventually selected General Paul von Beneckendorff und Hindenburg, who had retired as a corps commander in 1911 at the age of 65. Hindenburg was not as bizarre a choice as he might seem at first sight. He was no older than the three right-wing army commanders in France, his family came from Prussia and he knew the area well from his days as a staff officer at Königsberg.

Hindenburg and Ludendorff arrived in East Prussia on

Hindenburg and Ludendorff.

August 23 to find German redeployment under way and Samsonov's leading elements in action against 20 Corps. Hindenburg and his staff travelled down to visit the corps commander on the 24th and decided to throw their full weight against Samsonov. Radio intercepts, which suggested that Rennenkampf would pose no threat in the north and revealed that Samsonov proposed to press on and finish off 20 Corps, confirmed the German command in its decision, although August 25 was a day of tension at the Eighth Army's headquarters. The strain had already taken its toll at OHL, and on the 25th Moltke decided to send reinforcements to the east. Ludendorff, aware of the need for manpower in the west, doubted if the reinforcements were needed at the moment, but two corps and the 8th Cavalry Division duly set off.

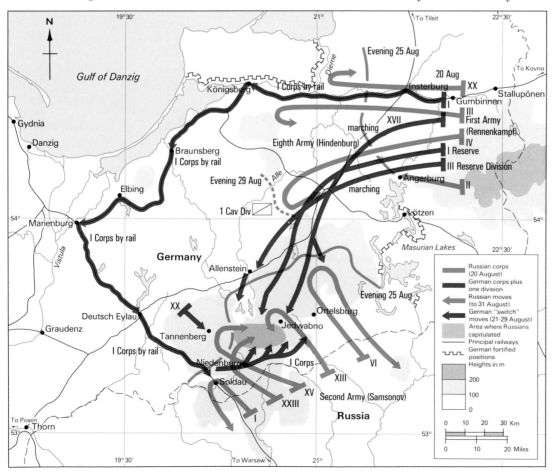

The German plan for Tannenberg was frightening in its simplicity. Radio intercepts had told the Germans that the advancing Russian armies would lack mutual support for some days, and the Germans planned to capitalize on this by concentrating against Samsonov's Second Army southwest of Tannenberg before Rennenkampf's First Army could intervene. Slick staff work and the comprehensive frontier railway system enabled the Germans to redeploy successfully, and Samsonov's army was cut to pieces. The Russian 23, 15 and 13 Corps, which had penetrated most deeply into East Prussia, were almost totally destroyed.

By the time they arrived the need for urgency had passed. On August 26 the Germans began their attack on the Second Army, and over the next five days, in the sand and pine forests around Neidenburg and Ortelsburg, the Russians were first fought to a standstill and then enveloped. Samsonov shot himself and about 100,000 of his men were captured: more than 30,000 died. It is easy to read too much into the battle of Tannenberg. The Germans deserve credit for their crisp redeployment, but the concept of the battle was simple enough and had been widely discussed before the war. Ludendorff's attempt to encircle Rennenkampf, which led to the battle of the Masurian Lakes in the second week of September, showed that it was difficult to repeat Tannenberg against an opponent who behaved with even moderate competence. The First Army was bruised but not crippled, and as the Germans pushed on across the Russian border they ran into a solid counteroffensive that brought them to a halt.

The Russians had little reason to congratulate themselves on their performance against the Germans in the north. In the south, however, they were considerably more successful. Their adversary there was the Austro-Hungarian army, smaller and less well-organized than the German, but, like its ally, already fighting on two fronts – against Russia and Serbia – and shortly to fight on a third, against Italy. The Austro-Hungarian army has received its fair share of criticism. It was too small and too ill-equipped for its wartime tasks, a fact which its general staff was slow to recognize; it was scarred by the problem of nationalities which characterized the empire itself; and it was the product of what was the least militarized of the Great Powers. Nevertheless, as Gunther Rothenberg has observed, it had "a truly remarkable capacity to absorb terrible casualties and recover from great defeat".

This capacity was well tested in the first months of the war. Conrad was eager to deal with Serbia first, and the early stages of mobilization were plagued by order and counter-order as he reluctantly recognized that Russia was the more serious foe. He finished with the worst of both worlds. The main Austrian concentration in Galicia was delayed and an inadequate force was dispatched against the Serbs, who soundly thrashed it in the battle of the Jadar in mid-August. Even when the concentration was complete, Conrad was uncertain what to do with it. He decided to mount a major offensive in the north, towards Lublin, with his First and Fourth Armies, while his Third Army moved eastward towards Tarnopol. Neither high command emerged with much credit from the fighting that swayed across Galicia over the next two months. Austrian successes in the north induced Conrad to let the Russian Third and Eighth Armies, which had driven back his own Third Army, advance towards Lvov (Lemberg), in the hope of taking them in the flank with the Fourth Army from the north. He was wrong to attempt map-board manoeuvres with troops who were already tired, and wronger still to risk sustained fighting around Lvov against Russians whose reserves far outnumbered his own. On September 11 he decided on a general retreat, and by the middle of the month his armies were back behind the rivers Dunajec and Biala. They had lost 400,000 men to the Russians' 250,000, and with them Galicia, although the fortress of Przemysl still held out.

Conrad had already asked for German help, and on

September 18 he met Ludendorff, who was under political pressure to support the Austrians and was also concerned at the Russian threat to Silesia. The reorganization of German forces in the east, under Hindenburg as commander-in-chief, added Mackensen's Ninth Army to the Eighth, now under General von Schubert. The first German drive was inconclusive. Mackensen came close to capturing Warsaw, but was driven back by a Russian counterattack in late October. In November a second attempt, with the Eighth Army's assistance, brought better results, and Lodz was taken. To the south, the Austrians flowed back into Galicia, briefly relieving Przemysl. After more heavy fighting around Lvov, the line stabilized along the Dunajec and Biala, through the central Carpathians and on through central Poland to the borders of Russia.

By Christmas 1914 a decision in the east seemed as distant as one in the west. But there were important differences between the two fronts. The sheer size of the Eastern Front and the harshness of much of its terrain meant that a continuous line took longer to solidify than it did in the west – and where the crust was as hard, it was rarely as deep. For all the massive size of the armies engaged, troops were thinner on the ground, and this, together with a shortage of defence stores, tended to make positions shallower than in the west: breakthroughs were easier and mobile operations more feasible. On both fronts, but especially in the west, it was evident that in order to fight the mobile war which alone could end the stalemate, a solution had to be found to the problem of barbed wire and machine gun: the locked front had to be opened.

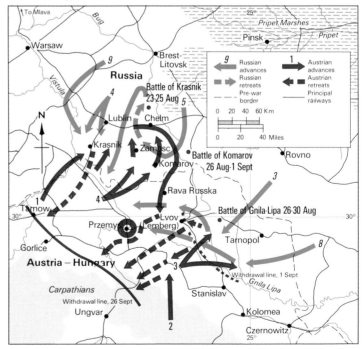

The Austrian First and Fourth Armies advanced on Lublin and initially made good progress. The Austrian Third Army was less fortunate east of Lvov, and was driven back by superior Russian forces. Elements of the Austrian Fourth Army, moving south from Lublin in an effort to take the advancing Russians in the flank, were defeated at Rava Rosska, and the Austrians withdrew from Galicia.

— chapter eleven —
— *World War I: Breaking the fetters* —

Richard Holmes

On January 2 1915 Lord Kitchener told Sir John French ". . . we must now recognize that the French army cannot make a sufficient break through the German lines of defence to bring about the retreat of the German forces from northern France. If that is so, then the German lines in France may be looked upon as a fortress that cannot be carried by assault, and also cannot be completely invested."

This assessment begged two crucial questions. First, was the German line really impregnable? This may have seemed clear to Kitchener, but it was less obvious to British commanders in France. Having set their hand to the task, they were reluctant to admit that it was beyond them, and personal ambition added another dimension to their judgments. As for the French, the pressures against acceptance of the status quo were massive and irresistible. The Germans occupied a broad swathe of territory containing much of France's industry and raw materials. Paris lay only just outside the German grasp. As George Clemenceau, elder statesman of the Left, reminded Frenchmen daily: *"Messieurs les Allemands sont toujours à Noyon."*

Even had the Allies been prepared to agree that the prospects of victory in France were poor, where else should they look? Opinions varied: Turkey, Italy and Salonika were but three of the options. Historians have tended to portray the Allied strategic debate as a dispute between "Westerners", who hoped for a decision on the Western Front, and "Easterners", who looked elsewhere. In fact there was more common ground than these labels suggest, and what was at issue was not whether there should be a Western Front, but the degree of emphasis accorded to it. The debate was sharpened by the fact that Easterners were often politicians, like Winston Churchill and David Lloyd George, dissatisfied by the way the war was conducted in the west. The Westerners were dominant among soldiers, who pointed out that the bulk of the German army was in France and would not be beaten by slick manoeuvring elsewhere. The dispute was partly a reflection of wider friction between soldiers and politicians. World War I, with its prodigious demands on national resources, so far outstripped what had gone before as to provoke a crisis in civil-military relations, especially among the Allies. War, as Clemenceau was to say, was too important to be left to the generals: but the generals resented interference in a sphere which they had long dominated.

Mud, blood and iron: the Western Front 1915

The question of the conduct of the war on the Western Front is itself complex. The numbing casualty lists, the dreadful conditions, and the gulf – often social as well as spatial – between senior commanders and their staffs and the men who did the fighting, provoke an understandable revulsion. Yet the school of historiography which depicts generals as feather-brained incompetents living securely in their chateaux glosses over some important points. Most generals who fought the war, whether for the Allies or the Central Powers, lacked recent relevant military experience. The Franco-Prussian War or colonial soldiering was as much as French or German generals could boast. The British had the Boer War, but the lessons of the veldt were little use in the hard school of Flanders. It was difficult to predict how weapons would affect tactics at a time of rapid technological innovation. The radio was in its infancy and generals depended on telephone and dispatch rider. Commanding from a nodal point where communications were

centralized and plans could be made without enemy interference was certainly not foolish. Nor was it cowardly. The British army lost more than 60 officers of the rank of brigadier general and above on the Western Front. There was some incompetence and much refusal to recognize the point at which resolution became obstinacy, but a good deal more honest endeavour to solve problems which had no easy answers. Yet there was too much politicking, especially among the various Allied commanders, and too many generals forgot that soldiers fight best for commanders they know and trust.

The central issue for the Allies on the Western Front was how best to break the German line and then to exploit the breakthrough. The Germans were not under the same obligation to attack: the status quo was not the national disaster to Germany that it was to France, and the poor performance of the Austrians meant that the Eastern Front was often a more acute concern. Furthermore, pre-war German doctrine had stressed the danger of frontal assault, and the creation of the locked front meant that no flank was available. For much of the war, therefore, the Germans were content to stand on the defensive in France, using limited attacks to keep the Allies in play. They were prepared to relinquish ground which was not tactically important and to take pains over the fortification of ground that was: German positions, with their growing emphasis on defence in depth, had a permanent character which Allied trenches lacked.

The capture of Neuve-Chapelle

The first attempts to break the German line were scarcely revolutionary. As Sir John French wrote in January 1915: "Breaking through the lines is largely a question of expenditure of high explosive ammunition. If sufficient ammunition is forthcoming, a way out can be blasted through the line." In March 1915 Sir Douglas Haig's First Army took the village of Neuve-Chapelle, his infantry advancing on a narrow front behind the heaviest bombardment yet delivered by British gunners. Neuve-Chapelle foreshadowed much of what was to come. First, shellfire did so much damage to the ground that the infantry found it hard to make their way forward. Second, it was impossible to get cavalry – the arm of exploitation and pursuit – through the crowded back areas and across the battlefield while the gap was still open. Horsed cavalry was not to gain much credit from the war, but we must acknowledge that for most of the time it was the only means of exploitation available. Third, the fact that the offensive was mounted on a narrow front enabled the Germans to strike at its flanks. Fourth, once the breakthrough had been identified, it was swiftly contained by reserves rushed to the threatened sector by railway. Finally, the limited British success consumed 100,000 shells, one-fifth of the BEF's total ammunition supply.

Expenditure of artillery ammunition in the first months of the war far exceeded expectations and, as commanders came to recognize the central role of artillery, the question

—TIME CHART—

1915		
Mar 10	British take Neuve-Chapelle	
Apr 22	Germans use gas at second battle of Ypres	
Apr 25	Large-scale Allied landings at Gallipoli	
May 23	Italy declares war on Austria	
June	Britain establishes Ministry of Munitions	
Sept 25– Oct 8	Artois offensive: battle of Loos	
Oct 5	Franco-British expeditionary force to Salonika	
Oct 6	Austro-German-Bulgarian invasion of Serbia	
Oct 13	Mass Zeppelin raid on London	
1916		
Jan	British introduce conscription	
Feb 21	Germans begin Verdun offensive	
May 31	Naval battle of Jutland: tactical draw	
June	Arab revolt against the Turks	
June–Sept	Russian Brusilov offensive drives back Austrians	
July 1	Allies mount offensive on the Somme	
Sept 15	British tanks used on the Somme	
1917		
Feb-Apr	Unrestricted German U-boat campaign	
Mar 12	Russian Revolution	
Apr 6	USA declares war on Germany	

Apr 16–20	French Nivelle offensive: mutinies in French army follow
June	Allies introduce convoy system
July 31	Third battle of Ypres (Passchendaele) begins
Oct 24– Nov 12	Austrians and Germans defeat Italians at Caporetto
Nov 7	Bolsheviks seize power in Russia
Nov 20	British use tanks at Cambrai
Dec 9	British take Jerusalem from Turks
Dec 15	Russo-German armistice signed
1918	
Mar 21	German Michael offensive opens on the Somme
April 9	Germans launch Georgette offensive on the Lys
May 27	German offensive on the Chemin des Dames
July 18	Allied Aisne-Marne counteroffensive begins
Oct 5	British breach Hindenburg line
Oct 24	Italians defeat Austrians at Vittorio Veneto
Nov 11	Germans conclude an armistice
1919	
June 28	Treaty of Versailles: German army limited to 100,000 men
1920	Russo-Polish War: Russians concede territory
1920–22	Greco-Turkish War: Greeks ousted from Asia Minor

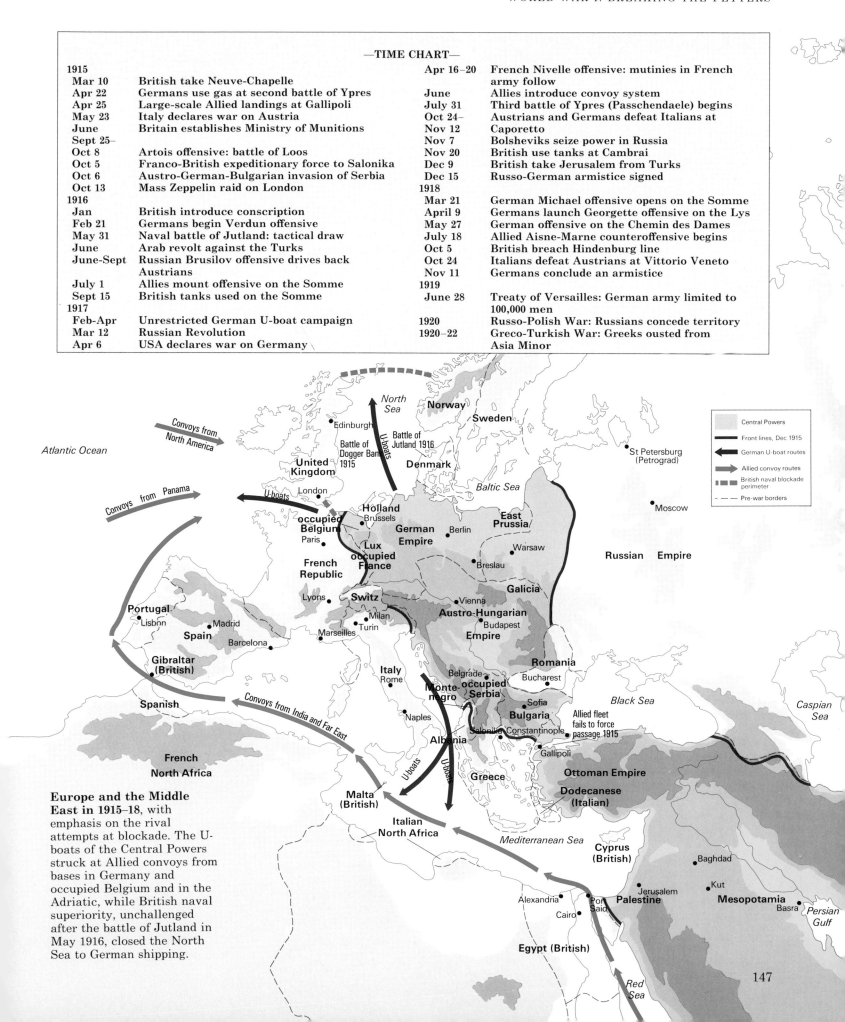

Europe and the Middle East in 1915–18, with emphasis on the rival attempts at blockade. The U-boats of the Central Powers struck at Allied convoys from bases in Germany and occupied Belgium and in the Adriatic, while British naval superiority, unchallenged after the battle of Jutland in May 1916, closed the North Sea to German shipping.

147

of ammunition supply loomed ever larger. Output was stepped up and procurement streamlined: in June 1915 Britain established a Ministry of Munitions, largely as a response to French's well-publicized complaint that lack of shells had deprived him of victory at Aubers Ridge on May 9. Allied domestic industry was unable to cope, and American factories made their own contribution to the war effort long before the United States entered the conflict. The quantity of shells fired was staggering. Even in 1914 the French army needed 100,000 each day. The British fired 1,723,873 shells in the preparatory bombardment on the Somme in 1916, and 4,282,550 in the early stages of the third battle of Ypres a year later. The latter represented a year's work by 55,000 munitions workers, underlining the effect that war production had on industry. With most fit young men away at the front, it was only by the increasing use of women in jobs hitherto restricted to men that the combatants were able to sate the hungry guns.

Plans for an autumn offensive

The British attack at Aubers Ridge was part of a combined offensive, with the French attacking on a nine-mile front north of Arras. Early gains proved delusive and the fighting dragged on until mid-June. The heavy casualties of the Artois offensive bruised the confidence of some French generals, one of whom warned President Poincaré that "the instrument of victory is being broken in our hands". Nevertheless, an Allied conference at Chantilly in June decided that a passive defence in the west was out of the question, and Joffre and Foch discussed their next move. The arrival of the British Third Army enabled the BEF to take over more of the line, and French, under pressure to cooperate more effectively, agreed to attack at

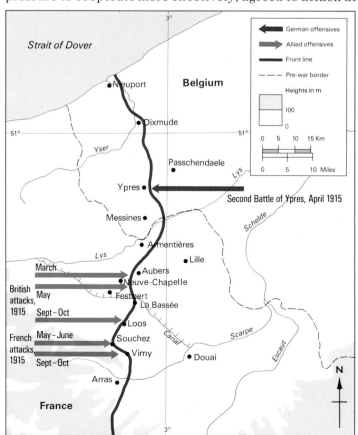

Second Battle of Ypres, April 1915

In early 1915 (*left*) French offensives foundered in Champagne and at St-Mihiel. The British attack on Neuve-Chapelle was to have been combined with a French attack at Vimy, but the Champagne battle prevented this. In May the Allies attacked simultaneously at Festubert and Souchez. In September their attacks north of Arras coincided with another French offensive in Champagne.

This British 6-inch gun, (*above*) in action on the Western Front in 1916, was derived from naval guns on improvised field carriages used during the Boer War, and was already obsolete. The gunner to the left foreground is setting fuses.

Loos as part of a two-pronged Allied offensive in late September. The French Tenth Army was to attack farther south in Artois, but the main blow was to be delivered in Champagne by the Second Army, whose new commander, Pétain, had been responsible for the initial successes in Artois in May.

The British had two useful cards to play. The first was gas. On April 22 the Germans had used chlorine gas in an attack north of Ypres. They were surprised by their own progress and were unable to exploit it properly: the second battle of Ypres resulted in heavy losses on both sides and ended with the British holding a much-reduced salient. The German example was quickly followed, the British planning to use gas in their attack at Loos. The other asset was the arrival in France of the first units of the New Armies, the enthusiastic volunteers who had flocked to the colours in 1914 in response to Kitchener's appeal for troops. Two New Army divisions, in 11 Corps, were on hand to support the attack.

A costly attack

The offensive, launched on September 25, proved sterile. The British made some progress, although their own gas caused them casualties, but 11 Corps arrived too late to widen the gap and lost 8,000 men in less than four hours. Pétain's offensive in Champagne was more lavish. He attacked with 20 divisions on an 18-mile front; his gunners fired nearly 5,000,000 shells. The French broke into the German first line and took 25,000 prisoners, but supporting positions on the reverse slopes of the open ridges were immune to the bombardment, and the scanty gains cost 145,000 men. As the second winter of the war set in, the situation on the Western Front was little different to that a year before. Some expensive tactical lessons had been learned: the Allies recognized that without proper artillery preparation an attack had no chance of success, and the Germans were improving defence in depth, which enabled a lightly held front line to delay attackers before counterattacking units moved up to restore the situation. Sir John French had been replaced by Haig, and Sir William Robertson, the BEF's Chief of Staff, was appointed Chief of the Imperial General Staff and was to have an important influence in British counsels.

Fettered to a corpse: the Eastern Front 1915

The fighting in the east in 1914 and the dangerous Russian offensive in the Carpathians in the early spring of 1915 convinced the Germans that they were "fettered to a corpse" in their alliance with the Austrians. In 1915 German strategy was dominated by "the Austrian emergency", and by remaining on the defensive in the west the Germans were able to throw more weight against the Russians. The arrival of eight German divisions from the Western Front, where they had learned invaluable lessons about the use of artillery and inter-arm cooperation, had a dramatic effect. On May 2 the Germans opened the battle of Gorlice–Tarnow and virtually destroyed the Russian Third Army, taking 150,000 prisoners and 125 guns. A month later Przemysl, taken by the Russians only in March, was recaptured, and by the end of June the Russians were back on the River Bug.

This substantial victory gave Falkenhayn pause for thought: like Schlieffen before him, he was alarmed at the prospect of burrowing ever deeper into Russia without reaching a decision. Nevertheless, after a conference at Posen on July 1, the offensive went on, but now

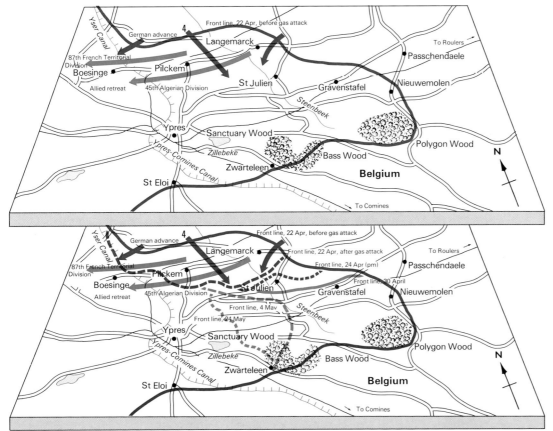

On the evening of April 22 clouds of chlorine gas rolled towards the two French divisions between Langemarck and the Yser Canal. The gas opened a gap over four miles wide in the Allied line, but the attacking German Fourth Army was insufficiently prepared to exploit its success. In ill-coordinated fighting, in which the Canadians suffered very heavily, the British failed to regain the lost ground. Smith-Dorrien, commanding the British Second Army, recommended withdrawal, only to be relieved of his command and replaced by the steady and methodical Plumer. Under Plumer's direction the exposed nose of the salient was given up, the British falling back to a line closer to Ypres.

149

Mackensen's Army group swung up from Galicia towards Brest-Litovsk, while the Twelfth Army thrust into Poland and the Eighth, Tenth and Niemen Armies pushed deep into Russian territory in the north. By the end of September the Russians had fallen back across the whole front. The campaign had restored the situation in the east and taken the pressure off the Austrians, but the Russians had managed to avoid being caught in a battle of encirclement. On September 1 the Czar himself took over from the Grand Duke Nicholas as commander-in-chief, with the hard-working and honest General Alexeyev as his chief of staff. The new team achieved an early success by mauling an ill-conducted Austrian offensive around Lutsk.

The Dardanelles

If the Germans felt moral and practical obligations to Austria, the Western Allies were no less obliged to Russia, and even before the retreat of 1915 the Russians had called for their assistance. In December 1914 the Grand Duke Nicholas requested Britain to make a "demonstration" against Turkey. He was concerned about the losses suffered on the Eastern Front and the shortage of artillery ammunition, and by a Turkish advance in the Caucasus. His appeal arrived at a time when several Allied leaders were already disillusioned with the Western Front, so that sending aid to Russia seemed attractive. It was ironic, though, that by the time the Grand Duke's request was received in London the Turkish army in the Caucasus was already in catastrophic retreat.

The Allied expedition, grandly conceived as an attack on Constantinople, got under way painfully slowly. On February 19 British warships shelled the outer forts at Gallipoli and a few marines landed unopposed. Almost a month later

Allied warships entered the Narrows and resumed the bombardment, but three battleships were sunk by mines and the British Admiral de Robeck lost heart. The first large-scale landings by General Sir Ian Hamilton's Middle East Force did not take place until April 25, by which time

"I was standing on the bridge in the evening when the Medjidieh *[steamship carrying wounded from the battle of Ctesiphon] arrived. She had two steel barges, without any protection from the rain, as far as I remember. As this ship, with two barges, came up to us, I saw that she was absolutely packed, and the barges too, with men. When she was about 300 or 400 yards off it looked as if she was festooned with ropes. The stench when she closed was quite definite, and I found that what I mistook for ropes were dried stalactites of human faeces. The patients were so huddled and crowded together on the ship that they could not perform the offices of nature clear of the edge of the ship, and the whole of the ship's side was covered with stalactites of human faeces."*

Major R Markham Carter, November 1915

strategic surprise had been lost. British troops seized Cape Helles but failed to capture the dominant Achi Baba hill, and the Anzacs (Australian and New Zealand Army Corps) wrested a toehold at Anzac Cove. The French put a brigade ashore at Kum Kale, on the Asiatic side, but it was withdrawn the next day and French troops were used alongside the British at Helles. A second landing on August 6 pushed inland from Suvla Bay and increased the gains at Anzac Cove, but the high ground of Sari Bair and Tekke Tepe remained in Turkish hands.

The landings themselves, bravely carried out under

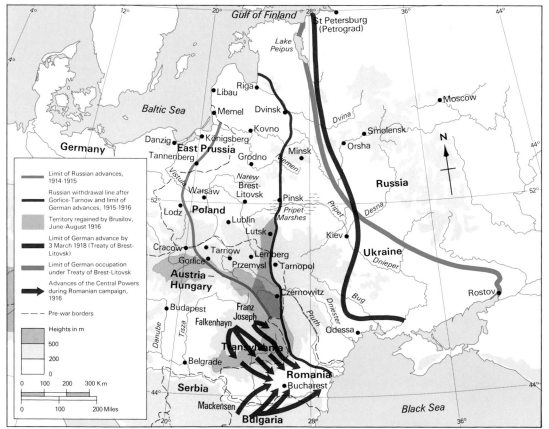

Austrian reverses in 1914 compelled the Germans to send more troops to the east, and in May 1915 Mackensen's Eleventh Army attacked at Gorlice-Tarnow, driving the Russians from Galicia. Warsaw fell on August 4, and Kovno and Brest-Litovsk by the month's end. The Czar took personal command in September and achieved brief success at Lutsk, but could not prevent further loss of ground. In March 1916 the Russians mounted an abortive offensive at Lake Naroch. In June, however, Brusilov's Southwestern Front put in a spectacularly successful attack, encouraging Romania to join the Allies. But a two-pronged offensive by Falkenhayn and Mackensen overran Romania, and further advances were made as Russian morale crumbled in 1917. An armistice was agreed in December, and peace was concluded at Brest-Litovsk in March 1918.

150

close-range fire, were costly, and the fighting on the peninsula was among the grimmest of the war: about 252,000 Allied soldiers and perhaps 250,000 Turks became casualties. Once the Turks had checked the initial attacks – with the aid of counterattacks organized by Mustafa Kemal Pasha, the future Ataturk, President of the Turkish Republic – the line solidified much as it had in France and what might have been a quick victory ground on into trenchlock. It became clear that nothing could be gained at Gallipoli, and in December 1915 and January 1916 the peninsula was evacuated in an operation which a German called "a hitherto unattained masterpiece". It was the only bright spot in the entire campaign.

The Balkans

The stabilization of the Eastern Front enabled Falkenhayn to look to the Balkans. Gallipoli had put great strain on the Turks, but they could not be effectively supported as long as Serbia stood firm. In September 1915 the Bulgarians were persuaded to conclude a secret alliance with Germany; on October 5 the British and French sent an expeditionary force to the Greek port of Salonika in the hope of assisting the Serbs. When the Germans and Austrians invaded Serbia two days later, Allied troops at Salonika, under the command of the French General Sarrail, could do little. Sarrail attempted to advance up the valley of the Vardar, but the Bulgarians forced him back. The Serbian army was overwhelmed, a fragment making its way through Albania to the island of Corfu, whence it was shipped to Salonika. The Allied force at Salonika was reinforced until nearly 500,000 men were sitting in what the Germans called "their largest internment camp". However, it prevented the Bulgarians from operating elsewhere

Italian *Alpini* pass an elderly gun in the Alpine foothills.

and, perhaps more significantly in the tangled web of Allied politics, gave employment to Sarrail, not only a well-known Republican but also a contender for Joffre's post.

Enter Italy

Italy, wooed by the prospect of territorial gains at Austrian expense, joined the Allies by declaring war on Austria on May 23 1915. Her army, with General Luigi Cadorna as its chief of staff, included 36 infantry divisions and comfortably outnumbered Austrian forces on the Italian front. But the nature of the front meant that this superiority was of limited use. Italy was committed to "liberating" the Italian-speaking territories across her frontiers. To attack the salient of the Trentino above Lake Garda, or to advance into Austrian-held Friuli, north of Trieste, the

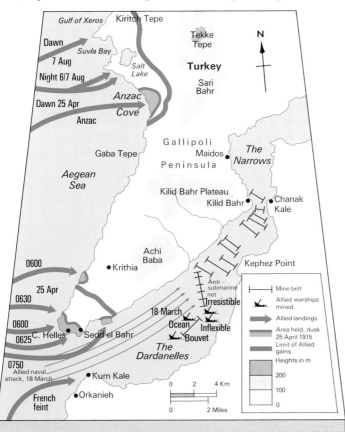

On March 18 1915 the Allied fleet bombarded the Narrows forts but withdrew after four ships were mined. Landings were made around Cape Helles and at Anzac Cove on April 25, but the attackers made little progress (*left*). A second landing on August 6/7 was equally disappointing. Anzac and Suvla were evacuated in December and Helles in January 1916.

Australian soldiers at Gallipoli (*above*). One is sniping with the aid of a periscope while his comrade observes the effects of his fire. The fighting was notably vicious, made more unpleasant by extremes of temperature and the closeness of opposing trenches.

151

Italians had to struggle uphill over some of the worst military terrain in the world, where the defender enjoyed all the advantage of the ground. Only on the narrow coastal strip leading to Trieste was the ground at all helpful. The sharp ridges of the Dolomites and the Carnic Alps separated the Trentino and Isonzo fronts, pushing Italian lateral communications through Venezia. The Austrians, however, looked down onto the Venezian plain from the Trentino, while in Friuli they held the barrier of the River Isonzo, with the Julian Alps behind it and the Friulian plain to its front. Terrain imposes its own discipline on any campaign, but events on the Italian front in 1915–18 were not merely influenced by the ground: they were dominated by it.

From late June until early December 1915 the Italians fought the first four battles of the Isonzo – there were to be eleven in all – and lost more than 278,000 men in making tiny gains. These fruitless attacks, so similar to what was going on on the Western Front at the same time, demonstrate that, whatever bad jokes were subsequently made about Italy's military performance, her soldiers could display valour and endurance for which "heroic" is no unreasonable description.

Verdun and the Somme: the Western Front 1916

An Allied conference at Chantilly on December 6–8 1915 agreed upon offensives on the Western, Eastern and Italian Fronts. It was the first serious attempt at a unified strategy and recognized the need to launch simultaneous attacks so that German reserves could not be switched to meet separate threats. Joffre and Haig discussed the offensive in the west over the following months. Joffre envisaged a "wearing-out fight" (bataille d'usure). Haig

had some reservations, but in mid-February the commanders agreed to mount a joint offensive astride the Somme, where the British and French armies joined, on or about July 1, 1916.

The Germans, too, had been considering plans for the coming year. Falkenhayn thought that the strain on France had almost become intolerable and, realizing that a breakthrough and break-out were beyond his resources, he decided to attack the French in a sector where, as he put it, "the forces of France will bleed to death – there can be no question of voluntary withdrawal – whether we reach our goal or not." Verdun, which already had popular significance to the French and lay close to the shell factories of Briey-Thionville and the rail complex of Metz, was the point selected. In the dreadful weather of January and February the Crown Prince of Prussia's Fifth Army prepared its attack, moving up guns and digging deep dugouts (stollen) in which the 72 assault battalions could shelter. By February 1 they had 1,220 guns in the sector, including 13 Krupp 420mm and 17 Skoda 305mm. The French had only 270 guns and 34 battalions; because of the disappointing performance of forts in the first months of the war, those around Verdun had had most of their guns removed for service with the field armies.

The German bombardment opened on February 21: 2,000,000 shells fell in the first two days and the early stages of the battle went much as Falkenhayn hoped, with the French being forced to offer up their infantry to the fury of the German guns. Fort Douaumont, key to the position on the east bank of the Meuse, fell on February 25, and the French considered evacuating the whole salient. Joffre appointed Pétain to command the sector and his arrival on the 27th heartened the defence. The line held, persuading

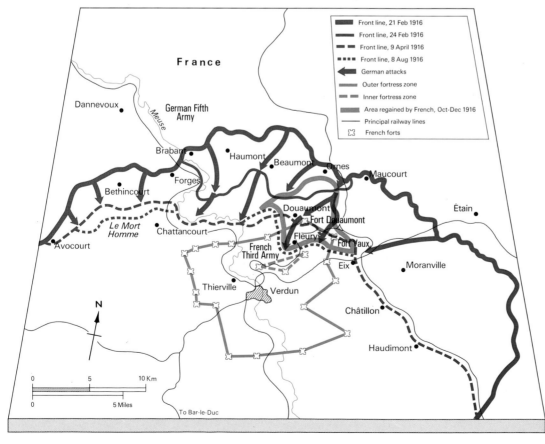

Verdun had long been an important fortress, and after the Franco-Prussian War Séré de Rivière encircled it with a belt of forts, subsequently modernized by the addition of "burster layers" of sand, and well-provided with steel cupolas – though by 1915 many of their guns had been removed. The Germans initially made good progress, capturing the commanding Fort Douaumont by a coup de main on February 25. They took the high ground of Le Mort Homme on the west bank of the Meuse, and on the east bank came almost within sight of Verdun itself, but were fought to a standstill by June. A French counteroffensive began in October, and much of the vital ground was recaptured.

Falkenhayn to step up the pressure on the west bank, where the high ground of Le Mort Homme and Hill 304 fell in early May. By this time the battle had little of the master stroke about it: the Germans were suffering almost as badly as the French in the murderous fighting on both banks of the Meuse. They took the stoutly defended Fort Vaux on June 7, and a month later managed to snatch a brief handhold on Fort Souville.

It was the high-water mark of German success. In October the French began a methodical counteroffensive, the work of General Robert Nivelle, who had taken over the Second Army in April, on Pétain's appointment to command Army Group Centre. Douaumont was successfully recaptured on October 24 in an operation that confirmed Nivelle as the French army's rising star. The fighting had cost the French at least 377,000 men and the Germans 337,000. In the words of Alistair Horne, "all the Germans had to show territorially for ten months of battle and a third of a million casualties, was the acquisition of a piece of raddled land little larger in area than the combined Royal parks in London."

The battle of the Somme

Verdun inevitably affected plans for the Somme offensive. Haig immediately agreed to take over more line, and he was urged to attack early to take the pressure off the French at Verdun. He was concerned about both men and ammunition. Conscription was not introduced in Britain until January 1916, and then only in a moderate form. The supply of ammunition was improving, but shortages – and defects which produced accidents at the gun and failures on the target – were frequent. Haig had a valuable ally in Robertson, but with the death of Kitchener – drowned in

HMS *Hampshire* on his way to visit Russia on June 4 – Lloyd George became Secretary of State for War, and Haig and he were increasingly to differ over strategy.

The Somme was above all a battle of men and material. General Sir Henry Rawlinson, whose Fourth Army was to bear the brunt of it, favoured a methodical bombardment to break the German defences on Pozières Ridge, the vital ground. His army was to attack the main position, with

The strain of Verdun shows on the faces of these French infantry.

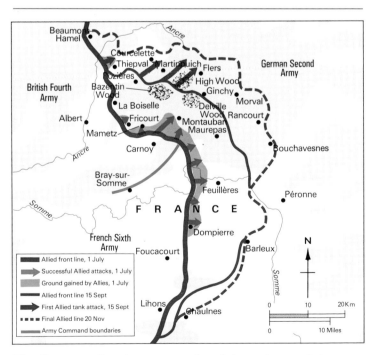

Field Marshal Sir Douglas Haig (1st Earl Haig) (1861–1928), attended Brasenose College, Oxford, before joining the 7th Hussars from Sandhurst in 1885. He failed the entrance exam to Staff College, but influential support secured him a place. After serving in the Omdurman campaign he became brigade major to French, and went to South Africa with him in 1899. He

ended the war in command of the 17th Lancers, and in 1903 was sent to India as inspector general of cavalry, returning to be Director of Military Training and influential adviser to R B Haldane, the reforming war minister.

In August 1914 he went to France in command of 1 Corps, later First Army. Dissatisfaction with French's performance induced him to send documents on Loos to his wife, a lady-in-waiting to the Queen. He replaced French in December 1915, and commanded the army in France for the rest of the war. His detractors portray him as a cold, unimaginative, self-seeking man who took little heed of the sufferings of his soldiers. John Terraine, however, suggests he was right in seeking to defeat the German army by attritional fighting. After the war he devoted himself to the British Legion.

The Somme offensive was planned as an Allied venture, though French losses at Verdun meant that the British bore the brunt of the battle. Opportunities for

breakthrough – on July 1 and 14 and September 15 – were missed, and it took months of bitter fighting to push the Germans out of their defensive system.

153

part of the Third Army mounting a diversionary attack at Gommecourt to the north and General Sir Hubert Gough's Fifth Army moving through to exploit. The bombardment began on June 24, and at half-past seven on the morning of July 1, after the explosion of mines under key German positions, the infantry moved forward: 17 divisions, 100,000 men, most of them volunteers of the New Armies.

The bombardment had not destroyed the German defences and casualties among the infantry, moving up steadily to follow the barrage, were frightful. One eyewitness reported:

> "The extended lines started in excellent order, but gradually melted away. There was no wavering or attempting to come back, the men fell in their ranks, mostly before the first hundred yards of No Man's Land had been crossed".

There were 57,470 British casualties on that memorable day and the gains were only slight, except in the south

---

"When the English started advancing we were very worried; they looked as though they must overrun our trenches. We were very surprised to see them walking, we had never seen that before. I could see them everywhere; there were hundreds. The officers were in front. I noticed one of them walking calmly, carrying a walking stick. When we started firing, we just had to load and reload. They went down in their hundreds. You didn't have to aim, we just fired into them. If only they had run, they would have overwhelmed us."

Musketier Karl Blenk, 169th Regiment.

---

around Mametz and Montauban. The French Sixth Army, on the British right, did better, achieving tactical surprise by attacking later than the British and doing so after a more concentrated bombardment.

THE SOMME: JULY 1 1916

British casualties on first day:

Time	Numbers committed	Casualties
0830 hours	84 battalions/ c.66,000 men	c.30,000 killed/ wounded
1200 hours	129 battalions/c.100,000 men	c.50,000 killed/ wounded
End of day	143 battalions/c.120,000 men	c.19,240 killed/ died of wounds, 35,493 wounded, 2,152 missing, 585 prisoners

German casualties on first day:
Estimated at c.6,000 killed/wounded; 2,200 prisoners

Haig was disappointed but not disheartened by the early results, and was determined "to press . . . hard with the least possible delay". On July 14–15, the second phase of the offensive, the British missed the opportunity of substantial gains around High Wood and the fighting degenerated into bludgeon-work in shattered villages and splintered woods. The weather was changeable that summer, and with the prospect of fighting into autumn Haig looked hard at all the available resources. These now included tanks, tested in England early that year and in France, albeit in small numbers, in August. A few were used on September 15 as part of the third phase of the offensive. Despite lack of preparation the results were initially good: one correspondent wrote of a tank walking up the main street of Flers with the British army cheering behind it. But the fleeting opportunity passed at Flers in September just as it had at High Wood a month earlier. The battle went on until November 19, ending with much of the German defensive system in British hands. Its cost had been terrible. The German method of calculating casualties makes estimates difficult and contentious, but it seems reasonable to assess losses at over 600,000 British and French to perhaps 600,000 German.

July 1 1916: the apotheosis of the machine gun. The Germans had dug deep into the chalk uplands above the Somme, making formidable defences in what had long been a quiet sector. The British plan relied upon the preliminary bombardment cutting the wire – its belts 20–30yds deep – and destroying dug-outs. Because they expected opposition to have been neutralized, British commanders planned a methodical advance by their infantry, largely volunteers of the New Armies. Although the bombardment, and mines exploded beneath the defences, caused casualties and damage, large sections of wire remained uncut and most dug-outs were intact. When the barrage lifted, surviving Germans emerged from their dug-outs and brought their machine guns into action from trenches or shell-holes. The attacking infantry, heavily-laden and moving at a walk in close order, suffered appalling casualties from the machine guns, and from the defensive fire of German artillery, registered on likely concentration points. Frontal machine-gun fire caught men trying to get through gaps in the wire. Fire from positions to the flanks was even more lethal, the gun's cone of fire making it most effective in an enfilade shoot. "Even so late as September," admits the Official History, "some battalions took pride in 'dressing' whilst advancing to the assault." Some officer showed more initiative: Brigadier General Jardine of 97th Brigade ordered his fir wave to creep forward under cover of the bombardment and then rush the German trenches: they were in them before the Germans got out their dug-outs.

Germans killed by shellfire on Pilckem Ridge, July 31 1917. The battle of Pilckem Ridge was the first component of Third Ypres. Fourteen British and two French divisions attacked, after bombardment which started on July 15, with over 2,000 guns and howitzers supporting them. Pilckem Ridge was taken, but gains farther south were disappointing and bad weather began to hamper operations.

Verdun and the Somme had taken a heavy toll and brought victory no nearer: it was inevitable that there should be growing criticism of their architects. Falkenhayn was dismissed in August, his place being taken by Hindenburg, with Ludendorff as "First Quarter-master-General" and the driving force in the combination. On the Allied side the crucial changes came in December, when Lloyd George replaced Asquith as Prime Minister of Great Britain and Nivelle took over from Joffre as commander-in-chief of the French armies.

The last gasp: the Eastern Front 1916–17

In March 1916 the Russians attempted to repair some of the damage of 1915 by going onto the offensive, but the resultant battle of Lake Naroch, despite masses of troops and ample artillery support, produced only a few briefly-held gains – and more than 100,000 Russian casualties. The failure plunged many Russian commanders into gloom and resulted in ever-increasing demands for ammunition. The Russian army's last glimmer of hope came with the appointment of General Brusilov to command the South-western Front. Not only was Brusilov himself a most competent commander, but his staff included officers who had studied previous operations and drawn useful conclusions. They suggested that earlier attacks had failed to achieve surprise because of the impossibility of concealing the preparations for them, and emphasized the importance of concealment, intimate cooperation between infantry and gunners and detailed planning.

Brusilov's offensive broke on June 4 and was at once spectacularly successful. The Austrian Eighth and Ninth Armies were demolished, more than 200,000 prisoners were taken and Austrian morale received a blow from which it never fully recovered. The ripples of the offensive spread wide: Falkenhayn was forced to switch troops from the Western Front and Allied morale was boosted on the eve of the Somme. Romania was encouraged to join the Allies, on August 27. The replacement of Falkenhayn by Hindenburg and Ludendorff was dictated by the need to restore German confidence in the wake of Verdun, the Brusilov offensive and Romanian entry into the war: Falkenhayn was sent off to conduct the campaign against Romania.

The Romanians themselves made his job easier. Operating with remarkable ineptitude, they invaded Hungary, leaving themselves open to a merciless counteroffensive. The Bulgarians and their allies of Army Group Mackensen jabbed up from the south, while Falkenhayn drove down

from the northwest with the German Ninth Army. Russian assistance was unavailing and the Romanian capital of Bucharest fell on December 7. In early 1917 the line stabilized with most of Romania in the hands of the Central Powers. The Romanian campaign was no mere sideshow. Although members of the British military mission had succeeded in sabotaging the oil wells around Ploesti, the Central Powers were able to extract large supplies of raw materials from Romania, a valuable asset at a time of growing shortage. Moreover, Russian support for Romania had consumed resources that might have been better employed elsewhere and losses in the south made their contribution to growing war-weariness in Russia.

Revolution broke out in Russia in March 1917. The war had not merely imposed a severe strain on an old-fashioned economy, provoked large-scale conscription into an army where incompetence was rife and casualties were heavy, caused substantial food shortages and reduced confidence in the Czar and his government. It had also helped fuel a great burst of economic activity producing what Stone has called "a crisis of economic modernization", while galloping inflation and strikes caused by the drop in the value of pay further unravelled the fabric of society. Losses amongst trained personnel, and the tendency to use the best units at the front and for those in key centres like Moscow and Petrograd to contain apprehensive conscripts, meant that the regime was short of loyal defenders when the crisis came.

The Provisional Government tried to continue the war, but the pressures which had brought revolution in March eroded the army's will to fight, and in July the Kerensky offensive crumpled as soldiers mutinied and deserted. The collapse was neither total nor uniform: many Russian soldiers tried hard to reconcile hatred of the war with recognition that the enemy was at the gates. Sometimes soldiers simply refused to move up into the line: elsewhere the disorder was more serious, with officers lynched and estates plundered.

In November the Provisional Government was itself overthrown by a Bolshevik coup. Among the new government's first decrees was one declaring Russia's readiness for peace: an armistice was agreed in December and in March 1918 the Bolsheviks concluded the treaty of Brest-Litovsk with the Germans, giving up a vast tract of territory and taking Russia out of the war for good.

The Italian Front 1916

Obedient to the Chantilly conference's decision to mount a joint offensive in 1916, the Italians attacked again on the Isonzo in March and again failed to make progress. However, the defeat of Serbia and the quiescent state of the Russian army encouraged Conrad to mount an offensive of his own. In May, after a diversionary attack which fixed Italian attention on the Isonzo, he struck hard on the Trentino front and broke deep into the Italian defences. Cadorna assembled a new army, the Fifth, and regained the lost ground, but the unexpected Austrian success brought down the Salandra government which was replaced by a broadly-based coalition.

With the danger in the Trentino past, Cadorna resumed his efforts on the Isonzo. On August 6 his troops began the sixth battle, taking the fortress-town of Gorizia on the 8th. This success, coupled with Romanian entry into the war

and the replacement of Falkenhayn, encouraged Italy to declare war on Germany on August 27. Stalemate followed euphoria: three more Isonzo battles achieved negligible results and the fighting died away in early November.

The Western Front 1917

The early part of 1917 was dominated by Nivelle's offensive. In December 1916 he outlined his plans to Haig, writing of "a vehement attack" which would produce a breakthrough, enabling a "mass of manoeuvre" to move through to fight the decisive battle. The British were to assist by freeing French forces for the mass of manoeuvre and by attacking to absorb German reserves. Nivelle also discussed the possibility of the capture of Ostend and Zeebrugge, although he believed that they would fall in any case if his plan succeeded.

The idea of an operation on the Flanders coast was not new. Churchill had proposed it in 1914 and Haig considered it in 1916. Not only could an offensive in the north take the rail centre of Roulers and imperil the whole German position in Flanders: it might also take Ostend and Zeebrugge. This was, in the winter of 1916–17, an especially attractive prospect. It was curious that for all Germany's pre-war efforts to build up her navy, the High Seas Fleet posed little real threat to British power. There were a few engagements between surface units – in the Heligoland Bight in August 1914 and off the Dogger Bank in January 1915, for example – but only one major fleet action, the inconclusive battle of Jutland on May 31–June 1 1916. Yet German submarines did increasing damage to Allied shipping: the British lost 104,572 tons of shipping in September 1916 and 182,292 tons in December: in February 1917 no less than 781,500 tons of Allied shipping went to the bottom. The eventual answer to German submarines was

The merchant ship *Parkgate* sunk by U-35. In the first years of the war German submarines usually surfaced to sink merchantmen by gunfire, allowing their crews to take to the lifeboats. This allowed submarine commanders to economize on their limited torpedoes: it is also clear that many did not relish sinking a merchant vessel without warning. The practice of arming merchant ships, and of using "Q ships" – armed vessels disguised as tramp steamers – made torpedo attack more common. Over 15,000 British merchant seamen died as a result of submarine warfare.

the convoy system, introduced in June 1917. But the British Admiralty, increasingly gloomy about future prospects and believing that Zeebrugge was the most dangerous submarine base, argued for an offensive to clear the Flanders coast, and Robertson was left in no doubt of the importance the War Committee attached to the operation. In early 1917, therefore, Haig found himself expected to collaborate in Nivelle's forthcoming offensive and also under pressure to mount an offensive in Flanders. Furthermore, Lloyd George was already contemplating making major efforts outside France. Haig was not heartened by an Allied conference at Calais in late February, which produced an ambiguous agreement placing him under Nivelle's command for the forthcoming operation.

Slaughter on the Chemin des Dames

The Nivelle offensive went wrong from the start. In March the Germans withdrew to the Hindenburg Line, a well-fortified position behind the Somme battlefield, which shortened their line and removed the salient which Nivelle had planned to attack. But the offensive went on. On April 9 the British attacked at Arras, taking a little ground at great cost, but with the capture of Vimy Ridge by the Canadians as a lasting achievement. The French, delayed by atrocious weather, attacked along the Chemin des Dames on the 16th. There was no surprise: the plans had been found on a captured NCO and the Germans were ready. Their position was skilfully laid out: about 10,000 yds (9,100m) behind the 21 divisions in the front line were 10 counterattack (*Eingreif*) divisions, with another 5 in reserve a further 10,000 yds (9,100m) back. Forward positions were thinly held and good use was made of natural caves in the chalk to enable the reserves of front line divisions to sit out the bombardment. Nivelle had

THE BLOCKADE OF THE CENTRAL POWERS

The British Admiralty initially hoped to impose close blockade of the German coast, but in 1911 it acknowledged that submarines and torpedo craft made this unworkable, and substituted distant blockade, effectively sealing off the North Sea. The blockade – and especially British wide interpretation of contraband – produced protests from neutrals. Some American vessels were stopped and their cargoes confiscated: the United States government protested, although President Wilson did not press the issue. In 1915 an Anglo-French conference agreed on a rationing system for European neutrals, who were to be permitted to receive a scale of imports based on pre-war levels.

The blockade bit deep. By 1916 complaints of shortages, profiteering and hoarding were rife, and emphasis on war production made consumer goods like clothing and footwear scarce. Rising prices generated industrial unrest: price controls brought in to appease workers fuelled the right's pressure for unrestricted submarine warfare. Those who could bargain for higher wages could cope, but small businessmen, the professional middle classes and pensioners were hard hit. In 1916 the potato crop failed, leading to the "turnip winter" of 1916–17, and there were strikes in January 1917 in protest at the reduction of the bread ration. There was a drought in the summer of 1917, and in 1918 morale was lowered by suggestions that a poor harvest meant another winter of privation. One authority suggests that 763,000 civilians died in Germany as a direct result of the blockade: others add starvation-related diseases to put the real figure much higher. While the blockade did not prevent the Central Powers from fighting on, it certainly lowered wartime morale and increased postwar bitterness. "What really depressed the mood of the people," wrote General von Kuhl, "was the British blockade, the impact of which was growing more and more severe."

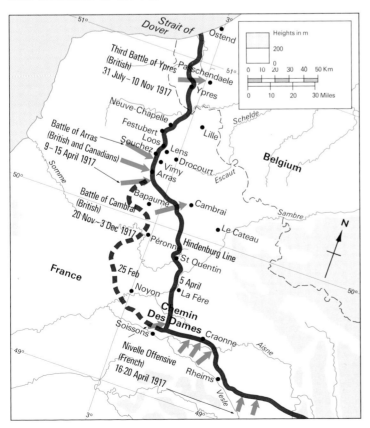

Nivelle's plan for the Allied offensive in 1917 (*left*) consisted of a British attack around Arras, to be followed by a massive French blow in Champagne. Although German withdrawal to the Hindenburg line left Nivelle facing well-prepared defences, he persisted nonetheless. Poor security compromised the operation, bad weather hampered preparations, and the bombardment was inadequate. The Germans broke the attack in its first 2 days.

A Sopwith "Ship Strutter" (*above*) takes off from a platform on a gun turret aboard HMAS *Australia*. This aircraft could only land ashore, however.

hoped to repeat his success at Douaumont by a heavy bombardment followed by a very rapid advance of 8,000yds (7,300m) in eight hours, but his estimate was utterly unrealistic. So deep was the German defence that penetration on this scale was too shallow to produce the intended breakthrough, and in the event little penetration was achieved: the Germans checked the assault in the forward zone without much need of assistance from the *Eingreif* divisions positioned behind the front line.

The slaughter on the Chemin des Dames nearly broke the French army. It was not just that the casualties were heavy – 117,000 men – but that hopes and expectations were dashed. Nivelle had promised a war-winning offensive and it had failed in mud and blood like all the rest. The French army was already sorely tried. Low pay, badly-organized leave, casualties, the spread of pacifist propaganda and the example of the Russian Revolution, all combined to strain the soldiers' loyalty. Mutinies began in late April and over the next few weeks there were 110 cases of "grave collective indiscipline" in 54 different divisions. Pétain took over from Nivelle and at once set about restoring the army's health. Food and accommodation were improved; leave for men and rest for units was made more frequent; decorations were distributed more liberally. More than 23,000 men were found guilty of indiscipline and more than 400 were condemned to death, but only 23 were officially executed, although others were undoubtedly shot. Offensives were suspended until the autumn, when a slickly-executed attack at La Malmaison inflicted 40,000 casualties on the Germans for the loss of 14,000 Frenchmen. The government tried to deal with civilian unrest, but was shaken by a series of scandals and fell in November.

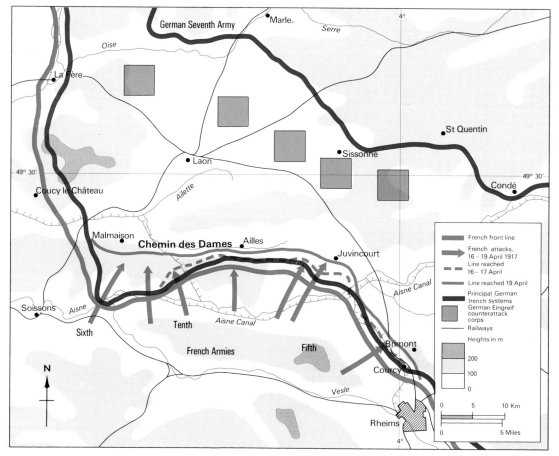

The Western Front, 1917. The Germans withdrew to the *Siegfried Stellung,* (Hindenburg Line to the Allies), a strong position an average of 25 miles behind their initial line, in March. In April the Allies attacked the shoulders of the old salient. The British attacked at Ypres in July–November, and at Cambrai in November.

Georges Clemenceau, the old "tiger" of French politics, took over as premier. Despite his 76 years he had lost none of his fire. He turned savagely on those who favoured negotiating with the Central Powers and firmly announced his intention of fighting on to victory. There were many, by no means all of them Frenchmen, who believed that victory for either side was remote, but a compromise peace was as far away as ever.

The one glimmer of hope in the dark months of 1917 was the United States' entry into the war. President Woodrow Wilson had tried hard to remain aloof, although news of atrocities in Belgium and the sinking of the Cunard liner

The loss of *Lusitania,* announced in the *New York American*. The German statement that she was "naturally armed" sits uneasily beside headlines like: "Mothers Clasping Dead Babies Picked Up in Water."

Lusitania on May 7 1915, with the loss of 1,198 lives, 128 of them American, helped harden public opinion. It took two German mistakes to end American neutrality. The first was the adoption of unrestricted submarine warfare on January 31 1917: all ships, irrespective of nationality, would be sunk on sight in the war zone. The second was the suggestion made by Zimmermann, the German foreign secretary, that if the United States did enter the war, Mexico should become a German ally in return for "generous financial support" and the return of "lost territory" in Texas, New Mexico and Arizona. Zimmermann's telegram to the German Ambassador in Mexico City, its code cracked by British naval intelligence, made national headlines in America on March 1. On the 16th two American ships were torpedoed; and on April 6 the United States declared war on Germany.

America's entry into the war was, as A J P Taylor has observed, "a promissory note for the future". Her army was small and largely inexperienced. However, Brigadier General John J Pershing had led a punitive expedition into Mexico in pursuit of Pancho Villa in 1916, and this made him, despite his lack of seniority, the natural choice for commander of the force sent to France. Raising and training troops would take time and America's vast industrial resources could not be quickly geared to the needs of war: orders were placed for 4,400 tanks, but only 15 ever reached France. Nevertheless, American entry into the war struck a great psychological blow: if the Allies could hold on until America's might was brought to bear,

their victory was certain.

There seemed little certainty of Allied victory in the wake of the Nivelle offensive. However, Haig had already sketched out his plans for an offensive in Flanders, and he met Pétain at Amiens on May 18 to discuss them. Pétain undertook to do what he could to assist the attack and to attract German reserves, but on June 2 Haig was told that the "bad state of discipline" in the French army would lead to the abandonment of the first of the supporting attacks. In the event, it led to the almost complete cessation of French operations during the forthcoming British offensive, officially the third battle of Ypres, but often known by the name of one of its component actions as the battle of Passchendaele.

The battle of Passchendaele

Few battles have proved as controversial as Passchendaele, and the arguments surrounding it can only be summarized here. Numerous factors combined to persuade Haig to attack in Flanders. We have already noted the demands for the capture of Zeebrugge and its submarine base, demands which intensified in June when Admiral Jellicoe, the First Sea Lord, made the announcement: "There is no point in discussing plans for next spring – we cannot go on." The state of the French army was another argument for a British offensive, although we cannot be sure how heavily desire to take the strain off the French weighed in Haig's mind. Elsewhere the situation looked bleak: Russian support could no longer be guaranteed and the Italians were fighting hard in the Trentino. Haig's critics would add to the list his desire to ensure the primacy of the Western Front, and would note the failure of his staff to appreciate the difficulties entailed in the operation.

On balance, given the information at Haig's disposal and the range of concerns which preoccupied him, his decision to fight at Ypres was not unreasonable. His resolve to fight

---

"A moment's pause to absolve a couple of dying men, and then I reached the group of smashed and bleeding bodies, most of them still breathing. The first thing I saw almost unnerved me – a young soldier lying on his back, his hands and face a mass of blue phosphorus flame . . . the first victim I had seen of the new gas the Germans are using. Good God, how can any human being live in this!"

Father William Doyle, 16th (Irish) Division, Zonnebeke, August 1917

---

on through the wet summer and into the muddy autumn may be more legitimately challenged. Yet if it was an error, it was one which sprang not from ignorance, but from a fixed opinion as to the nature of the war. Haig was to describe operations from the Somme onwards as forming one "great and continuous engagement", with the defeat of the German army as its aim. By persevering at Ypres when less determined men might have stopped, Haig undoubtedly did the Germans very great damage: what must remain contentious is whether the damage was inflicted at too great a human price.

On June 7 General Sir Herbert Plumer's Second Army took possession of Messines Ridge after the explosion of 19 mines containing a total of 957,000lbs of explosive.

This meticulously planned assault was the curtain-raiser for the main attack, by Gough's Fifth Army. On July 31, the first day of the battle, Gough took Pilckem Ridge, and Langemarck fell on August 16. Plumer edged forward on Gough's right with initial success but ultimately the same result: costly and demoralizing fighting in a wasteland of pulverized villages, blighted woods and water-filled craters, the whole permeated by foul and treacherous mud. The Canadians took the ruins of Passchendaele on November 6: the battle ended three days later.

It produced at least 244,897 British casualties. That total is disputed by those who accuse the British Official Historian of distorting the figures: he certainly overestimated German losses at 400,000. But for all their emphasis on defence in depth, their sturdy pill-boxes and their deft conduct of the defensive battle, the Germans had no easy time of it: their own counterattacks were often as costly as British attacks. German casualties were probably as heavy as the British, and Ludendorff admitted that "certain units no longer triumphed over the demoralizing effects of the defensive battle. . . ."

Enter the tank

Tanks were no use in the mud of Passchendaele: they were tried at St Julien, but to little effect. The Tank Corps found more favourable ground to the south, in the Third Army's sector in front of Cambrai. On November 20, 381 tanks attacked en masse, ripping a hole through two lines of the Hindenburg position on a front of six miles and taking 6,000 prisoners and more than 100 guns. It was a dazzling success and church bells were rung in England for the first

time during the war. However, many of the tanks broke down or became ditched and the attack bogged down in a struggle for the high ground of Bourlon Wood: when the Germans counterattacked on November 30 they recaptured most of the lost ground and part of the original British front line in addition. The year ended as it had begun, in stalemate.

Stalemate at the close of 1917 was by no means intolerable to the Allies. If they had failed at Ypres and Cambrai, they had at least recovered from the French mutinies and the Italian disaster at Caporetto. And the Yanks were coming, albeit slowly. There were three American divisions in France by the end of October: Pershing resisted attempts to persuade him to brigade his units with British or French formations and his battalions went into the line in Lorraine to gain experience. Across the Atlantic, fresh divisions completed their basic training and prepared for the voyage to Europe.

The Italian Front 1917

Largely on Lloyd George's initiative, the Allied conference at Chantilly in November 1916 considered sending artillery reinforcements to Italy to enable Cadorna to break through on the Isonzo. Cadorna was cautious about his prospects and the few British and French batteries which were sent made little difference. In 1917, as in 1916, the Italians honoured the spirit of inter-Allied agreements and attacked on the Isonzo in May, losing more ground to an Austrian counteroffensive than they took in the first place. In August the eleventh and final battle of the Isonzo went altogether better, ending with the Bainsizzia feature

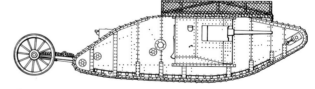

A British Mark IV tank (*right*) crushes wire at Cambrai. Steering involved the driver, the commander – sitting beside him and operating the brakes as well as a machine gun – and two gearsmen on the gearboxes controlling track speed. The remaining four crewmen manned guns. The tank was constructed by rivetting armour plate to butt straps and angle iron: missile fragments entered between the plates, and the plates themselves cracked – hence the crew's face protectors.

The map (*left*) shows the British advance in Third Ypres, but its woods and villages bear little relation to the strip of murdered nature created by three months fighting in appalling weather.

in Italian possession, albeit at the customary extortionate price in terms of human lives.

Costly though they had been for the Italians, the repeated attacks on the Isonzo had sapped Austrian strength, and General Arz von Straussenburg, Conrad's successor as chief of staff, feared that the Italians would eventually get through. The young Emperor Karl personally asked the Kaiser for German assistance. It came in the shape of General Otto von Below's Fourteenth Army, containing two Austrian corps and two seasoned German corps. Below concentrated on the Isonzo, and although Italian intelligence had some inkling of what was afoot, *Commando Supremo* failed to appreciate the strength or direction of the offensive.

The battle of Caporetto began on October 24 with a bombardment so intense that it surpassed anything so far encountered on the Italian Front. It did fearful damage to Italian positions, many of them unwisely sited on forward slopes; the gas which was mingled with the high explosive was especially effective against the poor quality Italian gas masks. By mid-afternoon there was a 15-mile gap in the Italian line and, as the Germans and Austrians surged through it, Italian morale, shaken by the ferocious bombardment, the myth of German invincibility and the

unfamiliarity of open warfare, collapsed. Cadorna, at the mercy of ponderous communications, reacted as best he could, trying to prepare a line on the Tagliamento onto which his Second and Third Armies could retire. It soon became clear that an orderly withdrawal was out of the question: much of the Italian army was streaming back out of control, along with more than 500,000 prisoners and deserters. Cadorna was dismissed and General Armando Diaz took over.

Disaster averted

Total disaster was close. It was averted partly by the Italians themselves rallying to meet the supreme crisis, although Allied assistance in the shape of six French and five British divisions helped solidify the new line on the Piave. The emergency was not without its benefits, for on November 5 1917 the Allied leaders, meeting at Rapallo to discuss sending aid to Italy, agreed to set up the Supreme War Council. This was far from being a unified command, but it provided a useful forum for discussion. Its military advisers met as an Executive Committee under Foch's chairmanship, foreshadowing more intimate cooperation.

In November and December the Germans and Austrians battered at the Piave line but could not break it. In 1918

most German troops were dispatched to the Western Front, while the Austrians concentrated for a final offensive: it was launched on June 16 and ended in failure a week later. Diaz methodically planned a counteroffensive – the battle of Vittorio Veneto – which began on October 24, a year to the day after Caporetto. The Austrians fought back hard for the first four days but were then bundled back across the Tagliamento in a calamitous retreat which bore witness to the overall collapse of the Central Powers' fortunes. An armistice came into effect on November 4, by which time the war in the west had only a week left to run.

The end in the West: 1918

The knowledge that the Americans would eventually intervene in strength was a major determinant of German strategy for 1918. Worsening food shortages, the consequence of the Allied blockade, had already produced rioting in German cities, emphasizing that Germany was running out of time. The armistice with Russia enabled Ludendorff to shift troops to the west, increasing his strength there from 152 divisions to 190; he planned to use them in an offensive which would end the war before American power could be brought to bear. The Germans were determined not to repeat past mistakes, and an OHL document of January 1918, *The Attack in Position Warfare*, encapsulated much German and Allied experience. A French officer, Captain André Laffargue, had sketched out infiltration tactics in 1915, and General Oskar von Hutier and his artillery expert Colonel Georg Bruchmüller had developed them on the Eastern Front. The events at Caporetto had also produced some useful lessons. Disrup-

tion rather than destruction was the essence of the new tactics. After a lightning barrage, assault squads of storm troops (*Stosstruppen*) penetrated the defence, bypassing strongpoints and exploiting weaknesses. Field guns moved up with the advance and low-flying aircraft harassed the defenders.

Ludendorff's offensive, codenamed Michael, came within measurable distance of winning the war. The British position was a half-hearted copy of the German, its strong-points patchily prepared and its reserves too far forward, at the mercy of the opening bombardment. When the Germans attacked on March 21 they had the added advantage of fog, which blinded most of the forward machine-gun posts. They took 140,000 sq miles (360,000 sq km) on the first day and rolled the British Fifth Army right back across the old Somme battlefield. The risk of the Allied armies splitting demonstrated the need for a unified command to enable reserves to be used most effectively, and on March 26 Foch was appointed to coordinate the Allied armies: on April 14 he was given the title "Commander-in-Chief of the Allied armies in France".

The worst of the danger was passing even as Foch was invested with his new powers. "Michael" ground to a halt at the end of March, and on April 9 Ludendorff launched a new offensive on the Lys; by the end of the month that too was stuck fast. On May 27 he struck again, this time on the Chemin des Dames. The French held the ridge too densely and Bruchmüller's gunners ripped the Sixth Army open, enabling the infantry to surge on across the Aisne and the Vesle, finally reaching the Marne. Here they were checked by reserves, among them two American divisions, whose

In the spring of 1918 the Germans launched three major offensives against the Allies. In March-April, Operation Michael drove the British Fifth and French First Armies back across the Somme battlefield. In April they attacked on both sides of Ypres, retaking many of the hard-won British gains of 1917 but failing to capture Ypres itself. Finally, in May the Germans lacerated the French Sixth Army on the Chemin des Dames.

Apparently expecting a gas attack in their sector, German airmen stand by their machine (*above*) on the Western Front in late 1917. Note the wing-mounted machine gun and the anti-personnel bombs in side racks: ground attack by low-flying aircraft formed an important part of the "shock tactics" developed by the Germans later in the war.

Map legend:
- Front line, 20 March, 1918
- Front line, 5 April
- Front line, 30 April
- German advances
- Pre-war borders

General Erich Friedrich Wilhelm Ludendorff (1865–1937) attended the *Kriegsakademie* in 1893 and in 1905 joined the mobilization section of the General Staff: patronized by Schlieffen and Moltke, he was earmarked to be chief of operations at OHL on mobilization. In 1913 his career faltered when he indulged in parliamentary lobbying to get the military budget increased. He was sent off to command a regiment, but on mobilization he became Deputy Chief of Staff of Bülow's 2nd Army, and played a leading part in the capture of Liège. On August 22 he was appointed Chief of Staff of the 8th Army, beginning his long relationship with Hindenburg. He remained in the east until August 1916, when he and Hindenburg were summoned to take over from Falkenhayn. Offered the title "Second Chief of Staff", he chose instead that of "First Quartermaster-General", and for the next two years helped exercise immense military and political power. Regarding the offensive of spring 1918 as a desperate throw, he became increasingly unstable when it failed. In August he acknowledged that the war must be ended, but could not stomach capitulation: he was dismissed on October 26. In 1923 he marched with Hitler in the Munich *putsch*, and subsequently became deeply involved in extremist politics.

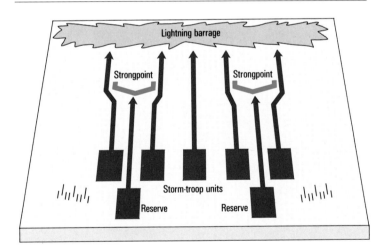

The Germans' tactics for the spring offensive of 1918 were based on their own and Allied experience. The linear tactics which had dominated operations in the first years of the war were jettisoned in favour of infiltration by parties of lightly equipped infantry who penetrated where the defence was weak, by-passing strongpoints, reinforcing success and avoiding costly frontal assault. The attack was prepared by a lightning barrage, containing a high proportion of gas so as not to destroy roads which would later be useful to the attackers. German training stressed flexibility and initiative, emphasizing that: "there must be no rigid adherence to plans made beforehand."

contribution – which included the actions at Cantigny and Belleau Wood – was valuable in both moral and material terms. With nearly 1,500,000 American troops in France there was little time left for the Germans. On June 15 Ludendorff attacked for the last time, jabbing down on either side of Rheims but making disappointing progress.

Although Allied casualties had been heavy, so too had the German, and they had fallen disproportionately upon the *Stosstruppen*. Morale was beginning to crumble. Allied counterattacks began on July 18 when Mangin's Tenth Army, making good use of Renault light tanks, boiled out of the forest of Villers-Cotterets into the shoulder of the German salient. The initiative was in Allied hands, and on August 8 the British Fourth Army attacked at Amiens. Ludendorff called it "the black day of the German Army", and his troops were back on the Hindenburg Line by early September. Pershing drew his army together, determined to use it as an intact formation despite objections from Foch and Haig, and on September 12 he nipped out the St Mihiel salient in Lorraine. The Americans then redeployed to attack in the Meuse-Argonne region on

"The atmosphere of intense excitement was amazing. Officers stood upright and shouted chaff nervously to each other. Often a heavy trench-mortar fired short and scattered us with fountains of earth; and no one even bent his head. The roar of battle had become so terrific that we were scarcely in our right senses. The nerves could register fear no longer. Everyone was mad and beyond reckoning; we had gone over the edge of the world into superhuman perspectives. Death had lost its meaning."

Captain Ernst Jünger, March 1918

September 26, their thrust forming the southernmost prong of a general Allied offensive. Pershing's progress was slow and his men paid heavily for their lack of experience and excess of enthusiasm. In the north, the British breached the Hindenburg Line, and although the German soldier could still fight back hard – the British lost 121,000 men in October – Ludendorff was convinced that the game was up.

Ludendorff's own confidence, shaken for some time, was worsened by news from other theatres of war. The Allied army in Salonika, now under Franchet d'Esperey, attacked the Bulgarians on September 15. The Bulgarians, deprived of German support, quickly collapsed, concluding an armistice on the 30th. In Palestine, Allenby defeated the Turks at Megiddo on September 19 and drove on towards Damascus. On September 28, after an emotional scene at OHL, Ludendorff telephoned Hindenburg to advise making peace on the basis of the Fourteen Points put forward by President Wilson in January 1918. Prince Max of Baden became Chancellor on October 4 and at once asked Wilson for an armistice. Exploratory discussions dragged on till October 23, when Wilson declared that the removal of the monarchy and German's military rulers were the prerequisite for negotiations. Ludendorff resigned on October 26 and on November 8 his successor, General Wilhelm Groener, told the Kaiser that he no longer enjoyed the army's confidence. William II abdicated the following day, appointing Hindenburg as Supreme Commander of the armed forces. The way was clear for the armistice, which came into force on November 11.

The sideshows: Mesopotamia, Palestine and Africa

The ill-starred Gallipoli campaign was not the only Allied attempt to overthrow the waning power of Turkey. The Russians made their contribution: a Turkish offensive into the Caucasus failed in the winter of 1914, and in January 1916 the capable Russian General Yudenich launched an attack which took Erzurum, Erzincan and Trebizond. However, these gains were swallowed up in 1917–18 when the Turks capitalized on unrest in Russia to capture the lost ground and some Russian territory into the bargain. The most telling blows were struck by the British in two campaigns, one in Mesopotamia, the other in Palestine. Both were marked by initial mismanagement particularly trying for the troops involved.

Britain's interest in Mesopotamia was generated by desire to protect the British-controlled oil wells in Persia and the Gulf. Troops sent by the government of India secured the head of the Gulf in November 1914; in 1915 they were reinforced for an expedition up the Euphrates towards Baghdad under the command of General Charles Townshend. In late September he took Kut-el-Amara, but on November 22 the Turks halted him at Ctesiphon, just 20 miles short of his objective. Townshend fell back on Kut and was besieged there: three attempts to relieve him failed and he surrendered with 10,000 men on April 29, 1916. Townshend's force had been too small for its task, especially in view of the problems of supply: the next British advance was altogether more methodical. General Sir

Stanley Maude set off from Basra with 50,000 British and Indian troops, manoeuvred the Turks out of Kut and took Baghdad on March 11 1917. Disease caused more casualties than Turkish bullets: Maude himself died of cholera.

The Palestine campaign had its origins in Britain's desire to protect the Suez canal, and there was indeed a half-hearted Turkish attempt to take it in February 1915. The failure at Gallipoli brought unexpectedly strong reinforcements, including a powerful Anzac contingent, to the garrison of Egypt, and just as the lure of Baghdad had taken the British up the Euphrates, so the prospect of advancing into Palestine drew them into Sinai. General Sir Archibald Murray took the Turkish outposts at Magdhaba and Rafa, but in March 1917 he was firmly checked at the first battle of Gaza. A second attempt to force the now much stronger Turkish position by frontal assault in April ended in failure: Murray was replaced by General Sir Edmund Allenby.

Allenby had not been a conspicuous success as commander of the Third Army in France, but he was the right man for Palestine. A cavalry officer, Allenby understood the importance of manoeuvre; he also radiated energy and magnetism which his predecessor conspicuously lacked. On October 31 he attacked again, but instead of repeating Murray's mistakes he pinned the Turks at Gaza while Chauvel's Desert Mounted Corps, with its Australian and New Zealand Light Horse units, turned their flank at Beersheba. Gaza fell on November 7 and Jerusalem on

Following the capture of Jerusalem on December 9 1917, General Allenby reads the Proclamation of Occupation from the steps of the Tower of David (*left*).

December 9. In 1918 many of Allenby's most experienced troops were taken to shore up the Western Front, but in September he inflicted a decisive defeat on the Turks at Megiddo, destroying most of Liman von Sanders' force and going on to take Damascus on October 2 and Aleppo on the 28th. Turkey was by now collapsing from within: on October 30 she made a separate peace with the Allies.

The war against Turkey naturally provokes re-examination of the debate between Easterners and Westerners with which this chapter began. For much of the war Mesopotamia and Palestine between them soaked up 1,000,000 troops, many of whom might have been used on the Western Front. British casualties outside Europe – not including those incurred in the Allied intervention in Russia in 1917–22 – numbered 850,000. The operations which caused them brought the defeat of Germany no nearer. The concept of "knocking away the props" from Germany made little sense because it was the Germans who propped up their allies, not the other way about. The Turks were hardy fighters, but they owed many of their successes to German staff-work: yet Germany's support for Turkey consumed only a fraction of her resources.

Long-term consequences

The sideshows were not entirely sterile. Although their contribution to Allied victory may have been marginal, they helped colour the globe in the inter-war years and beyond them. Subsequent conflict in the Middle East arose in part because the British attempted to foster Arab nationalism and to support the Arab revolt against the Turks – with a mission which included T E Lawrence – at the same time as they were encouraging the Jews to hope for the creation of "a national home for the Jewish people" in Palestine. The extent of British culpability remains in dispute, but the increase of Anglo-French influence in the Middle East after the war was undoubtedly to have important long-term consequences.

Other, smaller, sideshows were also to cast long shadows. Most German colonies were snapped up quickly. Togoland surrendered in late August 1914; German West Africa capitulated to a South African force on July 9 1915; and the last of the German garrison of the Camerouns gave up in February 1916. German East Africa, however, was defended by Colonel Paul von Lettow-Vorbeck, one of the most remarkable soldiers to emerge on either side throughout the war. He fought a vigorous campaign against a British force which often outnumbered his own by ten to one, surrendering only on November 23 1918. German possessions in the Pacific were also lost. The Japanese, who joined the Allies on August 23 1914, took the German enclave of Tsingtao in China in November, and went on to occupy German outposts in the Marianas, Carolines and Marshalls. The Australians seized Neu-Pommern and promptly rechristened it New Britain. These stirrings in the Pacific were, in their way, as portentous as any of the war's results elsewhere.

A detachment of the Desert Mounted Corps in the Jordan Valley (*above, right*). Allenby established this horsed and camel-mounted striking force from Anzac and British cavalry divisions in July 1917, immediately after taking command. The *Army Quarterly* reported: "In the heat of summer, Arabian camels will do about 250 miles comfortably between drinks; and this represented about three days' vigorous marching."

Mobility was all in desert warfare. Here (*above*), Lewis gunners of the 7th Indian Division, are seen with their transport mules during operations around Arsuf, Western Palestine, in June 1918.

Arab irregulars (*right*) and men of the Sherwood Rangers meet at Mouslimiié (Muslimie) Junction on the Baghdad Railway, just north of Aleppo, on October 28 1918.

chapter twelve

World War II: Blitzkrieg

Richard Holmes

In six weeks in the early summer of 1940 Germany overran Northern France and the Low Countries in a campaign which brought the word *blitzkrieg* – lightning war – to the lips of the Western world. Just over a year later the same torrent flowed into Russia, and if German achievements there were inconclusive, they were none the less astonishing: by the winter of 1941 the Germans were at the gates of Moscow, Leningrad seemed within their grasp and vast tracts of Russia had been overrun.

The achievements of blitzkrieg were literally dazzling, for they partially blinded both the Germans and their enemies to the flaws inherent in the technique. Moreover, the scale of German achievement induced many observers to see something utterly original in the form of warfare that burst across France and Russia in those two blazing summers.

The genesis of blitzkrieg

There was nothing new in the quest for rapid victory. Napoleon's Italian campaign of 1796 and Moltke's defeat of Austria in 1866 had shown that operations could be swift even in the age of black powder. Both French and German plans for war in 1914 had quick victory as their objectives: belief that the war would be over by Christmas was not confined to the rank and file. Yet for commanders of the 19th and early 20th centuries the dream of lightning war was often an illusion. Even after the railway enabled armies to mobilize and concentrate quickly, once soldiers left their railheads their mobility – and that of their logisticians – was poor. Furthermore, although some of the 19th century's technology had encouraged offensive operations by aiding movement, most of it – above all developments in firepower – strengthened the defensive.

It was the internal combustion engine in the armoured fighting vehicle, the ground-attack aircraft and the truck that made blitzkrieg possible. However, if it seems clear in retrospect that the mechanization of armies was desirable, this was less obvious in the 1920s and 1930s, and the debate over mechanization highlights the dilemmas faced by politicians and military bureaucracies alike in an era of technical and doctrinal innovation.

The immediate origins of blitzkrieg lie among the barbed wire and shell-holes of the Western Front. There were numerous attempts to break the stranglehold imposed by defensive firepower on offensive mobility, and blitzkrieg was born of the union between one of the mechanical devices used to overcome this deadlock – the tank – and the offensive doctrine developed by the Germans in 1917–18. As Charles Messenger writes:

"... by the end of the war ... the essential ingredients of blitzkrieg were already present on the battlefield. The Germans had contributed the art of infiltration and shock action, although the latter at this stage was in the shape of artillery. The Allies had introduced the tank. Both appreciated the value of psychological dislocation, and each had used aircraft in the ground support role."

Twenty years separated the signing of the Treaty of Versailles, which ended World War I, and the first use of blitzkrieg against Poland in 1939. The political and military climate during most of the period did not encourage radical military thought. The treaty included provisions designed to ensure that Germany could never again pose a military threat. Her armed forces were limited to 100,000 men and were deprived of the ingredients of aggressive war – tanks, heavy guns and aircraft – and her General Staff was abolished (to re-emerge under the innocuous title of

Truppenamt). It was easy for the wartime allies to assume that Germany was emasculated and to go their own ways. There had been divergent forces within the alliance even while the war was in progress, and these too helped ensure that the coalition which won the war did not long survive its conclusion.

Although it is too simplistic to blame the treaty that ended World War I for causing World War II, our understanding of the events of 1940–41 cannot be isolated from the political developments of the 1920s and 1930s. Not only did these result in a divided Europe – containing states like Poland and Czechoslovakia, products of the treaty, but vulnerable without collective security – as well as a resurgent Germany, but they also helped to determine the forces available to the combatants of 1940–41 and the doctrines they followed.

The most striking characteristic of foreign policy in the 1920s was the tendency for each of the former allies to concentrate on its own interests. Russia had made a separate peace with Germany and her communist government was hard at work strengthening its grip on a land ravaged by war. The Russian regime alarmed conservative politicians and at the same time excited idealists who hoped to see the old order swept away across the whole of Europe. If Russia was effectively excluded from European politics, so too was America. President Wilson had played an important part at Versailles, but his countrymen did not share his enthusiasm for foreign affairs, and in 1920 the Senate declined to ratify the treaty, leaving America outside the newly formed League of Nations.

The scars of war were livid on the face of France long after the peacemakers departed. France had lost 1,335,000 men killed – one-third of Frenchmen under 30 were dead or crippled – and the fighting had left a swathe of destruction across northern France. Frenchmen had gone to war in 1914 burning to recover the provinces of Alsace-Lorraine and sustained by the Russian alliance, but with the provinces regained and Russia lost, the prospect of another war was intolerable. Britain had escaped the worst of the war's destruction, but she and her empire had lost nearly one million men and her merchant navy had been mauled by German submarines. It was tempting for soldiers and politicians to see 1914–18 as an aberration which could never be repeated. In 1919 the "Ten Year Rule" was adopted, under which it was assumed that there would be no major war for the next ten years, and the demands of the empire dominated British defence policy.

Economic difficulties faced former allies and enemies alike. In 1923 Germany defaulted on the reparations she owed the Allies and the French and Belgians occupied the

The map (*opposite*) shows the political boundaries in Europe, the Middle East and North Africa immediately before the outbreak of World War II. The flashpoint was the German invasion of Poland and annexation of Danzig on September 1 1939.

—TIME CHART—

1919	Treaty of Versailles ends World War I
1920	Russo-Polish War: Russians concede territory
1923	Germany defaults on war reparation payments
1923	French and Belgians occupy the Ruhr
from 1926	German forces begin to use Enigma cipher machine
1929	France begins construction of Maginot Line
1933	Hitler becomes Chancellor of Germany
1934	German-Polish Non-Aggression Pact
1935	Franco-Soviet and Soviet-Czech pacts
1935	Germany forms first panzer divisions
1935 Mar	Hitler reintroduces conscription
1936 Mar	German troops reoccupy the Rhineland
1936–39	Spanish Civil War enables Germany, Italy and USSR to test new military techniques and hardware
1938 Mar	Germany annexes Austria
1939	
Mar	Germany occupies Czechoslovakia
Aug 23	German-Soviet Non-Agression Pact
Sept	Anglo-Franco-Polish alliance
Sept 1	German invasion of Poland
Sept 17	Russians invade Poland
Oct	British Expeditionary Force deploys in France
1939–40	Russo-Finnish War: effective use of ski-troops by Finns
1940	
Apr 8	Germany invades Norway and Denmark: airborne troops spearhead seizure of airfields
May 10	Germany attacks Holland, Belgium and France
May 14	Dutch army surrenders
May 16	Allied armies in Belgium start to withdraw
May 26	Evacuation of allied troops from Dunkirk begins
June 8	Allied evacuation of Norway
June 10	Italy declares war on Britain and France
June 14	Germans enter Paris
June 22	Franco-German armistice signed

167

Ruhr in an attempt to force payment. The episode soured Anglo-French relations and worsened Germany's already desperate plight, contributing to inflation and unemployment. Then, in 1929, came the Wall Street crash, which plunged the entire Western world into acute economic crisis. Extremism flourished in the atmosphere of the slump. In Germany the National Socialists, loud in their condemnation of the Treaty of Versailles and of the sinister influence of the Jews, became increasingly popular among Germans eager as much for scapegoats as for solutions. In January 1933 Adolf Hitler, the Nazi leader, became Chancellor. He reintroduced conscription in March 1935 and in the following year lent emphasis to his denunciation of Versailles by sending troops into the demilitarized Rhineland. The lack of response encouraged him, and in March 1938 he annexed Austria.

Many outside Germany had looked indulgently on Hitler's earlier activities, but by the summer of 1938 the spectre of German expansionism alarmed even those who had previously chosen to ignore it. In September the French and British leaders, Daladier and Chamberlain, met Hitler and his Italian counterpart Mussolini at Munich, discussed Hitler's intentions towards Czechoslovakia, and left assured that there would, as Chamberlain put it, be "peace in our time". The agreement did not save the Czechs, but it secured one year's breathing space for Britain and France. On September 1 1939 Hitler invaded Poland, and two days later Britain and France were at war with Germany. Marshal Foch had prophesied that Versailles was not a peace treaty but an armistice for 20 years: events had proved him right.

Foch's conviction that another war was inevitable was shared by few of his countrymen. The French army of the interwar years was, in Paul-Marie de la Gorce's words, "an Army without a Mission". Its morale was eroded by low pay and many of its officers were unsympathetic to the left-wing governments of the Third Republic. It had made great efforts to learn the lessons of 1914–18: its tragedy was that it had learnt the wrong ones. The official view, set out in the 1921 *Instruction provisoire sur l'emploi tactique des grandes unités*, was that artillery was the dominant arm: artillery neutralized, while infantry, supported by tanks and aircraft, occupied. Tanks were merely "a subdivision of the infantry arm" whose task was "to make it easier for the infantry to proceed". The value of fortifications seemed clear: the defence of Verdun epitomized French experience of the war, and it was easy, if misleading, to equate Verdun with its forts.

Fortifications on the frontier

The Maginot Line was the logical expression of this confidence in *matériel*. Begun in 1929, it ran along the Franco-German frontier, stopping short at the Ardennes. These were believed to be impassable to major military movement, and in any case Belgium was an ally. By the time Belgium declared her neutrality in 1936 there was no money available for a full extension – although some smaller defences were built – and there were technical problems in the construction of deep fortifications along the Belgian border. The Italian frontier was not neglected: forts blocked the Alpine passes and a fortified line ran from Mont Mounier to the sea.

Along the northeast frontier and the Rhine the Maginot Line consisted of 23 artillery forts, 35 smaller infantry forts, 295 interval casemates and blockhouses, and scores of minor defences. This artist's impression shows features found in the larger artillery forts. Ammunition and personnel entered the fort through separate defensible entrance blocks, usually some distance from the fighting part of the fort. The main barrack complex (*far right*), with three-tier iron beds, was near the entrance. The main gallery (*right*) housed an electric railway and an overhead monorail, the latter used to move ammunition in its square metal cages from the railway into the bays of the magazine (*bottom centre*). The fort produced electricity in its factory (*usine*) with diesel generators, and had transformer equipment and a ventilation system designed to purge the fort of foul air and cleanse incoming air in an anti-gas filter.

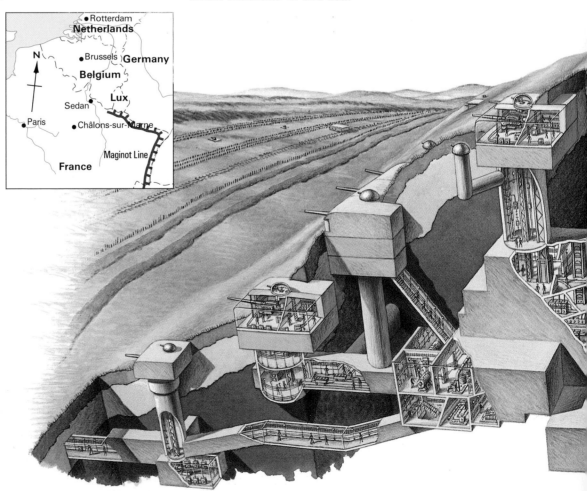

Major General J F C "Boney" Fuller (1878–1966) was commissioned into the infantry from Sandhurst in 1898, served in the Boer War, and went to Staff College in 1913. He arrived in France in 1915, and in late 1916 he joined the staff of the embryo Tank Corps, going on to become its chief of staff, and playing a leading part in the planning of tank operations in 1917–18. His visionary "Plan 1919" contained elements of future blitzkrieg.

After the war Fuller emerged as a prominent military writer, suggesting that a mechanical army would in the future replace the traditional combat arms. He became friendly with Captain Basil Liddell Hart; though the two differed on points of detail, they agreed that mechanization was essential. In 1927 he declined command of the Experimental Mechanized Force on the grounds that inadequate resources had been allocated to it. Promoted major general in 1930, he went on half-pay rather than accept command of the second-class district of Bombay. After retirement Fuller became involved in fascist politics, which dissuaded the War Cabinet from re-employing him in 1939. He wrote at length on military affairs, politics and the occult. At his best he was brilliant: the relative failure of his military career cannot obscure his stature as a prophet of modern war.

The Germans were well aware of the Maginot Line, but by 1939 they were concerned less with the problems posed by steel and concrete than with the potential offered by the tank and the ground-attack aircraft. During the 1920s some theorists in Britain, France, Germany, the United States and the Soviet Union had argued in favour of mechanized forces. Although their ideas differed in detail, Captain Basil Liddell Hart, Major General J F C Fuller, General Heinz Guderian, Marshal Mikhail Tukhachevsky and Colonel Charles de Gaulle had argued that the tank had a vital part to play. It was hard for them to convince their more conventionally-minded colleagues: personality clashes and inter-arm rivalry made matters worse. In Britain, France and the United States conservatism triumphed despite useful work by the advocates of mobility. The British Experimental Mechanized Force of 1927 was disbanded two years later, and it was not until 1938 that Britain at last got an armoured division. The Russians were more far-sighted: their 1936 Field Service Regulations noted that "only a decisive offensive in the main direction concluding with persistent pursuit, leads to complete annihilation of forces and means of the enemy", and by 1936 they had the largest mechanized force in the world. But Tukhachevsky and his ablest supporters perished in Stalin's purges, and incalculable damage was done to Soviet doctrine and operational efficiency, with many officers promoted to posts for which they had inadequate training and experience.

In Germany things were different. Hitler had written of winning a future war in "a single gigantic stroke", and when he saw Guderian's tanks training at Kummersdorff

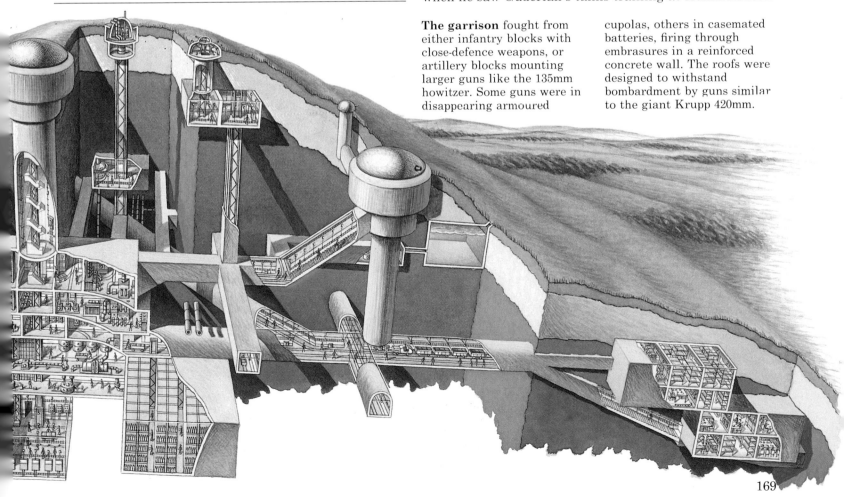

The garrison fought from either infantry blocks with close-defence weapons, or artillery blocks mounting larger guns like the 135mm howitzer. Some guns were in disappearing armoured cupolas, others in casemated batteries, firing through embrasures in a reinforced concrete wall. The roofs were designed to withstand bombardment by guns similar to the giant Krupp 420mm.

(probably in 1935) he was delighted, exclaiming: "That's what I need. That's what I want to have" The *Reichswehr* certainly had its conservatives, but it was less hostile to new ideas than most other armies, and collaboration with Russia had enabled research to be carried out even while Germany remained shackled by the Treaty of Versailles. Three panzer (armoured) divisions were formed in 1935, and there were five panzer divisions and four light divisions, with a tank battalion each, by 1939.

When the Spanish Civil War broke out in 1936, the Germans sent ground and air forces – including some *Panzerkampfwagen* (PzKw) Mk I tanks and Junkers Ju 87 Stuka dive bombers – to help Franco's Nationalists. While the effect of the German and Italian contribution must not be overrated, the war did provide evidence of the value of the ground-attack aircraft as "flying artillery". The evidence on tanks was equivocal. Some observers suggested that the war proved that anti-tank weapons had the edge, but the Germans noted that harsh terrain, obsolescent tanks and inexperienced crews made Spain an imperfect testing-ground. In March 1938 Guderian's panzers led the unopposed advance of German forces into Austria: they had difficulties with breakdowns and fuel, but the episode, in Guderian's words, "proved that our theoretical belief concerning the operational possibilities of panzer divisions was justified".

SS troops follow an armoured car, Danzig 1939.

Polish prelude

In 1914, the outbreak of war had come as a surprise: in 1939 it caused less astonishment. Hitler took Bohemia and Moravia – the as yet unoccupied part of Czechoslovakia – under his "protection" in March, denounced the 1934 non-aggression pact with Poland and the 1935 naval agreement with Britain, and signed a "Pact of Steel" with Mussolini. If none of this was unexpected, the Nazi-Soviet Pact of August 23 was unlooked-for and alarming. The threat to Poland was clear, and Britain's immediate undertaking to support her did little to deter the German Chancellor. The British government had reluctantly decided to send an expeditionary force to the continent in the event of war: on

March 29 it ordered the doubling of the Territorial Army and on April 27 conscription was introduced. All this promised but little to Poland. France, however, seemed more helpful. Marshal Edward Smigly-Rydz, the Polish commander-in-chief, had the word of his French opposite number, General Maurice Gamelin, that France would launch a major offensive against the aggressors within 16 days of mobilization – but this undertaking squared uneasily with France's defensive strategy and Germany's own fortress barrier, the Siegfried Line.

The Treaty of Versailles furnished the casus belli. It had separated East Prussia from the rest of Germany, although providing free access through the "Polish Corridor", while the German-speaking city of Danzig became a free state under League of Nations supervision. Hitler sought Danzig and secure communications with it, but the Poles stood firm. "Case White", the German war plan, had been produced even before the agreement with Russia. Hitler's directive for a surprise attack was elaborated into a double envelopment. Bock's Army Group North was to use its Fourth Army to clear the Corridor, while the Third Army swung down from East Prussia towards Warsaw. Rundstedt's Army Group South was to advance through Silesia, the Eighth and Tenth Armies on its left and centre making for Warsaw while the Fourteenth Army on its right pushed on towards Lvov. The bulk of the Polish army

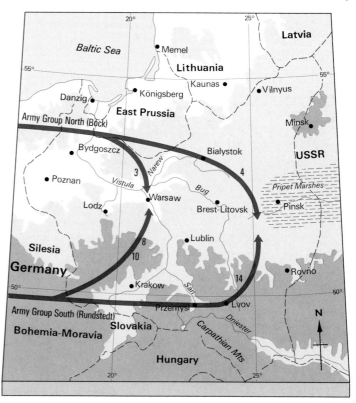

In concentrating against Poland, the Germans took the risk that the French would attack in the west. This risk was justified: the French launched only a limited offensive into the Saarland, advancing 5 miles on a 16-mile front. The plan for "Case White", shown in outline here, was for a classic double envelopment, two army groups advancing on narrow fronts to encircle the Polish armies. The operation did not display the stark simplicity of the initial project, due to Polish resistance, changes of plan, and Russian involvement.

would be cut off west of the Vistula-Narew line and dealt with methodically. Hitler briefed his senior commanders at Obersalzberg on August 21, stressing his determination to solve the Polish question and adding that he was sure that Britain and France would not intervene.

Germany's military might

Germany had forty infantry, six panzer, four motorized and four light divisions available for the attack, supported by 1,300 aircraft with another 400 in reserve or allocated to the army. The Luftwaffe was first to achieve air superiority – largely by destroying the Polish air force on the ground in a surprise attack – and then to interdict lines of communication and attack ground forces. Poland could muster six armies (and one reserve army) made up of thirty infantry divisions, eleven cavalry brigades, one mechanized brigade, two engineer brigades and eleven artillery regiments. The air command had a first-line strength of 388 aircraft, with another 950 under repair or in secondary roles. Though much of their equipment was obsolete, Poland's armed forces were not merely the lancers of popular legend. She had just under 900 tanks: many were small tankettes, but the 95 7TPjw tanks were a match for most panzers. Nevertheless, command and control arrangements were inadequate and operational mobility was poor: these defects would cost the Poles dear.

In 1920 the Poles had won the highly mobile Russo-Polish War and until the 1930s they had regarded Russia as their most likely enemy, although Stalin's purges and the Red Army's preoccupation with Japanese expansion in Manchuria suggested that the Soviet threat had diminished as the German danger increased. The Poles recognized that a two-front war plan was unrealistic, and in 1939 they implemented a version of Plan *Zacod* (West) for the forward defence of their frontiers. Another option, the defence of the shorter line of the Niemen, Bóbr, Narew, Vistula and San rivers, was rejected because it relinquished the industrial areas of Lodz and Upper Silesia and the agricultural zones of Poznan, Kutno and Kielce, and would have put the population of those areas at the mercy of a surprise attack. Smigly-Rydz hoped that defending the frontiers would enable him to mobilize and form reserve formations which would counterattack penetrations of the frontier cordon. His only real chance was to play for time, hoping that Anglo-French efforts in the west would check the Germans. By defending their frontiers the Poles risked the early defeat which made this strategy unworkable. Yet we may doubt whether even the most judicious defence of an interior position would have served Poland much better, given the fact that she had no long-term prospect of resisting the *Wehrmacht* without British and French help.

The Germans attacked at dawn on September 1. Their

Blitzkrieg depended very much on communications. This photograph of General Heinz Guderian in his command half-track in May 1940 shows the Enigma cyphering machine in the foreground. Behind the two Enigma operators sits the operator of Guderian's high-frequency command radio net. Just visible to the rear are motorcycle dispatch riders, ready to carry messages.

The heterogeneous character of the *Wehrmacht* is summed up in this photograph. In the foreground, half-tracks tow 88mm guns: to the rear are horse-drawn guns, more reminiscent of 1918 than 1940 but typical of much German World War II artillery.

initial air strike was not totally effective, because of fog and because the Poles, although caught half-mobilized, had dispersed many of their aircraft. Nevertheless, the damage inflicted on runways and hangars, the effect of interdiction on the flow of spares, and the attritional consequences of aerial combat, meant that the remaining Polish aircraft – approximately 100 – were flown to internment in Romania on the 17th. On the ground the Germans made rapid progress. Their advance was spearheaded by the new panzer divisions, boring their way through resistance with the assistance of their own motorized infantry and dedicated air support. Behind them marched the infantry, striving to remain in contact with the tip of the armoured drill, widening the penetration and bolstering up its flanks.

PzKw Mk I tanks in Poland, 1940.

Poland crushed

The Germans did not have it all their own way. The Poles fought back hard and sometimes their resistance shook the confidence of German commanders. Nevertheless, by September 6 both army groups had broken through and thousands of Poles had been cut off. Between September 9–12, seven Polish divisions were wiped out in the Radom pocket; by September 8 a panzer division was on the outskirts of Warsaw. On September 9 the Polish forces between Torun (Thorn) and Lodz counterattacked towards Warsaw, but the Germans swung divisions in from all sides and the battle of the Bzura river cost the Poles well in excess of 100,000 prisoners. The Germans completed their encirclement when the pincers of the two army groups met near Brest-Litovsk on September 17; on the same day the

Russians crossed the frontier to the east. The Polish government fled abroad and on September 27 its bomb-ravaged capital, with perhaps 40,000 of its population killed, fell to the Germans. In less than a month the Germans had defeated a nation of 33,000,000 and its armed forces of just over 1,000,000 at a cost of fewer than 46,000 casualties. To a world inured to the costly slogging-matches of 1914–18 this was indeed lightning war.

The ripples of the campaign spread wide. The French, whose own promised offensive inched laboriously into the Saar in early September, were well-informed of German tactics by their military missions in Poland. On September 21, General Gamelin announced that blitzkrieg could be countered by thinning out troops in the front line so as to

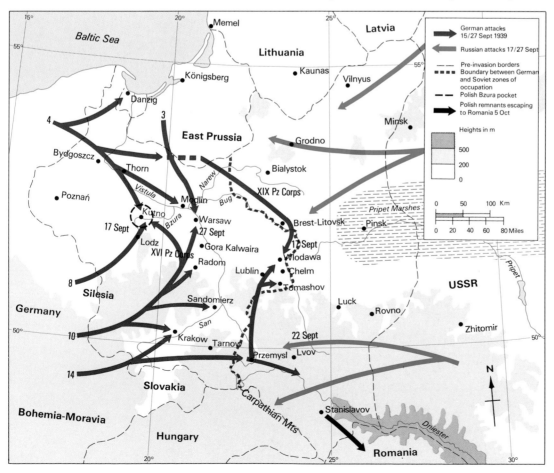

The German attack on Poland was the first example of blitzkrieg in action. Attack at dawn on September 1 was followed by a rapid advance, with battles of encirclement around Radom and on the Bzura river. The Russians entered Poland on September 17, and the country was divided into German and Soviet zones of occupation. Some Polish soldiers reached Romania, and a few ships and aircraft escaped the catastrophe.

provide maximum reserves for counterattacks, and pointed out that the potential separation of infantry from armour was a major weakness. In October the Germans, strengthened by the arrival of troops from the east, regained the ground lost the previous month, but a more general offensive failed to materialize. General Lord Gort's British Expeditionary Force deployed on the Belgian border in early October, but as the worst winter for many years tightened its grip on northeast France, lightning war seemed to have been replaced by what the French called *drôle de guerre*, the Germans *Sitzkrieg*, and the Anglo-Saxons, perhaps most aptly of all, the Phoney War.

There was nothing phoney about the fighting in Finland that winter. On November 30 the Russians, bent on rectifying their northwestern border, attacked Finland. The Finns had blocked the Karelian Isthmus, the 30-mile wide corridor between Lake Ladoga and the Gulf of Finland, with the Mannerheim Line, named after their commander-in-chief, but they were outnumbered in men and utterly outclassed in tanks and aircraft. Nevertheless, they did astonishingly well in the early stages of the fighting. Russian thrusts against the Mannerheim Line, north of Lake Ladoga, farther north opposite the White Sea and in the far north towards Petsamo, were checked with heavy losses, the Finns making good use of small ski-borne units which moved around the flanks of enemy columns. Russian air superiority was indecisive, although their ability to resupply forward positions by parachute was vital to their hard-pressed army.

But the Russians were not to be denied. Marshal Timoshenko took control of operations from Marshal Voroshilov and in February assaulted the Mannerheim Line with 13 divisions supported by an artillery concentration of World War I proportions. The tactics were those of Passchendaele rather than Poland, but on February 15 the line was pierced. Timoshenko exploited the breakthrough with fresh divisions and made good use of his armour – particularly the 47-ton KV-1 heavy tank – to knock out Finnish fortifications. Although some foreign volunteers fought for the Finns, British and French support for Finland was scarcely more use than it had been for Poland. An armistice was finally signed in mid-March and Finland agreed to relinquish control of the Karelian Isthmus and some other territory.

Repercussions of the conflict

The Russo-Finnish War is not one of the blind alleys of military history. Its events were studied closely in the west and gave comfort to the advocates of fortification and the critics of the tank. It showed that although the Red Army had undoubted strengths – robust and hardy soldiers not least among them – it lacked experienced officers, its tactics were cumbersome and its discipline was not always reliable. Timoshenko became deputy commissar of defence, and set about remedying many of the defects revealed by the war: the Red Army reaped many benefits from its costly victory over Finland. The most important of all the war's legacies was its effect upon Hitler. The German Chancellor was persuaded by the Red Army's lacklustre performance that it was no match for the *Wehrmacht*: thus the "winter war" was not least among the title deeds of Operation Barbarossa.

Finnish infantry and reindeer transport in the Petsamo area in the latter stages of the Russo-Finnish War of 1939–40. Although the Finns were badly outnumbered, they gave a good account of themselves: units like this earned a formidable reputation for their *motti* tactics, infiltrating to wreak havoc behind Russian lines.

Blitzkrieg in the West

Victory over Poland left the Germans curiously directionless. A sanguine Hitler ordered the preparation of an offensive in the west, but the Army High Command (*Oberkommando des Heeres*: OKH) was less confident. There was already tension between OKH and Hitler, springing from the only partial conversion of the army's senior officers to Nazism, the tendency of some to regard Hitler as a jumped-up corporal, and the General Staff's inclination to weigh its decisions carefully. Personalities also played their part. In 1938 Hitler had succeeded Field Marshal Werner von Blomberg (his reputation sullied by an unwise marriage) as supreme commander of the *Wehrmacht* and War Minister, setting up the Armed Forces High Command (*Oberkommando der Wehrmacht*: OKW), with the lightweight Major General Wilhelm Keitel as its professional head. Colonel General Werner von Fritsch, commander-in-chief of the army, resigned after a false accusation of homosexuality. His successor, Walther von Brauchitsch, an honest and capable man, found it hard to stand up to Hitler. Hitler had left the Polish campaign to OKH, but by the autumn of 1939 he was convinced that OKH was over-cautious, and came to regard OKW as the best means of giving substance to his wishes.

Hitler had written in *Mein Kampf* of "one last, decisive battle" against France. OKH's plan for an offensive, "Case Yellow", had little decisiveness about it. It was in essence a re-run of the Schlieffen Plan, its main attack delivered by Army Group B, thrusting through Belgium and northern France. The southern flank was to be covered by Army Group A, while Army Group C held the Siegfried Line. The

plan had critics, notably Lieutenant General Erich von Manstein, chief of staff of Army Group A. Fearing that the attack would not achieve surprise and would bring only partial victory, he argued instead in favour of an attempt "*to achieve decisive results on land*". His plan allotted the weight of the offensive to Army Group A, which would surprise the Allies by moving through the Ardennes – presumed to be impassable to a major military operation – turning the flank of the Maginot Line and pressing on to the lower Somme to cut the Allied armies in half.

Hitler axes Case Yellow

The Manstein plan was unenthusiastically received at OKH, but the capture of a copy of "Case Yellow", lost when its courier landed behind Belgian lines, made the high command more inclined to listen to alternative proposals. Manstein himself, regarded as "an importunate nuisance", was sent off to command a corps on the Polish frontier, but on his way there he lunched with Hitler and took the opportunity to expound his plan. The final operations order, issued on February 24 1940, overturned the OKH plan in favour of Manstein's. In the north, Army Group B (Bock) with 3 panzer and 26 other divisions, was to overrun Holland and northern Belgium. Rundstedt's Army Group A, with 45 divisions, 7 of them panzer, was to move through the Ardennes and force the crossings of the Meuse. A key role was allocated to General Paul von Kleist's Panzer Group: Guderian's 19 Panzer Corps was to cross at Sedan and Reinhardt's 41 Panzer Corps at Monthermé. Hoth's 15 Panzer Corps was to make its crossing farther north at Dinant. Army Group C (Leeb), with 19 infantry divisions,

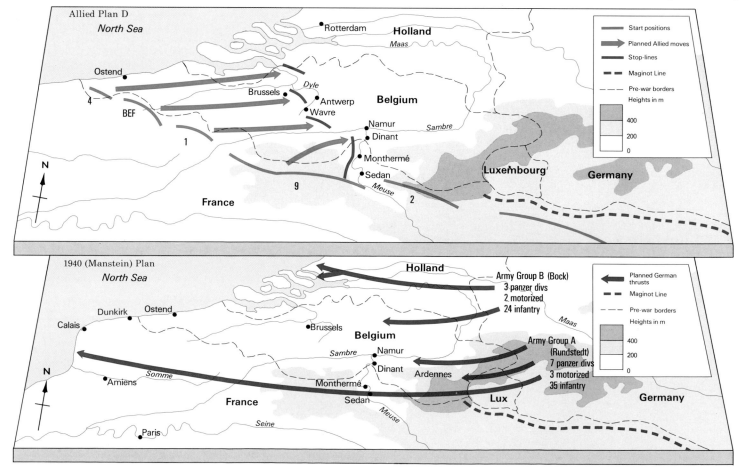

Colonel General Heinz Guderian (1888–1953), "The Father of the Panzer Divisions", was born at Kholm on the Vistula, son of a Prussian officer, and was commissioned into the infantry in 1908. In World War I he served as a signals officer, and later on the staff. After the war he raised volunteer forces in the Baltic states before returing to the *Reichswehr*. He went to the motorized troops department of the Ministry of Defence in 1922, and despite official discouragement championed the use of tanks. In 1931 he commanded a transport

battalion, re-equipped with dummy tanks and anti-tank guns as an armoured unit, and in 1934 a Motorized Troops Command Staff was set up, with Guderian as its chief of staff: he took command of 2nd Panzer Division in 1935. In *Achtung! Panzer!* he argued that the tank was a weapon of mobile, protected firepower, best used concentrated.

He led the advance into Austria in 1938, and he commanded a panzer corps in the 1939 and 1940 campaigns. He headed a panzer group – later 2nd Panzer Army – in Russia, but was dismissed in December 1941. In 1943 he was appointed Inspector-General of Armoured Forces, and in 1944 Chief of the General Staff. He stood up to Hitler, but perhaps inevitably, was dismissed in March 1945. "Hurrying Heinz" was a flawed genius: an inspiring leader but a difficult subordinate, whose expression *"Klotzern, nicht Kleckern"* (smash, don't tap) was the essence of blitzkrieg.

on February 16 caused a diplomatic furore – and Major Vidkun Quisling, leader of the fascist National Party, was in secret contact with the Germans.

Allied hesitation enabled the Germans to strike first. On April 8–9 British warships and Norwegian coastal defences inflicted some losses on the invasion fleet, but, assisted by Quisling and his supporters, the Germans took the capital Oslo and all the main ports with relative ease: Denmark, too, was swallowed up on April 8. The Allies replied with landings of their own, but only at Narvik, captured on May 18 by British, French and Polish troops, did they achieve any success. The Germans were less fortunate at sea, losing ten destroyers in two spirited actions off Narvik, but the eventual Allied evacuation of Narvik was marred by the loss of the aircraft carrier HMS *Glorious* on June 8. Allied

Wrecked German shipping in Narvik Fjord, April 10 1940.

evacuation left Norway, with its iron ore, naval bases and airfields, in German hands, and materially weakened Britain's hold on the approaches to the North Atlantic.

News of the withdrawal from Norway was overshadowed by more dramatic events elsewhere. German involvement in the north had led to the postponement of the offensive in the west. The Allies made little effective use of the lull. They remained convinced that the Germans would attack into Belgium, and four armies of General Billotte's First Army Group – the French First Army (Blanchard), Seventh (Giraud) and Ninth (Corap) and the BEF – were poised to move forward to join 14 Belgian divisions as soon as invasion jolted Belgium out of neutrality. Another ten Belgian divisions lay farther forward on the Meuse and the Albert Canal. Around Sedan, at the hinge between Billotte's mobile forces and the 50 divisions of the Second and Third Army Groups holding the Maginot Line, lay General Huntziger's Second Army.

The dimensions of the Allied catastrophe were to fuel the myth that the Germans enjoyed both qualitative and quantitative superiority. Both sides were roughly equal in manpower, with 136 divisions each, although Allied formations were of variable quality. There is no agreement among authorities on precise numbers of tanks, but the Germans deployed about 2,600 and the Allies around 3,000. Allied tanks were not universally hopeless: the French *Char B* mounted a useful 75mm gun and the British Matilda tank was so heavily armoured as to be immune to most German anti-tank guns. But while German tanks were designed, and their crews trained, for mechanized war, the baneful influence of interwar doctrine hung over

was to mask the Maginot Line.

The Allies, anticipating an attack into Belgium and Holland, planned to move forward to meet it with three French armies and the British Expeditionary Force of nine divisions and an armoured brigade. These could advance either to the River Schelde (Plan E), or to the line Antwerp–Dyle river–Wavre–Namur–Meuse river (Plan D). Plan D was preferred, but the discussion revealed disagreements within the French high command, for while Gamelin favoured Plan D, General Georges, commander of the northeastern front, feared that a German attack into Belgium might draw Allied strength away from the point of decision elsewhere.

The conflict in Norway

Ironically, despite preparation for war in France and the Low Countries, the first serious clash between the Allies and the Germans took place in neutral Norway. The Allies were interested in Norway because it supplied iron ore to Germany and also offered a base from which the Finns might be supported. Plans to aid the Finns from Norway were overtaken by the defeat of Finland, but the Allies nevertheless planned to lay mines to prevent German use of Norwegian territorial waters and to follow up by sending forces to Norway in early April. Hitler took a personal interest in Norway, prompted by Allied attention to the area and by Admiral Raeder's desire to use Norwegian naval bases to attack British shipping. The Norwegians were understandably anxious not to offend Germany – British capture of the German tanker *Altmark*, with British prisoners aboard, in Norwegian territorial waters

the Allies. About one-third of French tanks were parcelled out to support the infantry, and as many again were split up among five horsed cavalry divisions and three light mechanized divisions. By May 1940 the French had three armoured divisions, with a fourth planned, but these were recently formed and only partially equipped. The British were in a worse state, with the 1st Armoured Division still training in England. Thus, while Allied armour was spread out along the entire front, its best units ready to carry out Plan D, the Germans held three-quarters of their armour in the panzer corps of Army Group A, poised to administer a concentrated blow. The Germans enjoyed a clear superiority in aircraft, with some 3,000 to the Allied 1,800, and the Luftwaffe was much better trained than its opponents to cooperate with ground forces.

Neither high command was to emerge unblemished from the campaign, but the Allied apparatus was abrasively complex. Gamelin exercised overall control from his headquarters in the Chateau de Vincennes, on the edge of Paris. Georges' headquarters had been untidily split, Georges himself commanding the northeast front from La Ferté-sous-Jouarre, with an intermediate headquarters, responsible for the planning and preparation of orders, under General Doumenc at Montry. The front was divided into "Zones of Air Operations" which corresponded to the army groups: this seemed logical in theory, but in practice it was to prevent army group commanders from obtaining concentrated air support when they needed it.

They were to need it sooner than they thought. The Germans attacked on the morning of May 10, achieving almost total surprise. The Luftwaffe hammered airfields and paid particular attention to Holland, dropping paratroops and giving a foretaste of the raid on Rotterdam on May 14 that was to kill some 1,000 civilians. While the Allies swung forward into Belgium, obedient to Plan D, the long columns of Army Group A wound into the Ardennes, their progress screened by the Luftwaffe. Special forces, often in Allied uniform, jabbed at key points: the huge Belgian fortress of Eben Emael was taken by glider troops who landed on top of it. Over the next two days German armour continued to flow through the Ardennes, brushing aside sporadic resistance. The attention of the Allied high command was fixed on the north, where the Dutch were already in difficulties (they were overwhelmed on May 14) and French and Belgian units were in contact with German ground forces and, to their increasing peril, German aircraft. The Meuse barrier was quickly broken. On May 13, 7th Panzer Division, commanded by the enterprising Major General Erwin Rommel, secured a bridgehead north of Dinant, while Guderian, supported by an overwhelming 1,500 aircraft, burst across the river at Sedan, striking a blow which produced a psychological shock that jarred the French army terribly.

The relentless advance
The Germans spent May 14 consolidating and dealing with ill-coordinated counterattacks. Allied pilots made courageous attacks on the bridgeheads: of the 71 RAF bombers

A PzKw II leads a half-track towing an 88mm gun across a pontoon bridge over the Meuse under air attack in May 1940. The Mk II, with its 20mm gun, was too lightly armed for tank battles, but did well in the open fighting that followed the breakout. The 88mm originated in 1933 as the Flak-18, but was much modified subsequently: its high silhouette, clearly visible here, was always a source of vulnerability. It was probably first used in the anti-tank role in the west at Mercatel near Arras on May 21.

THE STUKA

The Junkers Ju 87 Stuka dive bomber and ground-attack aircraft was an indispensable ingredient of blitzkrieg, acting as "flying artillery" for armoured forces beyond the reach of conventional guns. It attacked in a steep dive to deliver its one 500kg bomb and four 50kg bombs. Experience in Spain, where some Stukas failed to pull out of the dive in time, had led to the fitting of an automatic device to help the pilot pull out, and red lines at 60°, 70° and 80° were painted on his side window. The Ju 87B, the Stuka's most usual version, had two wing-mounted machine guns and a rearward-firing machine gun, but was vulnerable to air attack and operated best beneath fighter cover. A later version, the Ju 87D, was built specifically for tank-busting on the Eastern Front, and mounted two 37mm cannon in underwing pods. The Stuka shrieked as it dived, and its banshee wail helped to make it one of the war's most terrifying weapons.

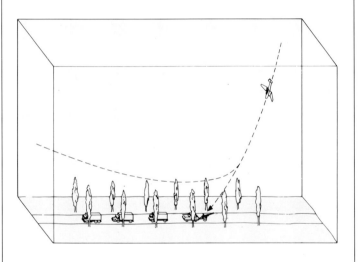

SINEWS OF WAR

The armoured fist of the panzer division is the popular image of the German army in 1940. But on the eve of war only 16 of Germany's 103 divisions were fully motorized, and this had been achieved at the cost of limiting the remainder of the army's motor transport. An infantry division had 942 motor vehicles but 5,375 horses: it consumed over 50 tons of fodder, and only 20 tons of fuel, each day.

There were several reasons for this two-tier structure. Hitler was more interested in new weapons than in wholesale modernization although some generals advocated a more equitable distribution of assets. Despite its much-publicized renaissance, German industry could not supply the army with the vehicles it needed. There was bewildering diversity, with over 100 different types of lorry in military use in 1938, and insufficient quantity: peacetime production was unable to keep pace with normal wastage. In 1939 the commandeering of 16,000 civilian vehicles helped, and by the winter of 1940–41 no less than 88 infantry divisions were equipped with captured French trucks.

German draught horses on the march.

sent into the valley of the Meuse that day, 40 fell victim to fighters or flak. Huntziger, deciding that the Germans were attempting to roll up the Maginot Line from the north, pivoted back south to prevent this happening: by doing so he widened the gap between his army and Corap's Ninth – the very gap through which Guderian lunged on May 15. That day saw the defeat of the French 1st Armoured Division by 7th Panzer and the mauling of the Ninth Army: it ended with General Corap replaced by the energetic Giraud, and German tanks within a dozen miles of his headquarters. By nightfall the danger was at last apparent to Gamelin, and to Paul Reynaud, the French premier, who pressed Winston Churchill, his British counterpart, for more fighters.

Churchill visited an increasingly edgy Paris on May 16 and was shocked to hear Gamelin admit that he had no strategic reserve to deal with the breakthrough. Yet the Germans were far from certain as to how to exploit it: Guderian had already clashed with Kleist, and Hitler, now more nervous than OKH, was worried that the French might counterattack. On the 17th Guderian was ordered to halt, but after a squabble, allowed to proceed with "reconnaissance in force", which he defined to include an advance by everything except rear echelons. On the same day Colonel Charles de Gaulle, recently given command of the part-formed 4th Armoured Division, put in a brisk but unsupported attack on 1st Panzer which caused a momentary check. Behind the lance point of the panzer divisions came the infantry, bare-armed and bare-headed in the heat, marching down the long poplar-lined roads of northern France, singing the songs sung by their fathers in 1914 and their grandfathers in 1870.

". . . we came under heavy artillery and anti-tank gunfire from the west. Shells landed all round us and my tank received two hits one after the other, the first on the upper edge of the turret and the second in the periscope.

The driver promptly opened the throttle wide and drove straight into the nearest bushes. He had gone only a few yards, however, when the tank slid down a steep slope on the western edge of the wood and finally stopped, canted over on its side . . .

The French battery now opened rapid fire on our wood and at any moment we could expect their fire to be aimed at our tank, which was in full view. I therefore decided to abandon it as fast as I could, taking the crew with me. At that moment, the subaltern . . . reported himself seriously wounded with the words: 'Herr General, my left arm has been shot off.' We clambered up through the sand pit, shells crashing and splintering all around."

Major General Erwin Rommel, Onhaye, near Dinant, May 14 1940

On May 16 the Allied armies in Belgium began to withdraw, under repeated and dispiriting air attack. There was talk of a counterattack by divisions brought up from the south, but the French high command, thinking at infantry speed, was hopelessly out of phase with the reality of the German advance. Reynaud decided to replace Gamelin with the 73-year-old General Maxime Weygand, commanding French forces in the Levant, and also summoned Marshal Pétain from his post of ambassador to Spain, appointing him deputy premier on May 18. Tinkering with the chain of command did nothing to halt the panzers. Early on May 18, 2nd Panzer entered St Quentin

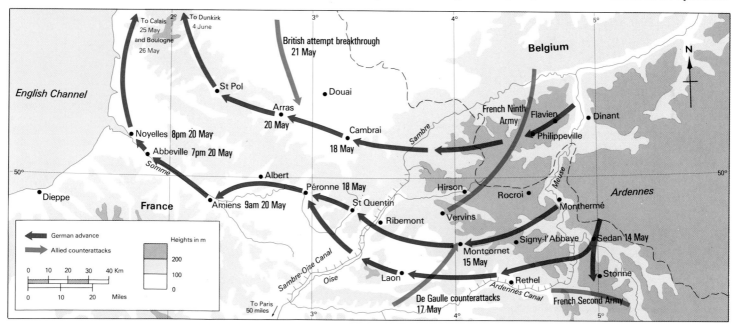

The German advance to the Channel coast, May 14–June 4 1940 (*above*). Shrugging off de Gaulle's counterattack from the Laon area on May 17, the leading German columns reached the sea at Noyelles on May 20, establishing a "Panzer corridor" that divided the British, French and Belgian forces to its north from the main French force to the south. Gort's attempt to break the corridor at Arras on May 21 contributed to the German decision to pause on May 24, and desperate delaying actions at Boulogne and Calais allowed a large part of the Allied force to reach Dunkirk for hurried evacuation.

German infantry under shellfire in Belgium, May 1940 (*right*).
A PzKw Mk I tank (*far right, top*) mounting two machine guns, follows a Mk III, armed with a 37mm cannon (right), through a French village.

and the direction of Guderian's thrust became clear to the French: he was making for the Channel, not for Paris, but little could be done to stop him.

Weygand took over on the 20th and set off to see the situation in the north, but events had spun on before he arrived. General Ironside, Chief of the Imperial General Staff, visited France to tell Gort – who had warned that he might have to consider evacuation – that he was to fight his way south to join the main body of the French army. Gort told him that the French were "tired", and Ironside travelled on to Lens to see Billotte, pressing him to counterattack from the south in concert with a British thrust from the north. On May 21 Gort mounted his counterstroke: a composite force of infantry and tanks swung into the flank of 7th Panzer at Arras, briefly persuading Rommel that he was up against five divisions. The Arras counterattack was imperfect both in planning and execution, but its impact was important, for it rever-

---

"Between Agny and Beaurains I met Major Fernie [who] told me to go and get the guns [firing from Mercatel]. They were not camouflaged, and their only cover was a fold in the ground. I got two of them before they realised I was into them – the range was about 200 yards. The survivors turned on me and one hit the gun housing . . .

I got my gun going [again] and returned to the attack. They must have thought I was finished for I caught the guns limbering up, and revenge was sweet."

WO3 "Muscle" Armit, 4th Royal Tank Regiment, Arras, May 21 1940

---

berated through the German command, and contributed to its subsequent failure of nerve.

Victory for Germany

British and German tanks were burning around Arras when Weygand arrived in Ypres to meet the commanders in the north. He spoke to Billotte and to King Leopold of the Belgians, but missed Gort, who arrived after he had left and was briefed by Billotte. It was Billotte's last official act, for he was mortally injured when his car crashed that night. Weygand's plan that the Belgians should hold the Yser while the French and British put in a two-pronged attack on the panzer corridor, their pincers meeting around Bapaume, was probably never viable, given the ragged state of French morale, but the demise of Billotte, one of the most capable of senior French commanders, who alone might have been able to implement the Weygand Plan, strangled the project at birth.

Weygand returned to Paris, apparently confident that his scheme would work. However, the German advance guard had reached the Channel on May 20 and behind it infantry were shoring up the sides of the panzer corridor. The journeys and conferences of May 20–21 wasted valuable time and with them evaporated the Allies' last chance of cutting off the panzers from their infantry: from May 24 the corridor was secure and the Germans had won the campaign. All that remained to be determined was the size of their victory.

Its scale was diminished by Hitler's decision to halt his panzers on May 24. Among the possible causes of this, the most significant probably were first, concern about tank losses – some 50 percent of Kleist's tanks were out of action,

The Fall of France: French troops surrender at Lille, May 1940.

many for trivial mechanical reasons; second, the desire to maintain an armoured reserve for use elsewhere in France; and third, the perceived danger of another counterattack (the legacy of Arras). The Allied defence of the Channel ports also slowed down the Germans: Boulogne fell on the 25th, and Calais, stoutly defended by the Rifle Regiments of 30th Infantry Brigade, on the 26th. The delays gave the British time to mount Operation Dynamo, the evacuation from Dunkirk, which began on May 26. By June 4, 338,226 personnel, 110,000 of them French, had been evacuated, at the cost of six British and two French destroyers and numerous small craft.

The triumph of blitzkrieg

Dunkirk marked the end of the campaign in the north. Belgium capitulated in the early hours of May 28. The remnants of the French First Army fought a brave and fruitless battle in defence of Lille, leading to suggestions that Frenchmen were fighting on while the British escaped. Although Churchill was to acknowledge that wars were not won by evacuations, the escape of the BEF, albeit without its heavy equipment, was of immense importance. As Alistair Horne was to write:

"Had the BEF been wiped out in northern France, it is difficult to see how Britain could have continued to fight; and with Britain out of the battle, it is even more difficult to see what combination of circumstances could have aligned America and Stalin's Russia to challenge Hitler."

The Anglo-French relationship was much discussed in the days that followed, as a separate peace lobby coalesced in Paris. There were complaints that the RAF had not appeared in the strength promised by Churchill, but these ignored the fact that fighters based in England were flying

about 200 sorties a day over France and had shot down 179 aircraft for the loss of 29 of their own between May 27–30. Weygand's men fought better than he or the Germans expected on the Somme and the Aisne, some of their "hedgehogs" of infantry and 75s holding out with a determination worthy of the men who had held Fort Vaux a generation before. But Paris was untenable: the government departed for Tours on June 10 and the triumphant Germans entered the capital on the 14th. It was no less ominous that Verdun fell the following day after 24 hours' hard and costly defence.

Mussolini declared war on France on June 10, but his intervention, not the brightest jewel in Italian diplomatic or military history, had little effect. The French government, with Reynaud under increasing pressure from the defeatists, moved on to Bordeaux. Late on June 16 the premier resigned. He was replaced by Pétain, who at once contacted the Spanish ambassador to arrange an armistice; it was signed on June 22 in the Forest of Compiègne, scene of the 1918 armistice. It left all France north of a line from the Swiss frontier at Geneva, and on via Bourges to St. Jean-Pied-de-Port, occupied by the Germans. The French government moved yet again, this time

General Charles de Gaulle (1890–1970), commissioned into the infantry in 1914, was captured at Verdun. He took command of the incomplete 4th Armoured Division in May 1940, and on the 19th attacked the panzer corridor with limited success. Promoted brigadier general and made under-secretary for war, he fled to England to champion Free France. He headed the Provisional Government in 1944 but thereafter withdrew from politics, emerging in 1958 to become president of the Fifth Republic until 1969.

to Vichy in the unoccupied zone. Charles de Gaulle, now a temporary brigadier general, broadcast from London that France had lost the battle but not the war. Over the next four years more and more of his countrymen would come to agree with him, but in the summer of 1940 it took unusual prescience to do so.

There was wider agreement on why the Germans had won. Weygand admitted that "we have gone to war with a 1918 army against a German army of 1939. It is sheer madness." Churchill saw things differently, arguing that "in the end it is certain that a regime whose victories are in the main due to machines will collapse. Machines will one day beat machines." The triumph of blitzkrieg was not merely that of superior technology. It was as much about morale as about machinery, and as the war went on the Germans would encounter enemies whose equipment was as good as or better than their own, and whose willpower was not undermined by the squeaky rattle of the panzer's tracks or the shriek of the diving Stuka.

British troops awaiting evacuation from the beaches at Dunkirk.

—THE WAFFEN SS—

The *Schützstaffel* – abbreviated to SS – originated in the "Adolf Hitler Shock Troops" formed in 1922 to protect Hitler. In 1929 they came under the leadership of the 28-year-old Heinrich Himmler, who had served as an officer-cadet in the Bavarian infantry. The SS was the instrument of Hitler's purge of the SA in 1934, and its three-battalion *Verfügungstruppen* (Special Disposal Troops) were given military training, not without criticism from the army. Himmler initially saw the role of part of the SS – the *Totenkopfverbände* – as guarding concentration camps while the rest, together with the police, which he also controlled, maintained internal security in Germany in wartime. He was at first hazy over the task of the *Verfügungstruppen*, the future *Waffen SS*, although by January 1937 they numbered 13 battalions, and a decree of Hitler's in 1938 described them as "a unit of the party, to be exclusively at my disposal".

In 1939 and 1940 Hitler allowed the *Verfügungstruppen* to take the field to gain experience and earn respect: he did not use the term *Waffen SS* until July 1940, and in 1942 told Himmler that he regarded its military status as a wartime expedient. Nevertheless, the *Waffen SS* grew steadily in size,

and by the war's end it comprised 38 divisions. Not all were of the quality of *Liebstandarte Adolf Hitler*, *Das Reich* and *Totenkopf*, and many were divisions in name only. Entry standards, at first ferociously strict, were gradually relaxed. Foreign volunteers were recruited, and the SS order of battle included the French *Charlemagne*, Belgian *Flandern* and *Wallonie*, and Balkan *Handschar* and *Skanderbeg* divisions.

Apologists maintain that the deeds which most redound to German discredit were not the work of *Waffen SS* field units but of *Einsaztgruppen* in the rear – and some hail the *Waffen SS* as a pioneer pan-European movement. It is fairest to judge the *Waffen SS* as what it was, a mixture of hard-core Nazis, policemen, foreign volunteers and German conscripts, its officers ranging from former regular officers like Paul Hausser to old bruisers like ex-sergeant "Sepp" Dietrich and killers like Oskar Dirlewanger. The *Waffen SS* was certainly to suffer by association with other elements of Himmler's labyrinthine black empire. On the one hand it produced some of the best armoured divisions of the war: on the other, some of its members carried out acts which brought them and their cause lasting shame.

The SS was, as the title of its magazine proclaimed, *Das Schwarze Korps* – the Black Corps. Its black uniform, with silver runic flashes and badges of rank on the collar, was replaced by field-grey for the *Waffen SS*, but the distinctive collar-patches and death's head cap-badge remained. The ranks of the *Waffen SS* differed from those of the army, breaking with terminology which links military rank titles to those developed in early modern Europe. The SS had no field marshals, but Himmler held the supreme rank of *Reichsführer SS*.

chapter thirteen

— World War II: The Eastern Front —

Richard Holmes

There is some truth in Russian complaints that western historians often do scant justice to the war on the Eastern Front. Its scale was gigantic. About 25 million Russians served in the armed forces, and perhaps 30 million soldiers and civilians died on both sides: the Germans advanced 1,200 miles into Russia and recoiled 1,500 miles to Berlin on a front never less than 2,400 miles wide. Its contribution to the defeat of Germany was no less massive. German and Italian forces lost some 59,000 killed, wounded or captured at El Alamein, October 23–November 4 1942: around 200,000 Axis soldiers, most of them German, were killed or captured at Stalingrad, three months later. About 250,000 Axis troops surrendered in Tunisia in May 1943: some 900,000 became casualties around Kursk that summer.

The importance of the Eastern Front to Germany was reflected in the intensity with which the war was waged. As General Reinecke, head of the General Army Office at OKW, observed: "The war between Germany and Russia was unlike any other war. The Red Army Soldier . . . was not a soldier in the ordinary sense, but an ideological enemy." This attitude generated at best a callous disregard for the fate of *Untermenschen* – more than 3,000,000 Russians perished while in German hands, mainly through overwork or neglect – and at worst inspired murderous atrocities and counter-atrocities. The corrosive effects of ideology and the bitterness produced by invasion and occupation combined to make the war one in which no holds were barred and the penalties of losing were catastrophic. If the dimensions of the disaster need emphasis, compare the map of Europe in 1939 with that of today. Nearly a quarter of the pre-war area of Germany, more than 44,000 square miles, was lost in the east. Germany is divided and the Reichs of 1871–1918 and 1933–45 are swept away as surely as that of Frederick Barbarossa, the medieval emperor whose name was the codeword for the German invasion plan of 1941.

The road to war

Despite Russo-German cooperation in the interwar years, the prospect of a German attack had long concerned senior Soviet officers. In 1935 Marshal M N Tukhachevsky suggested that Germany's aggressive designs lay towards the east. He pointed out that German doctrine relied upon paralyzing the enemy's will to fight, and would not succeed against an opponent who took the offensive himself. "In the final result," he declared, "all would depend on who had the greater moral fibre and who at the close of operations disposed of operational reserves in depth." Tukhachevsky did not live to see his prophecy proved correct: one of the many victims of Stalin's purges, which killed about 15,000 officers, he was shot in June 1937.

Military reform went on despite the purges. The Red Army's disappointing performance in Finland in 1939–40, and its more creditable achievements on the Manchurian–Mongolian border in 1939, were analysed; general officer ranks, abolished as relics of the old regime, were restored; discipline was strengthened and the powers of commissars were modified. Yet the reforms were incomplete and the shadow of the purges hung over an anxious and depleted officer corps.

Operation Barbarossa

Hitler's desire to obtain "living-space [*lebensraum*] in the East" made a clash with Russia inevitable. He had hoped to deal with Britain first, but the directive for "Case Barbarossa", issued on December 18 1940, decreed that his forces must be prepared to "crush Soviet Russia in a rapid campaign" even before victory in the west. It envisaged the destruction of the Red Army in Western Russia "by daring operations led by deeply penetrating armoured spearheads", and foresaw an advance to a line from the Volga to Archangel, later redefined to Archangel–Astrakhan. Three army groups would make the attack. Army Group North was to advance on Leningrad, Army Group Centre

---

—TIME CHART—

1937–39	Stalin purges Red Army	Nov 23	German forces trapped at Stalingrad
1939 Aug	Soviets use mass tank tactics against Japanese at Khalkin-Gol in Manchuria		(surrender Jan 31)
		1943	
1939–1940	Russo-Finnish War	Jan	Soviet attacks open corridor to Leningrad
1941		July 5–13	German Operation Citadel (Kursk) fails against Soviet deep defence
June 22	Germans launch Operation Barbarossa: invasion of Soviet Union	Dec 24	Soviets launch Ukrainian offensive
Sept (mid)	Leningrad encircled	1944	
Sept 30	Germans launch Operation Typhoon: assault on Moscow	Jan 14	Soviets launch northern offensive Operation Bagration begins: destruction of German Army Group Centre
Dec 5	Germans halted by Zhukov's counterattack	Sept 19	Finnish-Soviet armistice
1942		Sept	Romania and Bulgaria in Soviet hands
Jan–Mar	Soviet counteroffensive	1945	
Feb-Apr	German airlift maintains Demyansk pocket	Jan 12	Soviet drive for the Oder begins
June 28	German summer offensive opens	Feb	Hungary and East Prussia occupied
Sept	Street battle for Stalingrad begins	Apr 13	Soviets take Vienna
Nov 19	Soviet counteroffensive opens	May 2	Berlin taken
		Aug 8–22	Soviets defeat Japanese in Manchuria

Europe in November 1942, the high-water mark of German expansion in the east, was divided into four by Hitler's "New Order": "Aryan" areas like Austria, annexed and integrated into the German economy; occupied non-incorporated areas, such as France; the Balkan puppet and satellite states; and occupied Slavic areas earmarked for German colonization, the intended fate for the Ukraine.

The German attack on the Soviet Union in June 1941 had been followed by 14 months of German victory, interrupted by the Soviet counteroffensive of December 1941 which checked the advance and recovered some ground. In the summer of 1942 the Germans had embarked on Operation Blue, clearing the Crimea, taking

Rostov and pushing deep into the Caucasus. Although the initial plan had envisaged holding a defensive shoulder on the Volga, the capture of Stalingrad soon became a major objective, and in the late summer and early autumn the Germans fought hard for the city against a resolute Russian defence. But the tide of war turned against

them in November when the Russians counterattacked either side of Stalingrad, cutting off the Sixth Army. After an unsuccessful attempt to relieve them, the Germans surrendered on January 31 1943.

Legend:
- Border of Greater Germany (Grossdeutsches Reich)
- Axis territory
- Under German rule
- Under Axis military occupation
- Axis satellites
- Allied territory

on Moscow, and Army Group South on Kiev. The December 18 directive specified that Leningrad was a more important initial objective than Moscow, and that both objectives should be pursued only if Russian resistance collapsed. In this and in the subsequent debate over the goals of "Barbarossa" there was a dangerous lack of clarity in German planning. In one respect, however, these plans were clear enough. The demolition of Russia would be swift: General Halder, Chief of Staff at OKH, reckoned that the campaign would last only eight to ten weeks.

There was some reason for German confidence. The Red Army was ill-prepared, its war plan was for mere linear defence of the frontiers, and there were too few operational reserves. By June 1941 the Germans had 153 divisions, more than 3,000,000 men, concentrated in the east. There were 19 panzer and 14 motorized divisions, although of the 4,700 tanks on the German inventory, only 1,440 were modern PzKw III and only 550 the newer PzKw IV: the rest were obsolete Marks I and II or modified Czech types. The Red Army had about the same number of divisions on the frontier, but many of them were under-strength. Of more than 20,000 tanks, only 1,800 were modern KV-1, KV-2 or T-34 types. The 7,000 aircraft of the Red Air Force were inferior to those of the Luftwaffe and their pilots were not as well trained.

Hitler hoped to launch Operation Barbarossa on May 15, at the end of the spring thaw, thus allowing unrestricted campaigning until the arrival of the autumn mud in October. Winter lasted from November until the spring thaw – and the spring mud – in March. The spring mud persisted longer than usual in 1941, and German operations

in the Balkans may have contributed to delays. However, despite evidence of German preparations from agents of the "Lucy Ring", a German anti-Nazi organization operating through Switzerland, radio intercepts and the reports of forward units, Stalin refused to allow his forces to prepare to meet the attack.

German success

The offensive opened on the morning of June 22 1941. The advance of Field Marshal von Leeb's Army Group North was impeded by difficult country and hindered by Hitler's interference and disagreements between Leeb and his panzer group commander. Not until early September was it in a position to attack Leningrad. Hitler then forbade direct assault, ordering Leeb "to close in on the city and blast it to the ground". Field Marshal von Bock's Army Group Centre achieved more spectacular success. Its two panzer groups, under Hoth and Guderian, cut deep into the Soviet defences, linking up behind them to form huge pockets: some 500,000 Russians were swallowed up at Minsk in late June and another 300,000 around Smolensk in mid-July. Bock might have reached Moscow had he been allowed to do so; but Hitler halted his advance, declaring on August 21 that the Crimea and the industrial areas of the Donets Basin were more important objectives. Field Marshal von Rundstedt's Army Group South made slower progress, but in mid-September Guderian, diverted from the central thrust by Hitler's change of emphasis, hooked down to join up with Kleist's panzer group and seal off the Kiev pocket with some 665,000 Russians inside.

Uncertainty persisted. A directive of September 6

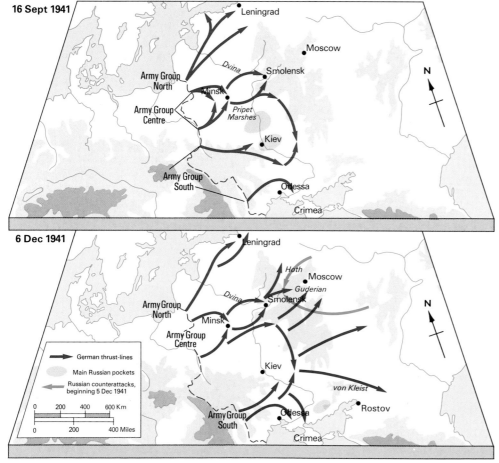

On June 22, 1941 the Germans mounted Operation Barbarossa, the invasion of the Soviet Union, with three army groups. The northern German thrust, over difficult tank country, made slowest progress, but by mid-September (*top*), had invested Leningrad. Army Group Centre, with greater resources, and moving over better terrain, was more successful, with Hoth and Guderian's panzer groups enveloping the main Russian concentration (west of Minsk) by the fifth day of the invasion, and the Russian reserve armies (west of Smolensk) by mid-July. Army Group South destroyed the southern front-line armies in early August. At this point Hitler halted the drive for Moscow and directed Guderian to thrust south and Kleist northeast, trapping the Russian reserves around Kiev. The resumed advance (*bottom*) faded in the face of adverse weather, fierce Russian resistance and German exhaustion.

sketched out Operation Typhoon, a plan for the destruction of Soviet forces east of Smolensk and a subsequent advance on Moscow. Its first phase went well: German armour encircled more than 600,000 Russians in the Vyazma–Bryansk pocket, and Guderian advanced 130

"This is a very grim war. And you cannot imagine the hatred the Germans have stirred up among our people, you know; but I assure you, they have changed our people into spiteful mujiks – zlyie mujiki – that's what we've got in the Red Army now, men thirsty for revenge. We officers sometimes have a job in keeping our men from killing German prisoners . . . And there's good reason for it. Think of the towns and villages over there, think of all the torture and degradation those people are made to suffer."

A Red Army Captain in conversation with Alexander Werth, 1941

miles to Orel in 4 days. But on October 6 snow fell and quickly thawed, turning the countryside into a sea of mud. The onset of the freeze in early November led the Germans to hope for increased mobility over the hard ground, and on November 15 Hitler ordered a final drive on Moscow. Although the Germans reached the Volga Canal, fewer than 20 miles north of Moscow, they could not take the Soviet capital, thwarted as much by the utter exhaustion of their troops as by dogged Russian resistance.

Russia fights back

Stalin played a central role in marshalling resistance. On June 23 the Stavka, "High Command Headquarters", was set up, and Stalin rapidly assumed the authority of commander-in-chief. New commanders arrived, some posted in from other areas and others released from the prison camps to which the purges had consigned them. While the NKVD dealt with the incompetent or the unlucky – Pavlov, defeated commander of the Western Front, was shot, as were his chief of staff and signals commander – Stalin and his associates busied themselves with containing the German advance, conjuring up new divisions and trying to cope with the damage done to the structure of the Soviet Union: almost half its population was behind German lines by December 1941, along with the facilities that had accounted for more than 50 percent of pre-war coal production and 66 percent of pig-iron, steel and rolled metals. The GKO (State Defence Committee) strained every nerve to set up arms factories east of the Urals, to inspire and control local administrations and, above all, to transform the conflict into a patriotic war by replacing the old political clap-trap with a genuine desire to throw back the Germans – a desire reinforced by the ghastly apparatus of the New Order which flowed in behind the invading armies. It is beyond doubt that the German tendency to regard all inhabitants of the Soviet Union as *Untermenschen* alienated many groups which had found the Communist yoke heavy. However, there was no attempt to unravel the Soviet Union by an early and consistent appeal to its anti-Soviet or anti-Russian elements: German attitudes helped turn the struggle into what Russian histories term "The Great Patriotic War".

There had been scenes of near-panic in Moscow as the

A German Klieff-type flame-thrower in use against a bunker. The flame-thrower was first used by Germans near Ypres in 1915. Very effective against bunkers, it was a horrific weapon that could shake the most resolute defence. Vulnerability and short range made it unpopular with operators, who were sometimes killed out of hand if captured.

A motorcycle reconnaissance battalion was included in the panzer division's establishment in 1941. These motorcyclists ride alongside a column of captured Czech 35t tanks, widely used by the Germans. They are from 6th Panzer Division, part of Reinhardt's 41 Panzer Corps, advancing on Leningrad in June-July 1941.

Germans neared the city, but although the government decamped for Kuibyshev, Stalin himself stayed on, while the tough and capable Zhukov, who had already stiffened the defenders of Leningrad before being appointed commander of the Western Front on October 10, fought resolutely to thwart Hitler's last thrust at the capital. On December 5 he counterattacked with divisions brought in from Siberia and the Far East, striking a blow whose surprise and impact jolted the Germans back, producing grim scenes as bearded and filthy soldiers, poorly prepared for winter, were assailed by an enemy they had believed to be at his last gasp. Hitler forbade withdrawal, ordering his troops to form fortified defensive positions: *Igelstellung*, or "hedgehogs". The decision had two important consequences. First, it stabilized the situation, despite heavy casualties and some loss of ground, thus enhancing Hitler's opinion of his own military judgement. Second, the fact that the German-held pockets of Demyansk and Kholm were supplied by air gave OKW an exaggerated idea of the Luftwaffe's abilities.

The situation in the south

Things had gone little better for the Germans in the south. Von Manstein, commanding the Eleventh Army, advanced into the Crimea and laid siege to Sevastopol, while the bulk of Army Group South pressed on to the Donets, taking Rostov on November 20. Rundstedt asked permission to withdraw to shorten his line, but was dismissed when he began to fall back. His successor, Field Marshal von Reichenau, reached much the same conclusion, and was allowed to pull back behind the Mius river. The dismissed

Georgi K Zhukov (1896–1974), Marshal of the Soviet Union, was an apprentice furrier in Moscow before joining the Novgorod Dragoons in 1915. He won the Cross of St George for bravery in World War I, fought with the Red Cavalry in the Civil War, and was a brigade commander in 1930. In August 1939 he defeated the Japanese at Khalkin-Gol in Manchuria. He became Chief of Staff of the Red Army in 1941 and, after the German invasion, was first sent to assist Voroshilov with the defence of Leningrad and then brought back to mount the counteroffensive before Moscow. Thereafter, as first deputy commander-in-chief, he played a major part in the planning of Soviet operations, and shares the credit for the Stalingrad counteroffensive of 1942 and the Kursk-Orel operation of 1943. In 1944 he coordinated the actions of 1st and 2nd Belorussian Fronts on Operation Bagration, and in April 1945 his 1st Belorussian Front took Berlin. Zhukov, fittingly enough, accepted the German surrender on the part of the Soviet high command. His high public profile attracted the suspicion of Stalin, who dismissed him, though he emerged to serve as Minister of War in 1955. Khrushchev dismissed him in 1957, and he was attacked for "mistakes and distortions" and "non-Party behaviour": he was rehabilitated in 1966.

The Junkers Ju 52 was the mainstay of Germany's pre-war Lufthansa. During the war this robust and reliable aircraft served mainly as troop transporter, freighter and for casualty evacuation. On the Eastern Front it was used to parachute supplies into pockets like Demyansk, and to evacuate most of the 42,000 wounded and specialists from Stalingrad.

Rundstedt was in good company: Brauchitsch, Bock and Guderian were victims of a purge rivalling in proportion the mass promotions which had followed the 1940 campaign in the west.

Early in 1942 the Soviet counterattack was expanded into a full-blown counteroffensive, based on Stalin's conviction that: "The Germans are in disarray as a result of their defeat at Moscow." It failed, and some of the attacking units were themselves overwhelmed: Vlasov's Second Shock Army was encircled on the Volkhov, and its captured commander became the focal point of the "Russian Liberation Army" recruited from anti-Communist volunteers in German prison camps.

The German offensive of 1942

Hitler's directive of April 5 1942 for the spring campaign, "Case Blue", emphasized that although Leningrad was to be captured, forces in the centre were to stand fast and "all available forces will be concentrated on the *main operations in the Southern sector*". The mopping up of resistance in the Kerch peninsula and the Crimea, and the destruction of the Izyum bridgehead south of Kharkov, was to be followed by a two-pronged advance, by forces moving from Orel and Kharkov in the north and Taganrog in the south, meeting in the Stalingrad area and cutting off the Russians around Voronezh and the Don. This would be exploited by an advance into the Caucasus and the oilfields of the Caspian.

Despite accurate intelligence, Stalin decided that the Germans would resume their attack on Moscow, and the Western and Bryansk Fronts were heavily reinforced. In mid-May Timoshenko's Southwest Front attacked towards Kharkov, but was counterattacked by forces poised for the Izyum operation and lost more than 200,000 prisoners. Farther south, Manstein cleared the Kerch peninsula and turned on Sevastopol which, battered by artillery that included the huge "Karl" mortars and the railway gun "Big Dora", fell on July 3.

By that time the main offensive was under way. Army Group South – 68 German divisions, supported by 50 Romanian, Italian and Hungarian formations – began its advance on June 28. The offensive did not produce the expected battle of encirclement, and again there was conflict between Hitler and his field commanders. Bock, restored to favour to command Army Group South, was dismissed again, and his force was divided into Army Group A, under Field Marshal List, and Army Group B, under General von Weichs. Halder argued that the Russians' withdrawal was deliberate: Hitler was convinced that they were in full retreat. On July 21 he wrote of denying them oil by advancing into the Caucasus. This would also reduce possible routes for Anglo-American aid, which would be further frustrated by the cutting of the Murmansk railway subsequent to the capture of Leningrad. On July 23 Hitler ordered Army Group A to complete the destruction of Soviet forces which had escaped from Rostov and then to press on into the Caucasus, while Army Group B was to "smash the enemy forces" concentrated in Stalingrad, taking the city so as to block land communications between the Don and the Volga.

Army Group A at first made encouraging progress, reaching the Black Sea in August and taking Maikop. But

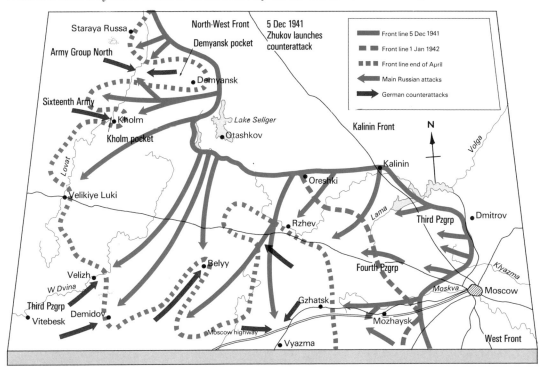

The German withdrawal through Belorussia. This 1944 photograph highlights not only the two-tier nature of the German army, with panzers at one extreme and horse-drawn transport at the other, but also the sheer misery endured through sustained operations in the varied and extreme climate of the USSR. Long and severe winters are followed by muddy springs, hot summers and muddy autumns. Even in the 1980s about 90 percent of the roads in the Soviet Union are unmetaled.

In December 1941 the Germans stood at the gates of Moscow: *Panzergruppe* 3 was 18 miles from the city. On December 5 the Russians counterattacked: early successes encouraged attempts to cut off Army Group Centre by attacks on both flanks, the Kalinin Front attacking from the north, and elements of the Western Front and the Bryansk Front from the south. German attacks eliminated some of the Soviet penetrations, and the fighting died away at the end of April with Army Group Centre in a large salient in front of Moscow.

when the advance gasped to a halt on the Terek in September, Hitler dismissed first List and then Halder of OKH, replacing them with Kleist and Zeitzler, the latter appointed largely because of his reputation as an obedient optimist. The first two years of the war had seen the total eclipse of OKH by an uncritical OKW, and the events of September 1942 were both an index of that eclipse and a portent of what was to come. We must guard against unquestioning acceptance of the view, so common in postwar German accounts, that all command errors sprang from the *Führer's* incompetence: there were times when Hitler's instincts were sounder than those of his generals. But by late 1942, German strategic direction of the war was desperately flawed – and a vivid demonstration of that fact was only months away.

The battle for Stalingrad

Army Group A closed in on Stalingrad, with Lieutenant General von Paulus's Sixth Army advancing directly on the city and Fourth Panzer Army swinging up from the south. Air attack reduced much of the city to rubble, but Stalin had decreed that it must be held. He dispatched Lieutenant General A I Yeremenko to command the Southeastern Front, with his headquarters at Stalingrad; Lieutenant General V I Chuikov was appointed to command Sixtysecond Army, whose sector included most of the city. On August 26 Zhukov was appointed Stalin's deputy commander and flew to Stalingrad to see things for himself. In mid-September the idea of a counteroffensive solidified in Stavka: as Sixth Army and Fourth Panzer Army were dragged into the grinding-mill of Stalingrad, under-

equipped divisions from Germany's satellites were stretched thinly on both flanks.

As Stalin recognized, the main business of the moment was to hold Stalingrad. It was held by the narrowest of margins. The sliver of defence on the west bank of the Volga was only 1,500yds (1,370m) deep in places, and Sixtysecond Army's divisions were reduced to a few hundred men apiece. Reinforcements and supplies had to cross the Volga under German fire. The fighting was ferocious even by the harsh standards of the Eastern Front: factories like *Barrikady* and *Krasnyi Oktyabr* became fortresses, and individual buildings like "Pavlov's House", held for 58 days by Sergeant Jacob Pavlov and a handful of men, featured on army commander's situation maps. The vicious close-range combat amongst the ruins wore down German technical superiority and the final surge of the offensive, staged in mid-October, ended with the Russians still hanging on.

With the battle for Stalingrad in the balance, the Russians perfected plans for Operation Uranus, their double envelopment of the Sixth Army. It began on November 19: on November 23 its pincers met at Kalach, leaving more than 250,000 Axis soldiers and auxiliaries in the pocket. Paulus requested permission to break out, but he was ordered to stand fast and to rely upon the Luftwaffe for his supplies. Now it was the Germans who fought to defend *Festung Stalingrad*, as the harsh winter drew in and supplies ran short.

Manstein, promoted field marshal after the fall of Sevastopol, was cast in the role of Sixth Army's saviour. He commanded Army Group Don, cobbled together from the

The map (*left*) shows the extent of German penetration on the Eastern Front up to November 18 1942. Note also the thrusts made in Operation Uranus, the Russian counteroffensive of November 19–23, which succeeded in encircling the German Sixth Army (General von Paulus) and part of Fourth Panzer Army, some 250,000 men in all, at Stalingrad.

Stalingrad (*left*), showing the progressive German penetration from September 12 to November 18 1942, when the Russians held only a narrow defensive strip west of the Volga, supported by artillery on the east bank. Paulus's attacks were made along a broad front: concentration on the flanks might have succeeded in interdicting Russian supply lines across the Volga.

remnants of Fourth Panzer Army, Sixth Army and other formations, with the task of halting the Russian attack and regaining lost ground. On December 12 Fourth Panzer Army struck out for Stalingrad and its leading elements fought their way to within 30 miles of the pocket. But Hitler would allow Sixth Army to move only if it continued to hold positions around Stalingrad: Paulus had too little fuel for a break-out and had begun slaughtering transport horses for food. On Christmas Eve the Russians attacked Fourth Panzer Army and pushed it back. Italian and

---

"Well, the time has come to be very honest and write a manly letter without trying to make things look better than they are . . . During the past days I was enlightened and saw very clearly that the end will be one about which nobody has spoken so far. But now I must express myself. There will be a day when you will hear about our battle to the last. Remember that the words concerning heroic action are merely words. I hope that this letter will come into your hands because it is the last letter which I can write . . ."

Gefreiter Schwarz to his wife, Stalingrad, January 13 1943

Romanian formations to the northwest were already in disarray, and there was no longer any hope of saving Sixth Army. Paulus, promoted to field marshal in the hope that he would commit suicide for honour's sake, surrendered on January 31 1943. Some 100,000 of his men went into captivity: only about 6,000 were destined to survive it.

The disaster at Stalingrad was an index of depressed German fortunes elsewhere. In mid-January the Russians tore a hole in Army Group B between Voronezh and Voroshilovgrad, while Soviet thrusts into the northern flank of Army Group Don threatened to cut off both it and Army Group A. Manstein wished to withdraw to concentrate his armoured forces for a counterstroke. He was reluctantly given permission to pull back to the Mius, but no sooner was he on the new line than the fall of Kharkov exposed his left flank. Hitler visited Zaporozhe on February 17, probably with the intention of dismissing Manstein, but he could not ignore the imminent catastrophe and not only left Manstein in command of Army Group South – created on February 13 from Army Groups B and Don – but increased reinforcements sent to the south. The stage was set for the last significant German victory in the east.

The Kharkov counterstroke

The plan was simple enough: First and Fourth Panzer Armies, the latter including the powerful SS Panzer Corps, would thrust up from between Zaporozhe and Krasnoarmeyskoye into the flank of the Soviet armies which had surged in between Army Group South and its neighbour, while Army Detachment Kempf, west of Kharkov, would attack northwestward as soon as the slackening of Russian pressure permitted. Execution of the operation showed Manstein at his most brilliant. Assisted by very effective air support, he first smashed Popov's armoured group around Krasnoarmeyskoye in late February and then went on to slice through the Soviet flank, mauling Vatutin's Southwest Front and Golikov's Voronezh Front and retaking Kharkov and Belgorod in

Savage close-range combat amongst ruined buildings in Stalingrad (*left*). Robust Russian infantry, making good use of the cheap, mass-produced PPsh sub-machine gun were at their best under these conditions. Russian propaganda emphasized the patriotic character of the struggle, but the Party maintained its firm grasp: a sailor and an infantryman (*above*) join the Party during the fighting. German prisoners endured a long march into captivity (*right*).

mid-March. Zhukov was flown down to coordinate emergency measures, but many of the winter's gains were lost and the operation left Army Group South in much the same position as it had been before the 1942 offensive. However, a large Soviet salient bulged out west of Kursk, between Belgorod and Orel. Manstein suggested an immediate attack on the salient, with Army Group South moving up to meet a southward thrust by Field Marshal von Kluge's Army Group Centre, but Kluge had no troops to spare and the spring mud soon rendered independent action by Manstein impossible.

The north and the centre

Between the opening of the German 1942 offensive and Manstein's recapture of Kharkov ten months later, the tide of war had washed far across the south. Although there had been less ebb and flow in the centre and north, the general picture was similar: growing Russian strength and skill, set against German superiority in armoured warfare, especially in the summer. Army Group Centre had sustained the main weight of the Soviet offensive in the winter of 1941–42. After desperate fighting it had re-established a cohesive front, with a Soviet salient around Toropets, south of Lake Ilmen. In October 1942 plans were made for an attack on the Toropets salient, OKH hoping that Army Group North might assist by moving south from the Demyansk pocket. But Army Group North had its hands full, first planning the assault on Leningrad and then coping with a Soviet offensive south of Lake Ladoga in August. The situation was not stabilized until mid-October, leaving too little time for an attack on Leningrad

Field Marshal Erich von Manstein (1887–1973) was commissioned into the 3rd Foot Guards in 1907, and fought on the Eastern and Western Fronts in World War I. In 1938 he was deputy to the Chief of General Staff, but was sent off to command a division after the dismissal of Fritsch. He was chief of staff of Rundstedt's Army Group A during the Polish campaign, and while serving in that capacity in the west he devised the plan for the invasion of France selected in preference to the OKH version. A corps commander in the 1940 campaign, he led a corps into Russia in 1941, and took the Crimea at the head of Eleventh Army, gaining his baton after the fall of Sevastopol. After service in Leningrad he returned to the south to command Army Group Don and later Army Group South. His plan for the relief of Stalingrad might have succeeded had it been better supported, and after the city's fall he stabilized the situation, defeating the Russians at Kharkov in February 1943. He advocated the Kursk offensive, and remained in command of Army Group South after its failure, fighting withdrawal battles with his customary skill. Dismissed by Hitler in March 1944, he was not re-employed. Manstein was one of the ablest practitioners of armoured warfare: his mobile defence in south Russia in 1943–44 is the paradigm of such operations.

Russian infantry in snow camouflage move through the outskirts of Leningrad. They are armed with sub-machine guns, and the Asian features of the two leading soldiers underline the racially heterogeneous character of the Red Army.

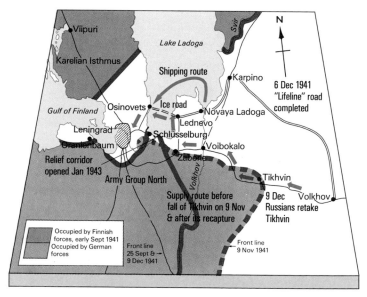

Hitler placed great emphasis on the capture of Leningrad, and in September 1941 Army Group North was on its outskirts, poised to attack. Alarmed at the prospect of squandering troops in house-to-house fighting in the city, with its unusually solid buildings and its canals, Hitler ordered that Leningrad was not to be assaulted in strength, but was to be tightly invested instead, battered by guns and aircraft, and starved into submission. The opening of the "ice road" across Lake Ladoga enabled the Russians to trickle supplies into the city when land routes were cut, and the morale of the population proved equal to the strain of privation and bombardment.

before winter. Failure to capture the city led to the cancellation of a projected offensive by the Finns and the German Twentieth Mountain Army in the far north.

In November 1942 the Russians attacked the Rzhev salient, south of Toropets. The operation was largely fruitless, although the Germans lost 7,000 men in January 1943 when Velikiye Luki was overwhelmed. In the same month a well-planned Soviet attack opened a narrow corridor to Leningrad along the edge of Lake Ladoga by taking Schlüsselburg (Petrokrepost). General von Küchler, commander of Army Group North, used the reverse to reinforce his request to be allowed to evacuate the Demyansk pocket, and Zeitzler persuaded Hitler to allow this. Evacuation of Demyansk rendered the Rzhev salient irrelevant, so Hitler yielded to the pleas of Kluge and Zeitzler, allowing Army Group Centre to collapse the Rzhev pocket in March. By the end of that month, the line ran south from Leningrad to Kholm and thence southeast to Orel. Between Orel and Belgorod the Kursk salient jutted into the German lines, which ran on along the Donets and the Mius. Army Group A held the Crimea and a small bridgehead around Taman across the Kerch strait.

As the campaigning season of 1943 approached, the balance of the war was gradually tilting against Germany. Industry and civilian morale suffered from the Allied bombing. The efforts of Dr Albert Speer, Minister for Armament and Munitions, helped ensure that Germany's war production continued to rise despite the bombs, but Russian and American production rose faster still. In May the last Axis forces surrendered in North Africa. Defeat in North Africa and at Stalingrad had shaken the confidence of Germany's allies, whose leaders trooped off to seek Hitler's reassurance.

Operation Citadel

Hitler proposed to recoup his fortunes with a decisive stroke. Operation Citadel would consist of a two-pronged assault towards Kursk by Army Groups Centre and South, with the aim of destroying Russian forces in the salient. "The victory at Kursk," announced Hitler, "must have the effect of a beacon seen around the world." The decision to attack was not an easy one. There was much debate on whether to allow the Russians to attack first and then to hit them "on the backhand", or to strike "on the forehand". An early forehand blow, as soon as the weather improved but before the Russians had recovered from Manstein's counterstroke, might have worked, but the postponement of Citadel until July allowed time for the Russians to reorganize. This time both operational logic and information supplied by the "Lucy Ring" left them in no doubt as to where the blow would fall, and Rokossovsky's Central Front and Vatutin's Voronezh Front worked frantically to prepare the defence, while Konev's Steppe Front, just behind the threatened sector, prepared for a counteroffensive.

The battle of Kursk began on the morning of July 5. On the southern flank, Fourth Panzer Army made a promising start north of Belgorod, but Army Detachment Kempf, south of the town, did less well in the face of solid defences and vigorous air attack. Ninth Army, on the northern flank, bit deep into Rokossovsky's first line, but its progress was slowed down by minefields, and determined

---

THE SIEGE OF LENINGRAD

War came gradually to Leningrad. In its first weeks thousands of children were evacuated, some into the path of the advancing Germans, because the authorities believed that air attack posed the greatest threat. As the Germans advanced, workers were mobilized to build the Luga line, running from Narva to Lake Ilmen, but Army Group North crashed through it in mid-August. On September 8 Schlüsselburg fell, and the Germans closed the ring: air raids rocked the city, and the warehouses at Badayev, containing vast reserves of food were burned. This underlined the real danger.

Rationing was introduced in July, but serious cuts were not imposed till September. Thereafter rations were reduced until even high-grade workers and soldiers were entitled to only twelve ounces of food a day. Black marketeering and theft became common. "I tell you plainly," said a Party official, "that we shot people for stealing a loaf of bread." There were rumours of cannibalism: rats, lipstick, and even leather dust from the tannery were certainly eaten. As winter set in, the miseries of cold were added to those of hunger: there was no electricity and central heating no longer worked. Soon the sight of a man collapsing, or a child's body being dragged off for burial in a sledge, no longer attracted attention.

Some food was shipped across Lake Ladoga before it froze in mid-November, but the fall of the railhead at Tikhvin on November 9 forced the Russians to construct a new road simply to get supplies to the lake's edge. As the ice thickened, supplies were brought across it in trucks, despite the attentions of German bombers. The "ice road" saved Leningrad. By late January 1942 food entered the city faster than it was being eaten, deaths and evacuations across Ladoga reduced the number of mouths to feed, and the recapture of Tikhvin had eased problems outside the city. By the time the ice melted in late April, nearly 53,000 tons of supplies, 42,500 tons of it food, had come across the ice.

In March strenuous efforts were made to clean up the debris in the streets: industrial production picked up, and a shipping route across Ladoga replaced the ice road. The second winter of the siege was infinitely more bearable than the first, and in January 1943 the Russians recaptured Schlüsselburg and opened a narrow corridor through the blockade. Nevertheless, it was not until January 1944 that the siege was fully raised.

---

defenders in well-sited positions. The Germans found it difficult to keep tanks and infantry together, experiencing mechanical difficulties with their PzKw V Panthers and PzKw VI Tigers. The Tiger, despite its 88mm gun, was not the hoped-for battle-winner, and the Porsche PzJg ("Ferdinand" or "Elefant") self-propelled 88mm gun was even less well suited to the close-range fighting that raged on the shoulders of the Kursk salient in early July. German difficulties on the ground were compounded by problems in the air. Although the Luftwaffe mounted 3,000 sorties a day, it was never able to guarantee superiority over the battlefield. Prowling Ilyushin Il-2 "Stormovik" ground-attack bombers struck at German armour, while Yakovlev Yak-9 fighters harried German ground-attack aircraft.

The Soviet counterattack

On July 13 Hitler cancelled the offensive, concerned at Soviet build-ups on both flanks of the salient, as well as Allied landings in Sicily. Manstein remained confident of achieving a breakthrough in the south, although his optimism brought no echo from Kluge. The Soviet counter-attack was swift and merciless. Kharkov was recaptured on August 23, and Hitler, although furious at the city's loss, agreed to permit Manstein to fall back to the Dnieper. By the end of September most of Army Group South had been pushed back across the river, with the Russians in pursuit: the Taman bridgehead was evacuated shortly afterwards. The Russians kept up the pressure, forcing crossings over the Dnieper, cutting off the Crimea and, on November 6, taking Kiev.

The four months and one day separating the opening of the Kursk offensive from the Soviet recapture of Kiev marked a crucial change of fortune on the Eastern Front. German losses were irreplaceable, while Russian tanks and aircraft arrived in growing numbers and the male popula-

---

"Captain Wesreidau often helped us to endure the worst. He was always on good terms with his men, and was never one of those officers who are so impressed by their own rank that they treat ordinary soldiers like valueless pawns to be used without scruple. He stood beside us during countless gray watches, and came into our bunkers to talk to us, and make us forget the howling storm outside. I can still see his thin face, faintly lit by a wavering lamp, leaning over, beside one of ours."

Gefreiter Guy Sajer, Autumn 1943

---

tion of areas reoccupied by the Red Army went straight into uniform. The Red Army still made occasional costly mistakes while the *Wehrmacht* showed flashes of its old brilliance, as when Manstein executed a smart backhand blow on the Dnieper bend in October. But as Manstein admitted: "When Citadel was called off, the initiative in the Eastern theatre of war finally passed to the Russians." Failure at Kursk sent shock-waves through Germany's allies, forcing the Germans to plan for the occupation of Hungary and Romania and straining German-Finnish relations.

Disaster in the south

In the south, Christmas 1943 brought the Germans no

Map legend:

	Orel axis Front lines
▬▬▬	4 July 1943
▬ ▬ ▬	10 July
▬▬ ▬▬	18 Aug
	Kharkov axis
▬▬▬	4 July
▪▪▪▪	12 July
▬▬ ▬▬	23 Aug
▨	Area held briefly by Germans in Operation Citadel
→	Russian counterattacks
——	Principal railways

0 25 50 Km
0 25 50 Miles

Map labels: Kirov, West Front 12 July 1943 offensive launched, Belev, Bryansk Front, Army Group Centre, Bolkhov, Bryansk, Second Pz Army, Novosil, Orel, Oka, Ninth Army, Sosna, Sevsk, Olkhovatka, Central Front, Kursk, To Voronezh, Seim, Korenevo, Belopol'ye, Voronezh Front 3 Aug 1943 offensive launched Steppe Front Prokhorovka, Sumy, Psel, Borodnya, Fourth Pz Army, Belgorod, Gadyach, Grayvoron, Army Detachment Kempf, Valki, Kharkov, Chuguyev, South-West Front, Donets, Army Group South, To Moscow

---

BATTLE OF KURSK	
Forces	**Losses**
German:	
900,000 men	70,000
2,700 tanks	c.1,500
10,000 guns	c.1,000
2,000 aircraft	c.1,400
Russian:	
1,300,000 men	Russian losses are not known: they were probably on a par with those above
3,600 tanks	
20,000 guns	
2,400 aircraft	

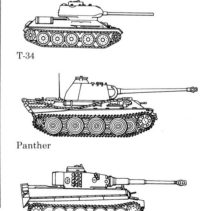

T-34

Panther

Tiger

Operation Citadel aimed at eliminating the Soviet salient which bulged out around Kursk. Many Germans had misgivings about the operation. The Russians received ample warning of the attack, and when it began on July 5 they were ready to meet it. Gains were disappointing, and when Citadel was called off the Russians counterattacked, the Central Front swallowing the Orel bulge, and the Voronezh and Steppe Fronts bursting out to take the Belgorod-Kharkov shoulder.

The T-34/85 tank (*top*) was outstanding, with good balance between speed, hitting power and armour. The PzKw V Panther (*centre*) was hastily produced, and those that survived Kursk required modification. Nevertheless, the Panther was superior to the T-34, and though it lacked the sophisticated features (including 88mm gun and 100mm armour) of the Tiger (*below*) it was more mobile.

respite. Vatutin attacked west of Kiev on December 24, and although Manstein riposted with an armoured counterstroke which mauled a Russian tank army, the fighting went badly for the Germans. Two corps were encircled in the Cherkassy/Korsun pocket: most of the soldiers escaped but their equipment was abandoned. Farther south, the Russians failed to achieve a similar envelopment of the bridgehead around Nikopol on the southern Dnieper bend, where the Germans withdrew before the claws closed. In early March 1944 the Russians slowed down, but the pause was the prelude to a massive operation aimed at the destruction of the Germans in the southern Ukraine. This hinged on First Ukrainian Front striking down from Rovno while the Second and Third Ukrainian Fronts stabbed westward. The attack began on March 4, but although it liberated the Ukraine, it narrowly failed to trap First Panzer Army. Manstein and Kleist were amongst its casualties, dismissed by Hitler on March 30 and replaced by Field Marshal Walther Model and Colonel General Ferdinand Schörner. The change of command made little appreciable difference: on April 14 the last German crossed the Dniester, and by the month's end a line had hardened across northern Romania, although Romanian troops, effectively subordinated to German command, were at their last gasp.

The defeat of Army Group South and Army Group A (the latter renamed Army Group South Ukraine in April) left the German Seventeenth Army increasingly isolated in the Crimea. Hitler had forbidden withdrawal, and Schörner, a vigorous commander and a committed Nazi, reported optimistically on the state of the Crimea's defences. The Russians attacked on April 8, quickly bundling the Germans back through the Gneisenau Line onto the defences of Sevastopol. On May 8 Hitler agreed to evacuation by sea, but 300,000 men were left behind.

The defeat of Army Group North

Army Group North had fought a relatively static war, preoccupied with the siege of Leningrad. However, its southern sector grew increasingly vulnerable, particularly after a Soviet breakthrough at Nevel in October 1943 opened a salient near the junction with Army Group Centre. Küchler's strength had been depleted by the removal of battleworthy divisions and their replacement by Luftwaffe field divisions and SS formations newly raised in the Baltic States. In September 1943 he had begun work on the "Panther" position – part of the projected East Wall, a defensive system running across the whole front. Küchler and his staff wished to withdraw to the Panther Line in January 1944, before the spring thaw, but Hitler was reluctant to move unless the Russians forced him to.

Hitler believed that the Soviets lacked reserves to mount an operation in the north while their offensive was under way in the Ukraine, but on January 14 a two-pronged offensive, with Govorov's Leningrad Front attacking around Leningrad and Meretskov's Volkhov Front south of Novgorod, proved him wrong. The Germans held well for the first few days, but soon ominous cracks appeared along both Soviet thrust-lines. Küchler tried without success to obtain Hitler's consent for withdrawal to an intermediate position preparatory to falling back on the Panther Line; on February 1 he was replaced by Model, who was waiting

Mines were widely used on the Eastern Front: in critical sectors on the Kursk front the Russians laid 1,500 anti-tank and 1,700 anti-personnel mines per kilometre. They could be detected by prodding with a bayonet or special probe, and by electromagnetic detectors like these. The Germans displayed great ingenuity in mixing mines and booby traps to make detection and removal hazardous, and non-metallic mines defeated mine detectors. The Russians sometimes used penal battalions to attack over minefields so as to clear them for following units.

PARTISANS

On July 7 1941 Stalin called on the population of invaded areas to rise. Partisan units were formed in areas likely to be occupied, though there was little time for them to be established. Many Russians were anxious to avoid activities which inspired reprisal, though some encircled Red Army men fought on. In 1942 officers and Party officials were parachuted in to organize units, air supply was stepped up and a chain of command established. German behaviour increasingly alienated civilians who might have tolerated peaceful occupation, but many Russians fought for the Germans either for ideological reasons or simply because it was hard to stay neutral and survive. In 1942 Himmler became responsible for antipartisan operations in territories under civil administration, and OKH in the zone of operations. Hitler's admission that military means alone would not defeat partisans did not transform policy: an order of December 1942 proclaimed that "anything that leads to success is proper . . ." Some officers disagreed. One wrote: "It is not Russian people who are our enemy but Bolshevism . . . our objective must be to win the trust of the population . . ." Partisan organization improved: a central staff controlled operations in the deep rear, while units near the front line came under the fronts opposite them. Antipartisan warfare focused on cordon and search operations, usually carried out by rear area formations. Often partisans managed to escape, but sometimes the Germans were more successful, especially when they used high-grade troops. Partisans were a source of irritation rather than a war-winner: most troops who fought them would not have been useful elsewhere, and the areas they dominated were not vital to German interests. For the Russians they were part of the necessary mythology of the war, proof of determination to eject the invader.

to take over from Manstein in the south. Model tried to patch together a counteroffensive but was forced to acknowledge that his army group was played out: the first week in March saw it behind the Panther Line.

Waning German fortunes persuaded the Finns to sue for peace. They rejected Soviet terms on April 18, but it was clear that Finland's days as a German ally were numbered. On June 9 the Russians mounted a full-scale offensive on both sides of Lake Ladoga. German assistance prevented the total collapse of the Finnish army, but after five weeks' fighting the Finns were back almost on the pre-1941 border.

The Belorussian offensive

The misfortunes of Army Groups North and South paled into insignificance before the catastrophe which overwhelmed Army Group Centre. German intelligence concluded that the Soviet summer offensive of 1944 would come in the south, although on May 12 it acknowledged the possibility of a secondary attack between the Pripet Marshes and the Carpathians. This misconception paid tribute to Soviet deception: Stavka persuaded the Germans that its offensive would indeed be in the south, while it concentrated 1.2 million men, 4,000 tanks and self-propelled guns, 24,400 guns and heavy mortars and 5,300 aircraft opposite Army Group Centre.

Operation Bagration employed four attacking fronts: First Baltic and Third Belorussian under Vasilevsky and First and Second Belorussian under Zhukov. After taking the strong points of Vitebsk, Orsha, Mogilev and Bobruisk, the Russians planned to move concentrically on Minsk, gobbling up chunks of Army Group Centre in the process.

The attack began on June 22, after preparatory work by partisans who cut the railway behind Army Group Centre in more than 10,000 places. The first day's fighting gave clear signs of what was to come. The Germans had too few reserves to check Soviet thrusts and were hamstrung by the customary "no withdrawal" order. Army Group Centre lost 28 divisions in 17 days. Model tried hard to establish a new defensive line, but it was not until mid-July, when the Russians were outrunning their supplies, that he was able to give some coherence to the wreckage of his army group. The collapse of Army Group Centre allowed the Russians to exploit to the flanks: they tumbled Army Group North Ukraine right back to the Hungarian border and compressed Army Group North into Estonia, between the Gulf of Riga and Lake Peipus.

Changes in command followed the attempt on Hitler's life of July 20 1944. Guderian took over from Zeitzler at OKH and, at his initiative, Friessner, commander of Army Group North, changed places with Schörner of Army Group North Ukraine. Guderian ordered these army groups to halt and counterattack, while Army Group Centre, to which he promised reinforcements, was to strike out to regain contact with its flanking formations. Guderian was unrealistically sanguine, but Soviet overstretch and the arrival of German reinforcements helped Model's line solidify in front of Warsaw. On August 1, General Bor-Komorovski's Home Army, the Polish resistance organization, rose in the city. The future of Poland had been the subject of a long and intense debate between the Russians and the Western Allies, and there have been suggestions that Stalin deliberately refrained from

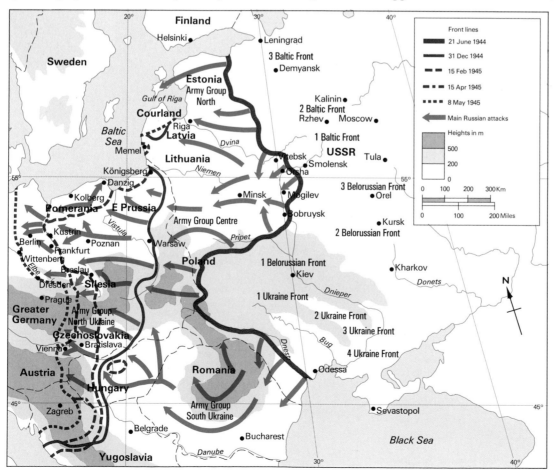

In 1944–45 the Russians expelled the Germans from Soviet territory, liberated Eastern Europe and pressed deep into the Reich. Operation Bagration, fittingly named after a hero of the Napoleonic Wars, destroyed Army Group Centre in June–July, driving the Germans from Belorussia. In July the Russians cleared the Western Ukraine, seizing a bridgehead over the Vistula south of Warsaw, and by October they had freed Romania. After the conquest of Estonia and most of Latvia, the Red Army moved on into Hungary and Czechoslovakia, taking Budapest in February 1945. In January the Russians attacked in the centre, capturing Warsaw and closing up onto the Oder and Neisse. In February–March they secured the flanks of their main drive against Berlin by invading East Prussia and Silesia, and in mid-April they crossed the Oder and Neisse in strength to take Berlin on May 2.

supporting a rising sponsored by the "London Poles" as opposed to his own "Lublin Poles", thus allowing the Germans to dispose of his ideological enemies. The rising was effectively crushed in a welter of appalling atrocities, although the Home Army did not finally capitulate until early October. Model himself was in no position to celebrate this ugly little victory, for he was sent off to command in the west on August 16: Colonel General Reinhardt took over his army group.

The loss of Romania and Hungary

Relative calm in the centre was followed by calamity in the south. In late August two fronts under Timoshenko's direction crashed into Romania. The pro-German Marshal Antonescu was deposed and Romania joined the Allies. Hungary, too, was wavering. Admiral Horthy, its regent, appealed for an armistice and was taken off to "asylum" in Germany: there were massive desertions from the Hungarian army as the war rolled across the Carpathians. Bulgaria declared her return to neutrality on September 4, but Soviet invasion spurred her into realignment and on September 8 she declared war on Germany.

The German High Command placed particular emphasis on the defence of Budapest, but the Russians encircled the Hungarian capital on December 26 and the plight of population and garrison grew increasingly grim. On January 1 1945 Army Group South mounted an abortive relief operation; on January 27 Soviet counterattacks began. The garrison commander ordered a breakout on February 11, but only a handful of his 30,000 German and Hungarian troops escaped. OKW's obstinacy had led to disaster at Budapest – and in early March it produced an ill-starred offensive in which Sixth SS Panzer Army, brought across from the Western Front, played a leading part. The Soviet counteroffensive pushed the Germans back across the Hungarian oilfields, which Hitler had been anxious to defend, and Vienna fell on April 13.

Into the Reich

Things went no better for the Germans in the centre and the north. On September 19 the Finns concluded an armistice, although the Germans managed to withdraw Twentieth Mountain Army from Finland into northern Norway. Army Group North fell back from Estonia into Courland and to positions covering Riga, cutting off elements of Soviet forces which attempted to split it. In October the Russians attacked across the border into East Prussia. Schörner hoped to counterattack into the Soviet left flank, but OKH sent Third Panzer Army back to Army Group Centre, and at the end of October he began to retreat into Courland. Although the offensive into East Prussia was checked, Army Group North remained isolated in Courland for the rest of the war.

By early January 1945 there was abundant evidence that Germany's plight was desperate. Although aircraft production peaked in September 1944 and armoured vehicle production in December 1944, allied bombers pounded industry and communications. Fuel shortages restricted everything from pilot training to the movement of panzer divisions. And Germany was simply running out of soldiers. Between June and November 1944 she had irrevocably lost nearly 1,500,000 men, almost two-thirds of them in the east. Stopgap programmes produced manpower of a

sort, but it was sketchily trained and poorly equipped: as the Allies closed in, the defenders of the Reich included battalions of sailors without ships and pilots without aircraft, auxiliaries from the east, and the striplings and greybeards of the *Volkssturm* ("People's Militia"). Guderian warned Hitler that the stage was set for an offensive which must overwhelm German forces in the East. Hitler and his cronies dismissed it as bluff – but on January 12 the bluff was called.

Zhukov's First Belorussian and Konev's First Ukrainian Fronts, with almost 6,500 tanks and nearly 2,500,000 men between them, were to thrust side by side to the River Oder. Chernyakhovsky's Third Belorussian Front was to

Tanks and infantry of the 5th Guards Tank Army, Belgorod, August 1943.

advance into East Prussia, while Rokossovky's Second Belorussian Front drove in from Warsaw to the Baltic around Danzig, cutting off East Prussia and protecting Zhukov's right flank. Stavka exercised its customary close supervision over planning, and Stalin himself monitored plans closely. The concentration was accomplished only with massive logistic effort, but was far enough advanced to permit the offensive to be launched on January 12, eight days earlier than planned, to assist the Western Allies in what Stalin termed their "difficult position" following the Ardennes offensive.

Konev broke out of the Baranov bridgehead quickly, pushing deep into the German rear and swinging north to threaten the troops facing Zhukov, who attacked on January 14, punching great holes in the German defence and taking Warsaw. A further German command reorganization – Army Group A was rechristened Army Group Centre; Army Group Centre became Army Group North; the former Army Group North became Army Group Courland – helped German defenders no more than it assists historians. The arrival of new commanders was scarcely more encouraging: Schörner took over the new Army Group Centre with his customary ferocity, and the SS leader Heinrich Himmler, unencumbered by any excess of military knowledge, assumed control of Army Group Vistula, holding the line between the new Army Groups North and Centre.

The offensive clattered on. Rokossovsky swung up into East Prussia, folding Army Group North against the Baltic. Zhukov seized bridgeheads over the Oder in early February, while Konev crossed the Oder to reach the

Neisse on February 15. A sudden thaw slowed the pace of the advance, giving Guderian the opportunity to mount a counterattack, Operation Solstice, in mid-February. It failed, but was not entirely sterile, for it helped inspire a change of policy at Stavka: the direct thrust for Berlin was replaced by flank-clearing operations in Pomerania and Silesia to prevent the threat of converging counterattacks. Rokossovsky jabbed up into Pomerania on February 24, and on March 1 Zhukov's forces joined in, taking Kolberg on March 18. Third Belorussian Front pinched out the Königsberg (now Kaliningrad) pocket on April 10, and Konev cleared upper Silesia.

The battle for Berlin

The last phase of the war was bound up in the tangled web of inter-Allied politics. In February 1945, Stalin, Roosevelt and Churchill met at Yalta, where they agreed that Russia would join the war against Japan soon after the defeat of Germany; that Germany would be divided into zones of occupation; and that a conference would be held to draft the charter for a "World Security Organization". Critics of Yalta have pointed out that it resulted in the extension of Soviet influence across Eastern Europe, and the forced repatriation of Soviet citizens in the postwar period has inspired passionate denunciation. However, Soviet military might underwrote the Yalta agreement, and Stalin, having stared ruin in the face, could hardly be expected to make generous compromises over future Soviet security. Subsequent negotiations on Poland and secret Anglo-American negotiations with German emissaries increased his suspicions. Roosevelt assured him that there would be no drive on Berlin from the West: Stalin replied that Berlin

had lost its strategic significance for the Russians and that the main offensive would be resumed in late May.

At a crucial meeting on April 1 Soviet planners claimed that an Allied attack on Berlin was imminent. "Well, now," asked Stalin, "who is going to take Berlin, will we or the Allies?" The question was rhetorical, for his staff unveiled the plan for the last battle. Zhukov would spearhead the drive on the German capital, with supporting attacks covering the flanks of the main thrust. Konev, Zhukov's personal rival, was to strike out for the Elbe between Dresden and Wittenberg, to join the Americans: his right flank armies were to advance northwest, hooking towards Berlin if the situation warranted it. Indeed, the question of whether Zhukov or Konev was actually to take Berlin was left undecided. Command of the whole operation was entrusted to Zhukov.

The Germans were already contending with an Allied advance which reached the Elbe on April 11 and were short of resources of all kinds. The wily Colonel General Heinrici now commanded Army Group Vistula which was responsible for the defence of Berlin, but Ninth Army, covering the direct approach to the city, had only 14 understrength divisions to face the 5 attacking armies of First Belorussian Front, and flanking formations were scarcely better off.

The Russians attacked before dawn on April 16. Zhukov's troops broke out of the Küstrin bridgehead while their comrades forced crossings over the Oder to north and south, but minefields, boggy ground and fierce resistance on the Seelow Heights checked the advance. Konev made better progress, crossing the Neisse in strength and breaking through the German defences. A furious Zhukov

The "Big Three" Allied leaders (*above*) – left to right, Winston Churchill, President Roosevelt, Joseph Stalin – dine together during the Yalta Conference, February 4–11 1945. At this summit meeting in the Crimea on the eve of victory, the postwar world took shape. The goodwill and military might of the Soviet Union were necessary for the speedy ending of the war, but Roosevelt has been much criticized for making too many concessions to the Russians.

Savage street fighting in April–May 1945 reduced Berlin to ruins (*right*). Here, the Soviet banner is raised over the Reichstag during the final stages of the battle (*above*). The Russians admitted to taking 304,000 casualties in the period April 16–May 8, but claimed that German losses exceeded 1,000,000.

discovered that Stalin, impressed by Konev's progress, had ordered him to turn his tank armies on Berlin. As Zhukov clawed his way across the Seelow Heights, Konev's armour slashed up from the southeast. Zhukov's gunners began to shell Berlin on April 20, on the 21st his men reached its suburbs, and on the 25th they met Konev's soldiers west of the city, completing its encirclement. On the same day Soviet and American troops joined hands on the Elbe, cutting the dying Reich in half.

Stalin decided on April 23 that the honour of capturing Berlin would, after all, go to Zhukov. The outcome of the last weeks of fighting was a foregone conclusion, but the battle was no less bitter for that. Fantastic schemes were produced as Hitler proclaimed that "the battle for the German fate" was being fought out. Some relief operations made initial progress but ground to a halt: others had no basis outside the Führer's fevered brain. Hitler himself refused to leave the city: he committed suicide on the afternoon of April 30, having appointed Admiral Dönitz as his successor. General Weidling surrendered Berlin on May 2, although sporadic fighting continued in and around the city for another two days. The battle had reduced the German capital to a wilderness of battered, stinking rubble, and, in John Erickson's words, "by the most conservative calculation. . .[it] cost half a million human beings their lives, their well-being or their sanity".

The end in the south
The fighting in Bohemia was on a far smaller scale, but had a poignancy all of its own. Vlasov, commander of the Russian Liberation Army, hoped to string together an alliance of non-Communist Czechs who would help him hold the Red Army off until the Americans arrived, but Allied recognition that Prague lay east of the demarcation line effectively killed the plan. Vlasov's men surrendered to the Americans, which did them little good, for most were later repatriated to face death or the labour camp.

Colonel General Jodl of OKW flew to Rheims on May 6 and signed an Act of Military Surrender the following day. There was some doubt as to how this could be enforced in the East, but Field Marshal Keitel, chief of OKW staff, was swept off to Berlin to enact another ceremony on May 8 and surrender orders were duly transmitted to German troops. Those who could fled to become prisoners of the Western Allies. However, about 1,250,000 men went into captivity in the Russia they had invaded with such confidence less than four years before.

Russia moves East
At Yalta Stalin had agreed to join the war against Japan within three months of the defeat of Germany, and by early August more than 1,000,000 Russians were facing the Japanese in Manchuria. On August 8, two days after the atomic bomb was dropped on Hiroshima, Russia declared war on Japan. The brief campaign that followed showed just how much the Red Army had learned over the past four years. Achieving almost total surprise, the Russians attacked with powerfully-concentrated forces moving on converging axes. In less than three weeks they utterly defeated the Japanese who, although short of tanks and aircraft, were no mean adversaries even at that late stage in the war. The Manchurian campaign was regarded by the Russians as the very epitome of Soviet blitzkrieg, and is closely studied even today.

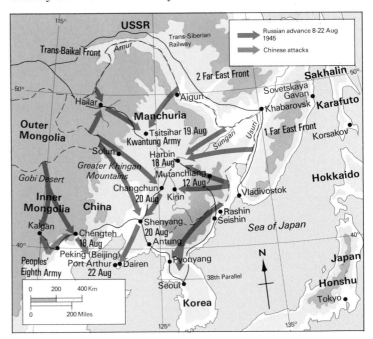

The Manchurian operation saw the Japanese Kwantung Army surprised in time and place: the attack came sooner than envisaged, was screened by elaborate deception, and included an unexpected thrust over the Great Kingan Mountains. Russian superiority was sharply focused: the Trans-Baikal Front advanced on a 930-mile (1,500km) front with 90 percent of its tanks and guns on the 250 miles of main effort.

— chapter fourteen —
World War II: The Desert War

Richard Holmes

Benito Mussolini attained power in Italy in 1922 and embarked upon a forward foreign policy. The prospects in Africa seemed most promising: Italy already possessed Eritrea and Somaliland and in 1935–36 she occupied Abyssinia (Ethiopia). Mussolini scorned the League of Nations' sanctions intended to force his withdrawal, and when they were lifted he showed his contempt for the League by resigning from it. By 1939 Italy had an army under the Duke of Aosta in Eritrea and Abyssinia, and another under Marshal Balbo in her colony of Libya. Her powerful navy seemed likely to dominate the Mediterranean. In May 1939 the "Pact of Steel" formalized the German-Italian alliance, although Italy did not enter World War II until June 1940, when Germany's panzers had dismembered France.

The defence of the Suez Canal and the oilfields at the head of the Persian Gulf were major priorities for the British. General Sir Archibald Wavell's Middle East Command was outnumbered by Italian troops in the theatre; the two British naval forces in the Mediterranean, under Admiral Somerville at Gibraltar and Admiral Cunningham at Alexandria, were similarly overshadowed by the Italian navy. General H M Wilson commanded British troops in Egypt, with the prime task of protecting Egypt against the 200,000 Italians in Libya. His chosen instrument was the 31,000-strong Western Desert Force, commanded by Major General R N O'Connor.

Correlli Barnett has described the desert war as being fought "like a polo game in an empty arena". The coast road ran from the Nile valley, round the great bulge of Cyrenaica, to Tripoli. To its south, a steep limestone escarpment led up to the rolling rock, sand and scrub of the desert itself. Although not the trackless and featureless sand-table of popular belief, the desert offered remarkably free play to armoured formations: it was, according to a German general, "a tactician's paradise but a quartermaster's nightmare".

Marshal Graziani, who took command in Libya after Balbo's death, set up bases on the Egyptian frontier in mid-1940. He wished to delay his offensive until Operation Sealion, the planned German invasion of Britain, had disheartened Wilson's troops, but orders from Rome jolted him into action. On September 13 1940 Graziani crossed the frontier, halting to establish camps around Sidi Barrani. O'Connor, disinclined to await Graziani's offensive,

planned instead a bold attack to destroy the Italian Tenth Army. On December 9 he thrust in between the Italian camps of Tummar and Nibeiwa in the north and Sofafi in the south, swinging in to take first Nibeiwa and then Tummar in the rear and subsequently encircling and capturing Sidi Barrani. Exploiting his success, O'Connor took Bardia and Tobruk before pushing across Cyrenaica to cut the road at Beda Fomm on February 5 1941. The retreating Italians failed to break through and finally surrendered: in two months O'Connor had taken 130,000 prisoners, 380 tanks and 845 guns for fewer than 2,000 British casualties.

In this prodigious victory lay the seeds of future defeat. The danger of air attack made dispersion necessary, and this tactic persisted even when the need for it did not exist. The need to conserve armoured resources encouraged the use of small composite units which militated against the development of proper inter-arm cooperation and contributed to the British tendency to fight dispersed.

O'Connor's command was eroded even before his victory was complete, when the experienced 4th Indian Division was sent to join the force which ejected the Italians from Abyssinia; it was replaced by the 6th Australian Division. Wavell, however, had an even greater problem.

Balkan interlude

Mussolini's invasion of Greece from Albania, launched on October 28 1940, had been sharply rebuffed, and on December 13 Hitler ordered preparations for an invasion of Greece. He also proposed to send German forces to

—TIME CHART—			
1922	Mussolini gains power in Italy	Nov 18	British offensive in Libya begins (Operation Crusader)
1935–36	Italy occupies Abyssinia		
1940		1942	
Sept 13	Italy invades Egypt	June 21	Tobruk falls to Germans after British defeat at Gazala
Sept 27	Germany, Italy and Japan sign Tripartite Pact		
Oct 28	Italy invades Greece from Albania	July	German advance stemmed at first battle of El Alamein
Nov 11	British raid Italian naval base at Taranto		
Nov 21	Greeks retake Koritsa	Aug 13–15	"Pedestal" convoy reaches Malta
Dec 9	O'Connor's offensive begins (to Feb 1941)	Oct 23–	British breakthrough at second battle of
1941		Nov 4	El Alamein
Jan–Mar	Greeks invade Albania	Nov 8	Allied landings in French North Africa
Jan–Nov	British conquer Eritrea, Somaliland and Abyssinia	1943	
		July–Aug	Allied invasion of Sicily
Mar 28	Italians defeated in naval battle off Cape Matapan	Sep 8	Allied armistice with Italy
		Sep	Allies land at Reggio, Taranto and Salerno
Mar 31	Germans attack in Cyrenaica	1944	
Apr 6	Germany invades Yugoslavia and Greece	Jan–May	Cassino battles lead to breaking of Gustav line
Apr 17	Yugoslavia capitulates	Jan 22	Anzio landings: contained by Germans
May 2	Last Allied evacuation from mainland Greece	May 23	Anzio breakout
May 20	German airborne invasion of Crete	Aug–Sep	Gothic Line broken
May 31	Last Allied evacuation from Crete	Oct	British landings in Greece and Albania
Aug	Anglo-Russian forces occupy Iran	1945 May	German forces in Italy surrender

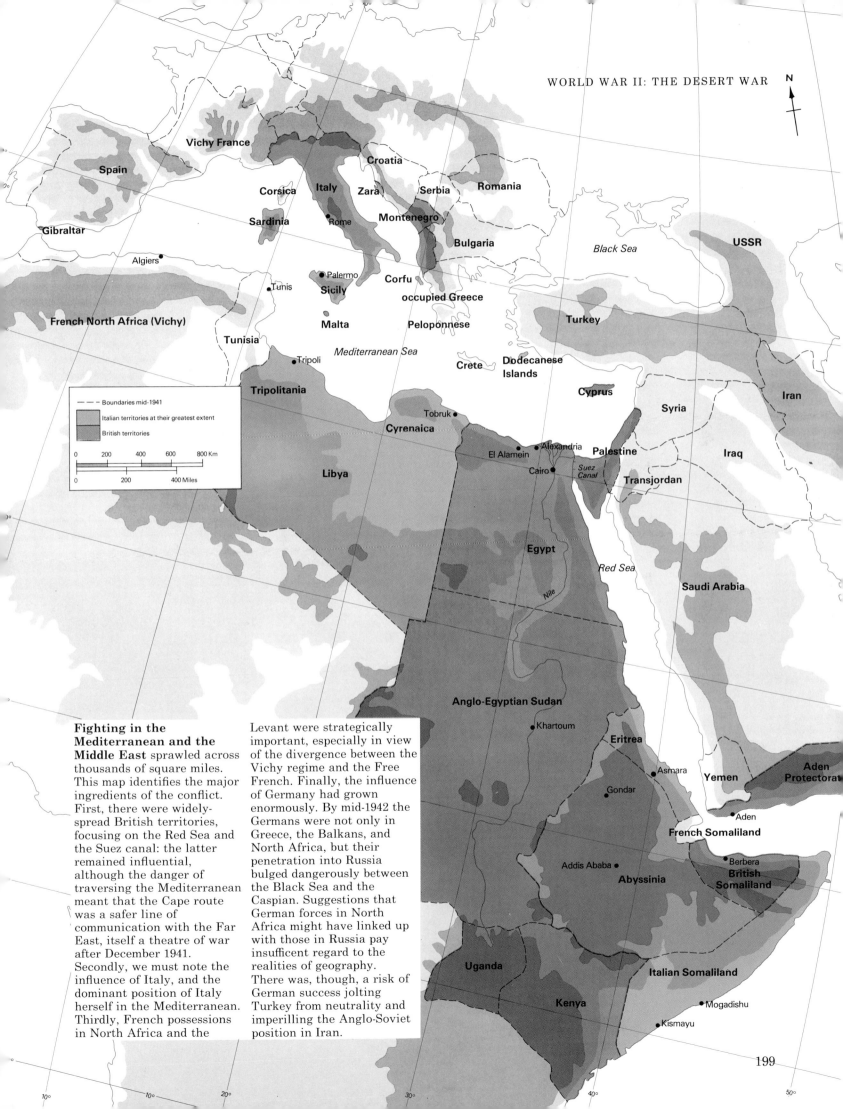

N

Vichy France

Croatia

Spain

Corsica **Italy** **Zara**

Serbia **Romania**

Sardinia • Rome **Montenegro**

Gibraltar

Bulgaria

Black Sea

USSR

• Algiers

• Palermo

Corfu

occupied Greece

• Tunis **Sicily**

French North Africa (Vichy)

Turkey

Tunisia

Malta

Peloponnese

Mediterranean Sea

• Tripoli

Crete **Dodecanese Islands**

Tripolitania

Cyprus

Iran

Tobruk •

Cyrenaica

Syria

Palestine

El Alamein • • Alexandria

• Cairo *Suez Canal*

Iraq

Libya

Transjordan

Egypt

Red Sea

Saudi Arabia

Nile

Boundaries mid-1941
Italian territories at their greatest extent
British territories

0   200   400   600   800 Km

0   200   400 Miles

Anglo-Egyptian Sudan

• Khartoum

Eritrea

Aden Protectorate

Asmara •

Yemen

Gondar •

• Aden

French Somaliland

• Berbera

Addis Ababa •

British Somaliland

Abyssinia

Uganda

Italian Somaliland

Kenya

• Mogadishu

• Kismayu

Fighting in the Mediterranean and the Middle East sprawled across thousands of square miles. This map identifies the major ingredients of the conflict. First, there were widely-spread British territories, focusing on the Red Sea and the Suez canal: the latter remained influential, although the danger of traversing the Mediterranean meant that the Cape route was a safer line of communication with the Far East, itself a theatre of war after December 1941. Secondly, we must note the influence of Italy, and the dominant position of Italy herself in the Mediterranean. Thirdly, French possessions in North Africa and the Levant were strategically important, especially in view of the divergence between the Vichy regime and the Free French. Finally, the influence of Germany had grown enormously. By mid-1942 the Germans were not only in Greece, the Balkans, and North Africa, but their penetration into Russia bulged dangerously between the Black Sea and the Caspian. Suggestions that German forces in North Africa might have linked up with those in Russia pay insufficent regard to the realities of geography. There was, though, a risk of German success jolting Turkey from neutrality and imperilling the Anglo-Soviet position in Iran.

Tripolitania, while the Sicily-based Tenth Air Corps was to attack British Mediterranean shipping. Mussolini persuaded Hitler to modify his plans: German troops arrived in Tripoli on February 12 1941, but the attack on Greece was left to the Italians. Even so, Hitler proceeded to isolate Greece. Bulgaria was persuaded to join the Tripartite Pact (the German-Italian-Japanese agreement signed on September 27 1940) but in Yugoslavia, a coup brought to power an anti-Axis administration, impelling Hitler to order the invasion of both Yugoslavia and Greece.

Although Wavell warned that no aid he could send would thwart a German invasion, he was peremptorily told to comply with the wishes of Prime Minister Churchill, who was convinced of "the massive importance . . . of keeping the Greek front in being . . .". The Greeks initially declined British support and by March the government was cautious of sending troops to Greece. But now Wavell was more optimistic, and he must share responsibility for the venture with Churchill and Eden.

Yugoslavia invaded

On April 6 the Germans invaded Yugoslavia, which surrendered on April 17. Most of the Greek army was deployed to resist an Italian attack through Albania, so the German drive through Yugoslavia and Bulgaria unhinged the defence. General Wilson's 56,500-strong force fell back first to the Aliàkmon line and then to Thermopylae. The RAF was driven from its forward bases, and the German

capture of the Corinth Canal bridge by airborne assault cut off the Peloponnese. Most of Wilson's men were evacuated by sea, leaving behind their heavy equipment, on April 24–May 2.

O'Connor believed that he could have advanced into Tripoli if his army had not been depleted for the Greek expedition, but it is unlikely that he could have maintained himself there for long.

In Yugoslavia, armed resistance to occupation, interspersed with elements of civil war, began almost immediately, and Germany's invasion of Russia brought the Yugoslav communists into the battle. The two major resistance groups were Tito's communist partisans and General Mihajlovic's *Chetniks*. The Allies initially supported Mihajlovic, who owed allegiance to King Peter II's government-in-exile, but eventually recognized that Tito's force was the more effective. Tito crushed the *Chetniks* and forged a formidable partisan army which, in 1944, assisted Russian columns to liberate Belgrade.

The communists also took the lead in Greece, where, by 1943, their ELAS (National Popular Liberation Army) was the largest resistance force. In the summer of that year ELAS attacked other resistance groups and established domination over most of the country. But this drove republicans into alliance with monarchists, besides alerting the British to the danger of a communist takeover. In 1944, when the communists inspired a mutiny in the Greek army-in-exile in Egypt, British troops accompanied the

The battle of Crete: May 20–31 1941. As the map shows, the German airborne invasion, committing an overall total of some 30,000 men and employing some 500 transport aircraft, concentrated on the three major airfields along the north coast: Maleme-Chanea, Rethimnon and Iraklion. The attacks on Rethimnon and Iraklion were repulsed with heavy losses, but the capture of Maleme allowed the

Germans to re-supply by air. The British retreated south, where some 15,000 men (about half the garrison) were evacuated by sea from the Sfakia area. Although the Germans secured the island, their heavy losses in men and material – some units took more than 50 percent casualties and more than 220 aircraft were destroyed – contributed to the fact that they never again made a major airborne assault. The

illustration shows German paratroopers debarking from a DFS 230 assault glider, which was capable of carrying ten fully-equipped men (including a combatant pilot).

monarchist government to Greece after the German withdrawal. When ELAS refused to surrender its arms and reform within a Greek national army, fighting broke out. The British in Athens were hard-pressed, but their reinforcement from Italy helped persuade ELAS to disarm. When an anti-communist purge provoked ELAS to take up arms again, they were defeated in a three-year civil war.

The fall of Crete

Some 27,000 British and Dominion troops and two weak Greek divisions had been evacuated from Greece to Crete. Major General Freyberg's force had inadequate fighter cover, few tanks and little artillery – and on May 20, General Kurt Student's paratroopers and glider troops attacked. The battle hung in the balance for some time, but the capture of Maleme airfield, allowing the Germans to reinforce by air, proved decisive. (At sea, two German troop convoys were lacerated by the Royal Navy, albeit at the cost of losses from air attack.) By May 28 it was evident that Crete was untenable: the navy suffered in evacuating 15,000 of the garrison; 13,000 British and 5,000 Greek troops could not be rescued. German casualties may have totalled 17,000 and losses of troop-carrying aircraft were also serious. German paratroopers were never again committed to a large-scale airborne assault.

German entry into the Mediterranean theatre was paralleled by waning Italian fortunes. O'Connor had trounced Graziani in Libya; on November 11 1940 the Fleet Air Arm successfully raided the Italian naval base at Taranto; on March 28 1941 an Italian naval squadron was mauled off Cape Matapan.

In East Africa, Lieutenant General Sir William Platt advanced into Eritrea from the Sudan in January 1941, defeating the Italians at Agordat before besieging Keren. Although stoutly defended by high-grade Italian troops, Keren fell on March 27 1941; Platt took Asmara on April 1 and Massawa on April 4. Meanwhile, Lieutenant General

---

"I did not care to let my fancy dwell too much on the previous fortnight or on the fact that we were clearly going back again. What we had seen and done had to be taken as part of the day's work, and as far as possible put out of mind. I was neither pleased nor ashamed at having killed a number of men, merely thankful that I had done so before they killed us; it was the memory of our own dead and wounded that had to be repressed. Nobody in 15 Platoon, I would say, was bloodthirsty in the sense of taking pleasure in maiming or killing fellow men, though we did it methodically enough; the kick came from success in achieving the object, not in the corpses we had made in the process."

Second Lieutenant Peter Cochrane, Keren, 1940

Sir Alan Cunningham moved up from Kenya into Somaliland, defeating the Italians on the Juba river, moving on into Abyssinia, and capturing Harar on March

25 after an advance of 1,054 miles (1,700km) in 30 days. The Duke of Aosta abandoned Addis Ababa and dug in at Amba Alagi; caught between Cunningham and Platt, he surrendered on May 18.

The Desert War

German troops arrived in Libya on February 12 1941, commanded by Major General Erwin Rommel. Although he initially had only a light armoured division, Rommel was not one to wait: on March 31 he struck eastwards with his one German and two Italian divisions. General Neame's Cyrenaica Command had only a thin screen forward at El Agheila, and Rommel soon knocked the British off balance. O'Connor, now commander of British Troops in Egypt, went forward to advise the inexperienced Neame and both were captured on April 6. Rommel drove the British from Cyrenaica and went on to invest Tobruk.

Three other campaigns competed for Wavell's resources in mid-1941. A pro-Axis revolution in Iraq threatened oilfields and pipelines: a British force relieved the base at Habbaniya and, on May 31, took Baghdad. The growing Axis presence in Vichy French Syria persuaded the British, assisted by de Gaulle's Free French, to invade: after determined resistance, the Vichy garrison agreed to terms on July 14. In August, Anglo-Russian occupation of Iran ended the danger of German penetration.

Wavell's first blow at Rommel, Operation Brevity, an attack towards Capuzzo and Halfaya Pass in May, was premature and achieved little. "Battleaxe", launched on June 15, also failed: the plan was pedestrian, many of the troops involved were inadequately prepared, intelligence was poor – and the effectiveness of the German 88mm gun was a rude shock, although British armour had first encountered it in 1940.

Churchill now replaced Wavell with General Sir Claude Auchinleck, who selected Cunningham, victor in Abyssinia, to command the new Eighth Army. Like Auchinleck, Cunningham knew little about armour or the desert – and, with preparations for a new offensive, Operation Crusader, already underway, he had very little time to learn. Eighth Army was comprised of two corps: Godwin-Austen's 13th, consisting of infantry and a brigade of infantry tanks; Norrie's 30th, containing the remaining armour supported by 1st South African Division and 201st Guards Brigade. "The alternative," Cunningham noted, "would have been mixed groups."

The real problem was more one of tactical doctrine than tactical grouping. Lord Carver tells how: "Time and time again tanks motored or charged at the enemy on a broad front until the leading troops were knocked out by enemy tanks or anti-tank guns: the momentum of the attack immediately failed. Such artillery as was supporting the tanks indulged in some spattering of the enemy . . . after which the tanks motored about or charged again with the same results as before . . . the infantry taking no part, their task being to follow up and occupy the objective after it had been captured by tanks." The Germans, as Major General von Mellenthin writes, emphasized inter-arm cooperation, seeking to concentrate "superior numbers of tanks and guns at the decisive point".

Rommel's *Panzergruppe* (later *Panzerarmee*) *Afrika* comprised the Italian 21 Corps of four infantry divisions, another Italian division in the frontier defences, a German

Field Marshal Rommel (left) with Major General Fröhlich, 1942. Erwin Rommel (1891–1944) was commissioned into the infantry in 1912. He impressed Hitler when commanding his escort, which helped bring him command of 7th Panzer Division. In 1940 his aptitude for blitzkrieg was a spectacular success. In February 1941 he took command of German troops in Africa. Defeated at Alamein, he withdrew into Tunisia, whence he was evacuated before the collapse. As an army group commander in France he advocated deploying armour to defeat invasion on the beaches. Wounded by air attack in July 1944, he committed suicide rather than face trial for alleged complicity in the July plot.

heavy artillery group, and Lieutenant General Crüwell's *Afrika Korps* of 15th and 21st Panzer Divisions and 90th Light infantry division. The Italian 20 Corps, which included the Ariete armoured and Trieste mobile divisions, was in support.

Cunningham had 724 tanks (excluding light tanks) to Rommel's 174 German and 146 Italian tanks – but Cunningham's armour was a scratch pack of old A13 and new Crusader cruisers, American Stuarts (Honeys) and Matilda and Valentine infantry tanks. The 2-pounder gun mounted in most British tanks was no worse than the German short 50mm; in armour, the infantry tanks were superior to any German tank and the cruisers to most German machines. The Germans held the better anti-tank guns in the 50mm and 88mm weapons, the latter devastating against any Allied tank. In fact, British disadvantages were serious: their tanks, often hastily modified for the desert, tended to be unreliable; inter-arm cooperation was patchy; too many commanders simply did not understand armoured warfare.

The plan for "Crusader", launched on November 18, was for 30 Corps to make for Gabr Saleh, where Cunningham hoped to bring German armour to battle; 13 Corps faced the frontier defences and would not be committed until the tank battle was won. The fitful British attacks soon met concentrated German armour. Cunningham failed to realize that 7th Armoured Division had wasted its strength in bitter little battles at Bir El Gubi and Sidi Rezegh, and Rommel hooked savagely around the British left flank. In an atmosphere of growing panic, Auchinleck replaced Cunningham with Major General Neil Ritchie, forbidding retreat and moving up to take personal control.

Auchinleck's determination paid off. The Germans lost heavily and on the night of December 7–8 Rommel began to

Dear Major-General Campbell,

I have read in the papers that you have been my brave adversary in the tank battle of Sidi-Rezegh on November 21–22, 1941. It was my 21st Panzer Division which has fought in these hot days with the 7th Armoured Division, for whom I have the greatest admiration. Your 7th Support Group of Royal Artillery too has made the fighting very hard for us and I remember all the many iron that flew near the aerodrome around our ears.

The German comrades congratulate you with warm heart for your award of the Victoria Cross.

During the war your enemy, but with high respect.

Von Ravenstein

---

fall back; early January 1942 saw him at El Agheila. But because reinforcements meant for the desert now went instead to the Far East, Auchinleck's strength was depleted at the very time he determined to press on and finish Rommel. Ritchie prepared an offensive into Tripolitania, but Rommel attacked first, on January 21, hammering 1st Armoured Division.

Eighth Army withdrew to a position running from Gazala on the coast to Bir Hacheim, with its infantry in "boxes" protected by mines and wire and its armour to the rear. Two factors dictated this: first, the Gazala line was a springboard for the offensive Churchill demanded; second, although ULTRA decoding revealed that the Germans planned an offensive, Auchinleck believed that Rommel was more likely to thrust straight along the coast for Tobruk than to swing round the desert flank. Ritchie had 850 tanks, including 167 new Grants, against Rommel's 560, among which the up-armoured Mk IIIH with its short 50mm gun predominated.

Rommel's offensive

Rommel began Operation Venezia on May 26, leading his armour south of Bir Hacheim, refuelling, then swinging north towards Tobruk. He caught 30 Corps unprepared, cutting up its isolated brigades but himself suffering losses. By dawn on May 28 both panzer divisions – the 17th depleted and short of fuel – were just north of "Knightsbridge", but it was not until midday that Ritchie realized how vulnerable Rommel really was and moved ponderously to crush him.

Rommel's plight was desperate. The Free French brigade at Bir Hacheim held out bravely, stretching his logistics around the British flank. He decided to break through the main Gazala line while the Afrika Korps concentrated to its east; when his tanks had been resupplied, he would resume the offensive. After a hard fight, the Germans overwhelmed 150th Infantry Brigade in the Sidi Muftah box. While Rommel regrouped in this area, "The Cauldron", a poorly coordinated British attack on June 5 failed with the loss of more than 200 tanks. The fall of Bir Hacheim, whose garrison broke out on the night of June 10–11, meant that Eighth Army was compressed into an L-shaped position along the remnant of the Gazala line and the Trigh Capuzzo. On June 11–12, the surviving British armour was lacerated around "Knightsbridge". Eighth Army was forced to retreat beyond Tobruk, which fell on June 21. Ritchie was awaiting attack at Mersa Matruh when Auchinleck arrived to take personal command. He

---

DESERT SPECIAL FORCES

Most fighting in Libya occurred just south of the coastal strip. Farther south stretched expanses of sand sea, which pre-war exploration had shown to be crossable. In mid-1940 Major R A Bagnold recruited volunteers: the Long Range Desert Group began operations in August.

The map (*below*) follows a mission of G and T Patrols under Captain P A Clayton. They left Cairo on December 26, 76 men and 24 vehicles strong. T Patrol entered Chad to meet a French force, and on January 8 both patrols and the French attacked the Italian airfield at Murzuk, destroying three light bombers but suffering casualties. Clayton was wounded and captured in further fighting, but the patrols reached Cairo on February 9 after 45 days and 4,500 miles in the desert.

The LRDG was expanded in 1941, and operated from Kufra and Siwa, moving back to Fayoum when the Germans advanced to Alamein. It reconnoitred and raided in the desert until Rommel was pushed out of the Mareth Line: it later fought in the Greek islands and the Balkans.

The Special Air Service was founded in 1941 by Lieutenant David Stirling. An initial attempt at parachuting proved disappointing, and in December the SAS carried out successful raids using the LRDG as a taxi service. It obtained its own vehicles, favouring the jeep mounted with Vickers K guns, and by the end of the Desert War had destroyed almost 400 Axis aircraft. The smallest desert special force was Popski's Private Army, commanded by Vladimir Peniakoff, which destroyed 34 aircraft and went on to fight in Italy.

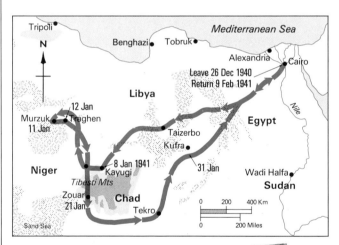

SAS patrol returns from action.

was at first forced back towards Egypt, but at El Alamein, where the front narrowed between the sea and the Quattara Depression, he halted. The newly promoted Field Marshal Rommel, now at the very limit of his troops' strength, failed to dislodge Auchinleck.

Montgomery takes over

The fighting around El Alamein throughout July convinced Auchinleck that although he could hold Rommel, he could not launch a swift counterattack. Churchill determined upon a change in command: General Sir Harold Alexander replaced Auchinleck and Lieutenant General B L Montgomery took over Eighth Army on August 13. There has been controversy over the relative roles of Auchinleck and Montgomery in paving the way for victory at El Alamein – but it is beyond dispute that Montgomery revitalized Eighth Army. When Rommel attacked the Alamein position on August 31–September 3 (the battle of Alam Halfa), he was sharply rebuffed.

Montgomery's offensive was carefully prepared. He had about 1,100 serviceable tanks, including 285 75mm-gunned Shermans. The Axis force, under General Stumme while Rommel was on sick leave, had 600 tanks, half of them Italian, of which only the 38 Mk IVs really matched the Shermans. British superiority in all other resources, from aircraft to anti-tank guns, was marked.

This strength owed much to the Allied seamen who ran the gauntlet to bring men and material to Alexandria. They also supplied the beleaguered island of Malta: in August, the "Pedestal" convoy shepherded the US tanker *Ohio* to Malta, enabling air and naval forces there to step up attacks on Rommel's supply lines.

Montgomery envisaged a battle with three phases. In

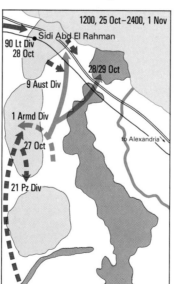

Field Marshal Sir Bernard Montgomery (Viscount Montgomery of Alamein 1887–1976) was a bishop's son, and joined the Royal Warwickshire Regiment after a Sandhurst career nearly ruined by rumbustiousness. He won the DSO as a platoon commander in 1914 and was a lieutenant colonel by the war's end, gaining valuable experience on the staff of Plumer's Second Army. Markedly efficient at the head of 3rd Division in 1939–40, he commanded corps in England until taking over the Eighth Army in 1942.

Montgomery instilled confidence into his "brave but baffled" army, and in October won El Alamein, a characteristically methodical battle. In 1943 he took the Eighth Army on into Sicily and Italy before returning to take an active part in the planning for the invasion of Europe, in which he was land force commander under Eisenhower. After the war he served as Chief of the Imperial General Staff and Deputy Supreme Allied Commander. Montgomery's critics note that he was a showman – witness his two-badge beret – inclined to be rude to superiors and subordinates alike, and that he had undue regard for the set-piece battle. However, he was right in recognizing the British army's limitations and in launching it into the sort of battle it could win, and for all his imperiousness he was an outstanding practical commander, identified by British soldiers and public alike as a winner at a time when winners were all too scarce.

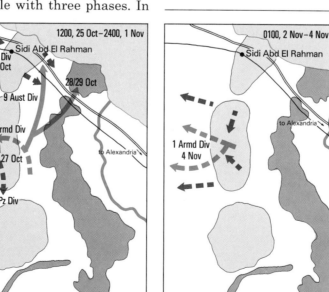

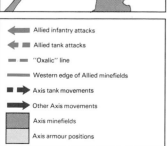

←	Allied infantry attacks
◄- -	Allied tank attacks
- - -	"Oxalic" line
———	Western edge of Allied minefields
■-■►	Axis tank movements
►►	Other Axis movements
▓	Axis minefields
░	Axis armour positions

The battle of El Alamein. The Axis army held a formidable position between the sea and Quattara Depression (off the map to the south), protected by deep minefields. Montgomery intended to open two corridors through the

minefields, reaching line "Oxalic" with 30 Corps, before 10 Corps exploited through the gaps. 13 Corps, to the south, was to mount a diversionary attack. The break-in went slowly, with fierce fighting and congestion around the corridors. Montgomery modified his plan for the next phase, using 9th Australian Division to "crumble" northwards, and pushing forwards in the centre with 1st Armoured and 51st Highland Divisions. Rommel counterattacked on October 27–28, throwing 90th Light and elements of 15th Panzer against the Australians and sending 21st Panzer against 1st Armoured and the Highlanders. The counterattack failed, and on the night of November 1–2 the breakout began, general Axis withdrawal starting on the 4th.

"Break-in", he would feed 10 Corps through gaps in the minefields opened by the infantry of 30 Corps, while 13 Corps mounted a diversionary attack to the south. In "Dogfight", infantry would crumble Rommel's positions while armour blunted his counterattacks. Then 10 Corps would make the decisive "Break-out".

The attack began with a prodigious bombardment on the night of October 23. By the time Rommel returned to take command on October 25 the Axis force had suffered severely and was dangerously short of fuel. Nor had the British an easy time: 30 Corps made uneven progress and there was a crisis of confidence as 10 Corps tried to move through. Montgomery was not ready to begin the break-out phase until November 1. On November 4, with total collapse imminent, Rommel ordered a general withdrawal. Helped by atrocious weather and British overcaution, the remnants of his mobile formations escaped. The British took Tobruk on November 13, Benghazi on November 20 and Tripoli on January 23 1943. On February 12 the rearguards of *Panzerarmee Afrika* fell back across the Tunisian frontier.

Operation Torch

El Alamein had not been fought in isolation. In summer 1942 the Allies had decided on Operation Torch, the seaborne invasion of French North Africa, and on November 5 the Allied Supreme Commander, Lieutenant General Dwight D Eisenhower, established his HQ at Gibraltar. The venture was contentious, for it was hastily planned and depended very much upon the Vichy French forces offering little resistance. In the event, the landings around Casablanca, Oran and Algiers on November 7–8 were sporadically opposed, but a ceasefire was soon agreed with the Vichy Commander-in-Chief Admiral Darlan.

The ceasefire persuaded Hitler first to occupy air bases in Tunisia and then to seize the unoccupied zone of France. This achieved little: the French fleet at Toulon was scuttled before the Germans could reach it, while the occupation caused French authorities in North Africa to cooperate more fully with the Allies. German reinforcements poured into Tunisia, halting the Allied advance at Medjez-el-Bab in late November and forcing the British First Army to abandon its advance on Tunis. Meanwhile, faced by Montgomery's inexorable advance across Tripolitania, Rommel sought to join with Colonel General von Arnim's Fifth Panzer Army in Tunisia, where they could cudgel the inexperienced First Army before Rommel turned back to strike at Eighth Army.

In January 1943 Arnim attacked the passes through the Eastern Dorsal, taking only one but badly shaking the French troops who held them. In February Rommel took armour northwards to help Arnim defeat the Americans at Kasserine Pass. But Rommel and Arnim got on badly, and both squabbled with Field Marshal Kesselring, Commander-in-Chief South. The Italian command was in increasing decline: Marshal Cavallero, its Chief of Staff, was dismissed on January 30. In late February Rommel swung his striking force from Kasserine back towards Mareth, but Eighth Army was alerted by radio intercepts and on March 6 Rommel's assault was broken by 30 Corps at Médenine.

In February, General Alexander set up Fifteenth Army Group to coordinate the battle in Tunisia, and his steady

---

LOGISTICS IN THE DESERT

The Desert War presented formidable logistic obstacles. Battle and manoeuvre over harsh terrain wore out engines and guzzled petrol, and the heat of the desert day made water more than usually indispensable. Local resources were negligible, and both sides had lengthy lines of communication, with perilous sea crossings followed by treks along the vulnerable coast road. While the British had the resources of Egypt behind them and believed the theatre to be of great importance, Rommel had to collaborate with the Italians – a task made harder by his sharp temper – and OKW regarded Africa as a low priority. Even when supplies reached Tripoli, they had to cover 1,000 miles to reach the front, and up to half Rommel's petrol was consumed getting supply trucks forward. For all his skill as a tactician, Rommel underestimated logistic problems, recognizing too late that "the battle is fought and decided by the Quartermasters." The war has left its mark on our vocabulary. The British sent petrol forward in a flimsy 4-gallon tin: Auchinleck estimated that it led to the waste of 30% of petrol between base and consumer. The robust stamped metal German petrol tin was eventually copied, and survives to this very day with the enduring nickname of Jerrycan.

Malta played a key part in the battle. Air and naval forces based there preyed on Rommel's supplies, but the island itself was under air attack and, in mid-1942, was dangerously short of supplies. The *Pedestal* convoy consisted of 16 ships including the US tanker *Ohio*, escorted by 3 carriers, 2 battleships, 7 cruisers and 24 destroyers. It passed the straits of Gibraltar on August 10, and was repeatedly attacked: one carrier, two cruisers and all but five of the merchantmen were lost, and many other ships were damaged. *Ohio* reached Valetta lashed between two destroyers. Nevertheless, *Pedestal* enabled Malta to hold out, its ships and aircraft doing increasing damage to Rommel's lines of supply.

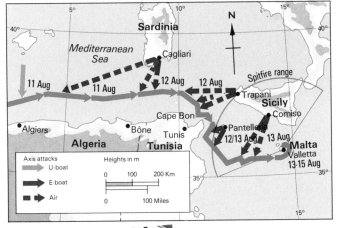

The map (*above*) shows the route of the Operation Pedestal convoy to Malta, and the various attacks mounted on it by the Axis forces. An Afrika Korps tank (*left*) is unloaded at Tripoli, February 1941. The vehicles had to cover 1,000 miles to reach the front.

guidance soon produced results. In March Montgomery levered Rommel (subsequently ordered back to Europe) out of the Mareth Line; on April 7 elements of the First and Eighth Armies met near Maknassy. Tunis and Bizerta fell; Arnim was trapped in the Cape Bon peninsula, and on May 12 Army Group Africa disintegrated with the loss of more than 250,000 prisoners.

The campaign in Italy

Allied leaders had agreed that victory in North Africa would be followed by the invasion of Sicily, Operation Husky, to secure Allied communications in the Mediterranean and to increase the pressure on the Axis. However, the Americans insisted that the cross-Channel invasion of Europe – "Overlord", then scheduled for May 1944 – should take priority over Mediterranean operations. The exploitation of "Husky" would depend very much on how the Italians fought.

Mussolini and his new Chief of Staff, General Ambrosio, were reluctant to accept German help in the defence of Italy. Kesselring, however, shrewdly increased German air strength in the Mediterranean and persuaded Ambrosio to accept German assistance in Sicily and on the mainland. Thus, General Guzzoni had 12 divisions, 2 of them German, to defend Sicily. For the Allies, Eisenhower was in overall command; Alexander, the land force commander, disposed of Lieutenant General Patton's US Seventh Army and Montgomery's Eighth Army; Admiral Cunningham commanded naval forces and Air Chief Marshal Tedder commanded air forces.

In preparation for "Husky", the Allies took the islands of Pantelleria and Lampedusa and bombed the Sicilian airfields. On July 10, Eighth Army landed in the southeast and Seventh Army on the south coast; these landings went well, but an attempt to drop paratroops inland proved costly. The British took Syracuse on July 10, but their advance was slowed until early August by German resistance around Mount Etna. Patton took Palermo on July 22, and on August 16 US forces entered Messina.

Sicily cost the Italians 130,000 men, and the defeat jolted Mussolini from power. His successor, Marshal Badoglio, agreed to an armistice on September 3, allowing the Allies to land unopposed on the mainland at Reggio and Taranto. Kesselring's immediate reaction to the armistice was to disarm Italian units and secure Rome. Rommel's Army Group B held Italy, the six divisions south of Rome forming General von Vietinghoff's Tenth Army. Failing a defeat of Allied landings on the beaches, the Germans planned four east-west defensive lines across Italy.

On September 9, Lieutenant General Mark Clark's US Fifth Army landed at Salerno, opposed by 14 Panzer Corps. The Germans counterattacked fiercely, the Luftwaffe bombing the beachheads and using radio-controlled glider bombs to damage Allied warships. Aided by naval gunfire and carpet bombing, the Allies held on, and on September 18 Vietinghoff disengaged and fell back to the Volturno. By early October the Germans had also abandoned Sardinia and Corsica.

On September 21 Alexander ordered an advance, with Eighth Army moving up the east coast and Fifth Army up the west. Montgomery took Bari on September 22 and Foggia on September 27; On October 2–3 an amphibious assault captured Termoli and Eighth Army crossed the Biferno and Trigno rivers. Fifth Army captured Naples on October 1, crossed the Volturno and came up to the Bernhard Line, an outpost of the main Gustav Line. Then bad weather slowed the Allied advance. Hitler ordered Kesselring – now Commander-in-Chief South West, with Army Group C (Vietinghoff's Tenth Army and Mackensen's Fourteenth) – to hold the Allies south of Rome.

Alexander's winter offensive comprised a right hook against the Sangro by Eighth Army followed by a left at the Bernhard Line by Fifth Army, supported by a landing at Anzio. Montgomery crossed the Sangro in November; on Clark's front, the Bernhard Line was broken with some difficulty. Eisenhower and Montgomery were now needed for "Overlord", so Wilson became Supreme Commander, Mediterranean Theatre, and Lieutenant General Leese took command of Eighth Army.

The battles at Cassino

Clark's assault on the Gustav Line began on January 12 1944. On his right the French Expeditionary Corps attacked the high ground north of Cassino, clearing the approaches to the Gustav Line but failing to penetrate it. In the south the British 10 Corps gained a bridgehead across the Garigliano, but the US 36th Division was cut to tatters on the Rapido. A corps landed at Anzio on January 22, but its commander feared a repetition of the Salerno counterattack and the landing was contained.

On January 24 the US 2 Corps attacked around Cassino but made little progress, although French attacks on the Colle Belvedere achieved minor gains at ghastly cost. For

The US Fifth Army in Operation Avalanche – the landings along the Gulf of Salerno, southwest Italy, on September 9 1943 – consisted of one American and one British corps. Here (*above*), men of the British Tenth Corps come ashore from an American landing ship in the Montecorvino area. Aimed at the capture of the vital port of Naples, "Avalanche" was delayed by determined opposition: British units of Fifth Army entered Naples on October 1.

The course of the Italian campaign (*right*) from the initial landings in Sicily in July 1943 to the surrender of German forces in May 1945. Note the parallel advances of the US Fifth Army and British Eighth Army up the west and east coasts respectively. Some of the fiercest fighting of the war was needed to break through the major German defensive positions on the Bernhard-Gustav, Caesar and Gothic Lines.

the second battle of Cassino, Alexander reinforced Fifth Army with Freyberg's New Zealand Corps. On February 15, after bombing had reduced the monastery above Cassino to rubble, Freyberg attacked the town and the northern hills, to little effect. New Zealand Corps tried again on March 15, following a massive air and artillery bombardment, but the German paratroops holding the town and the hills above it still resisted stoutly. In the fourth battle, beginning on May 11, the major effort was made in the Liri valley, south of Cassino. Aided by sound preparation and good weather, the British broke through, with the French and Americans fanning out to the south. General Anders's Polish troops at last took the ruined monastery on May 18.

The collapse of the Gustav Line took Kesselring by surprise, and the Allies were through the Hitler Line, just behind it, before a proper defence could be organized. The Anzio breakout began on May 23, and on June 4 the leading elements of Fifth Army entered Rome. The Allies pressed on quickly, but on June 20 they met cohesive defence in the Trasimene Line, whence Kesselring fell back slowly to the Gothic Line, north of Florence. Six Allied divisions, including the entire French corps, were now withdrawn from Italy for Operation Anvil/Dragoon, which began on August 15, when Lieutenant General Patch's US Seventh Army landed on the Riviera and advanced rapidly up the Rhône valley.

On the same day, Eighth Army began to concentrate for an attack on the Gothic Line, achieving a quick break-through on August 25–28 but slowing up as German reinforcements arrived. Fifth Army's attack began on September 10. Ten days' hard fighting convinced Kesselring that he could not hold the Gothic Line: late September saw Eighth Army on the southern edge of the Po valley and Fifth Army just short of Bologna. Bad weather, fatigue and artillery ammunition shortages slowed the advance. In January 1945 Alexander, the new theatre commander, was told that more divisions would be transferred to the west: he was to hold the ground already gained and employ limited offensives to pin down Axis forces.

German surrender

The Allies used the first three months of 1945 to refit and retrain. Vietinghoff, having taken over from Kesselring in March, was ordered to delay the Allies on the Po. In April, Eighth Army opened its final offensive, thrusting in between Lake Comacchio and Bologna, and Fifth Army attacked a few days later. Vietinghoff could not conduct a cohesive withdrawal: Army Group C was overwhelmed, and Vietinghoff surrendered on May 2.

Although the conduct of the Italian campaign has been hotly criticized, its contribution to the Allied victory cannot be doubted. Alexander wrote that he had tied down 55 German divisions in the Mediterranean "by the threat, actual or potential, presented by our Armies in Italy". The Germans lost 536,000 men to the Allies' 312,000. The valour of the fighting men on both sides lies beyond dispute, but it is sad to reflect that the sacrifices made by the Algerian *tirailleurs* and Polish riflemen at Cassino did not guarantee them postwar political recognition.

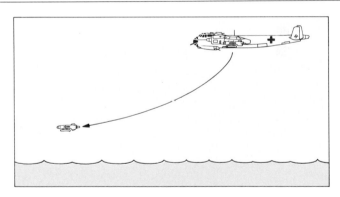

THE GLIDER BOMB

In December 1942 the Germans produced the Dornier Do 217K-2, designed specifically for the FX 1400 *Fritz X* glider bomb. The bomber mounted a *Kehl I* transmitter and the bomb a *Strassburg* receiver, enabling the bomb aimer, in a special crew station, to control the bomb. Do 217K-2s of *Kampfgeschwader* 100 commenced operations in the Mediterranean on August 29 1943, and on September 9 they sank the Italian flagship *Roma* as she steamed to join the Allies. During the Salerno landings KG 100 hit the US cruiser *Savannah* and several supply ships, going on to damage the battleship HMS *Warspite* and, in January 1944, to sink the British cruiser *Spartan* and the destroyer *Janus*. Although the glider bomb was unsophisticated by the standards of later stand-off systems, it represented a notable advance on free fall "iron bombs" and pointed the way ahead to new generations of "smart munitions".

chapter fifteen

World War II: The Pacific War

Eric Grove

The war between the Allies and Japan in 1941–45 was the greatest maritime conflict in history. Unprecedented naval power was deployed over a massive expanse of water, from the Aleutians in the north to Darwin in the south, and from Ceylon (Sri Lanka) in the west to Hawaii in the east. The dominance of the gun-armed surface warship ended and the major means of maritime strike became the aircraft and the submarine. The largest fleet actions in naval history were complemented by the most successful campaign of commerce interdiction ever waged. The USA emerged as the world's dominant naval power.

From the time of her initial "opening" to the West in the 1850s and her political revolution of 1868, Japan had steadily built up her economic and political strength. By 1920 she was a major power, with possessions extending from the Asian mainland in the west to the Marshall Islands in the east. Her naval ambitions received a temporary setback in 1922, when the Washington treaties set her ratio of capital ships at 60 percent of that of the United States or Great Britain (who acknowledged parity between themselves). This 5:5:3 ratio was extended to other classes of warship by the London Naval Conference of 1930, but by then the political tide was turning in Japan. In the early 1930s the Tokyo government was powerless as the Imperial Japanese Army (IJA) began its occupation of Manchuria and the Imperial Japanese Navy (IJN) began its own private war at Shanghai. "Independent" Manchukuo became an occupied satellite of Japan, while at home, extremists pursued a policy of "government by assassination".

The rise of militarism was associated with the effect on Japan of the economic crash of 1929. Trade dislocation, apparently permanent, was caused by the worldwide imposition of protectionist policies, while emigration prospects were curtailed by racialist laws that wounded Japan's pride. Thus there evolved the concept of a "Greater East Asia Co-Prosperity Sphere": a self-sufficient Japanese empire built at the expense of China and the European colonial powers. Expansion on the Asian mainland would provide room for Japanese colonists and sources of vital minerals: expansion southward would guarantee access to oil and rubber. Japan left the League of Nations in 1933 and in January 1936, after naval parity had been refused, walked out of the Second London Conference. A massive naval build-up, fuelled by an economic recovery whose fruits were increasingly monopolized by the armed forces, was planned to begin in 1937.

By that year, Japan's expansion in China had brought her increasingly into confrontation with the USA, where the Roosevelt administration moved towards a policy of economic sanctions backed by an increase in naval strength. In 1940, the defeats inflicted by Germany on France and Britain shocked the US Congress into appropriating funds for a "Two-Ocean Navy": a 70 percent increase in naval strength, to include 7 new battleships, 6 very large cruisers, more than 60 conventional cruisers, about 150 destroyers and some 140 submarines. Even more importantly, 19 aircraft carriers and 16,000 aircraft and pilots were authorized.

This legislation showed the Japanese that if they intended to carve out the Co-Prosperity Sphere, they must move fast. Their naval build-up had given them numerical parity in the Pacific and would give them superiority by mid-1942. However, by 1943–44 the USA would regain an unassailable lead. American economic sanctions and the weakness of the colonial powers in much of Southeast Asia and the East Indies give Japan both the motive and opportunity to strike south.

From July 1940 to June 1941, while the USA tightened its sanctions by embargoing strategic minerals and limiting oil supplies, Japan moved steadily south into French Indochina. With the need for self-sufficiency ever more apparent, and fearing that Germany's attack on Russia might tempt the IJA into realizing its long-cherished dream of war with the USSR, the leaders of the IJN exerted maximum pressure on the government to approve the strike south. Early in July 1941 a top-level imperial conference decided that Japan would first advance southwards and afterwards join Germany against the USSR: if this meant war with Britain and the USA, so be it. On July 21, when southern Indochina was occupied to provide forward bases, Admiral Osami Nagano, Chief of Naval General Staff, asked that war, if it came, should come sooner rather than later.

Roosevelt's next move caused many Japanese decision makers to agree with Nagano. All Japanese assets in the USA and in the British and Dutch empires were frozen, denying Japan access to all foreign goods, including oil. Japan had only six months' supply of aviation fuel. Effectively, she was faced with a stark choice: war or acceptance of America's terms. Japan felt unable to make the concessions demanded, notably withdrawal from China and Manchukuo – and as the last political attempts were made to stave off hostilities, she planned her offensive. Admiral Isoroku Yamamoto, Commander-in-Chief of the Combined Fleet, decided that the US Navy (USN) must not be given the chance to strike at Japan's widely deployed invasion forces. The US Pacific Fleet must therefore be destroyed in its base at Pearl Harbor, Oahu, Hawaii, on the first day of the war.

A superior naval air arm

Japan had the capability to do this because she possessed the world's most effective naval air arm – one result of a doctrine that emphasized qualitative superiority and novel technology to compensate for the quantitative inferiority imposed by the Washington and London agreements. During the 1930s, Japanese doctrine envisaged a decisive battle against a US fleet that had crossed the Pacific to relieve the Philippines, whittled down en route by submarines and land-based air attacks so that when the fleets met, numbers would be equal. Although this was not to be, the scenario had by 1941 resulted in a navy of considerable size and novelty. Large, long-range fleet submarines capable of carrying aircraft were already in service, as were midget submarines and their "mother" submarines and surface ships. The IJN's night-action techniques were unsurpassed in the pre-electronic age. Its secret weapon for surface battles was the

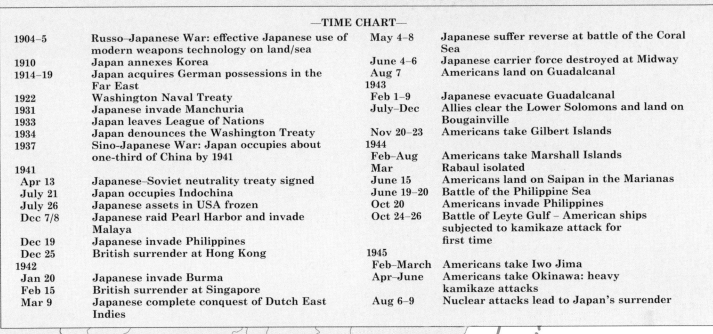

—TIME CHART—

1904–5	Russo–Japanese War: effective Japanese use of modern weapons technology on land/sea	May 4–8	Japanese suffer reverse at battle of the Coral Sea
1910	Japan annexes Korea	June 4–6	Japanese carrier force destroyed at Midway
1914–19	Japan acquires German possessions in the Far East	Aug 7	Americans land on Guadalcanal
1922	Washington Naval Treaty	1943	
1931	Japanese invade Manchuria	Feb 1–9	Japanese evacuate Guadalcanal
1933	Japan leaves League of Nations	July–Dec	Allies clear the Lower Solomons and land on Bougainville
1934	Japan denounces the Washington Treaty	Nov 20–23	Americans take Gilbert Islands
1937	Sino-Japanese War: Japan occupies about one-third of China by 1941	1944	
		Feb–Aug	Americans take Marshall Islands
1941		Mar	Rabaul isolated
Apr 13	Japanese–Soviet neutrality treaty signed	June 15	Americans land on Saipan in the Marianas
July 21	Japan occupies Indochina	June 19–20	Battle of the Philippine Sea
July 26	Japanese assets in USA frozen	Oct 20	Americans invade Philippines
Dec 7/8	Japanese raid Pearl Harbor and invade Malaya	Oct 24–26	Battle of Leyte Gulf – American ships subjected to kamikaze attack for first time
Dec 19	Japanese invade Philippines		
Dec 25	British surrender at Hong Kong	1945	
1942		Feb–March	Americans take Iwo Jima
Jan 20	Japanese invade Burma	Apr–June	Americans take Okinawa: heavy kamikaze attacks
Feb 15	British surrender at Singapore		
Mar 9	Japanese complete conquest of Dutch East Indies	Aug 6–9	Nuclear attacks lead to Japan's surrender

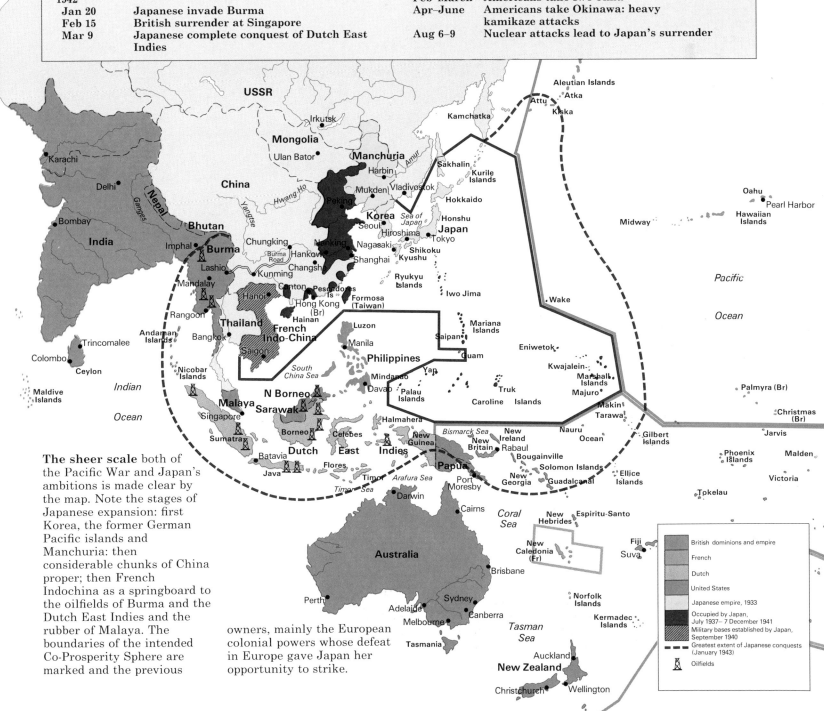

The sheer scale both of the Pacific War and Japan's ambitions is made clear by the map. Note the stages of Japanese expansion: first Korea, the former German Pacific islands and Manchuria: then considerable chunks of China proper; then French Indochina as a springboard to the oilfields of Burma and the Dutch East Indies and the rubber of Malaya. The boundaries of the intended Co-Prosperity Sphere are marked and the previous owners, mainly the European colonial powers whose defeat in Europe gave Japan her opportunity to strike.

Legend:
British dominions and empire
French
Dutch
United States
Japanese empire, 1933
Occupied by Japan, July 1937– 7 December 1941
Military bases established by Japan, September 1940
Greatest extent of Japanese conquests (January 1943)
Oilfields

Type 93 "Long Lance" torpedo, fitted in massed batteries in fast surface ships. Its newest battleships were equally remarkable: 64,000-ton monsters carrying 18.1-inch guns.

It was the naval air arm that gave Japan her most significant advantage. From April 1941 it consisted of the Eleventh Air Fleet, equipped with high-performance, long-range, land-based bombers, along with fighters and reconnaissance aircraft; and the First Air Fleet, with six major carrier air groups operating the finest combination of carrier-based fighters, dive bombers and torpedo bombers available to any navy. It was planned to use the carriers as a concentrated striking force so that their aircraft could attack en masse – a method considered to be especially effective against shore targets.

The First Air Fleet consisted of three carrier divisions. The First Carrier Division deployed *Akagi* (36,500 tons) and *Kaga* (38,200 tons), rebuilt respectively from a battlecruiser and battleship and extensively modernized in the mid-1930s. The Second Carrier Division contained two smaller carriers, the 15,900-ton *Soryu* and the 17,300-ton *Hiryu*, completed in 1937 and 1939 respectively. The Fifth Carrier Division was made up of the new 25,675-ton fleet carriers *Shokaku* and *Zuikaku*. Normal operational air groups comprised some 54 aircraft each for *Soryu* and *Hiryu*, 63 each for *Akagi* and *Kaga*, and 72 each for *Shokaku* and *Zuikaku*: all also carried between 16 and 25 reserve aircraft. The aircraft types were the fast, agile Mitsubishi A6M Zero fighter; the Aichi D3A dive bomber (later codenamed "Val" by the Allies); and the Nakajima B5N ("Kate") attack aircraft, carrying either an 18-inch

THE IJNS AIR ARM IN 1941

Mitsubishi A6M ("Zero")

Aichi D3A ("Val")

Nakajima B5N ("Kate")

Mitsubishi G4M ("Betty")

No navy – and few air forces – had better aircraft than Japan's in 1941. The **Mitsubishi "Zero"** fighter combined high speed with excellent manoeuvrability; its only weakness was vulnerability to battle damage. The **"Val"** dive bomber was as fast as many fighters but could also dive slowly and steeply enough to achieve pinpoint bombing accuracy. The **"Kate"**, an all-metal monoplane, was the most advanced carrier-based torpedo bomber of its time and was also intended for use as a high-level bomber against sea and land targets. The land-based bombers were also remarkable machines for their time. The newly delivered **"Betty"** combined very high speed for a large twin-engined aircraft with exceptionally long range. Like the Zero, however, it had little capacity to stand up to battle damage.

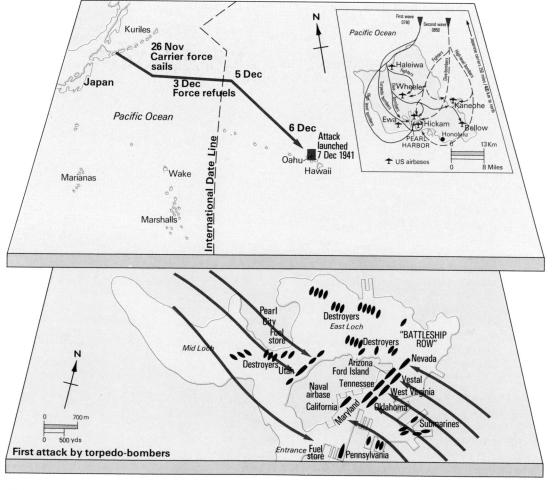

First attack by torpedo-bombers

Attack on Pearl Harbor: although the Americans had themselves rehearsed carrier air strikes on Pearl Harbor before 1941, they did not think the Japanese capable of doing the same. Great pains were taken to preserve secrecy. The Japanese tried to give the impression that the First Air Fleet was still in the Inland Sea and Nagumo kept radio silence as he advanced the 3,000 miles (4,800km) to the launching point 275 miles (440km) north of Hawaii. The Japanese had chosen a route that kept them clear of merchant shipping and sailed deliberately into seas notorious for bad weather, thus gaining the cover of fog and storms. The first wave was composed of 190 aircraft; 40 torpedo bombers, 50 high-level bombers, 50 dive bombers, and 50 fighters. The second consisted of 170 aircraft: 50 high-level bombers, 80 dive bombers and 40 fighters. The main target was "battleship row" to the east of Ford Island where the capital ships were moored.

torpedo or a 1,757lb armour-piercing bomb.

Yamamoto planned to unleash the First Air Fleet against Pearl Harbor, while at the same time a series of amphibious assaults supported by surface ships and shore based aircraft would be made from bases in China and Vietnam against Thailand, Malaya and Borneo; from Formosa and the Carolines against the Philippines; from the Carolines against the Gilbert Islands and Wake Island; and from Japan itself against Guam, America's foothold in the Marianas. The date chosen was December 7–8 1941 (depending on which side of the international dateline the target lay). Troops were ashore at Kota Bharu, northeast Malaya, one hour before the aircraft of Vice-Admiral Chuichi Nagumo's Striking Force hit Pearl Harbor.

The attack on Pearl Harbor
On the morning of December 7 1941 some 360 Japanese aircraft struck the US Pacific Fleet at Pearl Harbor in two waves. However, the Pacific Fleet did not then contain the USN's newest ships, which were engaged in the undeclared war waged since midsummer against the Germans in the Atlantic. All eight battleships currently at Pearl Harbor antedated the Washington treaties. Some had received limited modernization, but although well armed for surface action they were slow and deficient in protection. All were sunk or damaged, but only two were permanent losses: USS *Arizona*, sunk with the loss of 1,177 men, and USS *Oklahoma*. Nevertheless, US strength in operational battleships had been halved within a matter of minutes; two destroyers had been sunk and three cruisers and a

destroyer damaged; 188 US aircraft had been destroyed and 159 damaged. The Japanese lost only 29 aircraft, five midget submarines and one fleet submarine.

Although it was an awesome demonstration of the power of carrier aircraft, the Japanese victory was incomplete. Nagumo's refusal to allow his pilots to make a further attack against an alerted enemy had left the vital infrastructure of Pearl Harbor virtually unscathed. The base was able to begin repairs immediately and to continue operating its surviving ships – notably the Pacific Fleet's three aircraft carriers, the *Saratoga*, *Lexington* and *Enterprise*. At the time of the raid, *Saratoga* was on America's west coast while *Lexington* and *Enterprise* were on missions to Pacific outposts.

The Japanese had settled the debate between the USN's airmen and surface sailors as to which was the most potent capital ship, the carrier or the battleship. The USN now had no alternative but to use carriers as its primary weapon – and it was ready to do so, for by 1941 the USA had significantly developed its naval air power. The Washington treaties had caused the conversion of the incomplete battlecruisers *Lexington* and *Saratoga* into excellent carriers of some 33,000 tons, thus allowing the USN to explore the combat potential of large carrier air groups, both offensive and defensive, against targets at sea, on land or in the air. The 33-knot carriers, each with some 90 aircraft, had to operate at a different pace to the plodding 22-knot battle line. Treaty limitations and needless misgivings about the large size of the *Lexington* and *Saratoga* had led the USA to build two smaller and slower carriers, the

Devastation at Hickam Field; one of the four B-17 Fortress bombers destroyed on the ground on December 7. The US Army Air Force, based on Oahu, had over 200 aircraft damaged in the attacks, 77 of these being permanent write-offs. Their commander had been unwilling to disperse his planes for fear of sabotage.

A Japanese bomber scores a direct hit on the USS *Oklahoma* at Pearl Harbor. The battleship capsized after being hit with two more deadly accurate torpedoes, all before the attack was five minutes old.

unsuccessful *Ranger* and the more combat-worthy *Wasp*, in the 1930s. However, the preferred type was a carrier with *Lexington*'s capabilities if not her size: the *Yorktown*, *Enterprise* and *Hornet*, completed in 1937–41, put close on 100 aircraft in a 33-knot ship with a standard displacement of just under 20,000 tons. With such large air groups, US doctrine by 1941 was to form a task force around a single carrier screened by cruisers and destroyers.

US carrier aircraft in 1941 were efficient, although not as good as the Japanese planes. The best for its task was the Douglas SBD Dauntless scout aircraft and dive bomber. Between the wars the USN had concentrated on dive bombing as the most effective method of attacking both ships and ground targets. Torpedo bombing was considered not to be especially useful: the Douglas TBD Devastator could operate as a torpedo or level bomber, but its performance was in every way inferior to that of the IJN's "Kate" and it proved extremely vulnerable in action. The fleet fighter, the Grumman F4F Wildcat, was fast and manoeuvrable, and although inferior to the Zero in these respects, it was considerably more rugged.

There was no question of hazarding America's vital carriers in a premature offensive: indeed, the *Saratoga* was soon out of action, torpedoed by a Japanese submarine on January 11 1942. Until more carriers could be transferred from the Atlantic, only limited operations were possible. Thus, with only one or two temporary setbacks – notably at Wake Island, where two Japanese destroyers were sunk and conquest delayed for 12 days – the Japanese carried all before them. Britain, fully committed in Europe, had been unable to give Malaya sufficient air and naval cover to stem the Japanese advance. At Churchill's insistence, a reluctant Admiralty sent out the new battleship *Prince of Wales* and the old battlecruiser *Repulse* to operate from Singapore as a fast deterrent squadron. The Eleventh Air Fleet's Indochina-based torpedo bombers caught the British ships without air cover and sank them with little difficulty on December 10 1941.

Japanese troops in Malaya, although much outnumbered, outmanoeuvred the demoralized British imperial forces from the outset. Singapore surrendered on February 15 1942, and by then the Japanese had invaded the Dutch East Indies, perhaps their major objective. On January 8 they struck at Borneo and other northern islands. An ABDA (American-British-Dutch-Australian) Command had been set up at Batavia (Djakarta), Java, and a naval striking force formed under the command of the Dutch Rear-Admiral Karel Doorman. It failed to keep the Japanese out of Sumatra, and the showdown came when they moved on Java itself. Doorman sailed to intercept the invasion fleet, but his ill-coordinated force was almost entirely destroyed by Japanese surface ships in the battle of the Java Sea, February 27–March 1. On March 9, the Dutch East Indies fell.

The capture of the Philippines

Like the US fleet at Hawaii, US forces in the Philippines believed that Japanese airfields were too distant to constitute a serious threat. But just as the First Air Fleet had shattered American complacency at Pearl Harbor, so

General Yamashita's lightning campaign in Malaya (*left*) was one of the most remarkable of the war. Using only three divisions, the Japanese 25th Army defeated in detail the numerically superior forces of the British. Singapore's formidable shore batteries prevented a direct attack: instead the Japanese invaded northern Malaya, smashed the ill-handled 11th Indian Division at Jitra and rapidly moved south to outflank any hastily improvised defences.

The causeway connecting Singapore with the mainland was blown up once the last British troops were across. After a brief pause, the Japanese mounted an amphibious assault on the western part of the island on the night of February 8/9 which yet again wrong-footed the defenders. A week afterwards, a humiliated General Percival surrendered. Three days later on February 18 (*above*), Japanese infantry marched across the repaired causeway.

Map

Isthmus of Kra
100°
Singora
Imperial Guards Div (from Bangkok)
Patani
Changlun
Jitra
Alor Star — Kedah
1 Ind Div 15 Dec
Gurun
14 Dec Kroh
Butterworth
George Town
Penang 16 Dec
5°
Krian
Perak
Taiping 23 Dec
26 Dec
Ipoh 28 Dec
Malaya
Kampar 2 Jan 1942
1 Jan 1942
Kuala Lipis
1 Jan 1942 Telok Anson
Slim
2/3 Jan
100°
Kuala Selangor
10 Jan
Kuala Lumpar
Port Swettenham 11 Jan
10 Jan
Gemas
Port Dickson
Tampin
Malacca
Muar
Bakri
Batu Pahat

5 and 18 Div 8 Dec 1941
Japanese Twenty-Fifth Army (Yamashita)
N
Takumi Force
Kota Bharu
South China Sea
Gong Kedah
Kuala Krai
9 Ind Div
Kuala Trengganu
Kuala Dungun
Kuala Kuantan 30 Dec
GOC Malaya (Percival)
III Corps HQ
8 Aust Div
Batu Anam Endau 16 Jan
Segamat
19 Jan Mersing
16 Jan Kluang
25 Jan
7 Feb
31 Jan 1942
Last Allied forces withdraw to Singapore Island
Johore Bahru
Singapore 15 Feb 1942
Allied forces surrender
Strait of Malacca

Japanese attacks
Pre-war borders
Heights in m
1000
200
0
0   25   50 Km
0   20   40 Miles

Admiral Isoroku Yamamoto (1884–1943) was the son of a village schoolmaster. He joined the Japanese Navy in 1900 and was sent to Harvard University in 1917. A decade later, Yamamoto served as Naval Attache in Washington. Travelling widely in the USA, he developed a respect for, and an understanding of, American power that was unusual among his compatriots. He also specialized in the new branch of aviation and learned to fly; his first flag appointment was commander of First Air Fleet in the early 1930s. Yamamoto felt Japan had insufficient strength to oppose the USA and Britain, and although he led the delegation which walked out of the London Naval Conference, he put himself in great danger when he opposed Japanese alignment with Germany. Yamamoto was appointed commander of the Combined Fleet in August 1939. His tendency to overestimate American strength caused him to insist on the pre-emptive strike that brought a united America into the war, and the Midway (MI) plan, an operation that went disastrously wrong. He was, however, held in high regard throughout the service and his assassination in the air by the Americans came as a crushing blow to Japanese morale.

the Eleventh Air Fleet, flying from Formosa (Taiwan), did the same at Luzon. With US air power destroyed, major Japanese landings were made in the Lingayen Gulf on December 22 1941, after amphibious operations had secured airfields for shorter-ranged IJA aircraft in northern Luzon. The light carrier *Ryujo* supported a landing at Davao on the southern island of Mindanao on December 19. On December 24, US forces withdrew to the Bataan peninsula on the west side of Manila Bay, where they were to hold out for some three months. The American surrender began on April 8, and on May 6 the island fortress of Corregidor capitulated after heroic resistance. General Douglas MacArthur, Commanding General, US Army Forces in the Far East, had departed to Australia on March 11, promising that he would return to retrieve both the Philippines and his military reputation.

By the time Corregidor fell, the Co-Prosperity Sphere was virtually complete: the Dutch East Indies, Malaya and, by May 1942, Burma were secured. The only major unfinished business was the southern bastion of the newly acquired empire. On January 23 the Japanese occupied Kavieng, New Ireland, and Rabaul, New Britain; the latter was to be their major advanced base. The invasion forces were supported by the First Air Fleet. On February 19, four of Nagumo's carriers devastated Port Darwin in northern Australia, before moving westward to attack the Allied forces withdrawing from Java.

Thus, there was little air cover available for the Japanese landings at Lae and Salamaua, northeast New Guinea, on March 8. Unfortunately for the Japanese, two US carriers, the *Lexington* and *Yorktown*, were operating together in the area in the latest of a series of raids. The two carriers launched strikes at Lae and Salamaua over the

Owen Stanley Mountains, sinking 2 and damaging 13 of the 18 Japanese transports committed and thus preventing an amphibious move against Port Moresby. Although the Japanese continued to consolidate their position in the Solomons, Port Moresby's resistance exacerbated Japan's already complex problem of what to do next.

A question of strategy

There were three broad strategic options. The first was to continue westward into the Indian Ocean, taking Ceylon, bypassing India and cutting the vital Allied supply lines to the Middle East. The effects, both on a restive India and on the British position in the Middle East, were likely to be enormous: Britain might well be knocked out of the war. This was the closest the Japanese would come to a coordinated Axis strategy – but there were two major arguments against it. First, Japan's major enemy was the USA, not Britain; second, given her commitments elsewhere, the logistical requirements were beyond Japan's capabilities. Japan's overextension, especially her army's continuing war in China, also militated against the second strategic option: an advance farther to the southeast, occupying eastern Australia to prevent its use as a base for an Allied counterattack. The third strategic option was more purely naval and hence simpler: an eastward advance towards Hawaii to draw out and decisively destroy the remnants of US naval power in the Pacific. The increasing activities of the US carriers seemed to make this last option imperative, especially when US Army medium bombers launched from USS *Hornet* bombed Tokyo and other

American prisoners on Corregidor on May 7 1942. Although the Americans had surrendered the day before the Japanese continued to go through the motions of fighting until they had taken the whole fortress.

targets in the Japanese home islands on April 18. By the beginning of March the Indian Ocean option had been ruled out: the IJN contented itself with a major raid into the area, to prevent a British build-up from threatening the western flank of the Co-Prosperity Sphere.

By early April 1942, Britain had assembled a superficially imposing Eastern Fleet in the Indian Ocean, with 5 battleships, 3 aircraft carriers, 2 heavy and 4 light cruisers, a Dutch anti-aircraft cruiser and 14 destroyers (one Dutch), at Ceylon and a secret base at Addu Atoll. However, many of the ships were old, unmodernized and vulnerable to air attack, or in dubious mechanical condition. The total strength of the carriers' air groups was a mere 90 aircraft. The British naval air arm was more finely tuned to the battlefleet support role than those of the USA or Japan and the preferred type of Royal Navy fleet carrier, represented in the Indian Ocean by HMS *Indomitable* and *Formidable*, relied on armour and guns for protection rather than fighters. Air groups were limited in both numbers and capability: *Indomitable* carried 12 slow Fulmar fighters and 9 Sea Hurricanes, with 24 Albacore torpedo-dive bomber-reconnaissance aircraft; *Formidable* had only 12 American-built Martlet (Wildcat) fighters and 21 Albacores. The Albacores were slow biplanes, as were the 12 Swordfish torpedo-reconnaissance aircraft operated by the light carrier HMS *Hermes*.

The Eastern Fleet attacked

The Japanese attack on the Eastern Fleet at Colombo was planned for the morning of April 5. The First Air Fleet was deployed in almost full strength (*Kaga* was absent with engine trouble), so that 360 high-performance carrier aircraft opposed 90 inferior planes. Admiral Somerville, commanding Eastern Fleet, had the advantage that his aircraft could operate after dark. He also had high expectations of his general night-fighting abilities, so he divided his force into fast and slow squadrons and planned to manoeuvre for a night action. Given Japanese expertise in such combat, it was perhaps as well that he did not succeed. Although Somerville's two heavy cruisers were found at sea and sunk by "Val" dive bombers, and a number of minor warships and merchant vessels were destroyed, the attack on Colombo had little intrinsic success because of the absence of the main body of the Eastern Fleet. On April 8, when the Japanese struck at Trincomalee, *Hermes* was caught with her aircraft ashore and sunk by dive bombers – the first carrier sunk by carrier air strikes. To

Japanese dive bombers sink the carrier HMS *Hermes*.

add to Allied woes, Japanese raiding squadrons, one incorporating the light carrier *Ryujo*, were at the same time massacring shipping along the Indian coast. But the Japanese were unable to do more than raid. All their forces withdrew eastward for the series of operations whose combined codenames expressed the chosen synthesis of Japan's naval strategy: *MO-RY-AL-MI*.

MO signified a limited advance to the southeast in a further attempt to take Port Moresby from the sea, supported by two of First Air Fleet's carriers. It was to be followed by *RY*, the occupation of Nauru and Ocean Islands. These operations having secured the "defence perimeter", the full might of the IJN would be concentrated in a complex double strike on the Aleutian Islands (*AL*) and Midway (*MI*), in which the US carrier force would be destroyed. Unhappily for the Japanese, these plans were compromised by US code-breaking operations.

To counter *MO*, Task Force 17 – the carriers *Lexington* and *Yorktown* with supporting cruisers and destroyers – was deployed to the Coral Sea. The Japanese carriers engaged were *Shokaku* and *Zuikaku*, new ships but with

The Japanese carrier *Shokaku* under attack, May 8 1942.

inexperienced aircrews, and the light carrier *Shoho*. Thus, forces were approximately even in what would be the first major naval engagement fought without the ships actually sighting each other. Because of mistaken identification by reconnaissance aircraft, the Japanese traded the *Shoho* for USS *Neosho*, a fleet oiler, but when the carrier task forces finally launched strikes at each other the USN seemed to get the worst of it. The *Lexington* was sunk by internal explosions after torpedo and bomb hits had ruptured her fuel storage tanks and the *Yorktown* also suffered serious damage. On the Japanese side, *Shokaku* was damaged by dive bombers and *Zuikaku*'s air group suffered severe casualties: neither ship would be able to join the other two carrier divisions for *MI*. In fact, the American strategic victory was considerable. Fighting off a lacklustre attack by land-based aircraft, Allied surface ships formed into a separate striking force and so menaced the Port Moresby invasion fleet that it eventually turned back. *MO* was abandoned, as was *RY*: all would now depend on *MI*, the decisive battle farther north.

The battle of Midway

The Japanese plan for the battle of Midway was not a good one, mainly because forces were dispersed in a complex pattern of diversion and manoeuvre. The deployment to the Aleutians of Japan's less capable carriers, the converted liner *Junyo* and the *Ryujo*, depleted the main force to little advantage, since the Americans knew from their code-breaking that the Aleutians' operation was only a feint. When Nagumo's four fleet carriers approached the western extremity of the Hawaiian chain on June 4, they faced three US fleet carriers, the *Enterprise*, *Hornet* and *Yorktown*. Naval air strength was about even, but surprise gave the USN the initiative. Nagumo's aircraft hit Midway in the early morning, but it seemed that a second strike was necessary to complete the pre-invasion softening-up. Midway's aircraft were making brave but ineffective raids on the Japanese ships. The aircraft ranged on the Japanese carriers' decks in case US carrier aircraft should attack were struck down to the hangars for their ordnance to be changed. Then the US warships were belatedly sighted and the rearming was stopped. As Nagumo hesitated, it became too late to act before the Midway striking force returned – and shortly before it did, a Japanese reconnaissance plane spotted a US carrier. Nagumo had little choice but to complete the recovery of the Midway force before launching a full-strength strike against the American ships. As aircraft were rearmed and refuelled, the hasty crews left ordnance on the hangar decks rather than striking it below, creating a potentially dangerous mixture of fuel and explosives. Just as the Japanese aircraft were ranged

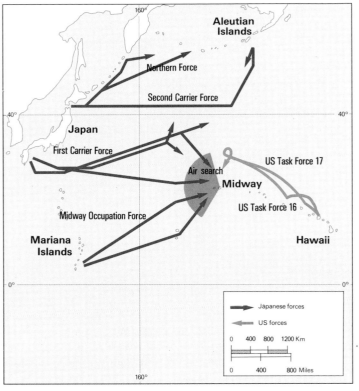

The *Essex*-class carriers (*below*), perhaps the most powerful and effective warships of World War II, were designed for the optimal operation of a 91-aircraft group: 36 fighters, 37 bombers and 18 torpedo bombers. In order to do this on a full-load displacement of just over 36,000 tons the flight deck and hangar sides were left unprotected. The hangar deck was, however, given $2\frac{1}{2}''$ armour against bombs and the hull had a 4″ armoured belt. The open hangar allowed aircraft to be warmed up before launch; essential for the launch of large strikes.

A consistent weakness in Japanese operational planning was a tendency to subdivide rather than concentrate forces (*above*). Vice Admiral Kondo's MI Invasion Force was subdivided into five: a Main Body of heavier units, a Close Support Group of cruisers, the Transport group with its close escort of the cruiser *Jintsu* and ten destroyers, a Seaplane Tender Group and a Minesweeper Group. Nagumo's Carrier Striking Force operated as a coherent group, but Yamamoto's main battlefleet was far behind and itself split into two to allow support to be given to AL if necessary.

on the flight decks for the strike at the US carriers, the American strike arrived.

Because the US squadrons had become dispersed, the slow Devastator torpedo bombers attacked first, unsupported. They were cut to pieces by Japanese fighters, but their sacrifice hindered Nagumo's launching of his

"With fire spreading and the cockpit filled with smoke, I had . . . to bale out, despite being at no more than 600ft at the time . . .

I hit the water a split second after the parachute had opened. It was a tremendous shock . . . I went down under the water and then, assisted by my life jacket, floated to the surface, only to see the cruiser Jintsu *passing me by, going at full speed . . . On the horizon I could see three black columns of smoke where I knew our fleet to be . . . I guessed it would take me all of 24 hours to make it – if I could . . . I began to think about the danger of sharks."*

Iyozo Fujita, Zero fighter pilot at Midway

own strike. Dauntless dive bombers now arrived and quickly reduced *Kaga*, *Akagi* and *Soryu* to blazing hulks. The survivor, *Hiryu*, claimed a limited revenge by putting the *Yorktown* out of action: a Japanese submarine finished her off two days later. However, *Hiryu* was crippled by dive bombers from USS *Enterprise* and sank on June 5. With the First Air Fleet removed from the board and his advanced cruisers suffering severe air attacks, Yamamoto, commanding from the super-battleship *Yamato*, felt constrained to abandon the operation. Battleships turned and

ran from carriers at 0255 hours on June 5 1942 – a significant moment in naval history.

The USA was still not ready to launch a full counter-offensive. The defeat of Germany was the Allies' agreed priority, but both the USN and General MacArthur urged some counteraction in the Pacific. Because of signs that

The "offensive defensive" begins: US Marines land at Guadalcanal.

the Japanese might still attempt to cut off Australia, an "offensive defensive" in the Solomons was decided upon, and US Marines landed on Guadalcanal on August 7, initiating more than one year's attritional warfare. Japan generally ruled the sea by night; the Americans dominated the air – and hence the sea – by day. Japanese submarines sank the *Wasp* and damaged the *Saratoga*, while two major carrier battles saw the *Hornet* sunk and the *Enterprise* repeatedly damaged. At one point the USN had only one operational carrier, and the British navy had to supply

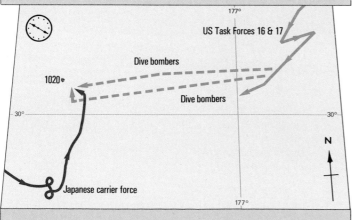

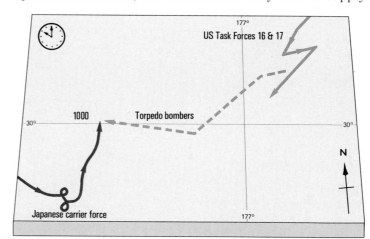

The battle of Midway began just before 0920 when *Hornet's* torpedo bombers attacked the Japanese Carrier Striking Force. However, they were without proper fighter support, and all were shot down by Japanese Zeros.

At 0930 another wave of torpedo bombers, this time from the *Enterprise*, spotted the Japanese and attacked, again unsupported. Out of the squadron of 14, 10 were shot down. As this attack ended, just after 1000,

Yorktown's torpedo bombers began their attack. Despite fighter support, all were lost. By about 1020, however, the defending Japanese fighters were at low altitude or being refuelled and rearmed on their carriers.

Dauntless dive bombers then arrived from two different directions. Five minutes later, the Americans were able to plant their bombs unopposed on the crowded decks of the Japanese carriers.

temporary reinforcement in the shape of HMS *Victorious* carrying American aircraft. But with the "Two-Ocean Navy" under construction and a massive training programme producing personnel to man it, a war of attrition was to America's advantage. *Ryujo* was sunk and severe casualties were inflicted on Japan's vital corps of trained aircrews. Early in February 1943 the Japanese abandoned Guadalcanal and moved to a new defensive line in the central Solomons. Their tide had begun to ebb.

The significance of the Guadalcanal campaign was reflected in the higher priority accorded to the Pacific theatre at the Casablanca Conference in Washington in January 1943 and, especially, at the "Trident" Conference in Washington in May 1943. This called for the Allies to exploit Japan's vulnerability to blockade and to subject her to sustained strategic bombing as soon as possible.

The Americans were to begin a two-pronged thrust. The South Pacific forces that had been under Pacific Fleet command would join MacArthur's Southwest Pacific forces in a drive on Rabaul and along the New Guinea coast. The other thrust, based on Pearl Harbor, would be an advance across the Central Pacific to the Philippines, via the Marshalls and Marianas. MacArthur tried to make his southern thrust the main effort, but Admiral King, the formidable US Chief of Naval Operations, prevailed, and in the months after "Trident" the Central Pacific Drive grew from a flanking operation for MacArthur to the main thrust. King saw the Marianas as the area where the Japanese would commit their main forces to a decisive battle that they would lose; he was supported by General "Hap" Arnold, the US Army Air Force chief, who realized the need for bases in the Marianas for his B-29

Superfortresses. The two-thrust strategy was confirmed at the "Sextant" Conference in Cairo at the end of 1943.

While strategic discussions progressed, the mincing machine ground on in the Solomons. On March 2–4 1943, in the battle of the Bismarck Sea, Allied land-based bombers destroyed a convoy carrying 7,000 troops from Rabaul to reinforce Lae. In a desperate attempt to regain air control in the area, Admiral Yamamoto went to Rabaul to direct a campaign against Allied air bases in which Japan's carrier air groups flew from land bases to supplement the conventional land-based squadrons. The campaign failed, and the losses sustained in the Solomons contributed to a sudden and steep decline in the quality of Japanese naval air power. The training system failed to produce new pilots of the required standard – aircrews began to be sent to operational squadrons to "learn on the job" – and many did not survive the process. Yamamoto himself became a casualty of the Solomons' air war when, its course revealed by a code-breaking operation, the bomber carrying him on an inspection tour was ambushed on April 18 1943 by long-range P-38 fighters from Guadalcanal.

The Allied advance

In the following months the Allied forces in the south advanced steadily. MacArthur's Southwest Pacific command concentrated on Papua-New Guinea and New Britain. Admiral Halsey's South Pacific force, subordinate to Admiral Nimitz at Pearl Harbor but operating under MacArthur's general strategic direction, covered the Solomons. Both amphibious offensives were to have come together at Rabaul, but in August 1943 it was decided to bypass this Japanese bastion. As the Japanese struggled to

The battle of production was one the Japanese could not win. Between 1941 and 1945, in the most remarkable naval shipbuilding programme in history, the Americans built a navy of enormous size. Twenty six 27,000-ton *Essex*-class fleet carriers were commissioned, 17 of which were completed before the war's end. This is the *Ticonderoga* being launched at Newport News on February 1 1943.

A vital role in the American amphibious offensive was played by the US Navy's Construction Battalions, nicknamed from their initials the "Seabees". Recruited from men with similar jobs in civilian life, the Seabees were given military training and worked in the front line, often under fire. They were usually among the first forces ashore. Their tasks ranged from building airfields and other base installations to repairing equipment. Admiral Nimitz summed up their contribution by his comment "In their hands the bulldozer became one of the instruments of victory".

217

maintain their seaborne communications, their skills in surface warfare at night still stood them in good stead, but the Americans were making steady improvements especially in radar capacity. When the Allies reached Bougainville in November 1943, Halsey's surface forces made good use of radar-controlled gunnery to repel a superior Japanese force that attempted to attack the landings during a night action in Empress Augusta Bay. The Japanese committed more ships and the best of their new, hastily trained carrier air groups in an attempt to retrieve the situation. The veteran *Saratoga* was now reinforced by the first of the new US carriers: in a series of raids on Rabaul and in repelling an aerial counterattack, Halsey's carriers inflicted heavy casualties on the IJN's ships and planes.

The drive for the Philippines

In New Guinea, with American and Australian forces supported by his own Seventh Fleet (basically an amphibious force), MacArthur advanced along the northern coast, taking Lae in September, Finschhafen in October and western New Britain in late December. An air offensive destroyed the aircraft based at Rabaul and neutralized its effectiveness as a base. In the first quarter of 1944, MacArthur and Halsey broke through the "Bismarck barrier", and MacArthur maintained the pressure by moving farther north towards the Moluccas and the major prize – the Philippines. The Central Pacific Drive, with the same eventual objective, had begun in November 1943 with the assault on the Gilberts.

During the latter part of 1943 the first products of the "Two-Ocean Navy" policy had reached Pearl Harbor. The first of the 27,100-ton fleet carriers, USS *Essex*, arrived at the end of May, to be joined later by the new *Lexington*, *Yorktown*, *Hornet* and *Wasp*, as well as *Intrepid*, *Franklin* and *Bunker Hill*: 16 *Essex*-class fleet carriers were in commission by the war's end. The nine *Independence*-class light carriers of 11,000 tons could each carry up to 45 aircraft, against 90-plus in an *Essex*; all were in commission by the end of 1943. Many small escort carriers were built on converted mercantile hulls; these acted as support carriers for the amphibious forces and, later, as vital components of the fast carriers' logistical fleet train. The new US carrier aircraft included the F6F Hellcat, which outperformed the Zero in almost every way. The TBF Avenger torpedo bomber had rapidly replaced the Devastator after Midway, while by 1943 the Dauntless was giving way to the higher-performance SB2C Helldiver. The carriers were screened and supported by a new generation of surface warships, ranging from the 33-knot, 45,000-ton *Iowa*-class battleships, especially built for fast carrier operations, through the mass-produced heavy and light cruisers of the *Baltimore* and *Cleveland* classes, to the *Fletcher*-class destroyers, of which no fewer than 175 were launched in 1942–44.

Japan could not compete in this production battle. Her only new fleet carrier, completed in March 1944, was the 29,300-ton *Taiho*, an armoured-flight-deck ship carrying 53 operational aircraft and 21 reserves. Six modified *Hiryu*-class carriers were laid down, but only three would be completed by late 1944. The third of the super-battleships, *Shinano*, was taken in hand for carrier conversion after Midway, but she would not be completed until November 1944. The submarine tender *Ryuho* and the seaplane

US CARRIER AIRCRAFT

Grumman F4F Wildcat

Curtiss SB2 Helldiver

Douglas TBD Devastator

Grumman TBF Avenger

Douglas SBD Dauntless

Grumman F6F Hellcat

The initial American carrier-based team of **Wildcat** fighters, **Devastator** high-level/torpedo bombers and **Dauntless** scout/dive bombers was not as good as that of the Japanese, the Devastator in particular proving to be a death trap. The latter was soon replaced by the **Avenger** which was much faster and longer ranging. The **Wildcat** was not the equal of the Zero and was replaced by the specially designed **Hellcat**. The first Hellcats entered combat in September 1943 along with the new generation of carriers. The Douglas Dauntless was the best of the first models in its designed roles and was preferred by many pilots to its Curtiss successor which perpetuated the pre-war name **Helldiver** (the Japanese nickname for all American dive bombers). The first Helldivers entered combat in December 1943.

The US fleet at sea in the Pacific showing the main instruments of US naval power in the second, victorious half of the Pacific War. In the foreground are representatives of the two carrier classes, an *Independence*-class light fleet carrier (CVL) leading a larger *Essex*-class fleet carrier. Although of only 11,000 tons CVLs carried 45 aircraft and were useful adjuncts to the larger Essexes with twice the aircraft complement that were the backbone of the fast carrier task forces. Behind the *Essex*-class ship are the *Iowa*-class fast battleships built specifically to operate with carriers. Their role was to strike and defend against surface threats when the carrier aircraft were unavailable, notably at night, and to finish off targets crippled by carrier strikes. However their most important role in practice turned out to be as anti-aircraft protection for the carrier striking forces.

tenders *Chitose* and *Chiyoda* were converted into light carriers, but the IJN's only other available carriers at the end of 1943 were the veterans *Shokaku* and *Zuikaku*, the converted liners *Junyo* and *Hiyo* and the pre-war light carrier *Zuiho*. Of the new aircraft, both the Yokosuka D4Y "Judy" dive bomber and the Nakajima B6N "Jill" torpedo bomber were fragile by US standards. They followed Japanese practice in emphasizing speed, range and manoeuvrability, but these high-performance characteristics made them difficult to operate from the smaller carriers, which retained the obsolescent "Kate" as their primary attack aircraft and used the Zero as a fighter bomber. In *Junyo* and *Hiyo*, "Vals" still constituted 50 per cent of the dive bomber force. Further, the ineptitude of the new, poorly trained aircrews meant that the limited investment in new aerial hardware was dissipated as soon as it was committed to combat.

In the early summer of 1944, having suffered further attrition during the US landings in the Marshalls in February, this deeply flawed naval air arm prepared for action once more. As the Allies expected, the Japanese planned to defend the Marianas with their full naval strength, in Operation *A-Go*. The Japanese had hoped that both attrition and superior technology would allow them to fight on favourable terms – but now both factors were working against them.

American submarine activity

From the beginning of the war US submarines had exacted a heavy toll in an unrestricted campaign against the shipping that was the lifeblood of the Co-Prosperity Sphere. For a long time, US submarine operations were hampered by the faulty Mark 14 torpedo, but by 1943 the faults had been corrected and new "wakeless" electric torpedoes, including a homing type, had begun to enter service. More than 100 large and habitable "fleet boats" were launched to replace older submarines between 1941 and the end of 1943. Their success rate steadily increased: the Japanese were slow to introduce convoys, and even when adopted they were usually poorly protected – easy prey for the skilled and ruthless US submarine commanders, who sometimes adopted the German system of hunting in "packs". In 1941–45, some 1,100 Japanese merchantmen of more than 500 tons were sunk by submarines. The tanker fleet was almost completely wiped out, denying Japan fuel for training pilots and other vital activities.

The deployment of the IJN's "First Mobile Fleet", which included both the major surface warships and the carriers, reflected US submarine activity. The base chosen in mid-May 1944 was Tawi-Tawi in the Sulu archipelago, within 200 miles of Tarakan, Borneo, from whose wells volatile oil could be pumped straight into warships' bunkers. When the Americans located the new base they concentrated submarines around it, confining the Mobile Fleet to its anchorage and preventing the inexperienced carrier pilots from receiving advanced training.

The Japanese failed to use their own submarines in their designed role as forward fleet pickets. Compromised by US code-breaking and hunted by antisubmarine warfare (ASW) systems tested in harsh Atlantic conditions, they suffered severely. Seventeen of some 25 Japanese submarines deployed in the Marianas were lost: the destroyer escort USS *England* alone sank 6 Japanese submarines in

Projecting carrier air power over the prodigious distances in the Pacific required enormous logistical back-up. The Central Pacific Drive was at first based largely on Hawaii, but as the fleet moved westward, mobile service squadrons were required. Service Squadron 10, the most important, set up headquarters at Majuro Atoll in the Marshalls, 2,000 miles (3,200km) west of Hawaii. It contained an enormous variety of auxiliary vessels, from self-propelled lighters and ammunition barges to destroyer tenders and hospital ships. Tankers were especially important and demand grew inexorably as the advance continued. Escort carriers were also vital to the support force, not only for protection but also as "jeep carriers", ferrying replacement aircraft from Majuro to the fast fleet carriers. After the Philippine Sea victory, Service Squadron 10 moved to Eniwetok in the Western Marshalls and then, in October 1944, shifted 1,400 miles (2,250km) southwest to Ulithi Atoll, a base for the Philippines operations. Here, obsolete tankers were moored as a floating oil storage centre. However, Ulithi was still too far away from the Philippines to give the carriers the required range, so an "At Sea Logistics Service Group" was created which, together with the "jeep carriers", shuttled between Ulithi and rendezvous points where replenishment of the task groups at sea was carried out. Thus the techniques that have ever since given fleets unprecedented freedom from shore bases were developed and refined.

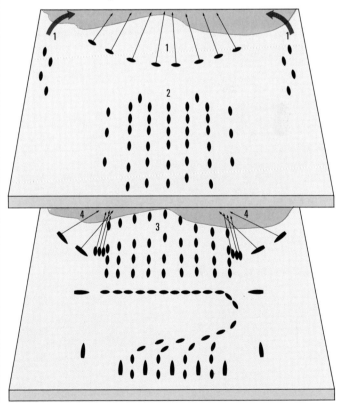

The US amphibious assault. 1. For weeks before "D-Day" the whole landing area would be bombarded by aircraft and surface ships. **2.** Before dawn on D-Day the transports and tank landing ships (LSTs) would assemble. **3.** About 90 minutes before the troops were due to land the leading LSTs would launch hundreds of tracked landing vehicles and amphibious tanks. **4.** Heavy and continuous gunfire support was crucial in defeating Japanese attacks on the new US beachhead.

the last 12 days of May 1944. Thus, the Japanese remained far from certain as to where the main US attack would be made. However, on June 12, Vice-Admiral Ozawa, commander of the First Mobile Fleet, was ordered by Admiral Toyoda at Combined Fleet HQ to begin *A-Go*: US carriers were striking at Saipan and Guam – so the Marianas were the target after all.

Ozawa's task was daunting, for he had only 430 aircraft in his 9 carriers against Admiral Spruance's 891 aircraft in the 15 carriers of Task Force 58. Ozawa hoped that land-based naval aircraft would inflict considerable attrition on the enemy before the fleets came into contact. He planned also to exploit the longer striking range of his aircraft and the availability of airfields on Guam in "shuttle bombing" missions beyond the reach of the US ships and aircraft.

In fact, Task Force 58 made short work of the land-based aircraft during its preliminary bombardments and fighter sweeps. However, Ozawa assumed the best when he located the US fleet on June 18 and ordered his carriers to strike at it next morning. Spruance, although informed of Japanese movements through submarine reconnaissance, feared to expose the Saipan invasion force to attack and decided to remain on the defensive: Vice-Admiral Lee, commanding "Battle Line", the 6 newest US battleships, refused a night action on June 17, while Spruance rejected the advice of his carrier commander, Vice-Admiral Marc Mitscher, to attack the Japanese at dawn on June 18. That may have been an error, for it would allow many Japanese ships to escape destruction, but such was US superiority that the battle of the Philippine Sea was in any case to mark the end of the operational capability of the Japanese carrier force.

The Japanese vanguard, the Third Carrier Division with

Admiral Raymond Ames Spruance (1886–1969) was not a carrier sailor but a gunnery specialist. He spent much of the 1930s on the staff of the War College at Newport, Rhode Island where he developed an enviable reputation as a quiet but brilliant strategist.

Appointed to command a cruiser division, Spruance took part in early carrier operations alongside "Bull" Halsey's task force. Placed hors de combat by a skin rash, Halsey chose his erstwhile subordinate to command Task Force 16 for the Midway battle.

Spruance's acceptance of the advice of his able chief of staff, Captain Miles Browning, in launching his strike to catch the Japanese aircraft while refuelling was a major contributory factor in the US victory.

After Midway, Spruance served as Chief of Staff to the Pacific Fleet Commander, Admiral Nimitz, and was then appointed to command the Central Pacific Force which became the 5th Fleet in April 1944.

Spruance's weakness was his virtue, an ability to weigh all facets of the problem and hence a tendency to err on the side of caution. He increasingly relied on the advice of his right hand aviator, Vice-Admiral Marc A Mitscher. Mitscher lacked the brilliance which might have led to a clash of personalities but willingly provided the air experience and instinct that Spruance lacked. Together, Spruance and Mitscher made the finest US fleet command team of the war.

The battle of the Philippine Sea, the biggest carrier battle of all time, was a crushing American victory. Admiral Spruance acted cautiously and stood on the defensive, allowing Ozawa to make the first moves. Admiral Obayashi, in command of the Japanese Van Force, launched his strike prematurely, about half-an-hour before the main body put its aircraft into the air. This made the job of the defending fighters a great deal easier. To add to Japanese woes, their second wave was also split into two separate raids launched between about 1000 and 1130. These were misdirected; the first went north of the actual position of the American carriers, the other south; most of the latter jettisoned their bombs without sighting the enemy, but this did not protect them from the US fighters which massacred many as they tried to land on Guam.

Situation 18-19 June 1944

Japanese First Mobile Fleet (Ozawa)

Third Carrier Division *Chitose, Chiyoda, Zuiho*
0830, 19 June
First strike launched

First & Second Carrier Divisions *Taiho, Zuikaku, Junyo, Shokaku, Ryuho, Hiyo*

1000-1015
First strikes launched
1005 *Taiho* torpedoed; sinks 1628
1220 *Shokaku* torpedoed; sinks 1624

1030
Second strike launched (misdirected)

2000

US Task Force 58 with 15 carriers (Mitscher)
1200, 18 June

1000, 19 June
US planes return from Guam raid. Japanese strike planes sighted

Guam

US fighters intercept Japanese aircraft

145°
15°

Situation 20 June 1944

Withdrawal towards Okinawa

1844
Hiyo sunk,
Zuikaku and *Chiyoda* damaged

2045
80 returning planes crash attempting night landing

1624
Strike launched

1200, 20 June

1600, 19 June

Guam

0   100   200   300 Km
0   100   200 Miles

145°
15°

the light carriers *Chitose*, *Chiyoda* and *Zuiho*, launched 45 Zero fighter bombers, 8 "Jills" and 16 Zero escorts, with 2 "Kates" as pathfinders, at 0830 on June 19. Almost immediately, *Taiho*, *Zuikaku* and *Shokaku* of the First Carrier Division put up 53 "Judys" and 27 "Jills" escorted by 48 Zero fighters; 2 "Jills" acted as pathfinders and a single "Judy" carried chaff in an attempt to baffle US radar. The Second Carrier Division launched a strike from *Junyo*, *Hiyo* and *Ryuho* at 1000–1015, flying off 25 Zero fighter bombers, 7 "Jill" torpedo bombers and 15 Zero escorts. At 1030 it launched 10 more fighter bombers and all its dive bombers – 9 "Judys" and 27 "Vals" – with 18 Zero escorts. Ozawa added his last aircraft, 9 Zeros and 9 "Jills" from *Zuikaku*, to this final wave. Depending on their fuel state after the attack, the Japanese aircraft were either to return to the carriers or land on Guam to rearm.

The Japanese erred in not concentrating their strikes and in losing the advantage of surprise because the first attacking wave was ordered to orbit in the vicinity of the US ships until the desired formations were achieved. The US carriers used radar fighter control to mount a near-perfect defence: some 450 Hellcats destroyed each of the four Japanese strikes in detail. The few Japanese aircraft that got through faced the new proximity-fuzed AA fire of the US ships. Only one bomb hit was made; the battleship *South Dakota* suffered some 50 casualties but little damage. A "Jill" crashed into the side of the *Indiana*, but its torpedo did not detonate and damage was superficial. Several carriers were near-missed.

"The Great Marianas Turkey Shoot"

An American pilot compared the slaughter of Japanese

aircraft to "an old time turkey shoot", and the battle of the Philippine Sea is sometimes called "The Great Marianas Turkey Shoot". Of some 330 aircraft committed to the 4 strikes, more than 220 were destroyed: AA fire claimed 19 aircraft; US naval fighters brought down the rest. Japanese losses were not confined to aircraft. Even as she launched her aircraft, *Taiho* was torpedoed by the submarine USS *Albacore*. At first she seemed little damaged by the single hit, but because of poor damage control the volatile fumes of bunker fuel and aviation spirit exploded and blew the carrier apart. The veteran carrier *Shokaku*, hit by three torpedoes from the submarine USS *Cavalla*, had a similarly explosive end.

Now it was the turn of the American carriers' strike aircraft, which had been sent away during the "Turkey Shoot" to bomb Japanese airfields on Guam. Late on the afternoon of June 20, they were launched at the Mobile Fleet, then estimated to be 220 miles away. It was a risky operation, for it would be dark before the aircraft could

The "Marianas Turkey Shoot" (*left*) was a convincing display of American superiority in carrier warfare, but Admiral Spruance was criticized for not having acted more offensively. His surface task force commander, Admiral Lee, was not at all confident about the force's capabilities against the Japanese in a night action. This caused Spruance to keep his distance from the enemy and he decided to turn back on the night of June 18–19.

The American counterattack was finally launched on the 20th, late in the day and at extreme range. Many of the aircraft committed to the strike, like the one pictured *above*, ended up ditched in the sea.

return. Following the launch of some 230 aircraft, the enemy's position was re-estimated at 280 miles distance and a follow-up strike was cancelled – but 54 Avengers (most armed with bombs), 51 Helldivers and 26 Dauntlesses pressed on with a formidable Hellcat escort. *Zuikaku* and *Chiyoda* were damaged by bombs and torpedo bombers sank the *Hiyo*. The Hellcats inflicted heavy losses on the Mobile Fleet's fighters: 65 Japanese aircraft were destroyed for the loss of 20 of the attackers, and by dusk Ozawa's carriers were down to only 35 operational aircraft. However, when the US aircraft returned to their carriers, in darkness and with fuel running low, some 80 were lost by ditching or in landing accidents. Most of the ditched crews were recovered, but this slowed the pursuit of the enemy, so that Ozawa's battered fleet at last escaped. Although it included five battleships and a strong force of heavy cruisers and destroyers, it could not stand against the might of the US fleet, triumphant in the air and now dominant even on the surface.

The long-expected, decisive "battle of the Marianas" had ended in crushing defeat for Japan. In the final major naval engagement, at Leyte Gulf in the Philippines in October 1944, the IJN had to expend virtually empty carriers as decoys in a vain attempt to allow its concentrated surface forces to destroy the US invasion armada. After coming close to success when Halsey's fleet carriers were lured away by Ozawa's decoy force, the plan failed, with the loss of the super-battleship *Musashi*, 2 older battleships, the carriers *Zuikaku*, *Zuiho*, *Chiyoda* and *Chitose*, 10 cruisers and 11 destroyers. Although US carrier planes and submarines inflicted much of the damage, the old battleships *Fuso* and *Yamashiro* succumbed to the guns

of the raised and refurbished Pearl Harbor battleships in the last major "battle line" action in naval history. However, the battle of the Philippine Sea, where the decisive losses were inflicted by submarines and air power, was a surer guide to the future of naval warfare than the gun duels of the dinosaurs in Surigao Strait off Leyte.

The Marianas campaign had also marked the use of sea power to acquire bases for a strategic air offensive, in which B-29 Superfortresses laid waste Japan's cities in incendiary attacks. Meanwhile, the Japanese had turned

---

"I will die the moment my torpedo hits an enemy ship. My death will be full of purpose. We have been educated here to forget all small things, and to think only of the one big thing . . . When you hear that I have died after sinking an enemy ship, I hope you will have kind words to say about my gallant death."

Yutaka Yokota, Kaiten ("human torpedo") operator

---

to their ancient tradition of self-sacrifice: first committed at Leyte Gulf, the kamikaze suicide aircraft constituted a formidable threat to the USN until the end of the war. The major kamikaze onslaught was made at Okinawa, from March 1945: even the super-battleship *Yamato* was sunk by US carrier aircraft while on a one-way "suicide mission" to Okinawa. But although the kamikaze inflicted significant damage, the USA possessed new ships and aircraft to replace its losses.

The Royal Navy returned to the Pacific early in 1945, thereby adding a carrier task group to the US forces. Maximum operational capability was conferred on the

The Japanese super-battleship *Yamato* under attack by US aircraft on October 25 1944 during the battle of Leyte Gulf. The 64,000-ton giant suffered only minor damage from bombs in the various stages of this three-day battle. Less fortunate was her sister ship *Musashi*, sunk by carrier aircraft on the 24th by no less than 20 torpedoes and 17 bomb hits. *Yamato*'s turn

came on April 7 1945 when she was sent on a one-way mission to Okinawa. Task Force 58's aircraft hit her with 13 torpedoes and at least 8 bombs. *Yamato* blew up with the loss of all but 269 of her 3,332 crew.

KAMIKAZE

A kamikaze Zero about to strike an American battleship off Okinawa in April 1945. The Japanese launched almost 1,700 kamikaze sorties against Allied forces off Okinawa. They hit almost 200 ships of which 16 were sunk, mostly destroyers on picket duties. This particular pilot is about to die in vain as a heavily armoured battleship was the least lucrative kamikaze target. American carriers with their unprotected hangars were much more vulnerable to serious damage; the *Bunker Hill*, flagship of Task Force 58, was hit by two, suffered over 650 casualties and had to return to the USA for repairs. Kamikaze pressure on this scale might have halted the Allied advance but there were not enough committed and trained suicide pilots. They were, however, a salutary warning about the kind of resistance to be expected when the home islands themselves were attacked.

British carriers by equipping them as far as possible with US "lend lease" aircraft and adopting such USN features as deck parks to cram on as many aircraft as possible. Even so, the British force was not strong enough to be placed in the forefront of the battle – and prudence and chauvinism combined to make US commanders keep the Pacific War as "American" as possible. Nevertheless, the British carriers proved that their armoured decks were an effective protection against kamikaze attacks. There were never sufficient resources, particularly in landing craft, to allow the liberation of former British possessions by British forces alone – with the exception of Burma, where the East Indies Fleet provided valuable assistance to General Slim's ground forces. In summer 1945, with the war in Europe over, Admiral Mountbatten's Southeast Asia Command at last had the resources to mount a major amphibious operation. "Zipper", the liberation of Malaya, was planned for late August 1945, but then put back to September.

Circumstances intervened: on August 6, a B-29 dropped an atomic bomb on Hiroshima, and, on August 9, a second was dropped on Nagasaki. Japanese industry was in ruins and shipping was at a standstill – 80 percent of Japanese tonnage had been sunk by submarines, aircraft or air-laid mines. Allied carriers ranged at will off the home islands and the remnants of the IJN were destroyed in their bases. The armies on the Asian mainland were collapsing in the face of the Soviet invasion of Manchuria.

In an unprecedented act of political intervention, Emperor Hirohito gave his support to the peace faction. On August 16 a ceasefire was ordered – and on September 2 1945 Japan's surrender was formally accepted by General MacArthur on the deck of the USS *Missouri*, in Tokyo Bay.

The devastation caused by the 13 kiloton nuclear bomb at Hiroshima (*above*). The Japanese were used to US B-29 bombers inflicting this scale of damage on their cities. What shocked them into surrender was the suddenness of the cataclysm and the fact that only one aircraft had been necessary.

The Japanese military code made no provision for prisoners-of-war, whose treatment was often brutal in the extreme. This emaciated British prisoner (*below*) survived the Burma-Siam "Death Railway" and some three years' imprisonment, including twelve months' solitary confinement.

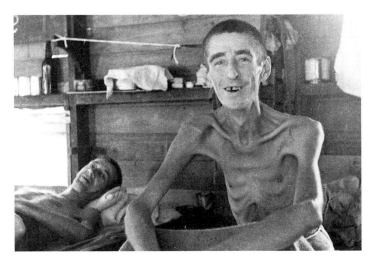

Burma was the one part of the British empire to be liberated from Japan by Commonwealth forces. After the dubiously successful Chindit operations in 1943 in the jungles of northern Burma, the unsuccessful Arakan offensive at the end of the year and the defeat of the Japanese move into India in 1944, General Slim's well-led 14th Army went on the offensive supported by the

American-led Chinese in the northeast. With the Japanese army defeated, Rangoon was finally taken by an unopposed amphibious operation.

223

— chapter sixteen —
World War II: The end in the West

Richard Holmes

On June 18 1940, in the aftermath of German victory in France, Winston Churchill told his countrymen: "Hitler knows that he will have to break us in this island or lose the war." A month later Hitler issued orders for Operation Sea-Lion, the invasion of Britain. OKW had already ordered the destruction of the Royal Air Force and the interdiction of the Channel: Kesselring's *Luftflotte* II (HQ Brussels) and Sperrle's *Luftflotte* III (HQ Paris) set about the task in early July with attacks on merchant shipping in the Channel, extending their reach to the airfields of southern England in August.

Few episodes of British history have so captured popular imagination as the Battle of Britain. Revisionist historians have pointed out that RAF estimates of German losses (2,698 in July–October 1940) were absurdly high compared with the real figure (1,733), and have noted that German maritime weaknesses cast doubt over the invasion. Nonetheless, German victory in the air would have increased Britain's peril, and it was averted by the skill and bravery of the RAF's pilots and, no less crucially, by its technology. Most British fighters were Hurricanes, useful against bombers but outclassed by German fighters. The Spitfire, however, was clearly superior to the Messerschmitt 109, while the twin-engined Messerschmitt 110 suffered heavily because of its lack of agility. Although Ultra gave an outline of some attacks, the crucial British advantage lay in radar. Radar stations on the coast enabled fighter controllers to plot incoming streams of bombers, "scramble" defending fighters and alert anti-aircraft defences. Britain's aircraft factories worked frantically to make good the RAF's losses, and although many German aircrew survived combat only to parachute into captivity, numerous British pilots and their crews baled out to fight once again.

Britain beleaguered

In September the Germans attacked inland airfields and aircraft factories, and on September 7 they mounted their first major attack on London, starting huge fires and killing 430 civilians. Although the change of emphasis was alarming for the civilian population – 12,696 Londoners

> "Of course, the press versions of life going on normally in the East End on Monday are grotesque. There was no bread, no electricity, no milk, no gas, no telephones. There was thus every excuse for people to be distressed ... There was no understanding in the huge buildings of Central London for the tiny crumbled streets of densely massed population. Here, people wanted to be brave but found bravery was something purely negative, cheerless, and without encouragement or prospect of success."
>
> Mass Observation report, London, September 1940

were killed in the first 3 months of the Blitz – it proved fatal to the Germans, for it took the pressure off airfields and enabled the RAF to take a heavy toll of bombers. In early October the Germans abandoned daylight bombing, thereafter attacking the capital and major provincial centres by night. The RAF, sorely tried though it was, had not been destroyed: on October 12 Hitler postponed Operation Sea-Lion, cancelling it for good in early 1941.

The ordeal of London under the Blitz was widely reported in the United States. Although most Americans still hoped to stay out of the war, on December 29 President Roosevelt told his countrymen:

"The British people and their allies are conducting an active war against this unholy alliance. Our own future security is greatly dependent on the outcome of that fight ... We must be the great arsenal of democracy ... We must apply ourselves to our task with the same resolution, the same sense of urgency, the same spirit of patriotism and self-sacrifice as we would show if we were at war."

America began her own economic mobilization in May 1940. That summer Roosevelt met urgent British requests for arms and ammunition by selling some World War I stocks; in September, 50 old destroyers were transferred to Britain in return for leases of Atlantic bases. The Lend-Lease Act of March 1941 gave the President wide powers to assist the Allies: throughout the war $31.6 billion worth of equipment went to Britian and her empire – about one-quarter of all British equipment came from the USA – with another $11 billion going to Russia, $3.23 billion to France and $1.6 billion to China.

The war at sea

Valuable though American support was, the uncomfortable fact remained that the North Atlantic separated "freedom's arsenal" from Britain. Despite the primacy traditionally accorded to sea power in British defence

Merchant ships in convoy.

policy, the parsimony of the interwar years had affected the navy. No fewer than 13 of Britain's 15 capital ships had been built before 1918; of 6 aircraft carriers, only one, HMS *Ark Royal*, was a modern purpose-built carrier, and naval aircraft were hopelessly outdated. The German fleet was smaller but more modern, but in the first months of the war it was not able to protect German merchant shipping, break the British blockade or interfere with the passage of the British Expeditionary Force to France.

After the evacuation from Dunkirk the navy was preoccupied with the threat of invasion and, quite aside from Germany's failure to secure command of the air, her

—TIME CHART—

1940	
July–Oct	Battle of Britain: effective use of ground-control and radar systems
Oct 12	Hitler postpones Operation Sea-Lion
end 1940	Enigma cypher broken
1941	
early 1941	Centrimetric radar first used to locate U-boats in battle of the Atlantic
Mar 11	Lend-Lease Act signed
Aug 14	Anglo-American Atlantic Charter signed
Sept 16	Beginning of American-escorted convoys
Dec 7	Japanese attack Pearl Harbor
Dec 11	Germany and Italy declare war on USA
1942	
June–July	Murmansk convoys decimated
1943	
early 1943	Intensive U-boat "wolf-pack" campaign
Apr–May	North Atlantic convoy ONS-2 withstands U-boat assault
June–Dec	Allied hunter-killer campaign against U-boats
end 1943	Allied U-boat kills exceed German replacement capability
1944	
June	Streamlined, high-speed Type XXI U-boats first deployed
June	Ultra intelligence: Colossus II digital programmable computer in operation
June 6	D-Day
June 13	V-1 flying-bomb attacks begin
July 25	Allied breakout from Normandy (Operation Cobra)
Aug	Proximity-fused shells used against V-1
Aug	British Meteor I jet-interceptor in service against V-1
Aug 15	Allies land in southern France
Sept	V-2 ballistic missile attacks begin
Sept 17–26	Operation Market Garden and battle of Arnhem
Dec 16	Ardennes offensive (Battle of the Bulge) begins
1945	
Mar 28– Apr 18	Allies envelop German Army Group B in Ruhr pocket
May 4–7	Unconditional German surrender
July 17	Anglo–US–Soviet Potsdam conference begins

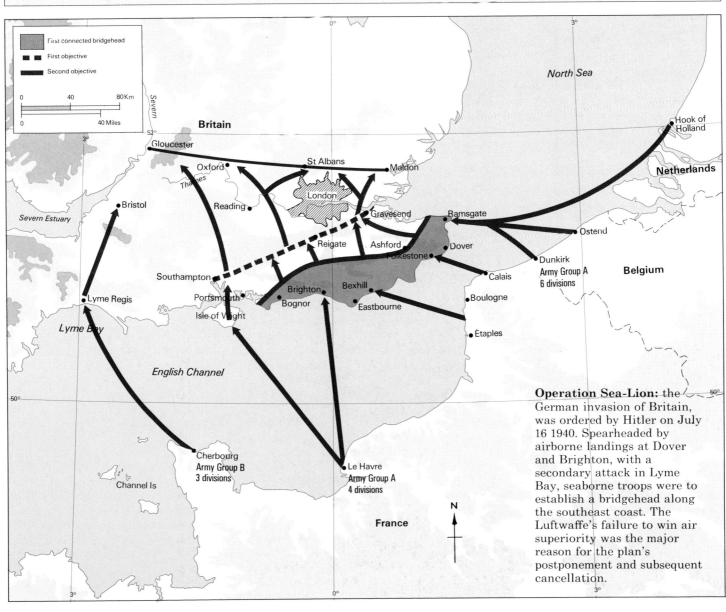

Operation Sea-Lion: the German invasion of Britain, was ordered by Hitler on July 16 1940. Spearheaded by airborne landings at Dover and Brighton, with a secondary attack in Lyme Bay, seaborne troops were to establish a bridgehead along the southeast coast. The Luftwaffe's failure to win air superiority was the major reason for the plan's postponement and subsequent cancellation.

225

inability to counter the Home Fleet made invasion a high-risk venture. The Admiralty was also anxious to ensure that the French fleet did not fall into German hands after the collapse of France, and an attempt in July 1940 to secure the neutrality of the French naval units at Mers-el-Kébir led to a tragic clash, with the sinking of the French battleship *Bretagne* and heavy loss of life.

German surface craft made damaging forays against trade routes. After a successful career of commerce-raiding, the pocket battleship *Admiral Graf Spee* was fought to a standstill off the River Plate, Uruguay, by a British flotilla and was scuttled outside Montevideo harbour on December 17. In early 1941 the battlecruisers *Scharnhorst* and *Gneisenau* made a destructive sortie, escaping into Brest at its conclusion, and in May the new battleship *Bismarck*, accompanied by the heavy cruiser *Prinz Eugen*, emerged from the Baltic and sank the battlecruiser *Hood* in the Denmark Strait. *Bismarck* was making for France when she was damaged by an aircraft from *Ark Royal*, and on May 27 units of the Home Fleet sent her to the bottom.

U-boats threaten

The real danger came, as it had in World War I, from German U-boats. In October 1939 the *U-47* penetrated the defences of Scapa Flow to sink the battleship *Royal Oak* at her moorings, and by March 1940, U-boats had sunk more than 200 ships. It was fortunate for Britain that the Germans had not devoted more of their resources to submarines: they had lost 31 of 56 operational U-boats by the end of 1940, and although the replacement rate was rapid, the British were able to build up their escort forces and improve their tactics.

The invasion of Russia relieved the pressure on Britain by compelling the Germans to devote some of their resources, especially shore-based aircraft, to supporting operations in the Baltic. Conversely, Britain's decision to send supplies to the Russian ports of Murmansk and Archangel exposed the crews of warships and merchantmen alike to the long and perilous voyage around the freezing waters of the North Cape, at the mercy of surface, submarine and air attack.

Growing Anglo-American cooperation centred upon the war in the Atlantic. In August 1941 Churchill and Roosevelt met off Newfoundland to sign the Atlantic Charter, and Roosevelt agreed that American ships and aircraft in the Western Atlantic would inform the British of any U-boat sightings. After a U-boat attacked an American destroyer in September, Roosevelt ordered his warships to attack submarines on sight. This was followed in October by the torpedoing of the destroyer USS *Kearny* and anti-German feeling in the United States increased. In November, an American naval squadron was sent to Iceland, where American troops had reinforced, and later replaced, the British garrison.

The outbreak of war with Japan altered the picture dramatically. America was now formally at war, but much of her naval strength went to the Pacific: U-boats did considerable damage to unescorted US merchantmen in the Atlantic before the Americans profited from British experience by introducing the convoy system. The year 1942 was the worst of the war for the loss of merchant ships: the Allies lost 1,664 of which 1,160 were sunk by sub-

---

THE FOOD FRONT

Food rationing in Britain began in January 1940, and by 1941 included meat of all kinds, sugar, butter and margarine, cooking fats, tea, jam, marmalade and syrup, and cheese. From December 1940, every ration-book holder had a fixed number of monthly "points" to spend on such goods as canned meat and fruit, fish and vegetables. The Ministry of Food could adjust demand where necessary by varying the cost of specific items.

Many foods became scarce. The flavoursome onion was in great demand, oranges were rare, lemons almost unknown, and most wartime children never saw a banana. For much of the war, adults other than young children and expectant mothers were limited to one egg per fortnight: dried egg became a universal standby. The very rumour of a rare item led to long queues at stores. Country dwellers could supplement their diets with home produce, while gardeners and allotment holders were exhorted to "dig for victory".

Lord Woolton, the able Minister of Food in 1940–43, called on housewives to win the war "on the kitchen front". "Woolton Pie" – potato, parsnip, carrot and turnip in white sauce, with a pastry crust – proved palatable, but the Ministry's experiment with whale meat was a failure. "British Restaurants" were set up to provide cheap, filling food, while many factories had their own canteens.

Although there was some black-marketeering, food rationing worked better than it had during World War I, mainly because of Woolton's policy of rationing items only when he could guarantee that the ration would be honoured.

---

marines. But by the autumn the Allies were imposing their control on the Atlantic. Their support groups fought it out with German "wolf packs" as technology tilted the balance in Allied favour. Radar was of limited use against submarines (although it did enable attacking aircraft to find the *Bismarck*) until the introduction of centimetric radar in 1941; thereafter, radar in ships and aircraft played a leading part in the war against U-boats. The development of the escort carrier gave intimate air support to convoys; the range of shore-based aircraft increased, narrowing the "air gap" in mid-Atlantic; improvements in asdic and sonar made the detection of submerged U-boats easier; and larger depth-charges and improved launchers made attacks on them more deadly.

This technology was not all one-sided. The Germans had made good use of magnetic and acoustic mines until effective countermeasures limited their value, and an acoustic torpedo achieved good results. The introduction of the snorkel enabled submarines to cruise just below the surface, using their diesel engines rather than electric motors which had only a limited battery life. The Germans went on to develop a formidable new fast submarine, the Type XXIII, but this was not available in large numbers before the war ended.

The battle of the Atlantic reached its climax in the first half of 1943. A total of 108 ships was lost in March; in May 41 U-boats were sunk for the loss of 50 merchantmen; and in October a double convoy lost only one merchantman while its escorts accounted for six U-boats. German submarines remained dangerous, in the Indian Ocean as well as the Atlantic, and even in the last five weeks of the war they sank ten merchant ships and two escorts. However, Allied victory in the Atlantic ensured that Britain would not be starved into submission and could become the springboard for the offensive against mainland Europe.

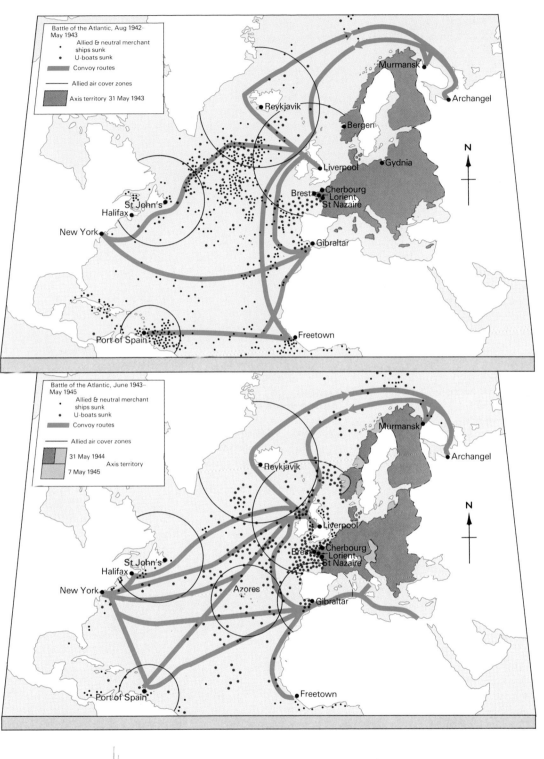

Battle of the Atlantic, Aug 1942–May 1943
- Allied & neutral merchant ships sunk
- U-boats sunk
- Convoy routes
- Allied air cover zones
- Axis territory 31 May 1943

Battle of the Atlantic, June 1943–May 1945
- Allied & neutral merchant ships sunk
- U-boats sunk
- Convoy routes
- Allied air cover zones
- Axis territory 31 May 1944 / 7 May 1945

The battle of the Atlantic.
An oil-soaked survivor (*above*) is taken aboard the U-boat that has sunk his ship. The map (*upper left*) shows major convoy routes and locations of merchant and U-boat sinkings earlier in the war. In 1941–42, U-boats sank 1,459 Allied and neutral merchantmen, totalling almost 7,619,000 tons. Merchant shipping losses during the later stage of the battle (*lower left*), June 1943–May 1945, fell considerably.

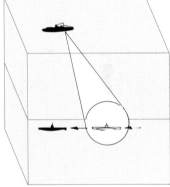

The underwater detection apparatus called "Asdic" (*above*) or "Sonar" (*sound navigation and ranging*) – transmits high-frequency pulses that "bounce" from a submarine, enabling its position and course to be judged.

The Type VII U-boat (*below*) was the mainstay of Germany's submarine fleet; of 705 in service in 1939–45, 437 were lost in action.

227

Towards "Overlord"

The attack on Pearl Harbor was followed four days later by a German declaration of war, providing the Americans with an immediate conflict of strategic priorities. Although the Japanese attack had caused great popular fury in the United States, Roosevelt and his Chiefs of Staff wisely recognized that Germany was their most durable adversary and determined to adopt the policy of "Germany first". This was affirmed at the first inter-Allied conference, "Arcadia", held at Washington in December 1941–January 1942, when America committed herself to "Bolero", the plan for the build-up of her forces in Britain. Over the next three years there was more than a little friction as the Americans focused their attention on cross-Channel invasion, suspicious of British concern for "sideshows" in the Mediterranean.

Some American enthusiasm was certainly premature. "Bolero" went slower than had been expected, and the experience of Dieppe, where the Second Canadian Division and British and American commando forces were landed at appalling cost on August 19 1942, showed that invasion would be complex and hazardous. Although the British managed to gain American support for Operation Husky at the Casablanca conference in January 1943, the "Trident" conference that May emphasized that the cross-Channel invasion, Operation Overlord, would take place the following year. The "Quadrant" conference at Quebec in August 1943 not only confirmed this, but also stressed American determination to mount Operation Anvil (later Dragoon) in southern France. The Brit-

ish, for their part, had been made cautious by early reverses and persistent shortages, and the fighting in Tunisia in 1942–43 gave them serious doubts about the combat effectiveness of the Americans.

That "Overlord" rode out the storms of inter-allied politics is a tribute to the energies of Lieutenant General Sir Frederick Morgan, Chief of Staff to the Supreme Allied Commander (Designate)—COSSAC—and his team, who did the important groundwork in 1943. The appointment of a supreme commander caused much debate, and it was not until they met at Cairo in December 1943 that Allied leaders decided on General Dwight D Eisenhower. Eisenhower's principal subordinates were British: Air Chief Marshal Tedder was his deputy and Montgomery was to command his land, Admiral Ramsay his naval, and Air Chief Marshal Leigh-Mallory his air forces. Montgomery detected flaws in the COSSAC plan, arguing for a heavier assault on a broader front, and on January 23 1944 Eisenhower accepted a revised draft.

The invasion of Europe

There was little enough time before the revised D-Day of June 5 1944 (bad weather was to cause a postponement of one day) for the mass of preparation and rehearsal upon which the assault would depend. "Overlord", with 150,000 men and 1,500 tanks to be landed in the first 48 hours, was smaller than the huge operations on the Eastern Front, but the fact that 5,300 vessels and 12,000 aircraft were involved made it the largest combined operation in history. Many of the skills acquired and equipment procured concentrated

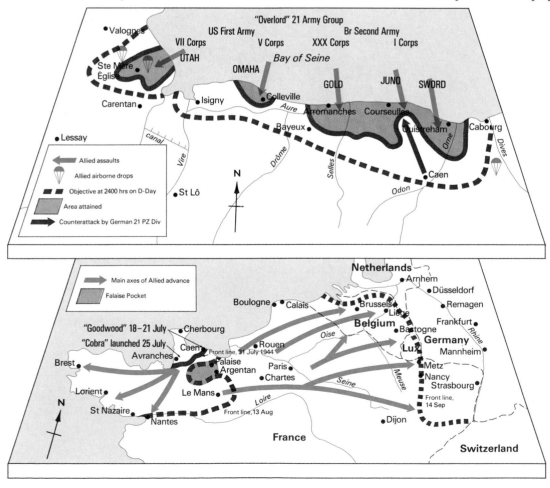

Normandy landings on D-Day (*upper left*), June 6 1944, involved more than 150,000 men in some 4,000 ships, escorted by approximately 600 warships and with more than 10,000 support aircraft. Following airborne attacks at 0200 hours, to secure high ground and communications routes and to silence enemy batteries, the first men of the US First Army hit Omaha and Utah beaches at 0630 hours; British and Canadian troops began landing on Gold, Juno and Sword beaches at 0700. Overall Allied casualties on June 6 totalled around 10,500 men and 114 aircraft.

The Allied "broad front" advance from Normandy towards the Rhine (*lower left*) took place July–September 1944. Some commentators maintain that a concentrated thrust on a narrower front, favoured for different reasons by the Allied commanders Montgomery, Bradley and Patton, might have shortened the war, but Eisenhower preferred not to expose his flanks to counterattack.

on specific amphibious problems: the specialized armoured vehicles of the British 79th Armoured Division were to prove invaluable in avoiding a repetition of Dieppe, and the "Mulberry" artificial harbours would compensate for the anticipated Allied failure to capture a port early on. No less important was the need to deceive the Germans as to the chosen site for the invasion in the Bay of the Seine. Radio transmitters helped create the impression of a fictitious First US Army Group in Kent, preparing to attack the Pas de Calais, and Allied airmen paid more attention to other areas than to the real invasion sector.

German failure to detect the direction of the assault meant that Rundstedt, re-employed to serve as Commander-in-Chief West, was unable to concentrate his not inconsiderable resources – 60 divisions, 11 of them armoured, in France and the Low Countries – to meet the invasion. He was handicapped by having numerous second-rate formations with no integral mobility, he lacked air cover, and he disagreed with Rommel, commanding Army Group B in northern France, over the best policy for dealing with invasion. Rundstedt favoured the classical solution of identifying the thrust and concentrating to counterattack it. Rommel, painfully aware of the practical effects of Allied air superiority, doubted if he would get his panzers to the coast unscathed and wanted to defeat the invaders on the beaches. Neither was helped by the fact that the OKH armoured reserves could be released only with Hitler's permission.

"Overlord" began in the early hours of June 6, when Allied parachutists were dropped behind the invasion

THE ENIGMA CYPHER MACHINE

Enigma was an electrical machine with a typewriter keyboard which transposed the typed characters according to a prearranged random setting. The basic Enigma used three rotors connected by electrical wiring to encipher letters. After a letter had been enciphered an operator would move the outer rotor round a step, modifying the connections, and deal with the next letter. The message receiver set up an Enigma in the same succession of states as the sender's, and fed in the cipher-text to recover the decrypted version.

The first approach to deciphering Enigma messages exploited their elements of repetition, introduced by the need for the message sender to let the receiver know the initial state of the machine. It was in sifting through these clues and checking the many possible deductions that could be drawn from them that the electro-mechanical "Bombe" proved invaluable. Matched to the electrical wiring of the Enigma (obtained from a pre-war spying coup), the original Polish device worked through all possible rotor positions and settings until one was found that matched the pattern of clues detected in the message traffic of a particular day.

This approach failed when the Enigma machine was made more complex. The British attacked the problem by *guessing* a word appearing in cipher in a message, and then checking every rotor setting and position to see whether any could produce the plain word from the cipher after logically possible letter-changes had been considered. Starting on May 22 1940, the high-speed British Bombe broke the current Enigma code and began to decrypt messages at the rate of 1,000 a day, providing an essential service to the end of the war.

General Dwight D Eisenhower (*above*) surrounded by his "Overlord" staff, shortly before D-Day. From America's entry into the war, Eisenhower had advocated a "Europe first" strategy, and in June 1942 he was appointed Commander, European Theatre of Operations, developing his high-command skills in the North African, Sicilian and Italian operations. In January 1944 he was appointed Supreme Allied Commander for the invasion of Europe. He proved an ideal choice: no great strategist, but a "manager" of genius, with the lucidity to recognize the decisions that needed to be made and the courage to make them. His general insistence on the "broad front" advance into Europe may be criticized: his overall achievement remains unassailable.

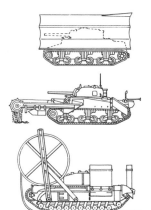

Men of the US First Infantry Division (*above*) shelter beneath the cliffs during the fiercest fighting of D-Day, at "bloody Omaha", where the 36,500 men landed took 2,500 casualties.

Specialized armoured vehicles (*left*), included, top to bottom, a "swimming" duplex-drive (DD) tank, a mine-clearing flail tank, and a "carpet-layer" with a reel of reinforced matting. These proved of great value.

beaches. Cloud and anti-aircraft fire hampered them, but the US 82nd and 101st Airborne Divisions caused confusion between Carentan and Valognes, while the British 6th Airborne Division seized objectives northeast of Caen, including Pegasus Bridge, taken by glider-borne assault. Success on the beaches was equally mixed. The leading American wave landed on the wrong part of Utah beach, but quickly secured a beachhead and pressed inland to join the parachutists. At Omaha beach things went badly wrong. American commanders had decided to lower their landing craft well offshore, and many were swamped or driven off course. Supporting fire did little damage to the defences, and the heavily laden men who reached the beach were pinned down. It took several hours, and untold bravery, before the survivors got off the beach, and by nightfall the perimeter was at best a mile deep. The British landings, on Gold, Juno and Sword beaches, were made easier by the specialist vehicles of 79th Armoured Division, and by nightfall the invaders were well inland, approaching Caen and Bayeux.

The battle of Normandy

The next few days saw sharp attacks delivered by newly arrived German armour and British attempts first to penetrate and then to outflank the city of Caen. The veteran but weary 7th Armoured division was rebuffed at Villers-Bocage on June 11–14, and at the end of the month an attack by 8 Corps made disappointing gains west of Caen. The Americans cleared the Cotentin peninsula, taking Cherbourg on June 27. By the month's close it was

clear that there would be no quick and easy breakout. The Germans fought with professional determination, their Panther tanks and *Panzerfaust* anti-tank rockets murderously effective in the *bocage* countryside with its hedge-

---

"Suddenly chunks of metal flew off the turret. Immediately a near solid column of dense black smoke spiralled vertically upwards for about 100 feet. He had expected to see a few sparse flames lick from the stricken tank followed by the hurried disembarkation of the crew, possibly wounded or burned, and after a short interval the thud of ammunition exploding inside. But the reality was quite different. The tank had without warning become an instant inferno."

Lieutenant Alastair Morrison, Normandy 1944

---

rows, sunken lanes and orchards. The battle for Normandy did not have the unsettling backcloth of Burmese jungle or Russian steppe, but it was an unrelenting struggle which wore down the men who fought it.

On July 18–20 the British Second Army mounted Operation "Goodwood", a concentrated attack east of Caen by three armoured divisions, preceded by a colossal strike by some 2,000 British and American heavy and medium bombers. Doubt remains as to whether Montgomery hoped that the attack would break out into the open country towards Falaise, or whether, as he subsequently maintained, he simply intended to attract German forces to the British sector to enable Bradley's First US Army to break

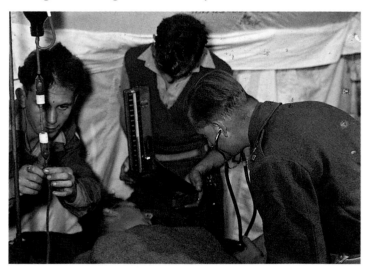

A field hospital in Normandy in the late summer of 1944 (*above*). Medical services in the field had improved greatly since World War I. The speedy treatment of wounded men in surgical field hospitals, the use of new antibiotics such as penicillin, and the development of blood banks and plasma (note the plasma bottle in the photograph), resulted in a reduction in deaths per 1,000 wounded

from 8.1 in 1917–18 to 4.5 in 1941–45 (US official figures). In the period from June 6 to September 15 1944, more than 2,000,000 Allied servicemen were landed in France: some 40,000 were killed during this time, with a further 190,000 wounded or missing. German losses during the same period are put as high as 700,000 killed, wounded or captured.

THE RED BALL EXPRESS

The logistics of "Overlord" were complex, and bad weather, German resistance, slow clearance of beach exits and the delay in opening Cherbourg meant that supplies arrived more slowly than planned. Since the Allied advance, too, was at first slower than expected, logisticians were able to build up vast stockpiles on the beaches. It had been expected that the Germans would withdraw gradually, allowing the advancing Allies to be supplied by conventional methods, but following Patton's breakout the American armies went farther and faster than their quartermasters had deemed possible. Throughout August 1944 emergency measures were needed to keep them supplied – and in September the taut chain of supply at last tugged them to a halt.

Because of damage to railway lines and rolling stock, the main burden of hauling supplies forward during August fell on road transport: the Red Ball Express, created by the US Army's transportation chief, Major General Frank A. Ross. The Express operated on a one-way road system which ran from Saint-Lô to Versailles, where one loop eventually branched north to Soissons and the US First Army, and another south to Sommesous and the US Third Army. On August 29 it reached a strength of nearly 6,000 vehicles.

Inevitably there were problems. Exhausted drivers sometimes crashed trucks or neglected maintenance; some resorted to malingering or even sabotage; a few sold their cargoes on the black market. Shortage of military police caused problems with traffic control, and repair facilities were unable to keep pace with the formidable demand. But overall it was a remarkable achievement: by hauling more than 400,000 tons in 81 days the Red Ball Express helped the US armies to sustain an advance that had seemed logistically impossible.

out farther west: the evidence suggests that Montgomery was more sanguine than the situation warranted. In the event, "Goodwood" proved a fiasco. The bombardment had not broken the defence and the attack stalled with the loss of 400 tanks. The American performance had initially seemed no more encouraging, with a slow advance through the *bocage* punctuated by German counterattacks and the dismissal of numerous senior officers. On July 25, however, on the heels of an aerial bombardment, Bradley launched Operation "Cobra", swiftly breaking out to Avranches, his Shermans ripping through the hedges with the aid of tusks made from steel salvaged from German beach obstacles. In the aftermath of "Cobra", Bradley's command was reorganized into the US 12th Army Group, with Courtney H Hodges and George S Patton commanding its First and Third Armies respectively.

The fighting of June and July had worn down German resources and the collapse of Army Group Centre meant that no help could be expected from the East Front. Rommel had been wounded in an air attack and Rundstedt was dismissed for displaying the realism that Hitler regarded as pessimism. Field Marshal von Kluge, Rundstedt's successor, was pressed by Hitler to counterattack the US First Army around Mortain on August 7, a battle which further reduced German strength and showed new American confidence. It was the prelude to the long-awaited breakout. Patton surged out to Mayenne, Le Mans and Alençon, before hooking north to take Argentan on August 20. The British and Canadians had been pushing away south and southwest of Caen against resistance which gave little clue to the exhausted state of the German army. The Canadians and Poles made some progress in Operations "Totalize" and "Tractable", and Falaise was at last secured on August 17. Its capture left much of the German army in Normandy in a shrinking pocket, harried from the air. The fact that many Germans escaped gave rise to a dispute among commanders and historians alike as to why the pocket was not closed sooner. Nevertheless, most of those who emerged safely from the cauldron did so on foot: the Falaise pocket was the culminating point of a battle which cost the German army 1,500 tanks, 3,500 guns, 20,000 vehicles and almost 500,000 men. The Allies lost 209,672 men, two-thirds of them British or Canadian.

Uneasy Autumn

The Allies made good progress after Falaise. In Paris the Resistance rose on August 19, before the liberators arrived, and it was to the Frenchmen of General Leclerc's 2nd Armoured Division (the famed *Deuxième DB*) that the German garrison commander surrendered on August 25. Kluge was already dead – by his own hand on August 19, on his way back to Germany under accusation of complicity in the July 20 "bomb plot" against Hitler's life. His successor, Model, could do little to stem the tide, and the Allies reached Brussels on September 4. On September 5 Rundstedt emerged from retirement yet again to become Commander-in-Chief West, leaving Model, as Commander of Army Group B, to grapple with the battle for Belgium.

Stiffening German resistance – a tribute to the astonishing resilience of the *Wehrmacht* – logistic overextension and a conflict of strategic priorities all contributed to Allied loss of momentum after the successful liberation of Brussels. Eisenhower was already involved in an argu-

ment over strategy before he took personal command of Allied land forces on September 1. Montgomery wanted Patton's Third Army to halt, in order to protect the flank of a thrust straight to the Ruhr by the British Second and US First Armies. But Patton was making good progress into eastern France and Eisenhower, weighing political and military considerations, was not disposed to stop him. On September 5 Patton resumed his advance, crossing the Moselle south of Metz and joining hands near Dijon with troops landed in southern France in Operation Dragoon. Montgomery still favoured a "knifelike thrust" to the heart of Germany, and saw an opportunity of using the Allied Airborne Army to secure the crossings of the Maas, Waal and Lower Rhine, laying a carpet over which the advance would roll.

"Market Garden" launched

Operation Market Garden began on September 17. The US 82nd and 101st Airborne Divisions took most of their objectives around Nijmegen and Eindhoven, but units of the British 1st Airborne Division, landing too far from the bridge at Arnhem, were involved in heavy fighting in their efforts to reach it. Bad weather and communications difficulties impeded reinforcement by air and prompt German reaction – aided by the fact that Model himself was close by – increased the odds against an already risky venture. The advance of 30 Corps from the south was slower than had been expected, and the survivors of 1st Airborne escaped across the Rhine on the night of September 25 26.

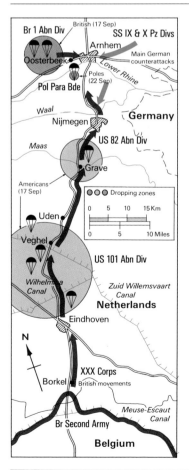

Operation Market Garden was favoured by Montgomery as a way of laying an "airborne carpet" along which the British Second Army would thrust into Germany. On September 17 1944, simultaneous airborne attacks were made to secure crossings on the canals north of Eindhoven and, farther north, on the Maas, Waal and Lower Rhine. The US 101st Airborne Division secured its objectives near Eindhoven, preparing the way for the advance of the British 30 Corps; the US 82nd Airborne Division captured the Maas crossing and advanced with 30 Corps to take the Waal bridge at Nijmegen on September 20. But thereafter 30 Corps' advance slowed, and at Arnhem the British 1st Airborne Division was encircled. After a heroic fight against odds, some 2,200 survivors managed to escape south; around 7,000 were killed, wounded or captured.

"Market Garden" may not have been a total failure, but its less than complete success left the Allies in an unsatisfactory position. A thrust into the Ruhr, probably technically feasible in September, was no longer possible against the hardening crust of German defence. The Allies were full extended, with only 54 divisions in Europe by the end of September and 600 miles of front to hold. Worse, they were still dependent upon ports in Normandy; although the Channel ports were taken in September, it took weeks to get them working. Much therefore, depended on the opening of Antwerp, captured on September 4 – although it was not until November 28, following the capture of Walcheren and minesweeping in the Schelde, that the first convoy arrived there. The rapid advance of late summer gave way to an autumn of attrition. Patton attacked in Lorraine but made slow progress, while Ninth Army edged forward east of Aachen in an advance which produced some of the theatre's heaviest fighting.

Hitler's last gamble

As early as October Hitler had considered a counter-offensive into the Ardennes, and despite the reservations of Rundstedt and Model he ordered an ambitious attack on a 75-mile front between Monschau and Echternach, with the aim of reaching Antwerp and cutting the Allies in half. The 3 armies involved, Sixth SS Panzer on the northern flank, Fifth Panzer Army in the centre and Seventh Army in the south, comprised 28 divisions, some newly raised but enthusiastic *Volksgrenadiers*, but others seasoned units refitted after fighting in Normandy or the east.

When the attack began on December 16 the Americans, thinly deployed in a quiet sector, were taken by surprise. The Germans pushed on quickly, profiting from the bad weather which kept Allied aircraft out of the sky. Nevertheless, American resistance, stubborn in some places if sketchy in others, delayed the Germans, and shortage of petrol halted several armoured spearheads. Failure to capture the communications centre of Bastogne, important in an area with few good roads, was particularly damaging. The Allies took several days to recover from the shock of the unexpected blow, but on December 19 Eisenhower took the important decision to place all forces north of the breakthrough under Montgomery and those south of it under Bradley. German penetration was contained, although not without anxious moments, and the flanks of the bulge were sharply counterattacked by Patton, moving up from the south, and "Lightning Joe" Collins of the US 7 Corps stabbing down from the north. As the weather cleared at the end of the month Allied aircraft flew 15,000 sorties in four days, taking a heavy toll of vehicles on the roads and pounding railheads across the frontier. A subsidiary German offensive into the Saar, beginning on January 1 1945, met with little success, and by January 8 it was evident even to Hitler that the gamble had failed.

The last act

It was not in Hitler's nature to capitulate, and Allied insistence on unconditional surrender gave the Germans no real alternative to fighting on to the last. Hitler cherished wild hopes that "secret weapons" might yet win

The German counter-offensive in the Ardennes (*above*), often called the "Battle of the Bulge", was launched on December 16 1944, when 28 divisions were committed to a thrust to Antwerp, to bisect the Allied armies and cut their supply lines. By December 24 the Germans had made a huge "bulge" in the Allied lines, only narrowly failing to break through.

The Messerschmitt Me 262 jet fighter (*left*), operational from late 1944, was not deployed earlier because of Hitler's insistence that it should be developed as a bomber. Powered by two 1,980lb static-thrust turbojets, it had a maximum speed of 542mph and, armed with four 30mm cannon and 24 5cm air-to-air rockets, proved deadly against bomber formations.

The Fieseler Fi 103 "flying bomb" (*above*), Germany's *Vergeltungswaffen Eins* (V-1), was a pulse-jet powered missile with a maximum speed of 400mph, carrying an explosive warhead of up to 1,870lb to a maximum range of 250 miles. Of some 9,000 "Doodlebugs" launched by catapult or air against Britain from June 13 1944 onward, around 2,400 fell on London.

the war. In 1944 attacks on England by V-1 flying bombs and V-2 rockets had come as a cruel shock, killing nearly 9,000 people, but the invasion overran their bases, and although 5,000 V-bombs fell in and around Antwerp they failed to check the flow of supplies. The Messerschmitt 262 jet fighter was an impressive weapon, but too few were produced to have much effect and their airfields were bombed as soon as they were identified. As far as the most awesome weapon of all was concerned, the Germans lagged far behind the Allies in research on atomic weapons. Even at this late hour Hitler hoped for a miracle, and when he heard of the death of President Roosevelt on April 12 he compared it to "The miracle of the House of Brandenburg", the death of the Czarina Elizabeth in 1762 and the accession of the pro-Prussian Czar Peter III, which had

saved Frederick the Great in the Seven Years War.

Hitler was hopelessly wrong. There was no pause after the Ardennes battle, and in February the British and Canadians fought their way through the Reichswald in a dismal battle of blighted woods, drenched fields and cratered towns. Kesselring took over the unenviable post of Commander-in-Chief West in March, but he could do little to halt the advance. The Siegfried Line was breached, and the Rhine was speedily crossed. As the US First Army closed up on to the river in early March it captured the bridge at Remagen intact. Patton forced a crossing south of Mainz, and on March 24 Montgomery seized a huge bridgehead around Wesel. On April 1 leading elements of the US First and Ninth Armies met at Lippstadt, with Model's entire army group inside the pocket. It fought on for 18 days, and Model shot himself rather than go into captivity with his 325,000 men. Ninth Army reached the Elbe near Magdeburg on April 11 and Allied columns swung up to the Baltic and down into Bavaria, where the rumoured "National Redoubt" proved to be an illusion.

On May 4 Montgomery took the surrender of German forces in Northwest Germany, Belgium and Holland in his headquarters on Lüneburg Heath. On May 7 Jodl and Admiral von Friedeburg, acting for Admiral Dönitz, appointed as Hitler's successor in the Führer's last hours, formally concluded the unconditional surrender of all German forces at Eisenhower's headquarters at Rheims. A telegram was sent to the Combined Chiefs of Staff, announcing: "The mission of this Allied Force was fulfilled at 3 am, local time, May 7th 1945. Eisenhower."

The German "V-2" rocket, more correctly called the A-4, shown here on its mobile launcher (*Meillerwagen*), was the first supersonic guided missile. The 13-ton rocket was 46ft long and carried an impact-fused, one-ton Amatol warhead. Its liquid-fuelled engine burned for around 70 seconds, reaching Mach 1 within 30 seconds and taking the missile to a maximum height of some 60 miles on a predetermined trajectory.

After a flight of about five minutes across a 200-mile range, the A-4 struck its target at a speed of 3,500fps (2,386mph), thus arriving totally without warning.

Between September 1944 and March 1945, Britain received 1,115 rockets, of which 518 fell on London, causing 9,000 casualties; Antwerp, its other major target, received 1,341 rockets and suffered 30,000 casualties.

— chapter seventeen —
Strategic bombing
John Sweetman

The term "strategic bombing" originated during World War I. At first associated with military targets beyond the immediate battlefield and out of artillery range, by 1918 it had come to mean raids on an enemy's homeland to destroy his industrial capacity to wage war and to erode the will of his civilian population to support it. In short, strategic bombing entails bypassing the front line to attack factories and cities.

In Germany, during the first decade of the 20th century, Count Ferdinand von Zeppelin developed aluminium-framed airships, while Dr Schütte and Karl Lanz built wooden-framed airships. Following the Wright brothers' success with heavier-than-air flight in 1903, aeroplanes also gradually evolved; although the military worth of any aircraft (airship or aeroplane) remained doubtful. Shortly before World War I, a senior British officer declared: "We are not convinced that either aeroplanes or airships will be of any utility in war." In 1914 Britain had 272 aeroplanes of which only 90 were serviceable; the USA had some 350, "all of inferior types". Exclusive of the United States, Britain and her allies (including Russia and Serbia) nominally mustered 1,235 aeroplanes and 8 airships; Germany, Austria and Turkey had 1,410 aeroplanes and 19 airships.

The first campaigns
The Germans were the first to undertake what was, in effect, a strategic bombing campaign. Early in 1915, German airships attacked East Anglia, and in March of that year the Kaiser relaxed his ban on raiding London. Meanwhile, in November 1914, the German military air service had formed a unit of Aviatik B aeroplanes at Ostend to bomb England. However, without possession of the French Channel ports, they lacked the range to do so.

Airships, which flew above anti-aircraft fire and defending aeroplanes, did have that range. During 1916, for example, 111 airships crossed the Channel and only five were shot down. Spectacular British successes, like the destruction of a Schütte-Lanz airship over Cuffley by a British fighter in September 1916, could not disguise the fact that some raids were most effective. During the night of September 8–9 1915, damage estimated at £530,787 was caused in London; one month later, another attack killed 71 people and injured 128. *Korvettenkapitän* Peter Strasser, commander of the Naval Airship Division, hoped that bombing would force Britain to her knees; and airship raids continued until August 1918, but with diminishing effectiveness as the defences improved. In all, between January 1915 and August 1918, German airships attacked Britain 51 times, dropping 5,907 bombs, killing 528 people and injuring 1,156. As Noël Pemberton-Billing MP observed: "We know what a disturbing effect air raids have in this country and how cheaply that effect is gained."

By 1917, the increased ranges of aeroplanes meant that London could be reached from German-held territory; and in May a unit based in Belgium began an aerial assault on the British capital. On May 25, 21 Gotha GIV bombers, prevented by cloud from reaching London, hit Shorncliffe and Folkestone instead, killing 95 people and injuring 195. Seventy-four defending aircraft took off, but only one enemy plane was claimed destroyed. Similar attacks took place on southeast England, with equally paltry defensive success, during the next three weeks. Then, on June 13, 17 Gothas dropped 118 bombs on London during a daylight raid, killing 160 people, injuring 408 and causing £130,000

worth of damage. Spent anti-aircraft shells added a further 20 casualties. Fifty-two defending aircraft had no success. Civic and national rage about "wanton, indiscriminate ferocity" and "deliberate war on the men, women and children of this country" led to angry calls for reprisals against German cities. A second daylight raid on July 7 killed 53 people and injured 190 (of whom 10 and 55 respectively fell victim to anti-aircraft shrapnel). As a result, home defences were vastly strengthened at the expense of squadrons in France.

During the winter of 1917–18, German aeroplanes mounted a series of night attacks on London and southeast England. Panic in shelters swelled the number of raid victims: on January 22 1918, for example, 14 people were killed and a further 14 injured in one stampede. Loss of sleep, interruption of manufacturing production and more general adverse effects on civilian morale were identified as direct consequences of enemy bombing. Altogether, German airships and aeroplanes dropped 280 tons of high explosives (HE) on Britain during World War I, killing 1,413 people and injuring 3,408.

Although no serious breakdown in either industrial production or civilian morale occurred, it has already been noted that German raids on Britain occasionally caused panic. During the first daylight attack on London, 70 people were treated in hospital for injuries and shock sustained in the melee at one factory that had not itself been bombed. During Zeppelin alerts at Hull "large numbers of people" streamed out of the city; elsewhere, coal mines and chalk pits offered safe refuge for the terrified. The War Cabinet was informed that "the public, and in particular the poorest classes, whose tenements are often of the flimsiest description, were tending to give way to panic". Despite this evidence, no general panic occurred and civilian morale did not disintegrate. However, whether subjected to direct attack or not, factories did stop work. During the night of September 24–25 1917, production of small arms ammunition at Woolwich Arsenal fell to 20 percent of the normal total and remained below average on the following day. The official air historian later estimated that 75 percent of workers in munition factories ceased work at the warning of an attack.

Britain's early bombing strategy
Like the Germans, the British drifted somewhat haphazardly into strategic bombing as the range and bomb-carrying capacity of aeroplanes increased. The Royal Flying Corps (RFC) initially used its planes to "influence local situations by bombing railway stations and junctions" in support of the Army. However, Major General Trenchard, commanding the RFC in France, soon ruled that "special squadrons" must be formed to carry out "sustained attacks" on targets beyond the front line, to draw off enemy aircraft which would otherwise intervene on the battlefield. In practice, this meant strategic rather than tactical action: Trenchard claimed that strategic

—TIME CHART—

1915 Jan 19–20	First Zeppelin raid on England
1915 Aug–Oct	Strategic offensive by Zeppelins
1916–17	Allies attack industrial targets
1917 May	First major daylight raid by Gothas
1917 Sept	First night attacks by Gothas and raids by Staaken (Giant) aircraft
1918	Introduction of British Handley Page 0/400 bomber
1919	RAF bombers in action in Afghanistan and Somaliland
1935	US B-17 (Flying Fortress) test-flights
1937 Apr 26	Destruction of Guernica by German bombers
1939	US B-24 (Liberator) test-flights
1940	London blitzed
1940 May 15	First RAF night raid (the Ruhr)
1942 Mar	Allied Gee radio navigation system deployed
1942 Mar	First widespread use of incendiary bombs in RAF raid on Lübeck
1942 May 30–31	"Thousand Bomber" night raid on Cologne
1942 Aug 17	First US daylight precision raid (Rouen)
1942 Sept	US B-29 (Superfortress) test-flights
1942 Dec	Oboe radio bomb-aiming system introduced
early 1943	H₂S terrain radar introduced
1943 July	Fire-storm raid on Hamburg
mid-1943	"Window" chaff in use
1943 Aug 17	First Schweinfurt daylight raid: heavy US losses
1943 Oct 14	Second Schweinfurt daylight raid: heavy US losses
1944 Feb	"Big Week": mass British and American raids on Germany
early 1944	Increasing range of escort fighters enables Allies to mount distant raids
1944 Nov	Major B-29 high-explosive bombing campaign against Japan begins
1945 Feb	Fire-storm raid on Dresden
1945 Feb 25	Incendiary campaign against Japan begins
1945 Aug	Nuclear attacks on Japan
1965–68	Vietnam: Operation Rolling Thunder
1972	Vietnam: Operations Linebacker I and II

The development of strategic bombing

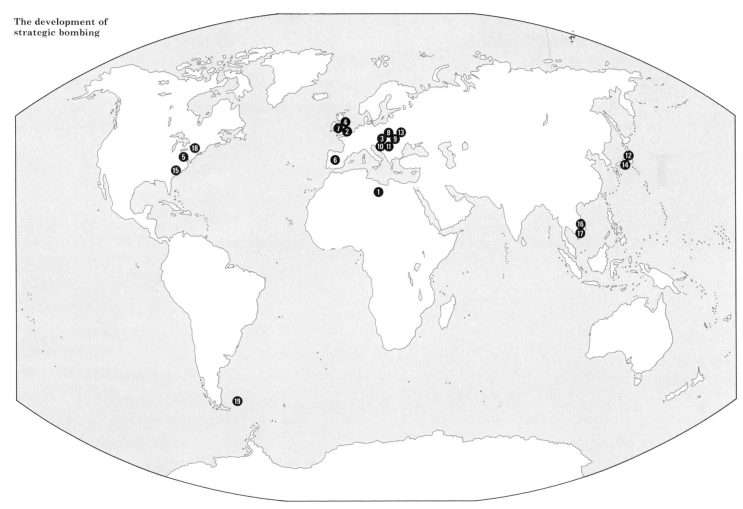

1. Italians bomb Turks in Libya, 1911
2. First strategic bombing campaign: Germany on England, 1915–18
3. Initial allied campaign against Germany, 1916–18
4. Handley Page v 1500: first purpose-built strategic bomber, 1918
5. Boeing B-17 Flying Fortress: self-defending bomber, 1935
6. Condor Legion in Spain, 1936–9
7. German "blitz" on Britain, 1940–4
8. RAF night bombing of Germany, 1939–42
9. 1,000 bomber raid on Cologne, 1942
10. Combined Bomber Offensive against Germany, 1943–5
11. Arado Ar 234 jet bomber developed by the Germans, 1943
12. American long-range bombing campaign against Japan, 1944–5
13. Grand Slam, 22,000lb bomb dropped by RAF, 1945
14. Atomic bombs on Nagasaki and Hiroshima, August 1945
15. B-36 inter-continental bomber, 1947
16. In-flight refuelling used by B-52s in Vietnam, 1965–72
17. Laser and TV-guided "smart bombs" in Vietnam, 1972
18. B-1 swing-wing bomber, 1974
19. RAF long-range bombing of Falklands from Ascension Island, 1982

bombing was originated by the RFC during the battle of Neuve-Chapelle (March 1915) when British machines attacked "German strategic centres".

The Royal Naval Air Service (RNAS), on the other hand, claimed that it had initiated strategic bombing in the opening months of the war, when its planes flew from Dunkirk to attack Zeppelin sheds on the Rhine. Following the delivery of Sopwith 1½-Strutter bombers to the RNAS early in 1916, plans were advanced to raid steel plants at Essen and Düsseldorf – the first industrial targets to be seriously considered. As a result, Captain Elder RN went to Luxeuil-les-Bains, near Belfort, to form No 3 Wing of "long-distance bombing machines operating against blast furnaces and munition factories in Alsace-Lorraine". During the winter of 1916–17, this British formation, together with the French 4th *Groupe de Bombardement*, attacked an impressive array of industrial targets. Apart from estimated direct damage, it was believed that the Germans had withdrawn aircraft from the front to re-inforce the searchlights, balloons and anti-aircraft guns grouped around sensitive industrial complexes. However, Sir Douglas Haig concluded that "it would . . . appear highly improbable that the output [of steel] had been seriously affected", and pressure for the deployment of aeroplanes in visible support of the trenches caused the disbandment of No 3 Wing in spring 1917.

"A most desirable thing"

Nevertheless, whatever the shortcomings of the Luxeuil experiment, the British War Cabinet believed that "a long-range offensive is in itself a most desirable thing". Thus, in October 1917, a second British bombing unit (41st Wing) was formed at Ochey, within comfortable range of German industrial and urban centres such as Koblenz, Mannheim and Cologne. Allegedly successful attacks during the winter months led to 41st Wing expanding into VIII Brigade early in 1918; by October 1918, 30 operational squadrons were planned "to attack, with as large a force as is available, the big industrial centres of the Rhine and in its vicinity". For this purpose in June 1918, Trenchard took command of an Independent Bombing Force (IBF) en-hancement of VIII Brigade. In a period of five months its bombers dropped 500 tons of HE, which *The Times* declared "extremely effective and profitable in actual results". Certainly, the IBF did achieve some striking results: one bomb on the Burbach works at Saarbrücken caused 400,000 marks worth of damage; 1,650lb bombs dropped on a small arms factory at Kaiserslautern caused damage estimated at 500,000 marks. The British Secretary of State for Air became "strongly convinced of the effectiveness of long-range bombing on German morale and on limitation of industrial effort". Production in Germany certainly suf-fered as a result of Allied bombing; for example, in one iron works at Völklingen, between September 1916 and Novem-ber 1918, work stopped 327 times with a resulting loss of 30,680 tons of production.

Nominally, although never in practice, Trenchard as-sumed command of an Inter-Allied Bombing Force during the closing weeks of the war. Its creation underlined the perceived importance of strategic bombing, as did the formation in September 1918 of No 27 Group RAF (the Royal Air Force [RAF] having come into being on April 1 1918, when the RFC and RNAS were merged). No 27 Group,

based in Norfolk, was equipped with the new four-engined Handley Page HP v 1500. This was the first purpose-built strategic bomber, but it did not see action before the armistice. By 1918, the early 20lb bombs had been replaced by 250lb and even 1,650lb bombs; incendiaries were in use and gas bombs were projected. Significantly, in view of its later adherence to strategic bombing theory, the United States Army Air Service launched a bombing offensive in June 1918 against industrial targets near Metz. Extending their range in the ensuing weeks, American bombers carried out 220 raids against 15 German cities, claiming to have inflicted damage worth 205 million marks, with 641 people killed and 1,262 injured.

During the course of World War I, strategic bombing had evolved on the Allied side from a combination of practice and theory. At first, the British and French sought to relieve pressure on the front line and to neutralize the

Handley Page v 1500, the first purpose-built strategic bomber.

During the night of January 19–20 1915, two airships bombed East Anglia causing 20 casualties. Only isolated minor raids then occurred until June, when attacks on London produced extensive damage and loss of life. By August, a mere air-raid warning was enough to create panic and stop industrial production throughout eastern and southeastern England. Anti-aircraft guns and aeroplanes proved "powerless"; only winter storms halted the airships. Raids restarted early in 1916 and continued intermittently until August 1918. But, by then, better defences had greatly reduced their effect.

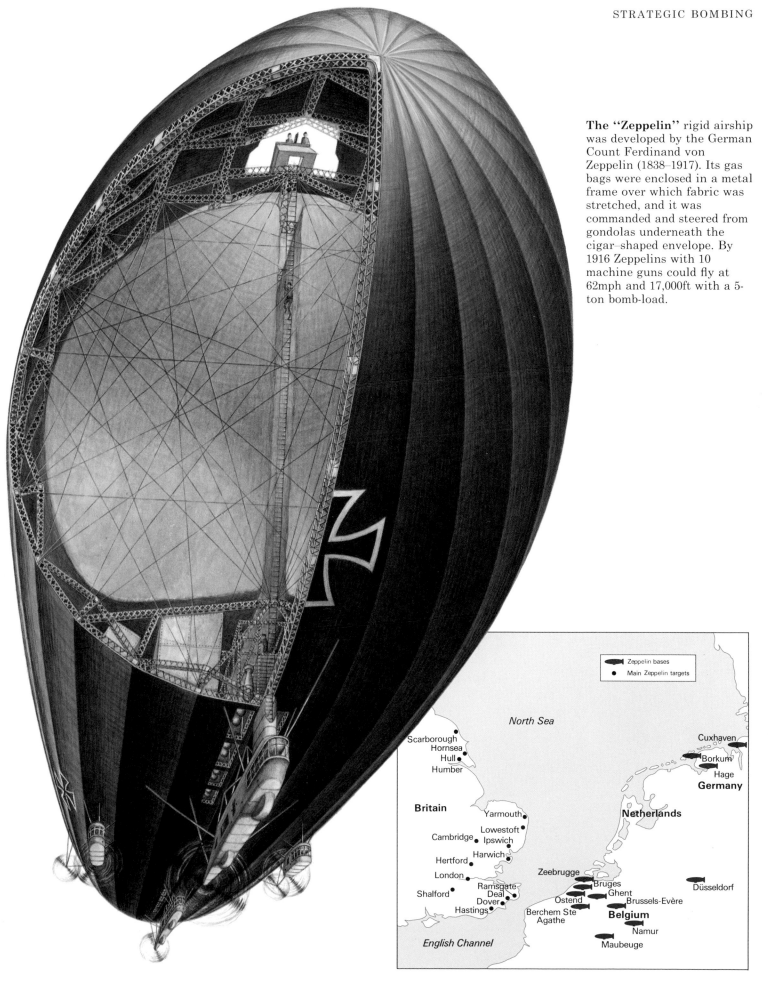

The "Zeppelin" rigid airship was developed by the German Count Ferdinand von Zeppelin (1838–1917). Its gas bags were enclosed in a metal frame over which fabric was stretched, and it was commanded and steered from gondolas underneath the cigar–shaped envelope. By 1916 Zeppelins with 10 machine guns could fly at 62mph and 17,000ft with a 5-ton bomb-load.

Zeppelin bases

Main Zeppelin targets

North Sea

Scarborough
Hornsea
Hull
Humber

Cuxhaven
Borkum
Hage
Germany

Britain

Yarmouth
Lowestoft
Cambridge
Ipswich
Hertford
Harwich
London
Ramsgate
Shalford
Deal
Dover
Hastings

Netherlands

Zeebrugge
Bruges
Ghent
Ostend
Brussels-Evère
Berchem Ste
Agathe
Belgium
Namur
Maubeuge

Düsseldorf

English Channel

Zeppelin menace by attacking airship sheds. As the ranges of aeroplanes improved, commanders became more ambitious in their choice of targets. Then their thoughts turned seriously to the destruction of the enemy's industrial base: the Germans had already pointed to attacks on urban concentrations as potential morale breakers. In 1916 the French drew up a list of German factories to be bombed (the first detailed plans for a strategic bombing campaign), revising it in 1917 in accordance with increased experience as well as technological advances.

An aerial knockout blow

Early in 1918, Sir Henry Norman officially proposed bombing attacks on six German towns to reduce them "to virtual collective ruins", ensuring that the rest of Germany would "approach a state of panic", that the enemy's military strength would be "largely paralyzed", and that "victory in the war would be in sight". He argued that such "bombing offensives on a really large scale" could produce "a quick and final victory". Norman's theory was not then put to the test; but the idea of an aerial knockout blow would foster much debate in the years to come. Of more lasting significance, perhaps, was Trenchard's contention that: "The moral effect of bombing stands undoubtedly to the material effect in a proportion of 20 to 1." The British Air Ministry claimed that "the effect, both morally and materially, of the raids on German territory . . . can hardly be over-estimated". On the contrary, as was the case with German raids on Britain, it certainly can. But the impact of air raids on the home front in both countries should not be entirely discounted. They helped, at the very least, to create a lasting fear of aerial bombing against civil populations.

Since the effect of aerial attack on civilian morale has been touched upon, it is worth noting also the effects of operational service on the airmen themselves. In 1914, when its military implications were largely unrealized, flying meant excitement and adventure: hence the men of the Royal Flying Corps crossed the Channel "gaily as to a dance". Two years later, the Canadian flyer J B Brophy could still eagerly await action; in 1917 an infantry officer, Lieutenant F M F West, fretted to join the aerial fray. In action, however, losses mounted. In 1916 pilots (admittedly not all engaged exclusively in bombing operations) lasted for an average period of three weeks at the front: Brophy's squadron suffered 19 casualties in five months – a high loss rate at that time. Of 55 British and French bombers that set off for Oberndorf in October 1916, 9 failed to return. During August 1918 the British Independent Bombing Force lost

---

". . . It is foolish to disparage the powers of the German aviator, for doing so must necessarily belittle the efforts of our own brave boys, whose duty it is to fight them. The marvellous fight which Voss put up against my formation will ever leave in my mind a most profound admiration for him, and the other instances which I have witnessed of the skill and bravery of German pilots give me cause to acknowledge that the German aviators, as a whole, are worthy of the very best which the Allies can find to combat them . . ."

Major James Byford McCudden VC, World War I

---

21 aircraft. These losses had their effect. Brophy noted that, in two months, six pilots were returned to England "as the result of nervous strain", which together with casualties incurred during the same period created a 150

Aerial reconnaissance – to identify troop movements and dispositions – was a major reason for development of air forces in World War I. At first, aircrew reported visual observations, then progressively hand-held and specially fitted cameras were introduced (like the one *left*, photographed in August 1918). Pictures were used to identify strategic bombing targets and to assess the damage caused by raids. During World War II and later, high-level Photographic Reconnaisance Unit aircraft operated similarly over enemy territory.

Marshal of the Royal Air Force Lord Trenchard (1873–1956) was Commander of the Royal Flying Corps in France in August 1915. He built a reputation for the forceful use of air power in support of the troops during the battles of the Somme, Arras, Passchendaele and Cambrai. He was also responsible for the deployment of bombers in long-range attacks against enemy targets beyond the battlefield, which he claimed marked the origin of strategic bombing.

A gruff, dominating personality (universally known as "Boom"), he originally resisted the establishment of a separate air force, but finally agreed to become the first Chief of the Air Staff of the new Service. In April 1918 he headed the British Independent Bombing Force set up specifically for the long-range bombing of Germany, and emerged from World War I convinced of the importance of strategic bombing.

As Chief of the Air Staff in 1919, he defended the RAF against attempts by the two Services to replace it with their own air arms. At his prompting, RAF bombers were used for "imperial policing", principally in the Middle East, and he insisted that a strong bomber force would deter potential enemies from attacking the homeland. His popular title of "father of the Royal Air Force" is inaccurate, since he had little to do with its birth and opposed its establishment in 1917, but he will always be associated with the successful struggle to retain an independent air force between wars.

bombers should make up two-thirds of its Home Defence Force – and strategic bombing, implicitly or explicitly, gradually became synonymous with the maintenance of an independent RAF. It did not seem extravagant for Stanley Baldwin (then a Cabinet Minister) to state in 1932 that "the bomber will always get through", for he was reflecting the views of influential theorists like the Italian Guilio Douhet and practical airmen like Trenchard and the American Brigadier General William Mitchell.

Developments in the USA

In the United States, Mitchell held that "the advent of air power, which can go straight to the vital centers and entirely neutralize or destroy them has put a completely new complexion on the older system of war"; while a senior Air Corps officer wrote: "I am convinced that the four-motored bombing plane is the weapon of hope for this nation." In 1935 test flights of the prototype B-17 (Flying Fortress) commenced, giving rise to the belief that a self-defending bomber – capable of flying at altitudes beyond the reach of anti-aircraft fire, while bombing accurately through the Norden high-level bomb sight – had become a reality. Four years later the B-24 (Liberator) prototype began test flights. The B-17 and the B-24 would carry out the bulk of American strategic bombing in Europe during World War II.

Meanwhile, the RAF had laid down specifications for four-engined bombers in 1936, although the first of these, the Short Stirling, would not become operational until 1941. With the death in 1936 of General Walther Wever, the first chief of staff of the new German air force (Luftwaffe), the impetus for four-engined bombers in Germany all but vanished, and the experience of the Luftwaffe's Condor

---

percent turnover in that one squadron. "I'm glad I don't develop nerves", he wrote, and welcomed an impending rest "before the whole squadron develops nerves". Increasing pride in the well-publicized (if exaggerated) bombing achievements of the RFC, RNAS, RAF, and especially the Independent Bombing Force, reinforced by the appeal of charismatic commanders like Trenchard, combined to minimize the effect of individual morale problems. Furthermore, in 1918, as in 1914, flying remained an adventure to the uninitiated.

The RAF fight for survival

In Britain, immediately after the end of World War I, the recently formed RAF had to face up to the other two services, which were determined to break it up in order to regain their own air arms. Evidence of successful bombing helped the RAF's fight to maintain its independent status. On May 24 1919, during the Third Afghan War, a Handley Page v 1500 bombed Kabul: four days later the Amir admitted that this had caused "great excitement and panic" and sued for peace. Trenchard, now Chief of the Air Staff, congratulated the bomber pilot on winning the war virtually single-handed. In the same year, RAF aircraft bombed rebels to defeat in Somaliland. Three years later the RAF assumed military control in Iraq: this policy of "imperial policing", whereby bombers rather than ground troops maintained law and order, nurtured belief in the decisive nature of bombing. In 1923 the RAF argued that

The German Condor Legion during the Spanish Civil War.

Legion in the Spanish Civil War gave the misleading impression that twin-engined aircraft were not only effective bombers but also could safely fly unescorted by day. The destruction of Guernica by the Condor Legion on April 26 1937 reinforced belief in the awesome effect of strategic bombing; but that target was undefended and the resulting carnage was more a testimony to inaccuracy than a practical demonstration of strategic theory. Elsewhere, in the Soviet Union, for example, the interwar years witnessed scant attention to heavy bombers, although the Italians developed three-engined bombers, and in Italy, as in Germany, lip service was paid to Douhet's advocacy of the effectiveness of massed air raids. Whatever the effectiveness of air forces, however, a universal fear of bombing had been created: in Britain, elaborate preparations were made to deal with the expected number of dead and injured caused by pre-emptive strikes before or immediately after a declaration of war.

The RAF went to war with nine types of bomber: none was four-engined and only two-thirds of them were able to reach the industrial targets in the Ruhr from England. At the time of Pearl Harbor, in December 1941, the United States had 159 four-engined bombers in service. Yet both nations had faith in the self-defending bombers that could fly by day to attack pinpoint targets.

The Luftwaffe's campaign against Britain

In 1939 the Luftwaffe could muster 1,505 bombers, of which 335 were Ju 87 Stuka dive bombers. Its campaign against Britain, which at first aimed to neutralize shipping, ports and airfields and then spread to attacks on urban concentrations, showed how vulnerable daylight bombers were, even with fighter escort, to single-engined monoplanes like the Spitfire and Hurricane. On August 15 1940, 16 out of 115 attacking bombers were lost, along with 7 German fighters: by the end of the month, the Luftwaffe was using as many as 65 fighters to protect a force of 15 bombers. In September 1940 the night "blitz" on London began: for 67 consecutive nights an average of 200 bombers each night pounded the British capital, assisted by a succession of increasingly sophisticated navigational aids and by special squadrons that flew ahead to illuminate the

target with incendiaries. London suffered heavy damage, with many of its inhabitants forced to live semi-permanently in underground shelters or to evacuate their homes and move to safer areas. Nor was the German bombing campaign confined to London. On November 14 1940, 503 tons of HE and 881 incendiary canisters were dropped during a 10-hour raid on Coventry. The city suffered 1,419

casualties. During the winter of 1940–41 heavy raids were mounted on other provincial towns, including Portsmouth and Southampton, and in February 1941, 1,200 sorties were flown in a night campaign which continued right through into May.

British scientists learnt how to counter the German *Knickebein*, *X-Gerät* and *Y-Gerät* navigational aids based upon radio beams, and in May 1941 night fighters ac-

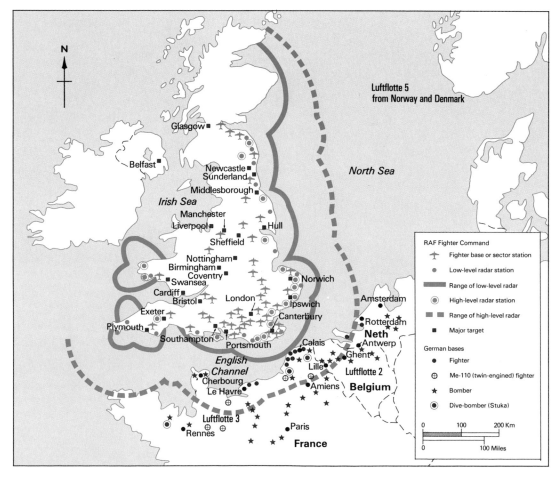

The German bombing attacks on England relied on the twin-engined Heinkel HE-111, Junkers Ju-88 and Dornier Do-17, with ranges of 1,000 – 1,500 miles. Daylight raids between July and September 1940 concentrated mainly on southeastern England, which was defended by No.11 Group at Uxbridge supported by a control system incorporating the coastal radar chain. Night raids, from September 1940 to May 1941, flew from bases in France, the Low Countries, Denmark and Norway, and brought more of the British Isles under attack. But British scientists learnt to counter German navigational aids; and night-fighters, equipped with A1 airborne radar, soon inflicted heavy losses.

counted for 96 bombers. Isolated raids continued, but after May 1941 no concerted strategic bombing campaign was pursued. However, in April–June 1942 the "Baedecker" attacks (so called from the well-known series of tourist guidebooks) hit cathedral cities like Canterbury and Exeter; in January–April 1944, in the so-called "little

---

"[A] harsh, rank, raw smell . . . came from the torn, wounded, dismembered houses; from the gritty dust of dissolved brickwork, masonry and joinery. But there was more to it than that. For several hours there was an acrid overtone from the high explosive which the bomb itself had contained; a fiery constituent of the smell. Almost invariably, too, there was the mean little stink of domestic gas, seeping up from broken pipes and leads. But the whole of the smell was greater than the sum of its parts. It was the smell of violent death itself."

John Strachey, London 1940

---

blitz", London and other major cities were attacked again. But by that time the British night fighter equipped with airborne radar had clearly established its superiority. No longer could the Luftwaffe's attempt to use strategic bombing to undermine the British war effort succeed. Nor had it ever seemed likely to do so.

Civilian morale

An enemy bombing campaign, as was shown in London in 1940, tends to act as a cohesive rather than a diffusive element. Throughout the Luftwaffe's raids on Britain,

60,595 civilians were killed and 85,000 injured. In London and elsewhere residents forsook their homes to sleep in tube (subway) stations, or in such areas as Epping Forest or Chislehurst Caves. Some more or less permanently evacuated to the countryside: the New Forest and Meon Valley provided refuge from Portsmouth and Southampton. Those staying in threatened towns faced not only physical danger but debilitating loss of sleep: one London survey showed 31 percent claiming not to have slept at all during air attacks, and only 15 percent claiming to have had more than six hours' sleep. Bombing did cause terror: "It's me nerves, they're all used up" . . . "I can't tell you the dread, every night it's worse". But morale was by no means at breaking-point: no undue lawlessness occurred and in London only 2 percent of psychological admissions to hospital were connected with air raids.

Nor was industry seriously affected. Lord Beaverbrook, Minister of Aircraft Production, told General H H Arnold, Commanding General of the US Army Air Forces, that the major German raids had reduced industrial production in the areas attacked by up to one-third. But that interruption was short-lived. After the Coventry raid in 1940, all the 21 factories damaged were again in virtually full production within five days. Birmingham's small arms production showed a temporary 5 percent drop as a result of air attack, but the output of Browning guns more than trebled in December 1940–March 1941 during heavy raids. The effect of Allied air raids on the enemy was not dissimilar. Despite the heavy bombing raids on Germany, munitions output increased twofold between 1942 and 1944, as did aircraft production. German civilians proved no less adept at

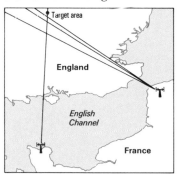

The X-Gerät system was used by German pathfinders against Coventry. Aircraft followed one radio direction-beam, which was intercepted by two others 30 miles and 12 miles from the target. A third beam crossed the target area.

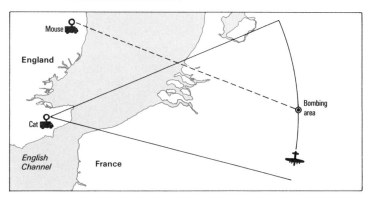

Oboe was first tested in December 1941 but did not reach service until 12 months later. The system relied upon signals from two ground stations in England. Flying along an arc, the aircraft kept a constant distance from one (CAT) and released its bombs or marker over a target on the signal of the second (MOUSE). Oboe's range was limited due to the curvature of the earth.

Air-raid damage was heavy in some areas of northern England. Merseyside for example, suffered 16 major attacks by a total of 2,417 bombers, Manchester was subjected to three full-scale "blitz" assaults, and in Hull only 6,000 out of 93,000 homes escaped damage. The morning after usually meant searching for trapped victims and clearing rubble.

survival than Britons and no widespread breakdown of morale took place.

Theorists may argue that the Luftwaffe did not have the means to mount an effective attack: its bombers lacked defensive firepower and were unable to carry adequate bomb-loads; its fighters lacked the range properly to protect the bombers. The RAF discovered these short-comings in its own operations in December 1939 when, even though flying in formation for mutual defence, Wellingtons and Blenheims incurred 50 percent losses. Hopes of a daylight strategic bombing campaign against German industry (outlined in pre-war plans) evaporated. Night bombing proved relatively less costly in aircraft, but it was inaccurate: attacks on German synthetic oil plants during 1940–41 were abandoned because of an average bombing error of 1,000yds. In August 1941, a study of aerial photographs showed that only one-third of RAF crews credited with successful attacks had bombed within five miles of their targets – in the Ruhr, that figure was one-tenth. Aiming at urban centres rather than specific industrial targets became policy in July 1941. However, because of the increased efficiency of German defences, aircraft losses rose: on November 7–8 1941, 37 out of 400 bombers were shot down. Although 1,046 bombers struck at Cologne in the celebrated "Thousand Bomber Raid" on May 30 1942 with minimal loss, during August–October 1942 RAF Bomber Command suffered a 5.3 percent loss rate. Improved radio aids to navigation (Gee and Oboe) were supplemented by H_2S radar-assisted bombing equipment, and the target-marking Pathfinder Force was formed in August 1942.

The Avro Lancaster, the most successful night bomber of WWII.

Night raids

In December 1943 the Chief of Air Staff, Sir Charles Portal, held that "if it had been tactically possible to concentrate a quarter of our total bombs dropped on Germany" on a specific group of targets associated with oil, ball bearings or aero engines "the war would by now have been won". However, experience showed that such a precise concentration was impossible. Air Chief Marshal Sir Arthur Harris, Air Officer Commanding-in-Chief Bomber Command, recognized that night "area bombing" was the only feasible tactic for the RAF. Yet in November 1943–March 1944, in 35 night raids, RAF Bomber Command lost 1,047

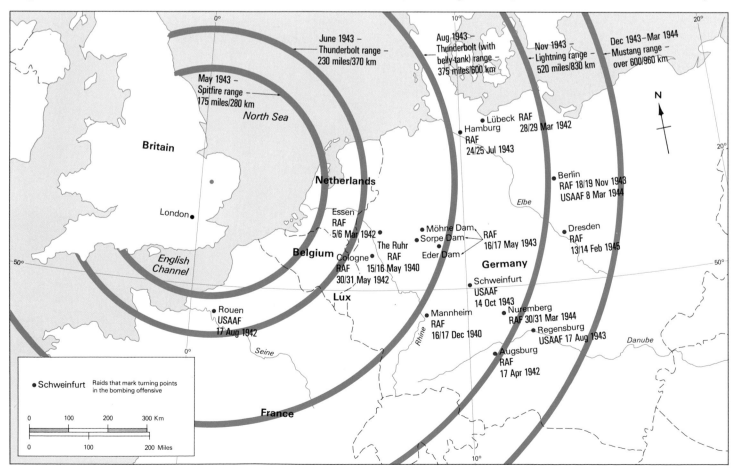

bombers and a further 1,682 were damaged. Each bomber carried seven men. During that period, German war production continued to rise as factories were dispersed outside major urban centres.

Harris bitterly but unsuccessfully resisted the plan to divert heavy bombers from Germany to attack targets in connection with the D-Day landings. However, following the success of the Normandy invasion in June 1944, the full might of the Allied Combined Bomber Offensive fell on Germany: enemy fighter bases were overrun; more accurate low-level marking techniques were developed; Mosquito long-range night-fighters appeared; the number of heavy bombers available vastly increased. In January–May 1945 the RAF mounted a total of 67,483 bombing sorties for the loss of 608 aircraft, dropping 181,740 tons of bombs. In the closing months of the war, strategic bombing undoubtedly played a major part in achieving victory for the Allies; but that was far from the pre-war conception of either a swift knock-out blow or a decisive independent role for air power.

During World War II, RAF Bomber Command lost 55,573 aircrew and 1,570 ground staff killed. In view of the heavy losses, it is not surprising that there were some problems concerning morale. The development of coordinated fighter, flak and radar defences caused mounting RAF Bomber Command casualties over Germany. During 1942, 1,235 bombers were lost and 2,839 damaged; while of 795 bombers dispatched to Nuremberg on March 30 1944, 95 aircraft (with 665 men) were lost and 71 damaged. That there was immense strain on aircrew is self-evident. "Waverers" (those with obvious morale problems) were

Marshal of the Royal Air Force Sir Arthur Harris (1892–1986) was decorated during World War I for organizing night-flying operations against Zeppelins over Britain. Harris held operational and staff appointments between the wars, including "imperial policing" duties in the Middle East.

In July 1941 he became Deputy Chief of the Air Staff at the Air Ministry, where he pressed for 4,000 heavy bombers as "an entirely practicable method of beating the enemy." Appointed head of RAF Bomber Command in 1942, he launched a number of devastating attacks on German cities such as Cologne (the spectacular "Thousand Bomber" raid) and Lübeck, partly to fend off demands for redeployment of heavy bombers both to other theatres abroad and also in a naval support role at home.

Stubbornly adhering to night area bombing despite rising losses, he deplored attempts to change the RAF to daylight bombing with concentration on specific industrial targets, and tried to persuade the Americans to abandon daylight attacks. He and General Spaatz attempted to prevent the diversion of heavy bombers from German strategic targets to tactical ones in northern France in preparation for D-Day, but they were overruled by General Eisenhower, backed by President Roosevelt.

Major targets in the strategic bombing offensive against Germany, and the combat radius of escort fighters throughout the campaign (*left*). Heavy losses had proved the "self-protecting bomber" a mistaken concept, but not until the advent of the P-38 Lightning and P-51 Mustang fighters in later 1943 could bombers be protected at all times during deep-penetration raids.

The first night-fighters (*above*) hunted independently in target areas, as "Window" chaff effectively blinded their radar systems. In 1944, improved airborne radar and firepower brought temporarily better results. The more sophisticated night-fighter, the Me-262 jet, had not reached service at the war's end.

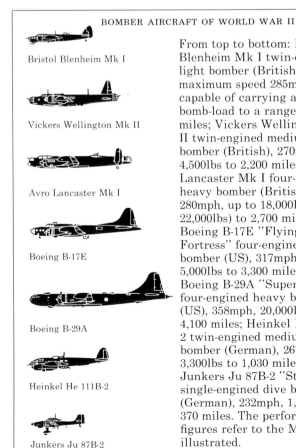

BOMBER AIRCRAFT OF WORLD WAR II

Bristol Blenheim Mk I

Vickers Wellington Mk II

Avro Lancaster Mk I

Boeing B-17E

Boeing B-29A

Heinkel He 111B-2

Junkers Ju 87B-2

From top to bottom: Bristol Blenheim Mk I twin-engined light bomber (British), maximum speed 285mph, capable of carrying a 1,000lb bomb-load to a range of 1,125 miles; Vickers Wellington Mk II twin-engined medium bomber (British), 270mph, 4,500lbs to 2,200 miles; Avro Lancaster Mk I four-engined heavy bomber (British), 280mph, up to 18,000lbs (later 22,000lbs) to 2,700 miles; Boeing B-17E "Flying Fortress" four-engined heavy bomber (US), 317mph, 5,000lbs to 3,300 miles; Boeing B-29A "Superfortress" four-engined heavy bomber (US), 358mph, 20,000lbs to 4,100 miles; Heinkel He 111B-2 twin-engined medium bomber (German), 267mph, 3,300lbs to 1,030 miles; Junkers Ju 87B-2 "Stuka" single-engined dive bomber (German), 232mph, 1,100lbs to 370 miles. The performance figures refer to the Marks illustrated.

usually identified and rested or medically treated, with a complete recovery rate of approximately 50 percent. But strict discipline had also to be maintained. Group Captain Leonard Cheshire VC, a veteran of 100 bomber operations, has recalled: "I was ruthless with 'moral fibre cases' . . . there was the worry that one really frightened man could affect others around him. There was no time to be as compassionate as I would have liked to be."

The USAAF in Britain

The US Army Air Forces faced similar problems. United States 8th Army Air Force aircrew flying from bases in Britain were committed to daylight missions in the face of vicious fighter activity and faced the additional irritation of an unfamiliar, damp, misty climate which frequently caused "scrubs" after briefing: crews were briefed for the combined raid on Schweinfurt and Regensburg in August 1943 seven days before they eventually flew. Like the RAF, the 8th AAF incurred heavy losses: six operations in the period July 24–30 1943 cost 8.5 percent of the attacking force. "Creep-back" (bombs dropped short), unsatisfactorily early returns from missions and last-minute claims of sickness before missions were symptoms of loss of nerves that were not confined to the RAF. Yet loyalty to fellow crew members, unit cohesion created by service routine and pride in concealing personal fears from others combined to ensure the overall preservation of morale. Too much emphasis can be placed upon those who cracked – the overwhelming majority did not.

The US Army Air Forces had entered World War II wedded to the idea of high-level, precision bombing in daylight by self-defending bombers – encouraged by excellent results during practice that led to claims that American bombers could place their bombs "in a pickle barrel". Combat conditions would soon destroy that fantasy. However, in 1941 the B-17 and B-24 seemed the ideal weapons for their envisaged role. During the spring and autumn of 1942, the 8th AAF was built up in England with its Bomber Command under Brigadier General Ira Eaker. Arnold, in the United States, fervently believed in the value of strategic bombing against Germany and intended to concentrate enough heavy bombers in England to make a cross-Channel invasion possible in 1942. However, the first USAAF B-17 did not reach England until July 1942. Once the idea of a cross-Channel invasion in 1942 had been abandoned, Arnold's target of 3,640 combat aircraft in England by April 1943 became unrealistic. In August 1942 he complained: "The air plan to win the war is completely abandoned in turning [heavy bombers] to Torch [the invasion of Northwest Africa in November 1942] and the South Pacific." The German aircraft industry, he insisted, must be attacked "relentlessly". Three months later, Winston Churchill noted the "pitifully small" American bombing results, and at the Casablanca Conference in January 1943 Churchill exerted strong pressure on the Americans to join the RAF in night bombing, which Eaker (by then commanding the 8th AAF) successfully resisted.

In February 1943, partly because of adverse weather and

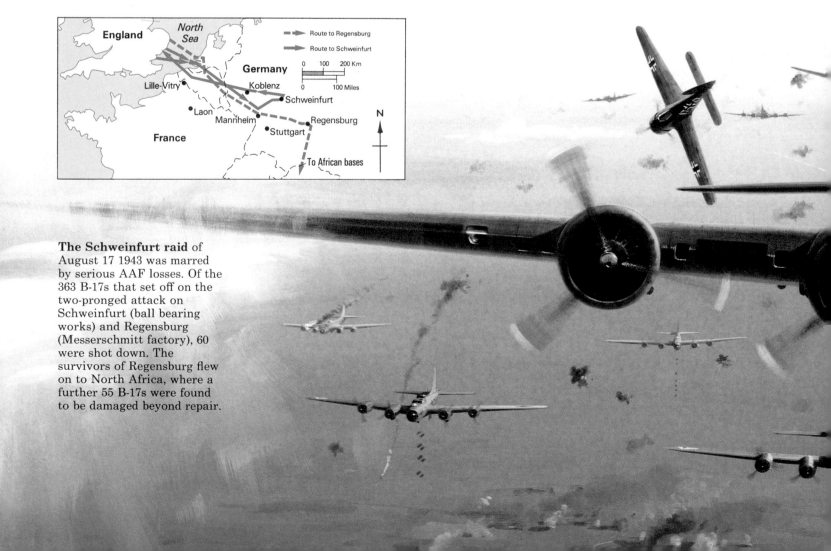

The Schweinfurt raid of August 17 1943 was marred by serious AAF losses. Of the 363 B-17s that set off on the two-pronged attack on Schweinfurt (ball bearing works) and Regensburg (Messerschmitt factory), 60 were shot down. The survivors of Regensburg flew on to North Africa, where a further 55 B-17s were found to be damaged beyond repair.

"In England monitors heard the German pilots gathering from all over France and Germany to ambush our homeward flight . . . All across Germany, Holland and Belgium the terrible landscape of burning planes unrolled beneath us. It seemed that we were littering Europe with our dead. We endured this awesome spectacle while we suffered a desperate chill. The cartridge cases were filling our nose compartment up to our ankles . . .

At last we came to the blessed sight of soaring Thunderbolts above the Channel coast . . . I felt exhaustion creeping in beneath the excitement like death beneath a fever."

Lieutenant Elmer Bandiner, the Schweinfurt raid, August 17 1943

partly because of the lack of available bombers, the 8th AAF was able to mount only two missions over Germany; and shortly afterwards Eaker admitted that "after sixteen months in the war we are not yet able to despatch more than 123 bombers towards an enemy target". The Pointblank Directive of June 1943 set target objectives for the British and American bombers that were aimed at weakening German fighter strength – the blueprint for the Combined Bomber Offensive agreed at Casablanca six months earlier.

Bombers without fighters
However, a major problem was that lack of fighter range

"simply ensured the bombers' safe delivery into the hands of the wolves", as escorts left the bombers at the German border and picked up the survivors again on the return leg. Without escort fighters, bombers had to fight their way to and from the target against a determined array of enemy aircraft equipped with cannon, machine guns and rockets and carrying out a bewildering variety of attacks from front, side, below and above at much faster speeds than any bomber could hope to achieve. Over most targets, predicted box barrages (anti-aircraft fire determined by the predicted course of the raiders) had to be negotiated, and elsewhere flak pockets (concentrations of anti-aircraft weapons) threatened the aircraft that strayed too close to them. Colonel (from 1943, Brigadier General) Curtis LeMay developed a complex system of stacking squadrons into defensive "combat boxes", thus providing a measure of mutual protection for the bombers; but heavy losses still occurred. In the combined mission against Regensburg and Schweinfurt on August 17 1943, 60 bombers (600 men) were lost, and in that week 8th AAF Bomber Command lost 148 aircraft. For ten days after the Schweinfurt disaster, the 8th AAF could not launch an operation. Inevitably, British pressure to switch to night bombing was renewed. On September 6, 45 B-17s were lost over Stuttgart; in operations on October 8–10 a further 88 bombers went down. A return trip to Schweinfurt by 291 bombers on October 14 cost another 60 B-17s – and unescorted, deep-penetration operations had to be suspended. Strategic bombing theory apparently could not be realized in practice.

However, improved navigational and bombing aids such as Oboe and H_2X (the American version of H_2S), together with the appearance of effective long-range fighters, at last gave hope: by 1944, P-51 Mustangs with additional tanks could fly 850 miles. During "Big Week" in February 1944, 8th AAF and 15th AAF in Italy sent an average of 1,000 bombers over Germany each day; and during March 1944 the loss rate of 3.5 percent compared most favourably with the 9.1 percent of October 1943. Arnold was convinced that the poor showing of the Luftwaffe on D-Day reflected the success gained by raids against aircraft factories and the losses sustained by German fighters attacking the bomber formations. In the eleven months that remained of the European War after June 1944, the 8th AAF, in association with RAF Bomber Command, concentrated their efforts on the German oil industry and transportation system. They succeeded in bringing both virtually to a standstill. During the course of World War II, the 8th AAF dropped a total of 701,300 tons of bombs (excluding 27,556,978 4lb incendiaries) on enemy targets. Casualties were heavy – including those among escorting fighters, the 8th AAF lost 43,742 killed and 1,923 seriously wounded.

Attack on Ploesti

From 1942, attempts had been made on industrial targets by bombers flying from the Middle East. The Ploesti oil refineries in Romania, which supplied high-octane aviation fuel, were one such target. A minor effort in June 1942 achieved little and served only to provoke strengthening of the enemy defences. On August 1 1943, 178 B-24s of

The North American P-51 Mustang was transformed into an excellent long-range, high-level escort fighter after a timely engine upgrade. It was used to defend US daylight bombing missions in 1944 and 1945.

the 9th AAF set off on a 2,700-mile round trip from Benghazi in Libya to try again: 46 bombers were lost, 8 landed and were interned in Turkey and 58 were badly damaged. The impact on Ploesti was visually spectacular but actually negligible.

The coordinated campaign in Europe

Bombers of the 9th and 12th AAFs stationed in the

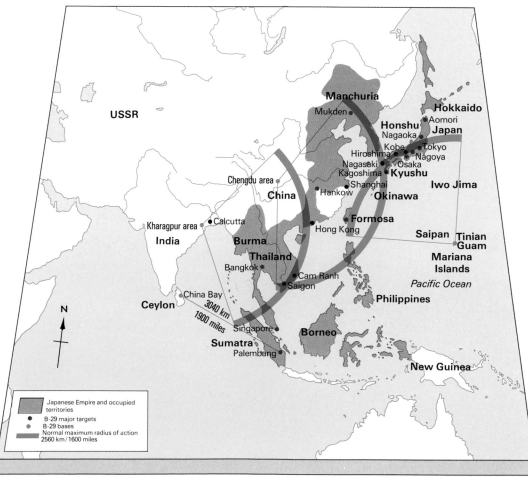

The strategic offensive against Japan from October 1944 involved B-29 crews flying a 3,000 mile return flight from the Marianas to Japan, with an additional dog-leg avoiding enemy-held Iwo Jima until March 1945. Initial bombing from high level proved disappointing, but a devastating low-level, night incendiary attack on Tokyo on March 9/10 1945, led to similar raids elsewhere prior to the dropping of the atomic bombs.

Mediterranean for the most part attacked airfields, convoys and other military targets closely associated with the North African battlefield or the invasions of Sicily and the Italian peninsula. In December 1943, after the 15th AAF became established in Italy, General Spaatz declared: "We don't need the Army to win this war. All we need you for is to capture airfields" – a throwback to the unreasonable pre-war belief in the decisive nature of bombing. Using airfields at Foggia, the 15th AAF bombed industrial targets in northern Italy, Austria, southern Germany and France, notably taking part in the February 1944 "Big Week" with the 8th AAF. In April 1944, it hit strategic targets in Bavaria, Austria and Hungary, beyond the range of bombers in England. In April–August 1944, 5,287 heavy bomber sorties were flown against Ploesti and 12,890 tons of bombs were dropped for the loss of 2,432 men (43 percent became prisoners of war), but the effect on oil production was not marked. Crude and synthetic oil plants in Czechoslovakia, Silesia and Poland were also attacked: 10,000 tons of bombs were dropped on four Silesian and Polish targets in the latter half of 1944, causing an estimated 80 percent reduction in output. During 1944, the 15th AAF struck 620 strategic targets with 192,000 tons of bombs. From the beginning of 1944, when Spaatz was appointed Commanding General of all American Strategic Air Forces in the theatre, Europe was seen as a strategic whole, with US bombers from the Middle East, Italy and England operating a coordinated campaign in close cooperation with the RAF.

Another strategic bombing campaign was carried out in the Far East, modestly signalled in April 1942 when 16 twin-engined B-25s were launched against Japan from the carrier USS *Hornet*. American admirals believed the bombing of Germany "of no use", and called for a maximum concentration of heavy bombers (although not necessarily all engaged in strategic bombing) in the Pacific. During 1943 bomber bases were established in China, with the aim of attacking Japan – ultimately with the B-29 Superfortress, the prototype of which flew in September 1942.

Air offensive against Japan

In April 1944 the first B-29s reached the Far East, and on June 15 a force of B-29s flew from China to attack a major steel works in Yawata, Kyushu. A single bomb hit a power house some three-quarters of a mile from the main target during a night mission in which 8 B-29s out of 75 scheduled to take part were lost and only 47 reached Yawata. However, the main campaign against Japan had begun, and Major General LeMay soon arrived in India to head 20th Bomber Command. Insisting on tight formation flying, he switched operations to daylight, high-level, precision bombing against industrial targets in Manchuria and Formosa.

On November 24 1944, 80 B-29s of 21st Bomber Command, commanded by Brigadier General Haywood Hansell and based on Saipan in the Mariana Islands, attacked Tokyo in a high-level precision raid, in accordance with a Joint Chiefs of Staff instruction to bomb urban areas and targets connected with the Japanese aircraft industry. In navigational terms alone, the task was daunting. The Marianas

The B-29A Superfortress, with a range of up to 4,100 miles, first bombed Bangkok from India in June 1944, then Manchuria, Formosa and Japan from China. Between November 1944 and August 1945, these aircraft attacked Japan from the Marianas, culminating in the two atomic raids.

Tokyo after B-29 incendiary attacks in August 1945. General LeMay opted for low-level, massed incendiary night raids in February 1945, after strong winds and cloud had proved high-level bombing inaccurate. The aim was to destroy countless scattered "shadow factories" in the major Japanese cities. In 10 March days, 32 sq miles of 4 cities were devastated.

were some 1,450 miles from Japan – far in excess of distances flown in Europe – across water and enemy-held islands. Moreover, thick cloud and high winds were usual over the targets, and these factors greatly affected bombing accuracy. In January 1945 LeMay took command of 21st Bomber Command on Guam and gradually increased the number and strength of raids on Japan. But results were still disappointing, so LeMay determined on a revolutionary tactic for the US AAF: low-level, area bombing at night. He reasoned that the poor conditions over Japan necessitated visual bombing from around 5,000–8,000ft – and that, if the B-29s flew unarmed, they could fly faster and carry a bigger bomb load. Since Japanese anti-aircraft guns were geared to high-level flying and Japanese night fighter defence was weak, speed would be the B-29s' main defence. The overwhelming bulk of the bomb-loads would consist of incendiaries, with flares to act as markers. On March 10 1945, 285 B-29s bombed Tokyo, where LeMay believed that 15.8 sq miles were "crammed" with shadow factories: two nights later, 1,950 tons of incendiaries fell on Nagoya, followed by similar raids on Osaka, Kobe, Oita, Omura and other towns. Within ten days, B-29s destroyed 32 sq miles of Japan's four most important cities – the first steps in a concentrated four-and-a-half month campaign. During the first two weeks of August, 25,000 tons of incendiaries and HE were dropped on 14 cities: on August 2 alone, 855 B-29s attacked 6 cities with 6,632 tons of bombs.

At the Potsdam Conference in July 1945, Arnold maintained that the B-29s could bomb Japan into submission. On August 6 and 9, by dropping atomic bombs on Hiro-

shima and Nagasaki, that is arguably what they did. But before the atomic bombs were dropped, the more conventional strategic bombing campaign had levelled an estimated 2,333,000 Japanese houses along with a large proportion of the business and industrial facilities in 60 cities. In the process some 240,000 Japanese were killed and 300,000 injured. Yet, perhaps significantly, as in England

---

"Beyond the zone of utter death in which nothing remained alive, houses collapsed in a swirl of bricks and girders. Up to about three miles from the center of the explosion, lightly built houses were flattened as though they had been built of cardboard. Those who were inside were either killed or, managing to extricate themselves by some miracle, found themselves surrounded by fire. And the few who succeeded in making their way to safety generally died about twenty days later from the delayed effects of the deadly gamma rays."
Japanese journalist, on the bombing of Hiroshima, August 6 1945

---

and Germany there was no general collapse in morale.

Postwar bombing campaigns
Since 1945, no strategic bombing campaign has been mounted. In Korea, Allied bombers attacked airfields and troop concentrations, partly because of the limited political aims of that conflict. No attempt was made systematically to bomb strategic targets, for these were mainly in Manchuria north of the Yalu river.

Vietnam saw similar constraints. Operation Rolling

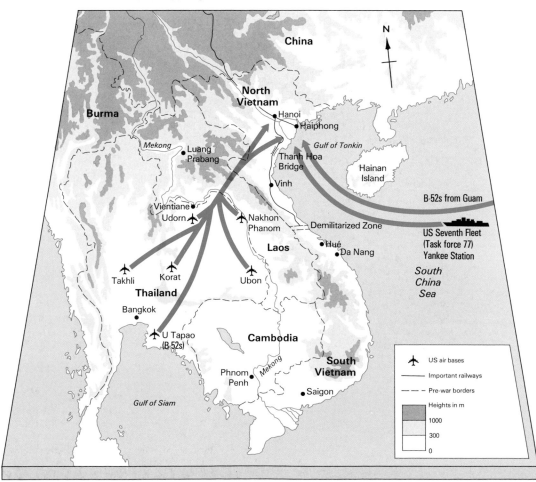

To bomb Viet Cong positions as part of Rolling Thunder in June 1965, B-52Fs flew the 5,500 mile return flight from Guam, which required in-flight refuelling from KC-135 tankers. More B-52s began operating from Thailand in April 1967 to reduce sortie time by one-third. For the Linebacker operation, B-52Ds, with their heavier bomb-loads, swelled the strategic bomber force engaged to over 200.

Thunder (1965–68) comprised "a program of measured and limited air action . . . against selected military targets" such as ammunition depots, storage dumps and lines of communication, specifically avoiding urban concentrations for fear of provoking direct Chinese intervention. Operation Linebacker I (May–October 1972), in response to the Tet (Easter) invasion of South Vietnam, again concentrated on lines of communication: in particular, Remotely Piloted Missiles (laser- and TV-guided "smart" bombs) were used to destroy important bridges. A massive 12-day assault by B-52s during Operation Linebacker II (December 1972) certainly extended the range of targets, but neither this operation nor its predecessors could accurately be classed as strategic bombing. The bulk of the bombing fell to F-4 Phantom and F-105 Thunderchief fighter-bombers; even the B-52s predominantly attacked tactical and interdiction targets. As the US Secretary of Defense, Robert F McNamara, observed, North Vietnam was an agricultural country with few significant industrial targets. Political considerations forbade the bombing of towns to undermine civilian morale: thus, the main objective of bombing was to prevent supplies from China and the Soviet Union reaching North Vietnam and being transported thence, down the "Ho Chi Minh Trail", to front-line troops in the South.

No strategic bombing campaign can be launched against an enemy who is devoid of important urban concentrations or vulnerable industries. In an age when modern industrial powers shelter under a nuclear umbrella, the possibility of a strategic bombing campaign in the future is debatable.

A Republic F-105 Thunderchief (*above*) unloads its bombs over Vietnam. The major burden of the bombing campaigns of 1965–73 was carried by the fighter-bombers of the US Air Force and US Navy: the F-105 and the McDonnell Douglas F-4 Phantom. The "Thud", as the F-105 was popularly known, is generally accepted to be one of the best combat aircraft of all time. The F-105 D/G had a maximum speed of 1,480mph and could carry up to 8,000lbs of bombs internally, with a further 6,000lb of ordnance – in Vietnam, often laser- or TV-guided "smart" bombs for precision strikes on bridges and similar targets; or "Shrike" anti-radar missiles – on external pylons.

A Boeing B-52 Stratofortress (*above*) drops its 750lb General Purpose bombs over a target in Vietnam. During the Vietnam War, the B-52 strategic bomber was used in a tactical role. In 1965–73, according to US official figures, B-52s flew around 126,500 sorties and dropped more than 2,600,000 tons of "iron bombs" on targets in southeast Asia. Thirty-one B-52s were lost during these operations.

The Rockwell International B-1 (*top*), a four-seat, low-level bomber, first flew in 1974. Opponents claimed that the manned bomber was obsolete, and the project was cancelled in 1977 but revived in 1981. The B-1, powered by four 30,000lb afterburning turbofans, has a maximum high-altitude speed of 1,451mph and carries up to 32 nuclear missiles or 115,000lbs of free-fall bombs. It is smaller than the B-52 but with twice the weapon load.

249

chapter eighteen

Guerrilla war

Charles Townshend

Guerrilla fighting is the prototype of all war. It was the original form of armed conflict and is still its archetype. It reveals the fundamentals of motivation, morale, discipline and tactics in a strikingly elemental way. Guerrilla warfare has always gone on, and there is no reason to suppose that it will ever disappear, as long as there are sharp disparities of power in the world. For guerrilla conflict is above all the response of the weak to those who are – for the moment at least – stronger.

Ancient and archetypal as it may be, guerrilla warfare has nevertheless changed dramatically in modern times. Technology has transformed its military potential and ideology has transformed its political significance. The precise point at which this transformation began is not easy to identify, but it certainly took place in the period of the French wars of 1792–1815. The fact that the Spanish word *guerrilla* (little war) entered the international vocabulary during the French attempt to subdue Spain after 1807 is significant. It helped to shape a distinct concept: before then, the distinctions between regular and irregular warfare were less clearly visible.

Irregularity, or the negation of regularity, is the key to the guerrilla concept. What has been called guerrilla, partisan (in Spain the term *partidos* was used more often than *guerrilleros*) or people's war, has been marked by its contrast with the dominant form of regular, conventional warfare. As that form has altered, so the perceived contrast has shifted. The pressure to regularize military forces seems to be inherent in the nature of states. The process was never consistent – there was a vast hiatus after the fall of the Roman empire – but in Europe, by the 17th century, the emergence of "absolute" monarchies was linked with the creation of permanent armies.

In the 18th century, these armies began to become professional in the modern sense of the term and to develop increasingly rigid operational conventions. Their strict discipline and their formalized fighting style restricted their flexibility, so that alongside them the more fluid art of *la petite guerre* developed, in which sharpshooters (*Jäger* or *tirailleurs*) with superior individual skills in weapon handling, fieldcraft and initiative protected the flanks of the regular formations. This form of "little war" was the subject of limited military interest – although Gerhard Scharnhorst (1755–1813), the Prussian army reformer, wrote a short essay on it – and it was always seen as contingent to regular operations. When, as in the American War of Independence, independent partisan operations appeared, they were still seen as marginal to the outcome of the war.

From the French Revolution to the Boer War

The idea of independent guerrilla activity, which might exploit the rigidities of regular forces and turn their apparent strengths into weaknesses, was hard to accept. Even though Revolutionary France flouted the 18th-century conventions, the main principles of international (or inter-state) war survived largely intact. As codified by Clausewitz, they boiled down to the concentration of force and the use of speed to achieve victory through a decisive battle. The rules of victory in battle might have changed somewhat – before 1792 armies often gave way with good grace, or prudence, when outmanoeuvred; afterwards they did not – but the notion of the decisive battle remained central. The fate of states would hang on the fate of their armies; victory in battle would produce peace. The state-

centred logic of this was obvious. Armies were expensive and war was disruptive of economic growth; its rapid termination was increasingly important for both social and political reasons.

Yet even Revolutionary France experienced opponents who ignored this logic: who refused to accept defeat as final, who would not offer battle at all, or who just kept on fighting, by one means or another, without any reasonable prospect of victory. The first such opponents appeared within France itself, where insurgency in the Vendée and *Chouannerie* in Brittany persisted through much of the 1790s. Other such struggles by irregular forces against Napoleon were made in the Tyrol and in Russia, as well as Spain. In dealing with them, the Republican and Imperial armies were hampered both by their own inflexibility and by political constraints. For these were ideological

The map indicates the locations of major occurrences of irregular warfare from 1793 (when the inhabitants of the Vendée department of western France and the *Chouans* in Brittany rose against the Revolutionary government) to the present day, when acts of terror inspired by beliefs that span the entire political spectrum trouble most countries of the world. Guerrilla activities may be associated with conventional warfare, in which partisans act as auxiliaries to regular forces, as did the Confederate irregular cavalry of the American Civil War and the *francs-tireurs* of the Franco-Prussian War; with nationalist aspirations, as in so many former colonial territories during the 20th century; or with socio-political ends, as in the case of such modern terrorist organizations as the West German Baader-Meinhof Group (Red Army Faction), on the extreme left, or, on the right, the anti-Communist "Contras", whose controversial campaign against the Sandanista government of Nicaragua has been waged from Honduras with American aid.

Indian resistan
wars 1850–98

USA

Confederate Shenandoah
operations 1864–5

Guatemala

Nicaragua
Contras
campaign
1980–

Cuba
Cuban revolu
1956–9

Cuban-inspired
guerrilla mover
1959–68

Colombia

Venezuela

Peru

Bolivia

Che Guevara
(d. 1967)

Brazil

Argentina   Uruguay

Monteneros   Tupamaros

Urban guerrillas
from late 1960s

—TIME CHART—

1808–14	Spain: anti-French insurrection
1864–65	USA: Confederate partisan operations
1899–1902	Boer War: British use blockhouses, barbed wire, concentration camps and scorched-earth tactics
1916–18	Arab rebellion: weakness of regular armies against insurgents shown
1919–21	IRA secures virtual independence for Ireland: demonstrates vulnerability of modern states to organized resistance
from 1930	Mao formulates rural guerrilla theories
1934–35	Long March of the Chinese Communists to Shensi illustrates importance of liberated base areas
1936	Civil war begins in Spain
1941–45	Yugoslav partisan campaign shows advantages given by rough and inaccessible terrain
1943–45	Guerrilla action in Yugoslavia and France weakens Nazi control
from 1945–54	Vietminh practising political assassination as insurgency tactic
1948–60	Malaya: resettlement successfully used by British to deny insurgents support
1952	Mau Mau insurgency begins in Kenya
1954	French defeat at Dien Bien Phu: effective manipulation of international opinion by Vietminh
1954–61	Algeria: French use helicopters, electronic defence lines and systematic torture in anti-nationalist campaign
1955	Grivas mounts guerrilla operations against British in Cyprus
1956–59	Cuban revolution: Guevara formulates *foco* guerrilla theory
1959–68	Cuban-inspired rural guerrilla movements across Latin America
late 1960s	Urban guerrilla movements in Argentina, Brazil and Uruguay
from 1970s	Urban terrorists in Western Europe

Communist and Muslim insurgency 1968
Philippines

China
Communist revolution 1927–49

Anti-colonial war 1946–54
Vietnam

East Timor
Nationalist rebellion 1976–

IRA campaign 1957–
N Ireland

IRA nationalist insurgency 1919–21
Eire

USSR

Partisan resistance 1939–45

Vendée uprising 1793
France

Malaysia (Malaya)

Basque (ETA) separatist insurgency 1975–

Yugoslavia

Anti-French insurrection 1808–14
Spain

Indonesia

Communist insurgency 1944–9
Greece

Communist insurgency 1948–60, 1975–8

PLO campaign 1968
Cyprus
Anti-British war 1952–9

Afghanistan

Nationalist rebellion 1945–9

Algeria
FLN war 1954–61

Israel

Arab revolt in Palestine 1935–9

Zionist-Arab struggle 1945–8

Hejaz revolt 1916–18

Mujahadin rebellion 1979–

Sri Lanka (Ceylon)
Tamil separatist insurrection 1983–

Eritrea
Secessionist war 1961–

Ethiopia

Kenya
Mau Mau insurrection 1952–9

Cameroon
Nationalist revolt 1955–60

Malagasy Republic
Anti-colonial uprising 1947–8

Angola
Nationalist rebellion 1961–74

Zimbabwe (Rhodesia)
Nationalist rebellion 1965–80

Namibia
SWAPO nationalist rebellion 1970–

South Africa

Mozambique
FRELIMO nationalist rebellion 1964–74

Boer War 1899–1902

ANC insurgency 1950–

struggles. As the most able of the Republic's generals in the Vendée, Louis Lazare Hoche, told his government, "For the twentieth time I repeat, if you do not grant religious tolerance, you must give up the idea of peace." And although the Republic gained a measure of military success through techniques which were to remain fundamental to later counterinsurgency efforts – the subdivision (*quadrillage*) and intensive search (*ratissage*) of insurgent territory – its record would give little confidence to other states faced by such challenges in the future.

Clausewitz and "Arming the People"

The potential significance of "people's war" was recognized by Clausewitz. Although the few pages of his chapter "Arming the People" are dwarfed by the rest of his great book *Vom Kriege* (On War), they contain some of his most perceptive and prophetic passages. Clausewitz said that a people's war in Europe was "a phenomenon of the nineteenth century" – a direct product of the French Revolution. France had politicized its masses, had given them a stake in the nation, and thus had released a military force that could only be matched by other similarly liberated nations. Such national mobilization was spiritual as much as organizational, and it was the spirit of the nation which could inspire the people to fight to the last.

Clausewitz remained cautious about the ability of popular forces to dictate the result of a major war. He saw them as auxiliaries to, not substitutes for, the regular armies of the state. (He thought always in terms of inter-state war, of course, not of internal revolution.) As a good professional soldier, he doubted whether civilians could continue to fight in face of the savage reprisals which the enemy might (as the French certainly did) inflict upon them. But he proposed that irregular operations should be deliberately restricted in scale, so as to ensure that the partisans would not suffer any serious defeat. This was a crucial point. The guerrillas should be "nebulous and elusive", slowly sapping the physical and psychological powers of the enemy rather than going for a knockout blow.

In a tentative way, and clearly in some awe at the prospect, Clausewitz identified the elements of what Mao Tse-tung's would call "protracted war". The central one was political motivation, which Clausewitz, again thinking in terms of inter-state war, assumed to be patriotism. Given the will to risk life for this cause, guerrillas also needed to believe that their resistance could be effective. Military skills would have to be gained, but the basic technology would be mental rather than material: not new tactics, but a reversal of conventional assumptions, putting diffusion in place of concentration, protraction in place of speed, harassment in place of battle. As Mao wrote: "The ability to run away is the very characteristic of the guerrilla."

Urban guerrilla warfare

These criteria for guerrilla warfare called for a kind of mental revolution; thus, it took time for their basic truth to be realized. In the century between the French Revolution and World War I, such ideas remained peripheral to the business of state-based war. Social revolutionaries, who were later to become the most committed advocates of people's war, ignored its potential. Obsessed with the urban industrial proletariat, they visualized revolution as

Giuseppe Garibaldi (1807–82) (*above*) became a popular hero throughout much of Europe after 1860, when he led his "Redshirts" in the struggle against Austrian domination of Italy.

The term "guerrilla" emerged from Napoleonic warfare in Spain. Francisco Goya (1746–1828) observed the cruelties practised by both sides, and in "Massacre of the Innocents" (*left*) he portrayed the fate of civilians caught up in irregular war.

a short, sharp catharsis fought out across city barricades. Even national liberation movements, without any prejudice against the peasantry, based their hopes on sudden, popular, urban uprisings. In Italy, for example, the patriot Giuseppe Mazzini (1805–72) wrote a substantial essay on "the conduct of guerrilla bands", but did little to put his theories into practice in his campaigns against Austrian and French forces. His compatriot Giuseppe Garibaldi (1807–82), hero of the patriotic campaigns of 1848 and 1860, was esteemed as a guerrilla leader, but his military operations were mostly conventional in the extreme. Following France's defeat by the Prussians at Sedan and the fall of the Second Empire, in September 1870, the French leader Léon Gambetta (1838–82) expended prodigious energy in an attempt to wage *guerre à outrance* in the spirit of 1793. But although guerrilla resistance flared up in many areas of occupied northern France, nearly all the Provisional Government's military resources were concentrated on the creation of new regular forces. When they were pushed into premature offensives by the urgent need to relieve Paris, the failure of these armies was practically a foregone conclusion. And although final defeat was staved off for five months after the elimination of the Imperial armies, guerrilla action played only a small part. Yet in France in 1870–71, the abyss of truly total war, in which combatants could no longer be distinguished from noncombatants, began to open up, and a fraction of the resources thrown away on the doomed armies of Orléans might, if devoted to a people's war, have sufficed to engulf even von Moltke and Bismarck.

Only outside Europe were guerrilla campaigns frequent

DISCIPLINE AND DEMOCRACY

Beyond doubt the basic motivation of most guerrilla fighters has been patriotism. Since Mao's time this has often been enveloped in a sophisticated Marxist vision of social revolution. Mao saw ideology as the force which would counteract the centrifugal tendency of guerrilla action. He insisted on rigorous discipline and high standards of conduct towards the people, as laid down in his famous "Three Rules and Eight Remarks". In his view, "With guerrillas, a system of compulsion is ineffective"; only when discipline is self-imposed "is the soldier able to understand completely why he fights and why he must obey".

Giap went on to refine further the paradoxical symbiosis of discipline and democracy. He reasoned thus: "Can we say that guerrilla warfare does not require severe discipline? Of course not. It is true that it asks the commander to allow each unit a certain margin of initiative in order to undertake every positive action that it might think opportune. But a centralized leadership always proved to be necessary. He who speaks of the army speaks of strict discipline." (*People's War, People's Army.*)

The Marxist idea of "democratic centralism" was applied with some success in Vietminh cells, party committees, and not least in "plenary meetings of fighting units" – as during the Dien Bien Phu battle. The Vietminh structured the individualistic "buddy system" into three-man cells which provided both psychological support for their members and a high degree of overall cohesion. "The facts have proved that the more democracy is respected within the units, the more unity will be strengthened, discipline raised, and orders carried out. The combativeness of the army will thereby be all the greater."

Defeat in the Franco-Prussian War was followed by insurrection in Paris where government troops and Communards battled in the streets and many buildings were burned (*left*).

An enduring legend of "Bloody Week" (May 21–28 1871) was the Commune's 8,000-strong brigade of *pétroleuses*, female incendiaries (*above*) tasked with the destruction of public buildings.

during the 19th century. In the American Civil War, for example, the operations of Confederate partisans in the Shenandoah Valley may have delayed Union victory by nearly a year. But guerrilla activity too easily degenerated into criminal banditry; and even where the partisans were of fair probity – as with Morgan, Mosby and Forrest's raiders – their military achievements carried an awesome price in the Union forces' destruction of their farmlands. The Confederate generalissimo Robert E Lee conventionally deplored "the whole system as an unmixed evil". In South America, several colonies successfully waged irregular campaigns for independence. Similar campaigns were fought in Haiti, Cuba and throughout Africa and Asia, where the inhabitants of the territories which would become Nigeria, Rhodesia, Algeria, Madagascar, Burma and Indochina struggled in vain to fend off European control. In these places, guerrilla warfare was not adopted for theoretical reasons: it was the traditional mode of warfare – and the only means available. Europeans consequently discounted its significance, concluding that while "savage warfare" might present problems, it offered no lessons to the civilized.

At the turn of the century, one war alone in Africa momentarily broke down this complacency. In 1899–1902, the *Tweede Vryheidsorlog* ("Second War of Independence") fought by the Boer republics of Transvaal and the Orange Free State gave the British, as Kipling said, "no end of a lesson". The British overcame the more-or-less regular Boer armies in about six months, but for nearly two years thereafter they faced an irregular campaign fought by resourceful commanders like Jacobus De La Rey, Louis

Botha, Christiaan De Wet and Jan Smuts. These "bitter-enders" maintained resistance through hit-and-run attacks by variably-sized *commandos*; that is, groups under command, but not under formal discipline. The ability of the commandos to elude or shake off British pursuit became legendary. Their capacity to cause real damage to British forces or installations was, perhaps, exaggerated.

In one sense, the Second Boer War was certainly a people's war. The unique way of life of the *burghers* was

---

"After three days of steady progress we were back on the open plains, within sight of the Bloemfontein–Johannesberg railway line, and we scouted round for a suitable crossing. This was becoming more and more difficult to find, now that the English were perfecting their block-house system. There were working parties of soldiers dotted along the railway track, engaged in putting up these block-houses, but we had no difficulty in galloping across the metals, despite a fairly heavy rifle-fire, and, having safely negotiated the line, we rode on. Towards dark we came on a field of maize into which we turned our weary horses, and there we spent the night."

Denys Reitz, commando in the Boer War

---

directly transmitted into the organization and fighting style of the commandos. The war demonstrated the Boers' deep will to independence, which even in defeat proved enough to thwart Sir Alfred Milner's dream of turning all southern Africa into a British colony. Yet the Boer struggle did not exactly prefigure the people's wars of the

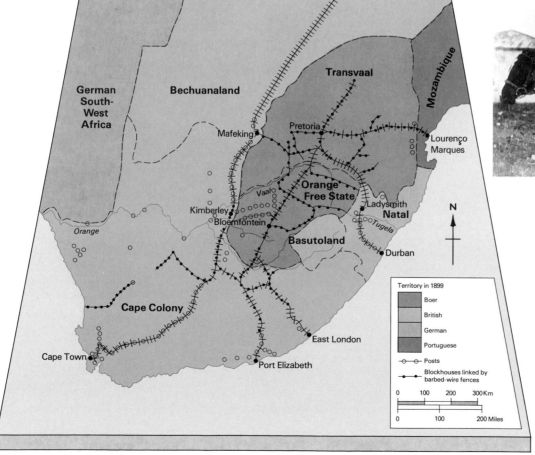

Horse, rifle, cartridge-belt and basic rations in a saddle-bag (*above*): this was the field equipment of the typical member of a Boer *commando*, whose mobility enabled them to tie down superior numbers of British regulars.

The answer to the commandos' hit-and-run tactics was to restrict their mobility by sectioning off the country with a chain of some 8,000 telephonically-linked blockhouses and 4,000 miles of barbed wire (*left*). The typical blockhouse was garrisoned by 7–10 men.

German South-West Africa

Bechuanaland

Transvaal

Mozambique

Mafeking

Pretoria

Lourenço Marques

Vaal

Orange Free State

Kimberley

Ladysmith

Natal

Bloemfontein

Orange

Tugela

Basutoland

Durban

N

Cape Colony

East London

Cape Town

Port Elizabeth

Territory in 1899
Boer
British
German
Portuguese
Posts
Blockhouses linked by barbed-wire fences

0   100   200   300 Km

0   100   200 Miles

20th century. Through the British policy of devastating the Boer farmlands and interning the families of fighters in concentration camps, the commandos were prised apart from their people. The momentum of their campaign declined, albeit imperceptibly at times, and their numbers dwindled. Attempts to raise the Cape Dutch in rebellion failed. The ability of the commandos to keep up their resistance was due above all to their extraordinary physical endurance in the empty veldt. The British had at last to resort to the crude and expensive method of blockhouses and barbed-wire lines to counteract Boer mobility; slowly but inexorably, this policy succeeded.

The first comprehensive statement concerning modern guerrilla warfare came from the pen of T E Lawrence ("Lawrence of Arabia", 1888–1935). The contribution of the Arab revolt to the outcome of World War I in the Middle East has, like Lawrence's own contribution to the masterminding of that revolt, been a matter for dispute. What cannot be disputed is the brilliance with which Lawrence conveyed his central message: "Granted mobility, security, time, and doctrine, victory will rest with the insurgents." This was a large claim, but Lawrence reinforced it by showing the weaknesses which regular armies were bound to display in the face of insurgents who did not obey conventional military logic. Armies were fixed to their lines of communication, while irregulars might be (echoing Clausewitz) a vapour, everywhere and nowhere. The very notion of military regularity involved reducing troops to the lowest common denominator, so as to ensure that their performance would be predictable; irregulars, by contrast, could exploit their individual qualities.

Lawrence identified the vital means by which guerrilla forces could achieve "security" and convey their "doctrine" to their people. The first was accomplished through intelligence; the second through propaganda. Insurgents with public support, however passive, could – with care and skill – achieve a virtual monopoly of operational information. Their enemies would be fighting in the dark. The insurgents could plan their operations in complete security: a military dream. Time would be on their side, provided they kept their public support. This would be accomplished in part by the self-publicizing nature of successful attacks, and in part by intelligent publicity. "The printing press is the greatest weapon in the armoury of the modern commander." (Radio, of course, was to be something else again.)

Independence for Ireland

Lawrence's works provided a powerful recipe for insurgency, and they had an enormous impact on their readers. That impact was further magnified by the astonishing achievement of the minuscule Irish Volunteer Forces (or Irish Republican Army) in securing virtual independence for Ireland between 1919 and 1921. At the time when Lawrence's first article on the Arab revolt was published, the IRA was feeling its way towards an operational repertoire which would paralyze the British administration in Ireland. It began with a civil boycott of the police and progressed from small-scale raids to secure weapons to night attacks on police posts. By the spring of 1920 the police had retreated into central positions and had visibly lost control of the countryside. The legal system came to a

Refugees, Washing Clothes, Springfontein Camp.

To break morale and to remove civilian support from the Boer partisans, the British burned their farms and incarcerated their families in "concentration camps" (*above*), where primitive accommodation, inefficient administration and poor hygiene resulted in an estimated 20,000 deaths (perhaps one-fifth of the total intake) from disease. Many

children died in measles epidemics: the British politician Lloyd George foresaw "a barrier of dead children's bodies . . . between the British and Boer races".

Michael Collins (1890–1922), the Irish revolutionary leader, served with the Irish Volunteers who seized the General Post Office in Dublin on Easter Monday 1916. He condemned the Easter Rising as "bungled terribly" at organizational level, and on release from internment he began the reconstruction of the Volunteers, later the Irish Republican Army (IRA). Collins recognized the cumulative moral effect which many small local

attacks would have on the credibility of the British administration. His prime target was the police special (political) branch, and during 1918–19 he substantially paralyzed the British intelligence system. The most dramatic point in his intelligence war was "Bloody Sunday", November 21 1920, when his Active Service Unit ("the Squad") assassinated twelve British officers. Collins recognized the limits of guerrilla action in the Irish case and accepted the compromise Irish Free State, becoming a leading figure in the Provisional Government. The civil war waged by Republicans who rejected his sober estimate of the possibility of forcing Britain to recognize an Irish Republic was a tremendous blow to him, and he was trying to arrange a settlement when he was killed in an ambush in his native County Cork on August 22 1922.

near standstill, as witnesses and jurors were persuaded (with no great difficulty) to stay away from the courts. An underground republican government was built up. Local government bodies broke off their contacts with the British administration; civil resistance appeared in many forms, such as a boycott of British goods. In reprisal, the police were rearmed and reinforced by the para-military "Black and Tans", and new repressive legislation was introduced. The IRA responded by turning its activists into full-time fighters in Active Service Units – more widely known as "flying columns". These units tried to perfect what is, together with the raid, the quintessential guerrilla operation, the ambush.

Their successes were patchy. Many Irish counties never produced a viable flying column or seriously menaced any British forces. Many flying columns were inefficient, never developing the support among the civil population which would give them ascendancy in the intelligence field. However, a few became quite formidable professional units. Their triumphs were essential to the republican publicity organization, which had little trouble in persuading Irish people that British repressive measures were brutal and unjustifiable, but which had a larger task in persuading a sceptical world that the self-proclaimed Irish Republic was a truly independent nation. The Irish Free State achieved limited independence in 1922; civil strife between its supporters and hard-line republicans then set former guerrilla comrades at each other's throats.

Overall, the Irish guerrilla campaign was a remarkable success. It showed the vulnerability of modern states to widespread and finely-controlled internal resistance, and

it sent shock waves throughout the European colonial empires. But it also embodied some warning signs, which might also have been signalled in Lawrence's remark that "at length we developed an unconscious habit of never engaging the enemy at all". Clearly, always to refuse to engage the enemy in open conflict was a deliberate defiance of conventional military thinking – but it raised the question of how, if such a policy was followed, guerrilla action could defeat a strong and determined enemy. The Turks against whom Lawrence had campaigned had not been strong; the British in Ireland had not been determined. Would attrition always work on the side of the irregulars? This question has never been finally answered – but the answer returned in the 1930s by the most famous of all guerrilla theorists was a qualified negative.

The theories of Mao Tse-tung

In *Yu Chi Chan* (Guerrilla Warfare) Mao wrote: "The concept that guerrilla warfare is an end in itself and that guerrilla activities can be divorced from those of the regular forces is incorrect." Guerrilla action, he wrote, was a phase in a war of national liberation or social revolution. It could raise the weaker side from a position of inferiority to a position of superiority, but it must always be superseded by the transition to open battle.

For this reason Mao, unlike many guerrilla writers, stressed the need for base areas – which might have to be defended by positional warfare. These bases, or liberated areas, would demonstrate the new social system and would provide the manufacturing capacity without which guerrilla forces could never hope to transform themselves into

In Dublin, 1921, an "Auxie" – a member of the Auxiliary Division of the Royal Irish Constabulary (RIC) – reloads his revolver, watched by British regulars. The "Auxies", recruited from among British ex-officers and with a maximum strength of around 1,500, were distinguished from the more numerous "Black and Tans" by their dark-blue uniforms and Glengarry caps.

Thomas Edward Lawrence (1888–1935), seen here in Arab dress, identified "mobility, security, time and doctrine" as the major criteria for a successful guerrilla campaign. Lawrence's contribution to the revolt of the Arabs against the Turks in 1916–18, has been questioned: however, his contribution to the theory of guerrilla warfare was considerable.

GUERRILLA DISCIPLINE

Che Guevara put the moral qualities of the guerrilla fighter above the ideological. For him, a guerrilla was made by personal probity and endurance. Not because a man carries a rifle, a back pack, sleeps rough and is hunted by the police, can he be called a guerrilla. Such things are equally true of bandits. Rather, he wrote, "Each member of the guerrilla army, the people's army *par excellence*, must embody the qualities of the best of the world's soldiers. The army must observe strict discipline. The fact that the formalities of orthodox military life do not correspond to the guerrilla movement, the fact that there is no heel-clicking or snappy saluting, no kow-towing explanations to superior officers, does not, by any stretch of the imagination, mean that there is no discipline. Guerrilla discipline is within the individual, born of his profound conviction, of the need to obey his superior, not only so as to maintain the effectiveness of the armed group of which he is a part, but also to preserve his own life. Self-control is the operative force."

M.P.L.A. troops train in Angola.

a real army. Mao was constantly suspicious of the tendency of partisans to degenerate into "guerrillaism", anarchy and banditry. Ideology, especially Marxism, might bind together disparate local units to some extent, but organization was also of vital importance. And, in fact, the Communist forces under Mao and Chu Teh, even at their lowest ebb during the Long March of 1934–35, always had regular units in divisional strength at the core of their military organization.

At the same time, Mao opposed the Clausewitzian tendency of many of his colleagues to distrust the military value of the people and to rely only on regular forces, even for "guerrilla" action. Guerrilla bands must, of necessity, spring from a people in arms; the discipline and skill called forth by a guerrilla campaign would make it the "university of war". There would develop between fighters and society a symbiotic relationship which even Mao's vivid metaphor of "the fish and the water" fails to convey in its full complexity.

Guerrilla strategy fused with social revolution

Viewed purely as a guerrilla theorist, Mao may not appear strikingly original. Indeed, many of his operational precepts were drawn from one of the most ancient writers on war, Sun Tzu (active c. 500BC) – which simply underlines the archetypal nature of guerrilla war. Mao's true significance is that he fused guerrilla strategy with social revolution. Before him, as we have seen, socialists looked to urban rather than rural struggle. Guerrilla campaigns were fought for nationalist rather than socialist objectives. Mao converted the Chinese Communist Party (CCP)

Mao Tse-tung (1893–1976) (*left*) is seen in 1966 with Marshal Lin Piao (1908–71), who holds the "Little Red Book" – the collection of Mao's "Thoughts" which embodied much of his doctrine concerning irregular warfare.

A rest during the "Long March" (*right*), the 8,000-mile migration from Kuomintang-dominated Kiangsi to the safety of the mountains of Shensi, made by the Chinese Communist "route armies" in October 1934 – October 1935.

to his vision of a socialist peasant revolution – although not without considerable difficulty, and only after the CCP had suffered crushing setbacks at the hands of the Kuomintang (Nationalist) government in Shanghai and other cities in the 1920s. Guerrilla methods alone could generate such a revolution.

Mao argued that a class struggle was "purer" than national liberation war, and so more effective: "One class may be easily united and perhaps fight with great effect, whereas in a national revolutionary war, guerrilla units are faced with the problem of internal unification of different class groups." The CCP's experience does not wholly bear this out. Some historians of the Chinese civil wars have concluded that only the upsurge of patriotism that followed the violent Japanese intervention in 1937 brought mass support for the Communists. The popular participation which made possible their final victory after 1945 was, in this view, fired by national spirit rather than class consciousness. But in any event, that victory was a world-shaking event and was seen as a stupendous vindication of the guerrilla concept. A whole spate of people's wars in the post-World War II period would draw from it not solely inspiration, but a kind of blueprint for action.

Evidence on which to decide the question of whether guerrilla methods could succeed against a strong and determined enemy was building up. Japan was ruthless (but, as Mao reassured his people, not strong enough for the gigantic task of subjugating China); the Kuomintang was strong and vicious. In Europe, the Germans were certainly all three. During World War II, partisan resistance to German occupation flared up in several countries, assiduously fostered by Allied agents. Only in Yugoslavia was such resistance wholly successful; elsewhere guerrilla activities were, overall, comparatively ineffective in damaging the German war effort, as distinct from raising the morale of the occupied peoples. Even in Yugoslavia, it is unlikely that the Germans would have been induced to quit the country if they had not been going down in general defeat. The forces of Marshal Tito, the partisan leader, were impressive, totalling some 800,000 by 1945. But the survival of Tito's movement in the dark days of the fifth German anti-partisan campaign (May–June 1943) was due in part to Yugoslavia's geography – mountainous, primitive and only on the margin of the German campaign in the East. The commonsense fact that certain sorts of terrain ("rough and inaccessible", as Clausewitz had stipulated) favour partisan resistance was underlined by both the Yugoslav and Chinese experiences.

Vietnam

The conflict which confirmed guerrilla warfare as the dominant, and apparently all but infallible, mode of the age was the Vietnamese resistance to the French colonial power after 1945. Here, Mao's blueprint seemed to be applied faithfully and triumphantly. The striking differences between the circumstances of Vietnam and China served, if anything, to heighten the impression that the fusion of guerrilla strategy and communist ideology was unstoppable. Certainly, Mao's precepts were practised by the Vietminh under leaders like Ho Chi Minh, Vo Nguyen Giap and Truong Chinh. Equally certainly, the Vietminh went beyond Mao in building up an even more comprehen-

Yugoslav partisans of both sexes (*above*) on a training march during World War II. The Spanish resistance fighters of Napoleonic times most often called themselves *partidos* (partisans), and in the 20th century the term has come to be associated with paramilitary bands operating in enemy territory and/or in support of regular forces. Led by Marshal (Josip Broz) Tito (1892–1980), and making expert use of Yugoslavia's mountainous terrain, the Communist partisans in Yugoslavia waged the most effective guerrilla campaign of the war.

Vo Nguyen Giap (b.1912), Vietminh leader against the French in 1946–54 and North Vietnamese commander-in-chief during the Vietnam War of 1956–75, was a "politician and strategist" rather than a professional soldier. He was an active Marxist revolutionary from his teens, and in 1941, having served with Mao Tse-tung's guerrillas against the Kuominang and Japanese, he was a founder member of the Vietminh under Ho Chi Minh. Giap took increasing responsibility for military organization, forming the first regular Vietminh unit, an armed propaganda brigade, in December 1944. For the next ten years he had overall direction of military policy (obscuring the contribution of others, such as Truong Chinh). Seeing protracted guerrilla war as inevitable, Giap organized the three-tier structure which gave coherence to a varied campaign involving guerrilla, mobile and main forces. He was a brilliant instigator of guerrilla action, but saw it primarily as a means of bringing the people to participate in the revolution and of training recruits for regular divisions. He made serious mistakes as a result of his eagerness to begin an open offensive against the French, but he was willing to admit them and to learn from them. Giap's supreme achievement was the victory at Dien Bien Phu in 1954, the result of a sophisticated assessment of French weaknesses.

sive popular organization, and in casting a grim shadow into the future by its deliberate use of terror.

Terror is to some extent inseparable from irregular war. The competition in brutality between the Spanish *partidos* and the Napoleonic armies reached such an intensity that both sides acknowledged the need for some restraint. Few governments have been able, even when they have wished to do so, to prevent their armed forces from taking reprisals against civilians who shelter elusive partisans. Few guerrilla bands have been able to survive without inspiring fear as well as admiration among the people. The IRA, for instance, for all its public support, felt impelled to carry out a number of exemplary killings of "spies and informers", in attempts (not altogether successful) to stop the flow of information to the British forces. Che Guevara was to take the same view of the need for "revolutionary justice". However, the Vietminh was, perhaps, the first revolutionary movement to take terror beyond angry vengeance or self-preservation, beyond the millennialism of the early anarchists, to a consistent long-term policy of assassinating all who collaborated in any way with its enemies, the French.

The Vietminh held that the instigators of terror were the French themselves: such allegations and counter-allegations are the stuff of ideological war and the truth is usually hard to discover. In this case, several historians have concluded that terror was first unleashed by the French, under the auspices of the British, in Saigon in September 1945 – the matter is by no means academic, for it reveals the attitudes behind the French decision to reoccupy Indochina – and the Vietminh merely organized

its counter-terror more efficiently and systematically than the French. It would be a serious mistake to think, as the French did, and as governments so often profess to do, that the power of the insurgents was maintained only by terror. That kind of misunderstanding crippled all French – and later American – efforts to thwart the Vietminh.

The Vietminh was an agent of social revolution. It put its roots deep into the structure of ancient village communities, dislocated by French-imported agricultural capitalism. For this reason, in part, its strength was always greater in Tonkin (later North Vietnam) than in the south. It is hard to think of a more vivid illustration of the guerrilla method than the penetration or, as a French writer put it, the *pourrissement* (addling), of the Red River delta, the only part of Tonkin to remain under anything like effective French control. Throughout most of northern and central Vietnam, the Vietminh built a set of organizations through which peasants could make the revolutionary movement part of their lives. A three-tier military structure allowed forces to emerge first at local level; second at regional level ("mobile forces" rather than guerrillas proper); and finally at national level, where regular formations operated either in the northernmost mountains of the Viet Bac, or from sanctuaries across the Chinese border.

The Vietminh and nationalism

The Vietminh was unmistakably a Marxist organization, but it was also a national movement. Vietnamese nationalism had persisted in face of fierce French repression from the 1860s to the outbreak of World War II. The defeat of

---

GUERRILLA LOGISTICS

Much of the impact of guerrilla forces results from their capacity to disrupt the hefty logistic "tails" of regular armies and tie up many troops in defending lines of communication. Their own logistic systems can be far looser. Initially, partisans have often armed themselves by capturing weapons from the police or army. In some cases, as with the Boer *commandos*, they may start with their own weapons, but come to depend on captures of ammunition for resupply. The scale of their operations is regulated by supply, but, at this level of conflict, little equipment beyond arms, ammunition and explosives is needed.

Transition to larger-scale operations raises different problems. Heavy as well as light weapons, bulky equipment and food supplies are needed. Unless these come from abroad, the insurgents must create their own munitions factories to manufacture them, and make careful preparations for the defence of their food production areas.

Perhaps the greatest epic of supply in guerrilla warfare was the battle of Dien Bien Phu – itself a classic siege rather than a guerrilla combat – during which the Vietminh was able to maintain 50,000 troops in intense fighting for several weeks, hundreds of miles from its support bases. Countless peasants were mobilized as porters over the routes from Thanh Hoa and Phu Tho to the northwest. Roads were built and some 1,000 trucks were used, but the bulk of the Vietminh supplies were moved on bicycles (carrying about 200lb; some specially strengthened to carry up to 450lb) across almost trackless terrain. Most crucial of all were General Giap's 105mm guns and their ammunition, whose presence utterly annihilated the calculations on which the French battle plan was based.

In their campaigns against the French and the South Vietnamese government with its American allies, the Vietnamese Communists developed a logistics system relying greatly on human muscles. Here, Viet Cong porters bring supplies down the Ho Chi Minh Trail, the clandestine supply route from North Vietnam.

France in Europe, and the Vichy government's agreement to the occupation of Indochina by Japan in 1941, brought about a transformation with the emergence of the Vietminh as the major resistance force. The reputation of its leader, Ho Chi Minh, ensured that the Vietminh had no plausible rivals in the nationalist movement, and in August 1945 it took governmental power in Hanoi. It was thus an incumbent regime, as much as an insurgent movement, when the French began their reconquest in 1946, and it never lost the political advantages that stemmed from this status.

"We are marching towards Hanoi. I am going to see my house again. Like an irresistible tide, the general counteroffensive has been launched. The colonialists are fleeing like smoke before the wind. We advance straight through jungle, forest, and rice fields, without even bothering to deal with those French outposts that are still holding out . . ."

Ngo Van Chieu, Vietminh combatant

Its victory was always likely, if not inevitable. French control was formally established through conventional military means in 1946–47, but it was undermined by constant attacks on outposts and road communications. The French were confined, by choice as much as geography, to the main roads, while the countryside passed to the control of the hydra-headed Vietminh. Although Vietnam is not a very large country, it has a variety of terrain – mountain, jungle, rice-paddy – unfavourable to regular,

especially mechanized, forces. Tireless guerrilla activity was countered only by sporadic army forays (the *tache d'huile* – "oil slick" – method). In his brilliant pamphlet "The Resistance will Win", Truong Chinh refined Mao's three-stage framework of protracted war. Shifting from the first phase ("phase of contention") to the second ("equilibrium") presented few problems; the real difficulty was the transition to the final phase ("general counteroffensive"). Giap, the presiding genius of the war, misjudged this point in 1951. He had forced the French out of the northern mountains, but his assault on their positions around the Red River delta was a costly failure. In fact, the Vietminh never developed the military capacity to expel the French bodily from Vietnam.

But the Vietminh's reverses of 1951 bore unintended fruit in reviving the overconfidence of the French army. The planning of the reoccupation of Dien Bien Phu in November 1953 only made sense on the assumption that the Vietminh could not operate in strength so far from their bases. Here the French failed to understand the Vietminh logistic structure. The severity of their resulting defeat was at least as much psychological as physical. In the final stages of the battle, after a long and patient siege from March to May 1954, Giap hurled his troops – with, he claimed, their active agreement – against the French fortifications without regard to losses, knowing the impact a victory would have on the international peace conference at Geneva. From first to last, this was a war of opinion.

By the early 1960s, when a renewed Vietminh campaign began in South Vietnam, the omnipotence of the guerrilla principle seemed to have been proved worldwide. One of

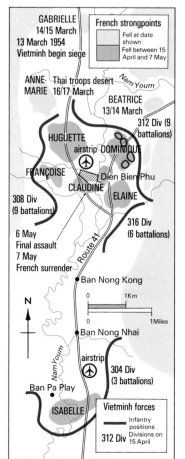

French Indochina up to July 1954 (*above*) Following France's defeat by the Communist Vietminh, the Geneva Agreements established independent regimes in Vietnam, Laos and Cambodia.

The battle of Dien Bien Phu, November 1953–May 1954 (*left*). The Vietminh suffered around 25,000 casualties before overrunning the French garrison.

A wounded French officer emerges from a bunker at Dien Bien Phu (*right*).

the few places where communist insurgency had been halted was Malaya, in 1948–60, where the British colonial authorities ran a fairly level-headed campaign to win the "hearts and minds" of the people. Improving the standard of living, resettling vulnerable civilians in defended villages and eventually mounting large-scale "food denial" operations, the government successfully cut off the insurgents, called "Communist Terrorists" (CTs) from their popular support. But the government's path was eased by the ethnic division which enabled it to mobilize Malays against the largely Chinese guerrillas, and still more by its readiness to grant Malayan independence. For this reason, the Malayan example offered no guide to the defeating of internal revolution, although many of its techniques (such as resettlement) were used or abused in other counter-revolutionary campaigns.

Algeria

Far more indicative was the gruesome French experience in Algeria in 1954–61. The French army drew from its humiliation in Vietnam a theory of *la guerre revolutionnaire* in which communism and nationalism were identified as two sides of the same coin, and psychology was the key to victory. In Algeria there were to be no concessions on the issue of independence. The National Liberation Front (FLN) was militarily more vulnerable than the Vietminh: never so thoroughly organized and always prone to internal factional struggles. New technology (especially helicopters) destroyed the immunity once conferred by the desert, while ruthless interrogation broke down the sanctuary of the *Kasbah*, the Muslim quarter of

Algiers, which had baffled many other governments. Electronic defence lines denied the rebels sanctuary across Algeria's borders. The French army won a military victory, but it proved politically sterile. Military toughness was justified, in the soldiers' own minds, by the indiscriminate terrorism of the FLN: as General Massu said, he looked on those who planted bombs in shops or restaurants as brutes ("comme des salauds") not deserving human treatment. But the use of torture, undeniably effective in a purely military sense, had an adverse effect on public opinion in metropolitan France, which increasingly disowned "the dirty war", its practitioners and its objectives alike.

The Cuban revolution

The most colourful triumph of the guerrilla idea came with the Cuban revolution. Small-scale it may have been, compared with Algeria and Vietnam, Indonesia or the Philippines, but Cuba's position on the doorstep of the USA guaranteed it world significance. The progress of Fidel Castro's tiny band, consisting of barely a couple of dozen fighters, after their disastrous landing in December 1956 which culminated in the total overthrow of the Batista dictatorship two years later, was spectacular enough. But its impact was enlarged by its charismatic leaders, Castro himself and Ernesto ("Che") Guevara (1928–67). Guevara wrote an instant classic, *Guerrilla Warfare*, which caused a sensation amongst revolutionary Marxists by declaring that "it is not necessary to wait until all conditions for making revolution exist: the insurrection can create them". This heterodoxy was labelled "the

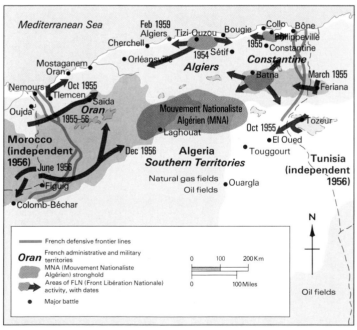

Operations during the French campaign in Algeria, 1954–61. By 1956, the *Mouvement Nationaliste Algérienne* (MNA), with its stronghold at Laghouat, had been largely absorbed by the *Front de la Libération Nationale* (FLN), although factional strife continued to

plague the guerrillas. Note the French defensive lines, particularly the "Morice Line", a massive electrified fence, strengthened with minefields and sensor devices, which ran the length of the Tunisian border.

French regulars examine the bodies of Algerian guerrillas (one of whom wears French paratrooper's boots, possibly looted from a dead adversary) in March 1956. They were casualties of Operation "Bigeard", named after the famous paratroop commander, Colonel (later

General) Marcel Bigeard, whose men were to establish a fearsome reputation during the battle of Algiers in 1957.

revolution in the revolution". A small armed band of dedicated men – for *machismo* pervaded Guevara's military ideas – could, like the eye of a hurricane (*foco*), concentrate the political consciousness of the people, demonstrating both the iniquity and the vulnerability of the government.

Guevara propounded this as a model for the whole of Latin America, and himself led a *foco* into Bolivia in November 1966. If ever a man died for his theories, it was Che Guevara. He refused to see how far the Bolivian government had eliminated the main peasant grievance by land reform. An Argentinian himself, believing in the unity of the Latin American revolution, he was unprepared for the hostility his Cuban veterans aroused among the Bolivian rebels. Finally, his aggressive use of the *foco* principle deprived him of support from the orthodox Communist movement, which had been crucial to Castro's success in Cuba. Isolated and betrayed, Guevara's little column was broken up and hunted down by the US-trained Ranger battalions of the Bolivian army. Guevara himself was captured and killed by Bolivian troops in October 1967. With shocking suddenness, the guerrilla idea plunged from apparently total triumph to almost unmitigated disaster.

Terrorists and urban guerrillas

Che Guevara denounced terrorism as "a measure that is generally ineffective and indiscriminate in its results". He contrasted it with sabotage ("a revolutionary and highly effective method of warfare") and "revolutionary justice", the use of individual punishments to enforce the authority of the *foco*. In view of Che's negative opinion, it is paradoxical that terrorist groups have been his principal legatees. The failure of the *foco* in Bolivia, and of several other rural guerrilla movements in the later 1960s, brought a recognition that the countryside could not invariably provide revolutionary potential. The vast, sprawling, ever-growing cities of South America came back into revolutionary focus. In Uruguay, for instance, the capital city was in effect the whole country. A new generation of revolutionary activists, mostly from the urban middle class, saw operations like bank raids, hold-ups, hijackings, and kidnappings, as ways of arousing public awareness. The disadvantages of the city as a theatre of war – the strength, proximity and rapid reaction time of government forces – might be balanced by the greater political impact of any armed challenge at the centre of power. The anonymity of the modern city might provide a sanctuary not so different, in a way, from that of the *sierra*.

The kind of force that can survive in the city is, however, very different from the rural band, not least in size. In his celebrated *Minimanual of the Urban Guerrilla* (1970), the Brazilian revolutionary Carlos Marighella put the optimum size of the "firing group" at four or five. This was ideal for what Marighella (himself killed in a police ambush in 1969) saw as the basic urban operation, the bank robbery. Such raids would teach fighters the vital operational skills of reconnaissance and coordination, and would have as their targets institutions with which the public had no great sympathy. Incidentally, they would generate the money needed to buy the expensive apartments and cars which would help to shelter the guerrilla groups from police suspicion.

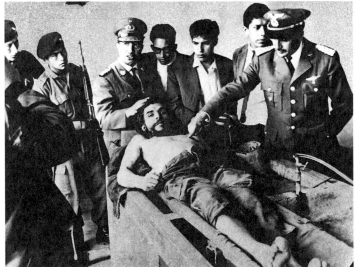

In Teheran, in 1984, an Iranian soldier prepares to depart for the front in the Gulf War (*left*). Although the Iranians are inspired by Islamic fundamentalism, they apparently also worship an icon of worldwide revolution: the Soviet-designed "Kalashnikov" AK-47 assault rifle – the favoured weapon of many guerrilla fighters – held aloft in a clenched fist.

Officials examine the body of the Argentinian-born revolutionary Ernesto ("Che") Guevara (*above*), captured and shot by US-trained Bolivian Rangers in October 1967. Guevara was Fidel Castro's right-hand man in the Cuban revolution of 1956–59.

The strengths and weaknesses of urban guerrilla strategy were vividly illustrated in the career of its most exciting practitioners, the Tupamaros of Uruguay. For a time in the late 1960s, these daring and resourceful partisans combined well-judged defiance of the state (for instance, by announcing that kidnap victims were being held in "people's prisons" in Montevideo and defying the police to locate them) with a form of social banditry or "Robin Hood" action, such as seizing food trucks and distributing their contents to passers-by. They achieved, as far as could be judged, considerable popularity. But the very nature of their cellular urban organization, the anonymity which protected them, prevented them from building a mass movement. They remained locked into a phase of colourful but hardly lethal displays. In common with other frustrated revolutionaries, they took to more indiscriminate terrorist action and were eventually overwhelmed by a popular authoritarian backlash.

The Tupamaros' experience, and that of many other idealistic groups, showed the flaws in the widely held belief that indiscriminate terrorism, by provoking governments into excessive repression, would generate public support. This belief has proved extraordinarily resilient, in face of repeated failures. In some cases, confusion about the function of public opinion has led to blurring provocation of the government with a generalized assault on the whole of bourgeois society, as in department-store bombings. The activism which leads to this form of conflict, for which "war" is only a rhetorical label, seems to be a structural product of post-industrial societies. It is linked, too, to technology. In the 19th century, the destructive power of dynamite (invented in 1866) inspired a whole generation of revolutionaries. Modern portable heavy weapons have given real military potential to groups which could earlier have carried only revolvers and small bombs. Combined with new ideological, especially religious, pressures, these factors are likely to ensure that terrorism becomes an inseparable part of modern life.

Counterinsurgency

To grasp the full extent of guerrilla warfare's divergence from regular war, one must examine the responses it has elicited from states. The primal reaction, as we have seen, is terrorism. Denied a target for their conventional military approach, armies have instinctively struck out against the civil populations who seem to be concealing the partisans. Reprisals are, at their crudest, a mere outlet for frustration, but they have also been seen to work. Information usually begins to flow back towards the authorities, sometimes in a trickle, sometimes a stream. But this assistance is bought at a political price that is too often prohibitive. Therefore, governments whose forces are not at the limits of desperation have tried to find a less brutal means of rallying the people to their cause. The crucial first step is to recognize that the allegiance of the people is the primary target for both sides in a guerrilla war. Some governments never manage to take this step; but once taken, several problems remain to be solved. These can be outlined under the headings of command, forces, operations, intelligence and propaganda.

Command becomes a problem in guerrilla wars because of the overlap between civil and military spheres. Even

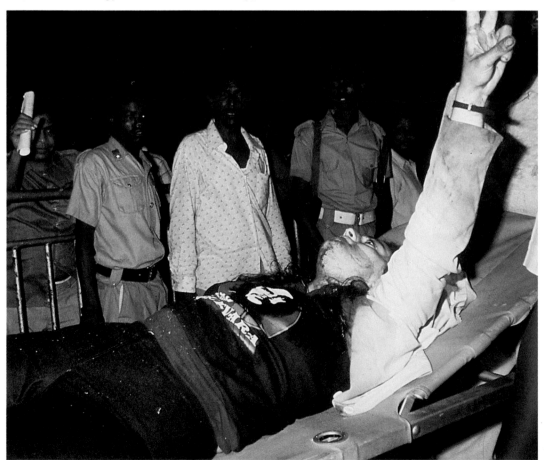

Another modern revolutionary icon, the face of Che Guevara, is seen on the sweater worn by a wounded member of the Popular Front for the Liberation of Palestine (PFLP) at Mogadishu airport, Somalia, on October 18 1977. It was indicative of the worldwide cooperation between terrorist groups that Palestinians should have undertaken the hijacking of a Lufthansa airliner in an attempt to force the release from prison of the leaders of the West German Baader-Meinhof urban guerrilla group. The West German anti-terrorist squad *Grenzschutzgruppe 9* re-took the airliner in a commando operation.

under martial law there can usually be no "free fire zone" in which pure military logic prevails. Instead, an uneasy coexistence between two quite different systems leads to misunderstandings, inefficiency, and sometimes outright conflict. Governments are concerned to preserve at least the illusion of civil authority. Admission that a state of war exists is often in itself a major political defeat. So a concept such as "the primacy of the police" limits the scope of military action. Organizational discrepancies can cripple every aspect of a counterinsurgency campaign, even where a single civil-military supremo has been appointed and a coherent overall strategy worked out, as was accomplished when General Templer became High Commissioner in Malaya.

Forces present an even more obvious problem. Sheer numbers, as Lawrence pointed out, are nearly always inadequate. Worse, existing forces are ill-adapted to the special demands of irregular warfare. Pure police and pure military techniques are both inappropriate; yet creating special forces is expensive, time-consuming and counter to the natural conservatism of existing forces. Some special forces may be actively dangerous to the state that creates them. In spite of these constraints, the development of *gendarmeries* and ranger units has been a marked feature of many successful counterinsurgencies. Popular militias, civil guards or special police have also played an important part, both in providing the manpower needed for close local control and in giving government supporters a way of participating actively in the struggle. A civil guard is, in a sense, a barometer of public opinion.

It is hard to reduce the whole field of anti-guerrilla

WOMEN IN GUERRILLA WAR

All guerrilla (though not terrorist) organizations have needed at least passive support from the mass of the people. Some have secured widespread active participation: perhaps the most accomplished of these was the Vietminh, which articulated – among other things – a women's liberation movement. "Women participate in the work of the revolution through the Women's Public Works Committee. This body centralizes and distributes all the work women can do, ensuring that they participate directly in the war effort; they are assigned appropriate military tasks, including sabotaging roads and trenches, carrying supplies and ammunition... washing the clothes of the sick and wounded, cooking, nursing, surveillance, keeping watch, and gathering information in areas occupied by the enemy." (Ngo Van, *Journal of a Vietminh Combatant*).

A female officer leads a Palestinian terrorist commando unit.

Using techniques that closely resemble the methods used by the British Special Air Service (SAS) and police in the storming of the Iranian Embassy, London, in April 1980 (*above*), gendarmes of the French R.A.I.D. anti-terrorist squad carry out a training exercise in Paris in 1987 (*left*). Because conventional forces are usually ill-adapted to the demands of irregular warfare, many countries use elite military or para-military bodies to combat terrorism.

operations to a single paragraph, but it is important to see the distinction between offensive and defensive measures. The balance between these is a matter of policy; military logic favours the first, civil logic leans towards the second. Soldiers have always tried to deal with irregulars by operations which resemble conventional warfare as nearly as possible. Thus, they favour concentration of force and the use of mechanized or air mobility to mount big sweeps (as in the French "oil slick" technique) or flying columns. Civilians argue for dispersion of force for local defence, small intensive patrolling and so on. In general, the civilian arguments eventually prevail. Against a skilful opponent, sweeps are usually a costly failure. Concentration of force hands day-to-day control of the country over to the insurgents. Only tight local control, if necessary by the construction of wire lines and minor fortifications (blockhouses), provides a reasonably predictable method of pinning down the irregulars.

Intelligence and propaganda

Locating or identifying insurgents has presented the most difficult of all technical problems. Only a sophisticated intelligence system has any chance of doing this. In the nature of guerrilla war, the police intelligence structure is often among the first targets for insurgent attack, and may well have been effectively eliminated. Regular military intelligence is ill-adapted to filling its place, partly because of a difference of outlook, partly through lack of expertise. Raids, searches, patrols, even defensive guards can be sources of usable information, but to gather, process and interpret such material calls for great dedication and

aptitude. The demands made on military officers are most lucidly analysed in General Sir Frank Kitson's remarkable tract *Low Intensity Operations*, which focuses on the techniques by which "background" information can be transformed into "contact" information. Shortage of the latter has vitiated countless anti-guerrilla operations. Patient interrogation of insurgents or suspects has proved the most consistent method, but under pressure of time the temptation to slip into brutality or systematic torture has not always been resisted. The dangers of this are obvious, as are those arising from the use of secret agents or "pseudo-gangs" to penetrate the insurgent organization.

The delicacy of these operational balances is dictated by the paramount need to preserve legality. "Propaganda" is an somewhat misleading blanket term for the whole range of action designed to show the people that the government is right – and that it will win. Credibility and legitimacy are at root inseparable, and governments have usually recognized that if they fail to protect people from insurgent activities, they lose the main justification for their own existence. No psychological warfare (or "psyops"), however skilful, can restore legitimacy once the complex fabric of the law has been undermined. This is the tightest of all the squeezes into which governments are forced by guerrilla action. The narrow line between insufficient and excessive force is hard to hold. Popular perceptions can destroy official calculations. Guerrilla movements, even those that are too weak to achieve any significant military advance, can crack the routine assumptions and behaviour on which complicated modern societies depend. That is why they are sure to persist in some form or other.

From Cromwell's time until the present day, Irish nationalism has relied on its "martyred dead" to evoke sympathy for its cause. In recent years, the funerals of IRA members killed in action (*left*) have often served as a flash-point for violence, when security forces have intervened in attempts to arrest "honour guards" like these armed, hooded and para-militarily-clad IRA members.

The climate of violence in Ulster (*above*): in a Belfast street, a child at play seems hardly to notice gunmen of the Irish National Liberation Army (INLA) – a breakaway group from the Provisional IRA ("Provos") – on a "training exercise".

chapter nineteen

Destroyer of worlds

John Pimlott

At precisely 0816 hours on August 6 1945, the Japanese city of Hiroshima ceased to exist. In a searing flash of heat, estimated to be 10 million degrees Fahrenheit at its core, the ground glowed red-hot, buildings vanished and people were vapourized. Within seconds, a massive firestorm had been generated as a seething mass of dust and ashes rose high into the sky in a strange mushroom cloud, sucking in the surrounding air to act like a gigantic pair of bellows. At the same time, a blast-wave rocked the remains of the city, advancing like "the ring around some distant planet" to destroy buildings and kill people more than five miles away, while a fog-like cloud of dust spread far and wide, depositing an invisible poison which would continue to take its toll for years to come. The atomic age had dawned.

To the people of Hiroshima, it was like a "peep into hell". Of the 320,000 soldiers and civilians in the city on August 6, about 78,000 died instantly. They were the lucky ones. Many others suffered appalling injuries as the heat-flash melted eyeballs and etched clothing on to skin; all the survivors of the initial blast unavoidably absorbed huge doses of radiation. Equally dramatically, the centre of Hiroshima was flattened – of 90,000 buildings standing at 0816, nearly 62,000 had ceased to exist a few milliseconds later – and all public utilities such as water, sewerage, electricity and telephones had been severed. There was little that the surviving rescue services could do to relieve the appalling suffering.

"Here it comes, the mushroom shape that Captain Parsons [a naval officer attached to the Manhattan Project] spoke about. It's coming this way. It's like a mass of bubbling molasses. The mushroom is spreading out. It's maybe a mile or two wide and half a mile high. It's growing up and up. It's nearly level with us and climbing. It's very black, but there is a purplish tint to the cloud. The base of the mushroom looks like a heavy undercast that is shot through with flames. The city must be below that."

Sgt George Caron, on the Hiroshima bombing mission, August 6 1945

A second bomb was dropped three days later on Nagasaki, killing a further 39,000 people. By then it was obvious that a new era of warfare had emerged. The idea of destroying an enemy's capacity to wage modern war by bombing his weapon-producing factories and killing his people was not new in 1945 – both the British and the Americans had only recently terminated a campaign against Germany with just such aims in mind – but the two atomic raids provided something which had not been available before: a virtual guarantee of instantaneous devastation through a single explosive device. The bomb which destroyed Hiroshima produced an explosion equivalent to 20,000 tons of TNT (trinitrotoluene) and was delivered by a single B-29 Superfortress bomber: for the RAF to have achieved the same impact over Germany only a few weeks earlier would have required a fleet of 10,000 Lancaster bombers all dropping their loads on the same spot at the same time. It was an awesome development, fraught with implications for the future of mankind.

The theory behind the atomic bomb had been known for some time. As early as 1938, the German chemist Otto Hahn put forward the idea that if the atoms of a heavy element such as uranium were bombarded with neutrons, the atoms would be split into fragments, creating new atoms of a lighter element. Because the new atoms would contain fewer neutrons, some neutrons would be left free to collide with other uranium nuclei, creating a "chain reaction" and producing an enormous burst of energy, equivalent to many thousand tons of TNT (the explosive yield equivalent to 1,000 tons of TNT is designated 1 kiloton; KT). The Germans under Adolf Hitler did not pursue the matter – but the Allies did. By 1945 scientists working in the United States under the overall codename "Manhattan Project" had succeeded in translating theory into fact. On July 16, at Alamogordo in the New Mexico desert, the first test-explosion was successfully achieved.

Intercontinental Ballistic Missiles (ICBMs) are maintained by both superpowers – the USA and USSR – in their homelands. The weapons are housed in specially hardened silos designed to withstand all but a direct hit by incoming nuclear devices. The Americans deploy their ICBMs in roughly a "cross" pattern, capable of reacting to threats from any direction, whereas the Soviets prefer to follow the line of the Trans-Siberian railway, with missiles ready to respond either northwards, across the polar route to the USA, or southwards into China. Each superpower also deploys Submarine Launched Ballistic Missiles (SLBMs) on board ocean-going Sub-Surface Ballistic Nuclear (SSBN) submarines. Each superpower normally has 24 SSBNs, each with up to 24 missiles on board, out in deep-ocean areas of the world. As they cannot be precisely located, they are virtually impossible to take out in a first (pre-emptive) strike, and will remain available as a second (retaliatory) strike arsenal, capable of inflicting enormous damage. In the case of the USA, these are supplemented by long-range bombers such as the B-1B, armed with nuclear weapons.

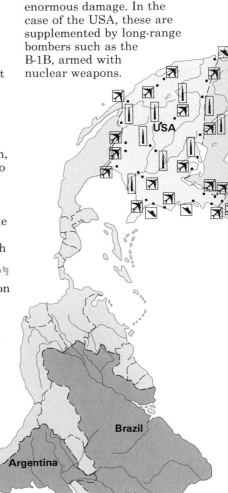

—TIME CHART—

1945 July	First atomic device exploded by US
1945 Aug	Hiroshima and Nagasaki destroyed by US atomic bombs
1946 July	US carries out first peacetime nuclear tests at Bikini Atoll
1949 Aug	Soviet Union explodes its first atomic bomb
1952 Oct	Britain explodes its first atomic bomb
1952 Nov	First thermonuclear bomb exploded by US
1954	US adopts "Massive Retaliation" policy
1954	US deploys Honest John tactical nuclear missile
1954 Sep	US commissions USS *Nautilus*, first nuclear-powered submarine
1956	US deploys intercontinental ballistic missiles
1957 May	Britain explodes its first nuclear bomb
1957 Aug	Soviet Union tests ICBM
1957 Oct	Sputnik satellite launched by USSR
by 1957	US and USSR experimenting with submarine-launched ballistic missiles
1958 Jan	Explorer satellite launched by US
1960 Feb	France explodes its first nuclear device
1962 May	US tests submarine-launched Polaris warhead
1962 Oct	Cuban missile crisis

1963–64	US adopts "Graduated Deterrence" strategy
1963 July	US and USSR sign atmospheric nuclear test ban treaty
1964 Oct	China explodes its first nuclear device
by 1965	American nuclear weapons delivery triad in place: ICBMs, submarines, and bombers
1966 Mar	France withdraws from the military structure of NATO
1967 Dec	NATO adopts "Flexible Response" doctrine
1972 May	US and USSR sign first Strategic Arms Limitation Talks treaty (SALT I)
early 1970s	US and USSR fit ICBMs with MIRVs, multiple warheads with individual guidance systems
early 1970s	US develops enhanced radiation weapons ("neutron bombs")
1974 May	India explodes a nuclear device
1974 Nov	US and USSR agree arms limitation figures (Vladivostok Accord)
1979 June	SALT II treaty signed by US and USSR
1983 Mar	US announces SDI ("Star Wars") research project
1987 Dec	US and USSR sign Intermediate Nuclear Forces (INF) treaty

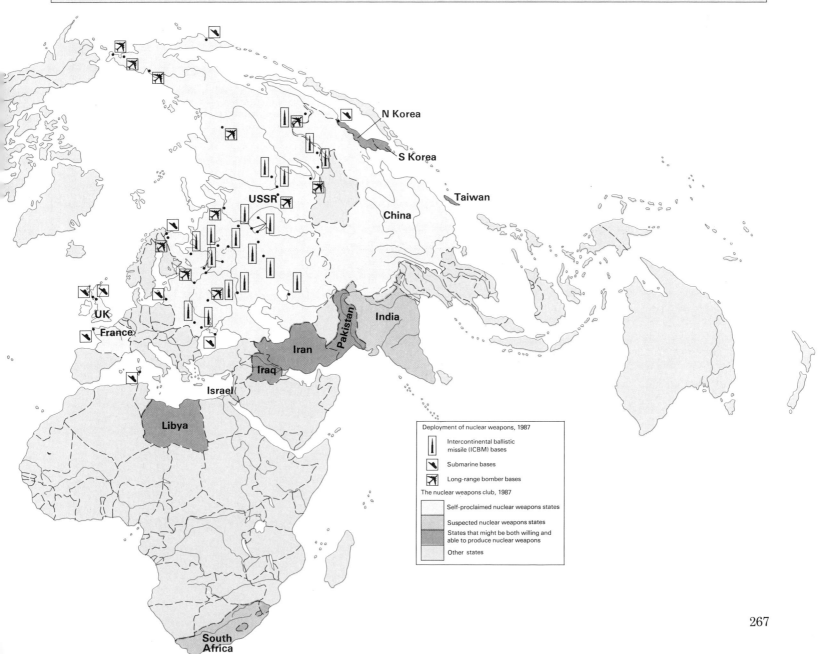

Deployment of nuclear weapons, 1987

Intercontinental ballistic missile (ICBM) bases

Submarine bases

Long-range bomber bases

The nuclear weapons club, 1987

Self-proclaimed nuclear weapons states

Suspected nuclear weapons states

States that might be both willing and able to produce nuclear weapons

Other states

The device exploded at Alamogordo was detonated by what was known as the "gun" method, in which two subcritical pieces of Uranium 235 were blasted together using conventional explosive; this method was repeated in the Hiroshima bomb three weeks later. The Nagasaki bomb was slightly different, in that the fissile material (plutonium) was surrounded by a casing of TNT which "imploded" to produce the chain reaction; but the results were the same. In less than a millisecond, enormous explosions, accompanied by blinding flashes of light and clouds of radiation, could be created.

Further deadly refinements

But this was only the beginning. By November 1952, the Americans had gone one stage further, producing weapons of limitless explosive potential. These were based on fusion rather than fission, whereby the hydrogen nuclei of deuterium and tritium were compressed with such force that they fused together to produce helium and release enormous bursts of energy. Because fusion could only take place at excessively high temperatures – an estimated 20 million degrees Fahrenheit – the process had to be triggered by means of an "ordinary" fission explosion, so the weapon which emerged comprised an atomic device inside a casing of deuterium and tritium: in other words, a hydrogen or thermonuclear device. Even more potential came from the decision to manufacture an outer casing of uranium, for this would absorb and react to the neutrons released by the fusion process. The result was a bomb equivalent to millions of tons (megatons) of TNT. If the outer casing was omitted, however, the neutrons would spill out, creating an invisible force of radiation which would be able to penetrate almost any protective screen to attack the central nervous systems of all vertebrate animals. By the 1970s, this particular refinement had led to the development of the Enhanced Radiation Weapon (ERW), more commonly known as the "neutron bomb" (it is in fact an artillery shell), as an integral part of the American nuclear arsenal.

Such weapons are awesome, but without the means to deliver them to their targets accurately and over long ranges, their value is meaningless. Hand in hand with the technology of creating the devices, therefore, there has taken place an equally dramatic development of delivery means. In 1945, these consisted of manned bombers such as the B-29 but, as the losses incurred by the Allies over Germany and Japan implied, aircraft were vulnerable to radar-assisted air-defence systems. Thus, although the bomber was not abandoned – by the mid-1950s the Americans were deploying the eight-engined B-52 Stratofortress while the Soviet Union was developing the Tupolev Tu–20 series, codenamed "Bear" – an alternative delivery system, capable of avoiding existing defences, was clearly needed. The answer was found in missiles which could carry nuclear warheads in an orbital trajectory. By 1954 the Americans had perfected the Honest John battlefield missile, with a maximum range of 25 miles, following this up two years later with the development of Thor, the first intercontinental ballistic missile (ICBM), capable of reaching targets in the Soviet Union from land-based silos in the United States. By 1957 the Soviets had followed suit, and both sides were experimenting with submarine-

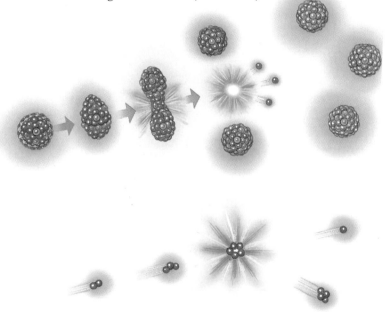

The Fission process (*top*) is based on the realization that when two pieces of subcritical Uranium 235 are blasted together (using ordinary high explosive), they will produce "lighter" elements, releasing neutrons to cause a chain reaction. The Fusion process (*bottom*) compresses together the hydrogen nuclei of deuterium and tritium to produce helium. In both cases, enormous amounts of energy are produced, with all the attendant by-products of heat, light, blast and radiation, capable of demolishing entire cities and killing vast numbers of people.

NUCLEAR JARGON

The confrontation between the superpowers which has characterized the post-1945 period has led to significant changes in the nature of war. These are reflected in (and often made more confusing by) the jargon used by both sides to describe their actions. Here is a short list of the most common acronyms, abbreviations and terms:

Arms Control: agreements to limit the numbers and types of weapons deployed by the superpowers

Cold War: hostilities between the superpowers which fall short of actual fighting

Detente: the easing of tensions between the superpower blocs

Deterrence: the ability to prevent aggression by persuading a potential enemy that the gains to be had from pursuing a particular course of action are outweighed by the losses that will be imposed if such action takes place

Disarmament: the physical removal or dismantling of weapons

ICBM: Intercontinental Ballistic Missile (land-based)

INF: Intermediate Nuclear Forces (those with less than intercontinental range)

MAD: Mutual Assured Destruction – the ability of one side to absorb a "first strike" while retaining sufficient weapons to hit back with a devastating "second strike"

SALT: Strategic Arms Limitation Talks

SDI: Strategic Defense Initiative ("Star Wars") – an American space-based defence system

SLBM: Submarine Launched Ballistic Missile

SSBN: Sub-Surface Ballistic Nuclear (missile-armed) submarine.

launched ballistic missiles (SLBMs). What emerged was a "Triad" of delivery means – by air, land and sea – which has remained the basis of the balance between the two sides ever since.

This balance is essential, for the true value of nuclear weapons lies not so much in their ability to destroy the enemy as in their threat to do so should aggressive moves be made. Because of the appalling destructive power contained in nuclear devices, any war involving their use would be devastating; thus, the aim has become one of preventing rather than conducting armed conflict at levels which might trigger their release. In other words, people must be deterred from using nuclear weapons for fear of the consequences. This may sound straightforward, but in reality it has produced a complex and inherently danger-ous strategy which, it could be argued, leaves the world on the edge of destruction.

Deterrence

Deterrence may be defined as "the ability to prevent aggression by persuading a potential enemy that the gains to be made from pursuing a particular course of action are outweighed by the losses that will be imposed if such action takes place". For deterrence to work, fear must be engen-dered and maintained at all times, satisfying the criteria of credibility, capability, communications and rationality. Of these, credibility is the key, for the opponent must believe that the threat being offered will be carried out if he provides sufficient provocation. Nuclear-capable states go to great lengths to create this impression, using both words and deeds to remind potential enemies that the option of

force as an instrument of state policy is not something that is likely to be ignored. Nuclear forces may be placed on alert during times of international crisis or tension – in October 1973, for example, the United States made just such a move in response to Soviet preparations to put airborne forces into the Middle East at the height of the Arab-Israeli Yom Kippur War – and leaders may go to considerable lengths to persuade observers that they would, in certain circumstances, "press the button". President Richard Nixon of the United States, in his deal-ings with the North Vietnamese in the late 1960s and early 1970s, even went so far as to propound a "madman theory", implying that, if sufficiently provoked, he would have no hesitation in using nuclear weapons to break the deadlock of the Vietnam War.

Credibility cannot stand alone, however, for it is equally important to show that, in addition to the will, a state leader also has the capability – the physical means – to carry out his threat. There are currently five states in the world with the proven ability to deliver nuclear devices – the United States, the Soviet Union, Britain, France and China – and all conduct periodic tests of their weapons and delivery means, in the full knowledge that these will be monitored and assessed by their opponents. Indeed, by 1987 the Americans and Soviets had gone one stage further, agreeing to allow such monitoring to take place "on site", for both superpowers recognize that completely secret tests could have a weakening effect on deterrence, by leading to the opponent's ignorance about the true nature of the threat. Frequent parades of the latest equipment aim to achieve a similar result.

The American atomic attack on the Japanese city of Hiroshima on August 6 1945 produced complete devastation (*top*), with only the strongest buildings surviving the blast and subsequent firestorm. The bomb itself (*above*) – nicknamed "Little Boy" – was detonated by means of a conventional explosion which hurled two subcritical pieces of Uranium 235 together to produce a chain reaction. Approximately 78,000 people died in the attack, but those who survived faced a blighted future: many were horribly scarred (*left*) by the blast and were subjected to doses of invisible, cancer-producing radiation – a long-term killer.

But this is only part of the picture. Regardless of political will or physical capability, deterrence would not be achieving much if the nature of both the threat and of the proscribed action was not communicated clearly to the potential aggressor. In 1950, for example, when United Nations forces crossed the 38th parallel into North Korea and pushed towards the Chinese border on the Yalu river, the recently installed communist regime in Peking tried to deter such a provocative advance by threatening to intervene with force. In this particular case, the threat was a purely conventional (non-nuclear) one – the Chinese did not test-explode a nuclear device until 1964 – but the fact that the message did not get through to the United States (the driving force behind the UN campaign), chiefly because Washington refused to recognize the communist regime, shows how easy it is to lose such an essential element of deterrence. If the action to be deterred is not described precisely, or the threat is vague, the aggressor will not understand the situation and will not have sufficient reason to avoid it. In Korea, the intervention of Chinese "volunteers" in October–November 1950, pushing UN forces back down the peninsula beyond Seoul, came as a considerable surprise.

Finally, throughout the exercise of deterrence, there is a presumption that both sides contain rational human beings who fear the consequences of nuclear attack. This will not always be guaranteed. Decision-makers may crack under the strain of an international crisis or may hold beliefs which transcend the fear of loss of lives or property. Indeed, in an age of strongly held religious or political ideologies, it is quite possible that leaders may emerge who believe in the need for sacrifice to achieve their aims, exchanging the physical elements of their state for the chance of furthering their ideals. Any threat against their cities or people would therefore be meaningless, and deterrence would ultimately fail.

Fortunately, this has not yet happened at a nuclear level, and deterrence, particularly between the superpowers (the United States and Soviet Union) has a well established record of success, based on the mutual satisfaction of the four criteria discussed above. But this does not mean that the process has been smooth or trouble-free: there have been times when the balance between the two sides has been uneven and the credibility of the threat less than convincing. An examination of the evolution of nuclear strategy since 1945 illustrates the point.

Nuclear strategy since 1945

The Americans retained a monopoly of atomic capability for just four years, during which they regarded the new weapons as little more than super-efficient conventional bombs, to be used (in conjunction with ordinary high-explosive devices) to destroy selected cities in the homeland of any future enemy. The "Age of Nuclear Innocence" began to be undermined in 1949, when the Soviets made their first atomic test-explosion, although in the absence of long-range delivery means, the threat to United States' interests was confined to the possibility of short-range attacks on targets in Western Europe. However, because most of the Western European countries were allied to the United States through the North Atlantic Treaty Organization (NATO), Washington had to start thinking, for the

Deterrence between the USA and USSR depends on Mutual Assured Destruction (MAD), whereby each is able to "absorb" the impact of a first (or pre-emptive) nuclear strike, while retaining sufficient weapons to carry out a second (or retaliatory) strike of devastating force. If, for example, the Soviets should attack first, they would aim to destroy US ground-based ICBMs, hoping to disarm their enemy in one go. The Americans would certainly lose many ICBM silos, but if the missiles were "launched on warning" this would ensure a second strike while still leaving bombers and submarines with the ability to launch devastating counterattacks.

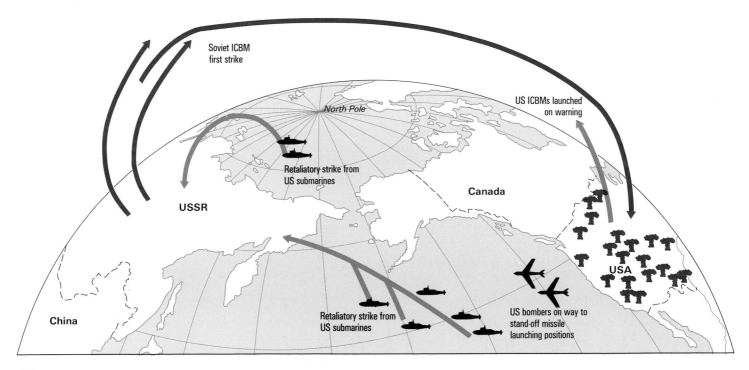

first time, in terms of deterrence – of preventing such attacks by posing threats the Soviets could not ignore. This was reinforced when Soviet aid to the North Koreans between 1950 and 1953 implied that force could still be used to back the expansion of communism in circumstances in which the use of atomic weapons was inappropriate or potentially wasteful. As the American stockpile of bombs was still relatively small – approximately 200 in 1952 – many strategists feared that a Soviet threat which resulted in their commitment to a peripheral theatre would merely be a prelude to a communist attack in a Europe now left without atomic protection.

Two interconnected strands of American policy emerged in response to such thinking. On the one hand, President Truman and his successor Eisenhower concentrated on the creation of a "containment wall" around the communist bloc, building up a network of alliances which stretched from Europe (NATO), through the Middle East (the Baghdad Pact–Central Treaty Organization) to the Far East (South East Asian Treaty Organization); on the other hand, they authorized the adoption of a strategy of deterrence. This emerged in 1954 and was crude but effective. Known as "Massive Retaliation", it made it clear to the Soviets (and, by implication, other communist powers), that any attempt to breach the containment wall would lead to an immediate and massive attack by the United States, using atomic and, by 1954, thermonuclear devices. As it was obvious that the Americans had the capability and, from the evidence of Hiroshima and Nagasaki only nine years earlier, the political will to carry out such a threat, credibility was assured. At the same time,

the nature of the threat and of the proscribed action was clearly laid down and the Soviet leadership, regardless of rhetoric, was rational to the extent of fearing massive civilian casualties. It was enough to ensure success.

But this was a one-sided strategy, dependent on a continued monopoly of capability which would allow the Americans to make threats without having to fear the consequences of a counterattack on their own population. This began to change as the 1950s drew to a close, for the Soviets, aware of their weakness, began to develop the means of hitting the United States. As early as August 1957 the Soviets tested a long-range rocket, and when this was followed only two months later by the successful launch of a space satellite (Sputnik I) – which implied a capability of delivering nuclear warheads in an orbital trajectory – the Americans suddenly realized their potential vulnerability. In the event, the Soviets still had to perfect their technology – it was to take them until the mid-1960s to achieve parity – but the writing was on the wall. Talk of a "missile gap" and the need to enhance America's strategic arsenal helped to ensure the election of John F Kennedy to the presidency in 1960 and led to a significant rethink of nuclear policy. Massive Retaliation lacked credibility if, in response to its threat, the Soviets could make counter-threats against American targets.

The Cuban Missile Crisis
The need for a new strategy was reinforced in October 1962, when the Soviets tried to accelerate the process of counter-threat by deploying medium-range ballistic missiles to Cuba, only 90 miles from the coast of the United States.

NUCLEAR TESTING

The testing of nuclear devices as a demonstration of capability is an essential part of the process of deterrence between the superpower blocs. But testing is also dangerous, producing all the destructive elements of a nuclear explosion – light, blast, heat and radioactivity. Finding somewhere safe to carry out such tests has been a constant problem.

At first, when the long-term effects of testing were not understood, explosions were carried out in the open, in remote areas. Such "atmospheric" tests were banned by international agreement in 1963, and this was followed by similar agreements governing experiments underwater or in space. This left little alternative except underground tests, carried out under strictly controlled circumstances in remote regions of the superpower homelands. Despite attempts to ban all testing, underground experiments continue today.

A Douglas SM-75 Thor nuclear missile (in RAF markings), 1961.

Although they were effectively deterred from delivering the warheads to Cuba by American resolve and by the imposition of a naval quarantine line around the island, the fact that Massive Retaliation was not automatically triggered showed that it was no longer a viable strategy. Indeed, Kennedy summed up the problems by now apparent when he pointed out that, when faced with a crisis, he had no choice between doing nothing (national humiliation) and Massive Retaliation (national suicide). If nuclear strategy is likened to a ladder, the President could stand on either the top or the bottom rung – there was nothing in between to cater for crises short of major war.

The rungs began to be filled in by Kennedy's Secretary of Defense, Robert McNamara, in 1963–64, when he introduced a new strategy, known as "Graduated Deterrence". It was based on the understanding that, because of new technology, nuclear weapons were no longer "big bang" devices, capable of destroying nothing smaller or more select than a city. By the early 1960s, the Americans were already deploying battlefield nuclear weapons for use over short ranges against enemy armed forces. This alone implied that the decision to "go nuclear" did not mean the automatic destruction of cities and civilian populations, and when it was coupled with the fact that increased accuracy over long ranges opened up the possibility of picking more precise targets, a new degree of flexibility emerged. McNamara drew a distinction at strategic level between counter-force targets, covering the enemy's war-related industries or force concentrations, and counter-value targets, covering his cities and people.

Out of all this emerged an increased range of options, avoiding the automatic escalation to all-out strategic exchange which, once the Soviets perfected the means to retaliate in kind, represented a policy of suicide. In the event of a conventional war in Europe, McNamara argued, the US President could now respond with tactical nuclear weapons if required, making it clear to the Soviets that he had no intention of moving to strategic attacks unless provoked. If that provocation arose, the threat of further escalation was still there, but even then, value targets need not necessarily be top of the agenda. Before they were hit, force targets could be distinguished and taken out in attacks which, once again, would signal to the enemy that the Americans were not using their full capability and were prepared to hold back if the war was brought to a close. At each level of escalation, "pauses" would occur during which the Soviets would be forced to make a "cost-gain calculation" ("Is what I am after worth the punishment that will be suffered in return?"), and during these pauses the chances of a negotiated settlement of differences might be opened up.

Graduated deterrence

In terms of the four criteria of deterrence, this strategy had more in its favour than its predecessor. No one doubted the Americans' capability – by the mid–1960s the Triad of ICBMs, SLBMs and bombers was firmly established – and since each level of projected escalation would be in proportion to the level of threat from the enemy camp, credibility would be enhanced. Communication between the superpowers was good – one of the immediate results of the Cuban missile crisis had been the creation of a secure "Hot Line" between Washington and Moscow, designed to make direct contact much easier – and it could be presumed that the Soviets had no more desire to die than they had had ten years earlier. In fact, the main criticism of Graduated Deterrence came from within NATO, where European countries interpreted the American desire to delay escalation to all-out nuclear attack as an early attempt to distance the United States from the potential flash-point of the Inner-German border. Such an interpretation was unjustified, but to countries used to living under a nuclear umbrella which seemed to guarantee deterrence by threatening Soviet cities, the United States appeared to be modifying its commitment. In the event, France responded

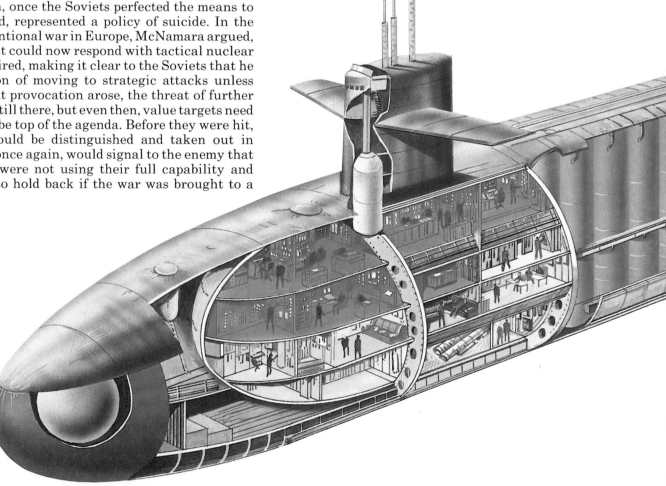

by withdrawing from the military structure of the NATO alliance in 1966, determined to create her own massive retaliatory force (the *force de frappe*; later called the *force de dissuasion*), and when McNamara's strategy was extended to conventional levels and incorporated in the new NATO doctrine of "Flexible Response" a year later, it was touch-and-go as to West German acceptance. The requirement for NATO to offer some sort of conventional war-fighting response to communist attacks at that level did little to please the countries of Western Europe which would, inevitably, be the battleground.

The Single Integrated Operational Plan

But considerations such as these were relatively minor when compared to the credibility of Graduated Deterrence, and the strategy, constantly modified and updated, has remained in force since its official adoption in 1964. Ten years later, Nixon's Secretary of Defense James Schlesinger widened the scope of the strategic level of targeting by means of his "Limited Nuclear Options" policy, and information is constantly fed into the system to ensure that all targets are covered in what is known as the Single Integrated Operational Plan (SIOP). The only real problem is that the Soviets, with their traditional view that maximum force achieves the best results, may not stick to the "rules", but it is generally recognized that the Politburo, being rational men, will not risk the destruction of their own state by provoking uncontrolled escalation. It

seems a reasonable presumption, reinforced by the fact that no major crisis between the superpowers has occurred since the early 1960s.

The main reason for this is that a situation of stalemate, known as Mutual Assured Destruction (MAD), has emerged as the two sides have matched each other in terms of technology. MAD is based on the ability of each superpower to survive, or "ride out", a nuclear attack, retaining sufficient weapons to ensure a devastating response. If, for example, the Soviets should decide to hit the United States in a "first strike", their aim – quite logically – would be to try to destroy nuclear forces which threaten the Soviet Union. Their targets would therefore be the ICBM silos, bomber bases and submarine pens which contained such weapons, and although enormous damage would be inflicted – in 1981 President Ronald Reagan was informed that more than 90 percent of the ICBM silos would be lost in a first strike of this nature – they would not cover the full range of American nuclear forces. Bombers could already be in the air, ICBMs could be "launched on warning", leaving empty silos to be hit as they travelled

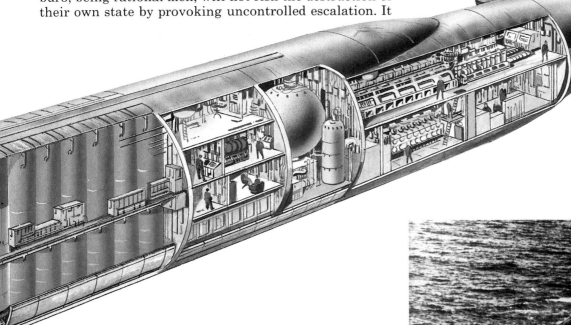

The latest American SSBN USS *Ohio* (*above*). Armed with 24 Trident I (C-4) SLBMs, each with eight 100KT MIRV'd warheads, the *Ohio* was laid down in 1976, joining the fleet in 1982. By 1988, eight "Ohio" class SSBNs were in service, with a further 16 planned. The Soviet equivalent is the 'Typhoon' class SSBN (*right*) – the largest submarine design ever built – although it is armed with only 20 SS-N-20 SLBMs, each with 6–9 MIRV'd warheads. The "Typhoon" has a number of distinctive features, not least the deployment of its SLBMs forward of the fin.

273

towards targets in the Soviet homeland, and, most important of all, the nuclear-armed submarines – known in nuclear parlance as sub-surface ballistic nuclears (SSBNs) – would still be available, hiding in deep-ocean areas of the world, invulnerable to a first strike. Together, these forces would constitute a devastating second-strike retaliatory response which the Soviets would be foolish to ignore. As their first strike would, in effect, constitute a policy of suicide, they are deterred from carrying it out. The same applies to an American first strike, presuming that such a move was ever contemplated.

This situation of mutual deterrence has remained in force since the mid–1960s but it is, in the celebrated words of the nuclear strategist Albert Wohlstetter, "a delicate balance of terror". Nuclear forces do not remain static in terms of development: they are updated and modernized constantly as both superpowers pursue new technology which, by its nature, threatens to give one side a sudden advantage, opening up a "window of opportunity" in which a successful first strike could become feasible. If this should ever happen – if one side should feel able on the one hand to hit its opponent with such force that retaliation would be less than devastating, or, on the other, to defend itself effectively against a second-strike response – then deterrence would clearly fail.

Nor is this unthinkable. Despite the advantage of a central balance, both superpowers have actively pursued research into new capabilities, arguing that if they did not, they might be left behind in the arms race and rendered vulnerable to developments made by an opponent they distrust. One of the most worrying areas is anti-submarine warfare (ASW), for if either side could perfect the means to track SSBNs, even in their deep-ocean hiding places, the submarines would become vulnerable to a first strike. If the SSBNs were taken out, the state under attack would lose the mainstay of its second strike, tempting the ASW-capable state to use its nuclear arsenal to ensure victory. Satellite technology, coupled with underwater seismic devices, could conceivably produce this capability: if it was ever perfected, the balance would disappear.

Star Wars

Equally worrying is the American obsession with creating a space-based defensive system – the Strategic Defense Initiative (SDI or, more popularly, "Star Wars") – which, if deployed, would be able to destroy all warheads fired towards the United States on an orbital trajectory. First announced as a long-term research project in March 1983, SDI has been advertised as a system designed primarily to negate nuclear threats, thereby opening the way to the dismantling of all ICBMs and SLBMs as obsolete weapons. In theory, this is an attractive proposition – if warheads cannot reach their targets, they obviously lose their value – but in reality it is fraught with danger, threatening to upset the central balance which prevents nuclear war.

The idea of defending a state against nuclear attack is, of course, nothing new. Some countries, both nuclear-capable like the Soviet Union and non-nuclear like Switzerland, have devoted huge sums of money to what is known as "passive defence", building nuclear-proof shelters for their people and laying down contingency plans for a post-nuclear world. This is not really a destabilizing factor –

An MGR-1B Honest John short-range tactical battlefield support missile (*above*) is hoisted aboard its wheeled truck transporter-launcher, December 1954. The first tactical nuclear missile to be deployed, it is now an obsolete design.

SS-13 "Savage" Intercontinental Ballistic Missile in Red Square, May 1965 (*right*). A three-stage inertially guided missile, the SS-13 was the first solid-propellant Soviet ICBM. It usually carried a single 600KT warhead and had a range of 8,000km (4,970 miles): examples were still deployed in the Yoshkar Ola missile field in the mid-1980s.

1-kiloton neutron device

All die | Many die

Incapacity and coma within 5 minutes. Death within 24 hours

Permanent incapacity in five minutes. Death within 48 hours

Incapacity within 5 minutes and brief recovery. Death within 48-96 hours

Serious functional impairment within 2 hours. Most survivors remain impaired until death within weeks

Radiation sickness within 2 hours. Many survivors dead within months

Probable radiation sickness. Survivors liable to cancer, cataract, immune deficiency, etc

Whole body dose in rads (inside tanks)

100,000
10,000
1,000
100
10

Knockdown zone

Latent lethality zone

Latent damage zone

500 metres | 1,000 | 1,500 | 2,000

The awesome effect (*above*) of a 1KT Enhanced Radiation Weapon (ERW), commonly called the "neutron bomb". Fired by an artillery piece, the ERW explodes over its target, dispersing invisible neutron radiation over an enormous area.

even if the majority of people survived, there would be little left for them to inherit, so the deterrent aspect of a nuclear strike is not undermined – but the same cannot be said of "active defence", involving the physical destruction of incoming warheads. In the 1960s, both superpowers seemed to be on the brink of developing viable anti-ballistic missile (ABM) systems – rockets which could be fired into the path of the incoming warhead, destroying it by means of a small nuclear explosion which would create a chain reaction inside the warhead (a process known as "fratricide"). By 1969, however, both superpowers had recognized the potential for destabilization and had decided that no such system could destroy more than a certain proportion of the attacking warheads. In addition, the costs of developing and deploying systems such as the US Sprint and Spartan and the Soviet Galosh were likely to be crippling.

In such circumstances, the superpowers decided that it would be safer (and cheaper) to negotiate over ABM systems and, as an integral part of the first Strategic Arms Limitation Talks (SALT–I), a special treaty was signed. Coming into effect in 1972, after three years of negotiation, the treaty confined the deployment of ABMs to two sites only in each superpower's homeland – one to protect the ICBM central control complex and the other covering a major city. By 1974, when the next stage of SALT–I – the Vladivostok Accord – was finalized, neither side had taken up the option of two sites and it was agreed to restrict deployment to one site only. The Americans placed theirs around the ICBM control centre at Grand Forks, North Dakota, while the Soviets defended Moscow; however, by

NUCLEAR DEPLOYMENT

The nuclear "Triad" of land-, air- and sea-based delivery means is designed to enable each superpower to absorb a sudden attack while retaining sufficient weapons to mount a devastating retaliatory strike. Missile-armed submarines and bombers, which are difficult to target, will make up the "second strike" retaliatory force, but the intercontinental ballistic missiles (ICBMs), in fixed land-based silos, are vulnerable to surprise attack. Both superpowers are constantly studying new methods of deployment which will enhance ICBM survivability.

One of the most obvious solutions would be to make the missiles mobile, for this would deny the enemy the opportunity to target precise, predictable locations. But this can create problems. The American MX Peacekeeper ICBM, for example, was initially intended to be deployed on an underground "race-track", moving constantly between a number of silos, but when the scale and cost of this was realized, the idea had to be scrapped.

An alternative was to put all the MXs together in a "dense pack", acting as a huge target which would relieve pressure on other ICBMs, but this failed to receive Congressional approval. In the end, it was decided to put the new missiles in existing silos: the problems had not been solved.

the end of the 1970s the former system had been dismantled and the latter allowed to fall into obsolescence. It began to seem as if the ABM threat had been contained.

SDI renewed that threat, particularly when it became apparent that President Reagan was intent on its deployment, regardless of cost or potential destabilization. To a certain extent, he could argue that if the Americans failed to pursue new technology, there was fair chance that the Soviets would gain the advantage, developing their own version of SDI. But the fact that the Americans were clearly willing to devote enormous sums of money (an estimated trillion dollars over 20–25 years) to the technological problems of SDI – sums the Soviets could never hope to match – worried First Secretary Mikhail Gorbachev. His constant reference to curtailment of research or delays in deployment as necessary prerequisites to arms control agreements with the United States implied that he was unwilling to devote funds to his own research programme. The chances of the Americans gaining a decisive advantage seemed high.

Obliterating multiple warheads

The practicalities of SDI are complex. If the Soviets should ever fire their missiles, the United States would have approximately 20 minutes in which to activate its defences. During the first three minutes, when Soviet missiles and warheads were still integral, it would be possible to inflict telling damage, but thereafter, as the warheads left the missiles and produced a shower of targets for SDI systems, the task would become infinitely more difficult. The reason for this is that since the 1960s both sides have developed multiple warheads, allowing each missile to carry large numbers of nuclear devices. Initially, using multiple re-entry vehicle (MRV) technology, this comprised no more than three or possibly five warheads, intended to fall in a "claw" or "footprint" pattern around a target such as an ICBM silo to increase the chances of inflicting damage. By the early 1970s, however, each individual warhead had been given its own guidance system to produce multiple independently targeted re-entry vehicles (MIRVs), and the number of warheads carried by each missile had been increased to at least ten. Thus, if the Soviets should launch 1,000 ICBMs – a distinct possibility in a first strike – the Americans would have to cope with 10,000 incoming warheads, to say nothing of the myriad of "decoys" (dummy warheads) which would accompany them. The possibility of developing manoeuvrable re-entry vehicles (MARVs) – warheads which contain on-board computers, capable of monitoring their primary target and, if activated by a high level of radiation (implying that the target has already been hit), of spinning off to a secondary target – merely makes the situation worse. If SDI is to work, every one of the incoming warheads would have to be destroyed.

First priority must be given to early warning, giving the Americans the maximum possible time in which to activate their defences, and for this purpose special geo-stationary satellites would be deployed to keep a constant watch on Soviet missile silos. At the first clear sign of a launch which threatened the United States, attempts could be made to hit the missiles before they dispersed their warheads, at a time when they would still be within the earth's atmosphere over the Soviet Union. This is fraught with problems,

The theory of "Star Wars" (Strategic Defense Initiative), based on the various stages of ICBM trajectory. Surveillance satellites pick up the launch of Soviet ICBMs and call in space-based battle stations to hit the missiles while they still have their warheads on board. ICBMs which escape will then deploy their warheads, increasing the targets to be hit by further battle stations, some using mirrors to reflect laser beams from the ground. Any which remain will be destroyed in the terminal phase by ground-based ABMs or high-flying aircraft.

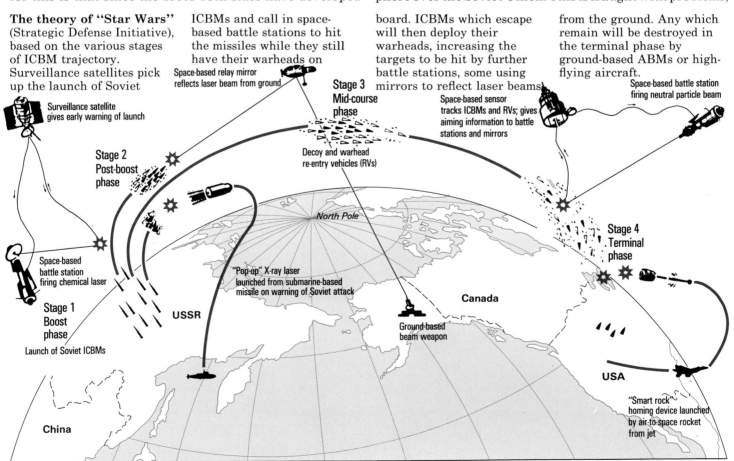

CIVIL DEFENCE

Defence against nuclear attack is not easy to organize, chiefly because the destructive nature of present-day warheads seems to make any attempt at protection meaningless. But this has not prevented some states from doing all they can to ensure the survival of at least a proportion of their populations.

In the West, there are few countries which provide shelters for anyone other than selected "high value citizens", chosen for evacuation from threatened cities as soon as a crisis begins. In the United States and Britain, this would include members of the government, whose services would be essential if the decision to retaliate was to be made: the remainder of the population, unless fortunate enough to own their own nuclear shelters, would be left to face the full horrors of a nuclear strike. Collectively, such people are known (in the United States) as "megabods".

Only in neutral countries such as Sweden and Switzerland are public shelters provided as a matter of policy. Interestingly, a similar scheme is available for citizens of the Soviet Union, although in all such cases it is difficult to see how survivors would cope in the aftermath of a full-scale nuclear exchange. With cities in ruins, food stocks destroyed, water contaminated and radioactive dust falling, the prospects are not attractive.

particularly in terms of weapons systems which would be capable of responding effectively within the 3 minutes available, so the emphasis necessarily has shifted to using defensive measures against the dispersed warheads during the 14 or so minutes of their orbital trajectory or the 3 minutes of their re-entry over the United States. A number of measures have been suggested. Those systems actually in space, capable of hitting warheads during the trajectory phase, would be mounted on board "killer satellites", controlled from high-flying command aircraft or, ideally, by the satellites which originally monitored the launch. Some would be armed with laser weapons, capable of destroying the sensitive electronics of the warheads by means of solid "pulses" of concentrated light; others would use "charged-particle" technology to project beams of nuclear-activated neutrons in an attempt to create fratricide. There is even the possibility of special "rail guns", activated by means of parallel electro-magnetic circuits, being linked to "throw" solid projectiles towards the warheads. In all cases, a constant barrage of fire would need to be sustained for the full 14 minutes.

Assistance from ground-based systems

If this system was ever deployed, it could, theoretically, destroy over 95 percent of the warheads, but the 5 percent which survived – a total of 500 nuclear devices from a launch of 1,000 IBMs – would still devastate the United States. In an attempt to destroy them before impact, therefore, further ground-based defensive systems, including laser or charged-particle weapons sited on mountains in the United States, as well as air-launched anti-satellite

Women peace protestors (*left*) sit down in the road outside Greenham Common airbase in Britain to prevent the deployment of Ground Launched Cruise Missiles (GLCMs). Introduced to Britain in 1983 as part of a NATO force-modernization programme, GLCMs became a focus for protest until it was agreed that they should be withdrawn as part of the 1987 INF Agreement.

A demonstration in London (*above*) focuses on the role played by Prime Minister Margaret Thatcher in agreeing to the deployment of GLCMs to Britain. NATO countries agreed in 1979 to provide bases for Cruise and Pershing II missiles on a "twin-track" policy: modernization of NATO theatre nuclear forces coupled to "meaningful" arms control talks.

(ASAT) missiles, would need to be added, with the capability of destroying warheads as they re-entered the earth's atmosphere. By then, the decoys would have been burned up on re-entry – but it would need only a small proportion of the activated warheads to survive for damage to be inflicted. And since each warhead is capable of producing an explosion equivalent to the Hiroshima bomb, a number of American targets could still be devastated.

SDI has not yet approached reality. Some experiments have been conducted with ASATs fired from high-flying aircraft into space, and in June 1984, in what was known as the Homing Overlay Experiment (HOE), a ground-based missile successfully intercepted and destroyed a dummy warhead high over the Pacific, but the system is nowhere near completion. Sustained firing of laser or charged-particle weapons would require an enormous deployment of killer satellites and fuel, particularly since the Americans are constrained by the Outer Space Treaty of 1967 from using controlled nuclear explosions in space to produce the necessary energy. Even so, the mere chance of SDI success must worry the Soviets and threaten the central balance.

A sudden technological breakthrough of this kind would constitute vertical proliferation – the increase of capability among existing nuclear powers. Equally worrying is the possibility of horizontal proliferation, spreading nuclear capability to states hitherto excluded from the "nuclear club". By the late 1980s, this club still contained only five proven members – the United States, Soviet Union, Britain, France and China – with most commentators agreeing that Israel, despite denials, should be in-cluded as an "associate member". That alone is worrying enough, for if nuclear weapons should ever be used in the Middle East it would be difficult for the superpowers to remain aloof – but it may be no more than the tip of the iceberg. Iraq, currently involved in a long-running and bitter war in the Gulf, is generally accepted to have the potential to produce a nuclear device (hence the Israeli airstrike on the reactor outside Baghdad in June 1981), while Pakistan, with its long history of conflict with India, is also close. India itself test-exploded a device as long ago as 1974, and although there is no evidence of a delivery system having been developed, the potential for horizontal proliferation in the sub-continent is clearly high. What may happen, of course, is that a balance of capability mirroring that between the superpowers will be created, but if newly-capable nuclear powers should ever use rather than merely threaten to use their weapons, local wars could easily draw the superpowers into a confrontation not of their making, negating the central balance. This is particularly worrying if the weapons should reach the hands of "irrational" leaders of smaller states – men whose ideological beliefs leave them unmoved in the face of deterrent threats to their cities or people.

This particular problem highlights a third area of potential weakness in the deterrent balance: the whole process is controlled by human beings who, by their very nature, are fallible. The chances of the balance being upset through human error or miscalculation are, and always have been, high. On June 3 1980, for example, the Americans went on to alert when the computers at NORAD – the North American Air Defense system – insisted on tracking

An AGM-86B Air Launched Cruise Missile (ALCM) is dropped from a Boeing B-52G Stratofortress bomber during testing in the mid-1970s. The ALCM (up to 20 of which may be carried by a B-52, 6 under the wings and 14 in a bomb-bay dispenser) is normally fired some 1,550 miles (2,500km) from its target. Small wings then fold out and the missile glides down to its operating height of 50ft (15m). A special Terrain Contour Matching (TERCOM) guidance system ensures extremely high accuracy.

incoming Soviet missiles. In this particular case, after just over three minutes of heart-stopping action, the fault was traced to a short-circuit in a silicon chip which cost less than 50 cents to manufacture. During the subsequent Congressional Inquiry, it was reported that this was by no means an isolated incident, for a total of 3,703 alarms had occurred in the previous 18 months, including the inadvertent use of a "simulated attack" computer tape during a routine exercise, which led one ICBM Bombardment Wing to assume a high level of preparation for war. The problems in all cases were solved easily and quickly, but the potential for disaster is clearly constant. Similar problems in the Soviet Union are not reported, although the apparent ease with which the Chernobyl disaster occurred in April 1986 does little to allay public fears.

Peace movements

All this has led, understandably, to the growth of peace movements demanding an end to the "nuclear nightmare", pointing out the basic immorality and practical dangers of a dependence on weapons of assured destruction. Their arguments are attractive and, to a certain extent, valid, but the fact remains that deterrence has worked, preventing a superpower war for more than 40 years. Furthermore, although both horizontal proliferation and human error must be causes for deep concern, the United States and Soviet Union have displayed growing responsibility as nuclear powers, maintaining the central balance and even negotiating to ensure its continued effectiveness. The ABM Treaty and the subsequent adoption of a "common ceiling" of deployed delivery means – the SALT–I "pack-

STRATEGIC NUCLEAR FORCES 1987			
		USA	*USSR*
ICBMs	Missiles	1018	1398
	Warheads	2118	6420
SLBMs	Missiles	616	979
	Warheads	5536	2787
Long-range	Aircraft	180	170
Bombers	Nuclear Weapons	2520	680

A comparison between the nuclear arsenals of the two superpowers. These figures reflect the situation before the INF treaty of December 1987, and may well represent the high-water mark of nuclear stockpiling.

age" – was a major step in this direction, and although SALT–II (1979) failed to bring that ceiling down in practical terms – by then, MIRV technology had made the counting of delivery systems meaningless – and the threat of SDI remains, there is no reason to presume an imbalance at the present time. Indeed, the Intermediate Nuclear Force (INF) Agreement, signed by the superpower leaders on December 8 1987, could lead to a more substantial dismantling of weapons (disarmament) rather than just their control. After all, enormous "overkill" is unnecessary; so long as both sides fear the consequences of a nuclear strike, deterrence will persist. There is no evidence to suggest that such a fear has disappeared.

The superpowers' leaders meet in Washington, December 1987. As Mikhail Gorbachev of the USSR is greeted by US President Ronald Reagan, they prepare to sign an Intermediate Nuclear Force (INF) agreement which will lead to the elimination of land-based theatre nuclear forces in Europe: Soviet SS-4s and SS-20s; US GLCMs and Pershing IIs (the "double-zero" option). Although this was an historic occasion – the first at which the superpowers agreed to get rid of nuclear weapons – it was seen as only the beginning of more dramatic arms control, affecting the strategic arsenals on both sides. Such a policy of arms control/disarmament is extremely valuable if the two superpowers wish to avoid an uncontrollable (and hideously expensive) arms race, but the maintenance of a "balance of capability" is essential to mutual deterrence.

— chapter twenty —
An unquiet peace
John Pimlott

Since 1945, there has been no major war between the great powers of the world. The trauma of World War II, with its emphasis on total political and military aims, has not been repeated, and although this has not prevented conflict at lower levels, between states intent on using force as an instrument of restricted policy, the nightmare of full-scale, global war has been avoided.

The main reason for this lies in the development and deployment of nuclear weapons, the existence of which effectively prevents fighting wars at the "no-holds-barred" levels of World War II. Indeed, since 1945 one of history's periodic "revolutions in strategy" has taken place, characterized by a shift away from war-fighting to war-prevention among those states with nuclear capability, as they realize that it is equally productive (and infinitely safer) to threaten with rather than to use weapons of such awesome destructive potential. The situation which exists between the major superpowers, the United States and Soviet Union, illustrates the point. Despite the presence of many of the traditional causes of conflict between them – a deep ideological divide, territorial rivalry, mistrust and misperception – they have refrained from overt use of force to achieve their respective aims in the long-running confrontation known as the "Cold War". Defined as "hostility short of actual fighting", this has not been without its dangers – the Cuban Missile Crisis of October 1962, for example, came perilously close to a "shooting war" – but the realization that any recourse to nuclear exchange would be in the interests of neither one side or the other has put a "lid" on the unrestrained use of violence among the major powers. Under the "rules" of Mutual Assured Destruction (MAD), such recourse would be unquestionably suicidal.

But this does not mean that wars have ceased to occur in the modern world. At a rough estimate (and depending on how strictly one defines the term "war"), some 130 wars have taken place since 1945, affecting the lives of more people than the two world wars together and inflicting many millions of casualties. It is obvious, then, that the use of force as an instrument of state policy is still an attractive option beneath the level of superpower confrontation, and there is little evidence to suggest that this will change. Even among the nuclear-capable states, restricted violence, consciously avoiding escalation to levels at which weapons of proven mass destruction are used, has its place. The world may have avoided a nuclear holocaust, but it is still a violent and dangerous place.

After Hiroshima and Nagasaki
At first, as the full implications of the atomic attacks on Japan in 1945 sank in, a feeling arose – in the United States at least – that war had suddenly become too horrific to contemplate. Any crisis or confrontation would be defused as soon as an opponent was threatened with the frightening destructive capability of new weapons, precluding the need for large armed forces, whose task in future war would be to act as little more than a "tripwire" to atomic release. Even before such thinking had been rationalized into the strategy of Massive Retaliation (1954), the Americans had shifted the emphasis of their force levels away from large conventional armies and had concentrated much of their potential in a separate Strategic Air Command (SAC), the bombers of which would pose a constant threat – in conventional as well as atomic terms – to would-

be enemies. War Plan "Trojan" in 1949, for example, envisaged a situation in which Soviet cities would be destroyed with a mixture of conventional and atomic air-delivered weapons, with virtually no land or naval operations taking place.

This doctrine was severely tested in June 1950, when forces from communist North Korea, backed and equipped by the Soviet Union, swept across the 38th parallel in a conventional invasion of South Korea. Such was the confidence in SAC that the Americans had reduced their force levels to a dramatic extent; despite their political commitment to the protection of South Korea, no US units were available to prevent a communist advance through Seoul and as far south as Pusan. Indeed, when President Truman had created what was known as his "containment wall" around the communist bloc a few months earlier, South Korea had appeared on the maps on the "wrong" side, implying that it had been written off already as territory impossible to defend.

The world is not a peaceful place for the majority of its inhabitants. Since 1945 few areas have avoided the effects of war, although the scale of conflict has varied considerably. Some regions – notably the Middle East and Southeast Asia – have felt (and continue to feel) the full weight of conventional fighting, with armed forces using the entire range of non-nuclear weapons in a bid to gain victory. Elsewhere, as in Africa and Latin America, the scale of the fighting may, with certain exceptions such as the Falklands, have been less, but the damage in political, social, economic and human terms can be just as serious.

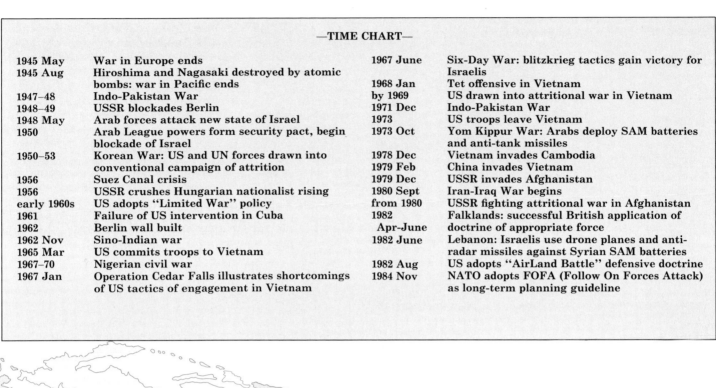

—TIME CHART—

1945 May	War in Europe ends
1945 Aug	Hiroshima and Nagasaki destroyed by atomic bombs: war in Pacific ends
1947–48	Indo-Pakistan War
1948–49	USSR blockades Berlin
1948 May	Arab forces attack new state of Israel
1950	Arab League powers form security pact, begin blockade of Israel
1950–53	Korean War: US and UN forces drawn into conventional campaign of attrition
1956	Suez Canal crisis
1956	USSR crushes Hungarian nationalist rising
early 1960s	US adopts "Limited War" policy
1961	Failure of US intervention in Cuba
1962	Berlin wall built
1962 Nov	Sino-Indian war
1965 Mar	US commits troops to Vietnam
1967–70	Nigerian civil war
1967 Jan	Operation Cedar Falls illustrates shortcomings of US tactics of engagement in Vietnam
1967 June	Six-Day War: blitzkrieg tactics gain victory for Israelis
1968 Jan	Tet offensive in Vietnam
by 1969	US drawn into attritional war in Vietnam
1971 Dec	Indo-Pakistan War
1973	US troops leave Vietnam
1973 Oct	Yom Kippur War: Arabs deploy SAM batteries and anti-tank missiles
1978 Dec	Vietnam invades Cambodia
1979 Feb	China invades Vietnam
1979 Dec	USSR invades Afghanistan
1980 Sept	Iran-Iraq War begins
from 1980	USSR fighting attritional war in Afghanistan
1982 Apr–June	Falklands: successful British application of doctrine of appropriate force
1982 June	Lebanon: Israelis use drone planes and anti-radar missiles against Syrian SAM batteries
1982 Aug	US adopts "AirLand Battle" defensive doctrine
1984 Nov	NATO adopts FOFA (Follow On Forces Attack) as long-term planning guideline

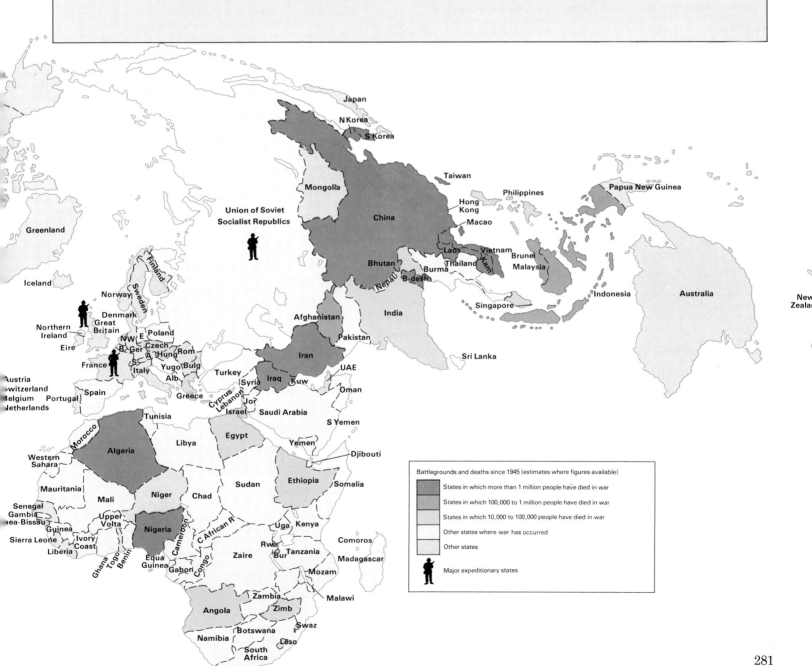

Battlegrounds and deaths since 1945 (estimates where figures available)

- States in which more than 1 million people have died in war
- States in which 100,000 to 1 million people have died in war
- States in which 10,000 to 100,000 people have died in war
- Other states where war has occurred
- Other states
- Major expeditionary states

Realizing the mistake, Truman hastily committed US forces to Korea, under the command of General Douglas MacArthur. Units were withdrawn from occupation duties in Japan and sent to bolster up the defences around Pusan, but it soon became obvious that they were incapable of achieving much on their own. According to prevailing doctrine, the Americans should have released their airpower in a crushing display of force, using both conventional and atomic weapons to punish the enemy for his actions. However, the situation was not straightforward enough to merit such a response. Although the Soviets were undoubtedly backing the North Koreans, none of their forces was directly involved in the fighting, so attacks on the Soviet Union could hardly be justified, while the commitment of what was still a relatively small atomic stockpile to a peripheral area like Korea would leave US interests elsewhere unprotected. As some commentators noted, Korea might be nothing more than a feint, forcing the Americans to expend their atomic weapons preparatory to a Soviet invasion of Western Europe. For these reasons, Truman could not use his full power – in 1951 he was even forced to sack MacArthur when the latter tried to insist on atomic attacks – and had to look elsewhere for instruments of military pressure. Regardless of the arguments, atomic weapons, as they existed in 1950, were incapable of preventing war at every level: conventional forces still had a vital role to play.

Once this was recognized, US armed forces were expanded and, in conjunction with units drawn from anti-communist members of the United Nations, committed to a conventional war in Korea. Their potential was shown as early as September 1950, when MacArthur used US Marines – the only available strategic reserve – to carry out an audacious "left hook" at Inchon, liberating Seoul and forcing the North Koreans into a precipitous retreat. Pursuing them across the 38th parallel, UN units looked set to reunify Korea by force, but in November the Chinese, fearful of the apparent threat to their border on the Yalu river, intervened. By sheer weight of numbers, they pushed the UN contingents back beyond Seoul; UN counter-attacks in 1951 regained the city, but the war quickly degenerated into an attritional slogging match around the 38th parallel. It was to continue until July 1953, ending in an armistice which did no more than preserve the *status quo ante bellum*. Fearing events elsewhere, the Americans had fought at significantly less than their full potential, avoiding escalation but sacrificing victory.

The strategy of Massive Retaliation

Few countries enjoy wars with a less than decisive outcome, and the United States is no exception. Far from accepting that Korea might be a pattern for warfare beneath the atomic level, President Eisenhower, Truman's successor, looked for ways to strengthen the atomic threat and thus make the use of force an unattractive option to America's enemies anywhere around the globe. The result was the strategy of Massive Retaliation, stating unequivocally that any future assault on the "containment wall" would elicit an atomic (or, by 1954, a nuclear) response. It was a threat which could not be ignored so long as America retained a monopoly of global delivery means.

This began to change in the late 1950s, when it became

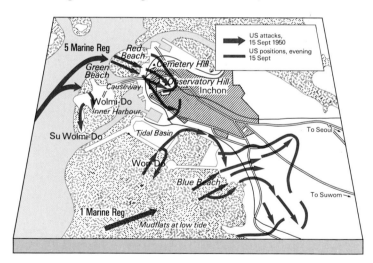

The North Korean invasion of the South (*left*) in June 1950 triggered a war which was to last for three years, drawing in the United States, a number of United Nations' members and the newly communist Chinese. Despite mobile battles in the early stages, the conflict was characterized by attritional stalemate.

The US landing at Inchon (*above*) in September 1950 was a decisive "left hook", designed to outflank the North Koreans besieging Pusan. A beachhead was established despite enormous problems, enabling US troops to advance inland to liberate Seoul. The North Koreans withdrew across the 38th parallel, closely pursued by UN forces.

obvious that the Soviets were developing their own intercontinental nuclear capability: once they gained the ability to hit back, destroying American cities in turn, Massive Retaliation ceased to be a credible strategy. If this was not to lead to a nuclear stalemate, beneath which Soviet expansion would be impossible to stop, a significant rethink about the use of force had to take place. It was provided by a number of American theorists who collectively advocated a policy of "limited war" in which conventional force could be used, subject to conscious restraints designed to prevent escalation. While the existence of nuclear weapons would continue to prevent all-out nuclear war, they argued, enhanced conventional capability would ensure that attacks at lower levels were contained and the communists kept in check.

One of the most influential of these theorists was Robert Osgood, whose book *Limited War* appeared in 1959. In his search for evidence to show that conventional warfare was still possible in the nuclear age, he turned to the events in Korea a few years earlier. Despite the lack of a clear-cut victory by 1953, it was apparent that the tide of communist expansion had been stemmed, and, according to Osgood, this had been achieved by the Americans adopting a number of conscious constraints. With the exception of a two-month period in 1950, when MacArthur's forces had invaded North Korea, the aim of the war had been to keep the fighting contained to a relatively small area around the 38th parallel. At no time had the Americans tried to defeat China or the Soviet Union – indeed, US pilots had been under strict orders not to fly close to the border of either state for fear of provoking a response. Moreover, the Americans had physically prevented the movement of Nationalist Chinese forces from Taiwan to the mainland, for although this would undoubtedly have diverted communist attention from Korea, it would have widened the war considerably. Within Korea itself, there had been restraints on the use of particular weapons, not just in terms of atomic bombs but also of long-range aircraft: despite the unquestioned ability of the Americans to flatten North Korean cities using conventional air weapons, such a policy had not been adopted, again for fear of the possible consequences. In short, limitations of aim, geography, force commitment and weapons had been deliberately adopted and an atomic confrontation avoided.

US policy in Vietnam

These limitations were formed into a conscious strategy in the 1960s, when the Soviets clearly began to match the American capability to wage nuclear war on a global scale. They were put into effect again in Vietnam, when US forces faced what appeared to be a situation very like that in Korea. Between 1965 and 1973, US main-force units in Vietnam operated under considerable restraint: the political aim was restricted to the preservation of South Vietnam against a communist threat from the North; at no time was the destruction of the Hanoi regime actively sought, in case its communist allies intervened; US aircraft flying over the North, particularly during Operation Rolling Thunder (1965–68), were deliberately kept away from the Chinese border as well as the population centres of Hanoi and Haiphong; atomic or nuclear weapons were not used and, despite an eventual commitment of more

General Douglas MacArthur (*above*), American commander of UN forces in Korea, addresses the people of newly liberated Seoul, October 1950. His belief in the need for all-out victory threatened to escalate the Korean conflict, drawing in the major communist powers. He was sacked by President Truman in April 1951.

US troops (*right*) move through typically rugged terrain in Korea, 1951.

than 500,000 US troops, the full potential of American armed might was not brought to bear. Hardly surprisingly, the results were the same as in Korea: by 1973 a political compromise had been hammered out which seemed to guarantee the integrity of South Vietnam while allowing the Americans to withdraw.

"Limited war" viable?

There the similarities ended, however, for within little more than two years the South had fallen to invasion from the North, with the Americans, sick of war, doing nothing to prevent it. This implied that "limited war" was no longer a viable strategy: regardless of the power of the Americans, they could not withstand the frustration and attritional cost of a conflict which the enemy was both willing and able to prolong. Nor was the United States the only nuclear-capable country to experience this, for in the years after 1979 the Soviet Union, caught in a very similar situation in Afghanistan, faced the same kinds of problems. What should have been a short, sharp intervention designed, in Soviet eyes, to preserve a friendly regime in Kabul, soon developed into a prolonged attritional war which they dared not use their full potential to resolve. The Moscow Politburo might not have talked of limited war as such, but they too were forced to recognize the limitations of their power: to use maximum force (including nuclear weapons) was out of all proportion to the nature of the threat and was fraught with the danger of undesirable escalation; thus, conventional armed services, operating under restraint, were left to fight a difficult and essentially unwinnable war.

Superpower experience in these wars seems to suggest that restraint is self-defeating, but it is interesting to note that other members of the "nuclear club", perhaps less fearful of provoking a major confrontation, have managed to use force effectively at conventional levels. In 1979, for example, the Chinese invaded northern Vietnam and, despite a number of problems, succeeded in fighting a short, sharp campaign which did not degenerate into either nuclear war or attritional stalemate. More dramatically, in 1982 the British fought and won a war in the South Atlantic, liberating the islands of the Falklands and South Georgia from Argentinian control in a campaign lasting less than 14 weeks. In this particular case, all the restraints associated with limited war were apparent: the British pursued the precise aim of regaining possession of the islands and at no time even contemplated the overthrow of the regime in Buenos Aires; the weapons and forces used were restricted to those needed to gain victory according to this limited aim, with no attacks on the Argentinian mainland or deployment of nuclear capability; the campaign was confined to a relatively small area of the South Atlantic and the chances of escalation were minimized by a successful diplomatic offensive to isolate the enemy preparatory to military action. If this had been a war involving one of the two main superpowers, it might have followed the pattern of Vietnam or Afghanistan; as it was, Britain was able to manipulate events to produce a situation in which highly professional, well motivated forces gained unquestioned victory in a short, sharp display of appropriate power beneath the nuclear level.

But if it is wrong to dismiss limited war as an unwinnable

Royal Marine Commandos (*left*) march across East Falkland, late May 1982. The British landing on the west coast caught the Argentinians by surprise: the subsequent breakout by 45 Commando and 3 Para in the north across difficult terrain trapped the Argentinians in Stanley, enabling the British forces to conduct a final, victorious offensive in mid-June.

HMS *Coventry*, hit by Argentinian bombs, May 25 1982.

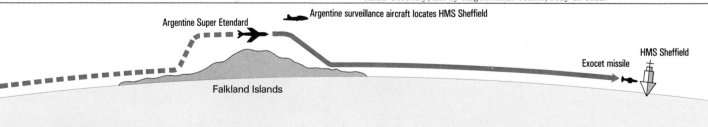

Argentine surveillance aircraft locates HMS Sheffield

Argentine Super Etendard

HMS Sheffield

Exocet missile

Falkland Islands

The Argentinian air attack on the British Type 42 destroyer HMS *Sheffield* on May 4 1982 came as a profound shock. Despite an array of radars and detection devices, the warship was hit by a French-manufactured Exocet sea-skimming missile, fired from a Super Etendard attack aircraft. Flying from bases in Argentina, the Super Etendards picked up their target, thought to be a British aircraft carrier, and launched their missiles at low altitude some distance away, before turning rapidly for home. Twenty British sailors died and *Sheffield* was lost.

method of fighting, it is equally wrong to presume that all wars since 1945 have been conducted under conscious restraint. If the ultimate aim of limited war is to avoid escalation to levels of totality, the strategy can only be adopted by states which enjoy the capability of conducting total war on the global, massively destructive scale associated with World War II. In the modern age, this means states which possess and deploy nuclear weapons: an exclusive "club" which contains only five proven members. The rest of the world – a further 162 recognized states – does not have the capacity to escalate its conflicts to this level and so, regardless of its aims or methods of fighting, it cannot engage in limited war as such: indeed, so long as nuclear-capable states are not dragged into its conflicts, they may be fought at levels which, to the countries involved, are as unrestricted as they care to make them. This helps to explain why so many "local" wars have taken place since 1945, despite the creation of the nuclear "lid": to the overwhelming majority of states, nuclear weapons are totally irrelevant.

Wars since 1945
Such a lack of conscious restraint has meant that conventional wars may be fought along traditional lines, using military forces to their full potential within the context of their locality. This has been reflected in the plethora of wars which have occurred since 1945, caused by a variety of traditional grievances. The most violent and prolonged have been triggered by disputes over territory, particularly if it contains a valuable resource or commercial commodity, and by ideological or religious differences. In the Middle East, for example, the territory held by Israel has provoked a seemingly endless series of conventional wars, waged against Arab powers which not only lay claim to the land in the name of the Palestinians but also resent the intrusion of Judaism into a Muslim area of the world. Similarly, in the Arabian (Persian) Gulf, the long-running Iran–Iraq War may have started in 1980 over possession of the Shatt al Arab waterway but it has been fuelled by deep ideological divisions between the Shiah and Sunni Islamic sects. In Africa, wars have been based on ethnic or tribal divides – the bitter conflict between Nigeria and secessionist Biafra between 1967 and 1970 is a case in point – while in the Indian subcontinent the territorial and religious disputes between India and Pakistan have led to wars in 1947–48, 1965 and 1971. Elsewhere in Asia, wars have occurred between Indonesia and Malaysia (1963–66) and Vietnam and Kampuchea (Cambodia) (1978–   ). In Central America, El Salvador and Honduras even came to blows in July 1969 over a World Cup football match, which brought to a head deep-seated grievances over nationalistic and territorial disputes. In every case, the fighting was not consciously constrained, following patterns of warfare familiar throughout history: as political or diplomatic manoeuvres failed, countries were quite prepared to use force as an instrument of national policy.

What has changed, however, is the nature of such wars, for the period since 1945 has seen a massive leap in weapons technology at the conventional level, coupled to a considerable increase in the availability of such weapons. Much of this has been triggered by the superpowers and their immediate allies in response to the demands of the Cold War

SUPPLIES TO THE FALKLANDS

The Argentinian invasion of the Falklands and South Georgia in early April 1982 caught the British by surprise. Although a Task Force was put together immediately, with the aim of regaining control if diplomacy failed to achieve a settlement, it faced a nightmare logistic chain, stretching some 8,000 miles from the UK to the South Atlantic.

However, contingency plans for just such an emergency did exist, and one of the first actions of the British government was to order the requisition of merchant ships. Eventually, more than 50 merchantmen, including container ships, ferries and oil tankers, were provided, maintaining a constant flow of supplies throughout the war. One container ship, *Atlantic Conveyor* was lost, destroyed by an Argentinian Exocet missile on May 25 1982, but the rest survived. Their contribution, together with that of the Royal Fleet Auxiliary, was crucial to British victory.

The Iran-Iraq War (see map, *right*) entered its eighth year in September 1987, with casualties estimated at over a million dead. Despite Iranian numbers and fervour – epitomized by the Revolutionary Guards (*above*) – Iraq has survived using massive firepower, but has failed to achieve victory.

confrontation, but the field-testing of new weapons, out of which further development occurs, is necessarily left to countries actively engaged in conventional war. Since 1945 the methods of waging war may not have changed significantly – the tendency in conventional terms is still to aim for mobility and hitting power on the battlefield, using armour and airpower as the major instruments – but the capabilities of individual systems and their impact have been progressively improved. Tanks have been better armed and armoured, with more reliable engines; aircraft have become faster and their weapon loads have increased dramatically in terms of both accuracy and destructive power; individual weapons in the hands of ordinary soldiers have been characterized by increased range and kill capability. The result has been a much more dangerous battlefield, with few targets remaining immune to fire, as well as a much wider battle area, within which weapons are effective. Add to this the concurrent improvements to back-up systems such as reconnaissance, surveillance, transport and communications, and it quickly becomes apparent that the physical environment of battle has changed considerably.

One of the trends which emerges is that the pendulum of advantage between the offence and defence in conventional war has swung rapidly from side to side, depending on the availability and capability of individual weapons systems. By 1945, despite the Allied victory over the Axis powers, most commentators agree that, in technological terms, the advantage lay with the defender: tanks such as the German King Tiger were designed specifically to withstand attack, being fitted with thick armour plate and large-calibre main guns which increased their weight and prevented fast movement over difficult terrain; anti-tank guns had been made more powerful and anti-aircraft systems, both ground-based and in terms of interceptor fighters, had been improved. At the same time, the issue of automatic weapons to ordinary soldiers had improved their ability to lay down defensive fire, while such technological innovations as infra-red enhancement – introduced by the Germans in 1944 in the shape of night-vision equipment – made the build-up of enemy forces prior to an attack easier to monitor. The full effects of such developments were disguised by the fact that the Allies maintained offensive capability by sheer weight of numbers, but the trend was there all the same. It was to be shown more clearly during the attritional battles of the Korean War a few years later.

Conflict in the Middle East

But the pendulum did not remain static – indeed, since 1945, as technology has continued to improve, it has begun to swing wildly from one side to the other, until it is sometimes extremely difficult to say with any accuracy where it stands at any one time. To illustrate this, it is best to look at a series of related conflicts in the modern period. As the longest-running dispute and the one which, sadly, has produced the most frequent armed clashes, the conflict between Israel and her Arab neighbours provides the most reliable evidence, significantly enhanced by the fact that both superpower blocs have used the various wars in the Middle East to field-test new equipment through their respective allies.

An Israeli Super Sherman (*left*) in Sinai, June 1967. Israeli success in 1967 led many to believe that tanks could achieve victory on their own; by October 1973 infantry in armoured personnel carriers (*above*) had to be provided to suppress Egyptian Sagger anti-tank missiles.

The Israeli assault on Sinai, June 1967 (*right*). As soon as Egyptian defences had been breached at Khan Yunis, Abu Ageila and Kuntilla on June 5/6, Israeli armour pushed deep into Sinai; the east bank of the Canal was reached on the 8th.

In order to survive, Israel has had to adopt a distinctive strategy. The country is small and surrounded by potential or actual enemies, its economy is not strong and its population is considerably outnumbered by that of its neighbours. At the same time, Israel can make use of such latent strengths as internal lines of communication, Arab disunity and the Israeli will to survive. In response to the demographic and economic problems, Israel has become a "nation in arms", training nearly all its young men and women in military affairs and keeping them ready, as reservists, to be mobilized in time of crisis. Because any mobilization unavoidably disrupts the running of the state and its economy, however, it is essential that wars be as short as possible, inflicting decisive defeats at minimum cost. Moreover, given the small size of Israel, it is imperative that all fighting should take place on Arab soil. These considerations have meant that Israel has to be constantly on guard, monitoring Arab moves to ensure that it is not caught by surprise, before mobilization has been ordered: indeed, the aim is one of pre-empting any perceived Arab threat, hitting the enemy before he crosses the borders of Israel. As speed is clearly of the essence, an Israeli emphasis on armour and airpower, combined to produce the impact and mobility of blitzkrieg, is inevitable, backed by a ready acceptance of UN-imposed ceasefires before the enemy can recover and at a time of maximum Israeli territorial gain.

The Six-Day War
This strategy was shown in its purest form during the Six-Day War of June 5–10 1967, fought in response to a build-up

of Egyptian forces in the Sinai desert and an apparent attempt by President Nasser of Egypt to coordinate the actions of the Arab powers along Israel's other borders. As soon as the threat was clear, Israel mobilized her reserves and mounted a highly effective pre-emptive airstrike against Egyptian airfields. Going in at 0845 hours (Egyptian time) on June 5, when dawn patrols had returned and the aircraft were being refuelled, Israeli fighter-bombers destroyed nearly 250 Egyptian machines on the ground. This was followed by similar strikes against Jordanian, Syrian and Iraqi airfields, ensuring that the Israeli Air Force (IAF) enjoyed complete air superiority from the start of the campaign, something that was essential to the success of the armoured assaults which then went in. By the end of June 5, three armoured columns had broken into Egyptian border defences in Sinai, and although the fighting was occasionally hard, enough had been done to ensure blitzkrieg success. On June 6 and 7, Israeli armour moved deep into Sinai, with a small tank force racing through Egyptian lines to take up blocking positions at the eastern end of the Mitla Pass, the only easy line of retreat across the Suez Canal. Caught between the hammer of the columns advancing from Israel and the anvil of the Mitla Pass, Egyptian units broke and fled. By June 8, Sinai was in Israeli hands.

Israel gains territory
By then, a similar campaign had secured Jerusalem and the West Bank (Judaea and Samaria) from the Jordanians, and on June 10-11 the Golan Heights were seized from Syria. At that precise moment a UN ceasefire, sponsored by the

Ariel ("Arik") Sharon (b.1928) a brilliant if somewhat reckless commander, was born in what was then British-ruled Palestine. As a member of the Haganah (Jewish Defence Force), he fought in the guerrilla campaign against the British and in Israel's "War of Independence" (1948–49) against invading Arab powers. In October 1956 he spearheaded the successful Israeli attack on Egypt in Sinai, although he was criticized for exposing his soldiers to an unnecessary

battle in the Mitla Pass. This did nothing to curtail his military career. During the Six-Day War (June 1967) he commanded one of the armoured divisions which defeated the Egyptians in Sinai; in 1973, by which time he had entered politics as a founder of the *Likud* Party, he led a reserve armoured division to victory in the Yom Kippur War, crossing the Suez Canal to encircle an entire Egyptian army. By June 1982 he was Minister for Defence and, as such, was responsible for Operation Peace for Galilee, Israel's invasion of southern Lebanon. However, in September 1982, he was sacked in the aftermath of the Sabra-Chatilla massacres in Beirut, despite the fact that these were actions carried out by Israel's Lebanese allies, who should have been under stricter control. In 1984 Sharon was rehabilitated as Minister of Trade and Industry.

287

United States, came into force, leaving Israel in possession of 26,000 sq miles (67,000 sq km) of additional territory, much of which enhanced the long-term security of the state. With Sinai as a buffer and the Suez Canal as an effective barrier, attacks from Egypt could be contained, while possession of the West Bank pushed the Jordanians

---

"We went into the Old City [Jerusalem] and from then on it was hand-to-hand and house-to-house. That's the worst thing in the world. In the desert, you know, it's different. There are tanks and planes and the whole thing is at a longer range. Hand-to-hand fighting is different, it's terrible. I killed my first man there . . . All of a sudden I saw this man coming out of a doorway . . . We looked at each other for half a second and I knew that it was up to me, personally, to kill him, there was no one else there."

Israeli paratrooper, the Six-Day War of June 1967

---

back, out of positions from which they had threatened the thin coastal strip of Israel. Seizure of the Golan prevented Syrian observation of the northern Galilee area and placed the Israelis in a position to monitor activity on the Damascus plain. It was a remarkable achievement and a complete vindication of the strategy so carefully thought out by the Israelis.

But the defeated Arab powers were unlikely to accept the results and, in the years immediately after the 1967 War, they searched for ways to blunt Israel's offensive capabilities. By 1969 the Egyptians had realized that the Israeli strategy depended on wars of short duration and,

during the "War of Attrition" (which lasted until August 1970), they mounted raids into Sinai, designed specifically to force Israel to maintain at least a partial mobilization. The Israelis hit back, even attacking Cairo, and it was out of this escalation that a more permanent answer to the Arab problem emerged: the use of new technology to counter the specific Israeli advantages of armour and airpower. The pendulum, hitherto inclined firmly in favour of the offensive in the Middle East, was about to swing to the opposite extreme.

The key to Arab recovery lay in a variety of Soviet-supplied weapons systems. In air-defence terms, these included both surface-to-air missiles (SAMs) and radar-controlled cannon (ZSU-23/4s), deployed initially to protect Cairo against Israeli retaliatory raids during the War of Attrition. By the early 1970s, these had been combined to produce an air-defence "umbrella" of significant potential. Any hostile aircraft flying towards Egypt at high-to-medium altitude would be picked up on the SAM radars and intercepted by heat-seeking SA-2 or SA-3 missiles – static systems which, because of their "slant range" if fired at less than the perpendicular, extended their coverage to take in most of western Sinai from positions inside Egyptian territory. If the aircraft chose to approach at medium-to-low altitude, more mobile systems – the SA-6, SA-7 and ZSU-23/4 – would create a similar barrier, denying to the IAF the ability to fly close-support missions in order to open the way for an armoured assault. At the same time, infantry-operated anti-tank guided weapons (ATGWs) such as the AT-3 Sagger would give the Arabs the ability to disrupt armoured momentum, forcing the Israelis

French-manufactured Mirage III-C fighters (*left*) of the Israeli Air Force, March 1967. Acquired by the Israelis to counter Soviet equipment in Arab hands, the Mirage, with a maximum speed of 863mph (1390kph) and a weapons kit of two 30mm DEFA cannon, proved its value during the Six-Day War.

Sinai, June 1967 (*above*). One advantage enjoyed by the Israelis was their ability to depress the tank guns of their Super Shermans, allowing them to hide behind sand-dunes in ambush positions. By comparison, Soviet-supplied Egyptian tanks had to mount the sand-dunes to fire.

to fight attritional rather than blitzkrieg campaigns. By 1973, with all these systems available, the Egyptians and the Syrians had decided that they were in an ideal position to go to war on more equal terms, forcing Israel to give up the territory gained in 1967.

On October 6 1973, the Arabs attacked across the Suez Canal and onto the Golan, catching the Israelis by surprise (it was the day of Yom Kippur – an Israeli national holiday) and seizing vital ground before the Israelis could order a mobilization of reserves. The Arabs deployed both anti-tank and anti-aircraft systems to disrupt Israeli counter-attacks, and it began to look as if they would gain an unprecedented victory, especially on the Golan, where Syrian armour and mechanized infantry maintained a relentless pressure. In the event, however, it was the Egyptians who posed the more worrying long-term threat, for once across the Canal – the crossing in itself was a remarkable achievement, made possible by the use of high-pressure hoses to destroy the Canal banks and open the way for Soviet-supplied pontoon bridges – they deliberately created a "brick wall" of defences, hoping to force the Israelis to batter themselves against it until lengthy mobilization or high casualties gave them no alternative but to accept peace talks on Arab terms. As Israeli aircraft fell to SAM/ZSU defences – 14 percent of the IAF's front-line strength was lost in the first 48 hours alone – and armoured counterattacks failed to breach the Sagger line, the strategy seemed close to success.

But the Israelis were swift to respond, evolving a package of effective countermeasures while the war was going on. In the air, the IAF changed its attack profile, so that instead of flying in at high altitude before sweeping down onto the target and then escaping high and fast – a technique virtually guaranteed to attract every SAM for miles around – the pilots went in low, attacked low and escaped low, often at less than 50ft (15m). This placed the aircraft beneath the effective cover of the SAM/ZSU radars, which could not distinguish between attacking aircraft and "ground clutter" below 50ft (15m), and when low flight was combined with such technical aids as heat flares (designed to divert heat-seeking missiles from the exhausts of fast jets), "chaff" and electronic countermeasures (ECM) pods (to disrupt SAM/ZSU radars), it was quite possible for aircraft to survive over the battle area. At the same time, new tactics on the ground, including the close coordination of armoured and mechanized infantry formations, effectively countered the Saggers, the kill-capability of which had been found to be limited. By October 15, when the Israelis mounted their main counterattack across the Canal into Egypt, infantrymen in M113 armoured personnel carriers (APCs) were laying down suppressive machine-gun fire on any position that might hide a Sagger operator, opening a route for the armour to recreate the impact and mobility of blitzkrieg. Within a matter of days, the pendulum had been forced back in favour of the offensive.

Operation Peace for Galilee

This was reinforced by the events of June 1982 when, in response to Palestinian artillery and rocket attacks on Jewish settlements in northern Galilee, the Israelis carried out an invasion of southern Lebanon (Operation Peace

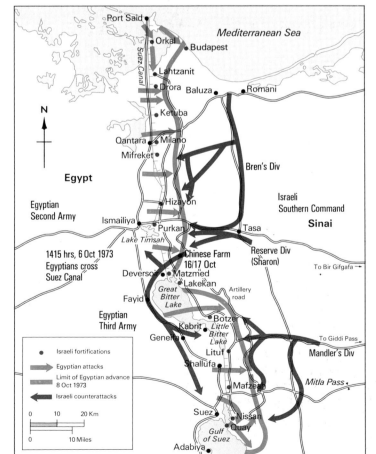

The campaign in Sinai, October 1973 (*left*). Egyptian attacks across the Canal on October 6 came as a surprise to the Israeli defenders and, in the early stages, their reaction was poor. On October 14/15, however, they counterattacked at Deversoir, crossing the Canal and encircling the Egyptian Third Army.

The common denominator of war (*above*) – an Israeli soldier, well dug in, watches for Arab attack, Sinai, June 1967. In the 1967 War, with armour seeming to win victories on its own, infantry had little part to play in Sinai; in October 1973, they returned to the forefront of the battle.

for Galilee) which took them to the outskirts of Beirut in less than five days. Armoured mobility, along the coast against the Palestinians and towards the Bekaa Valley to prevent Syrian intervention, was ensured by a continuation of all-arms coordination (including the provision of mobile artillery and combat engineers) and by enhanced armour protection. The latter took the form of "add-on" ceramic plates to cover the vulnerable parts of Centurion and M60 main battle tanks (MBTs), while the Israeli-produced Merkava MBT incorporated "spaced" armour designed to absorb the impact of most anti-tank projectiles. There were even rumours of "reactive" armour being used, whereby small explosive charges detonate to divert the shock of incoming rounds as soon as the latter touch the tank's protective shell.

In the air, developments were equally dramatic. Using remotely piloted vehicles (RPVs) to fly over SAM sites to trigger the radars – the precise frequencies of which would be monitored and fed into appropriate ECM pods – and to pinpoint the exact location of mobile SAM systems by means of video cameras which relayed "real-time" information back to Israeli command posts, the IAF mounted highly successful strikes into the Bekaa Valley, destroying most of the Syrian systems in a single day. In the event, the fighting around Beirut was to drag on (chiefly because the UN, diverted by the concurrent conflict in the South Atlantic, failed to impose a ceasefire), but in purely military terms, it was obvious that offensive success had been repeated, principally by means of new technology.

Technology versus tenacity

But technology alone cannot guarantee victory in modern war, as the Americans found to their cost in Vietnam between 1965 and 1973. When the decision was taken to deploy US Marines to protect the airbase at Da Nang in March 1965, the American armed services were the most sophisticated in the world. Ground forces were equipped with all the latest technology, from MBTs to automatic small arms, and were backed by aircraft and naval units of enormous offensive capability in conventional terms. Nor was this a static capability, for as the Americans were drawn deeper into the war, conducting "search and destroy" operations against the Viet Cong (VC) in the villages of South Vietnam as well as "main-force" campaigns against the North Vietnamese Army (NVA) along the borders, their involvement fuelled the process of weapons development. By 1972, the last full year of American commitment, an entire generation of new weapons had emerged, including such innovations as "smart" bombs – deadly accurate air-delivered devices which were guided onto their targets by means of laser designation – and purpose-built helicopter gunships, capable of laying down devastating bursts of rocket and machine-gun fire. Indeed, in set-piece engagements with the NVA or in ambushes triggered by the VC, American ground units could call down an unprecedented weight of fire – from attack aircraft, helicopter gunships, field artillery and even B–52 bombers or battleships – while enjoying enhanced mobility through helicopter transportation. In such circumstances, victory should have been assured.

Yet that victory did not materialize, for a variety of reasons. The communists, far from being intimidated by US technology, soon learnt to avoid its effects, exploiting their

knowledge of the ground to mount unexpected attacks which often prevented a concentration of US force by diverting resources away from offensive operations. Even when sustained battles did take place on American terms – as in the Ia Drang and A Shau valleys in 1965–67 or around Khe Sanh in 1968 – the Americans found it difficult to inflict decisive defeats, particularly when, in what soon became an attritional war, the communists proved both willing and able to make enormous sacrifices. Far from "bleeding the enemy white", as General William Westmoreland, commander of Military Assistance Command Vietnam (MACV) from 1964 to 1968, intended, the constant drain on US resources led to domestic opposition to the war, unacceptable political costs and, increasingly, unit demoralization in Vietnam. Furthermore, such "main-force" battles often diverted attention away from communist activities in the villages of South Vietnam, where subversion and intimidation were preparing the way for a

---

"Two hundred metres away, facing the Marine trenches, there was an NVA sniper with a .50 calibre machine gun who shot at the Marines from a tiny spider hole. The marines fired on his position with mortars and recoilless rifles, and he would drop into his hole and wait. Gunships fired rockets at him, and when they were through he would come up again and fire. Finally, napalm was called in, and for ten minutes the air above the spider hole was black and orange from the strike, while the ground around it was galvanized clean of every living thing. When all of it cleared, the sniper popped up and fired off a single round, and the Marines in the trenches cheered. They called him Luke the Gook, and after that no one wanted anything to happen to him."

Michael Herr, war correspondent.

Maoist-style revolution, and when VC guerrilla attacks triggered the same scale of American response as the NVA advances – as during the Tet Offensive of early 1968 – the inevitable destruction involved did little to persuade the people of the South to support their government or its allies. In the end, the Americans used a bludgeon of crude power to fight the war, and although on occasion this could inflict terrible damage on NVA or VC forces, it was not subtle enough to ensure a lasting victory. American soldiers came to depend on firepower and technology, avoiding any attempt to fight the enemy at an appropriate level of response.

"Cedar Falls"

This was shown throughout the war, but one particular operation will illustrate the nature of the problem. By late 1966, it was apparent to the Americans that one of the most dangerous areas of VC infiltration was the notorious "Iron Triangle", less than 40 miles (64 km) to the north of Saigon, the capital of South Vietnam. Bounded by the Saigon and Thi Tinh rivers, with the Than Dien forest reserve to the north, the Triangle was anchored on the villages of Ben Suc, Phu Hoa Dong and Ben Cat. It was thought to contain the "Headquarters, Viet Cong Military Region IV and its base camps and supply bases as well as the 272nd Viet Cong Regiment", and the aim of the American operation – codenamed Cedar Falls – was to clear the area of enemy forces. Beginning on January 5 1967, it was a multi-division, two-phase affair in which the borders of the Triangle would be secured prior to a sweep by armoured, mechanized infantry and heliborne units, backed by artillery and airpower. Once in possession of the region, the Americans would destroy the villages, forcibly remove the surviving population to more secure (government-controlled) locations and turn the area into a "free strike zone", within which all human movement would be presumed to be hostile and reacted to accordingly.

In purely tactical terms, the operation ran smoothly, despite the fact that an American public relations officer in Saigon inadvertently announced the plan to the press 20 hours before it began. Elements of the US 25th Infantry

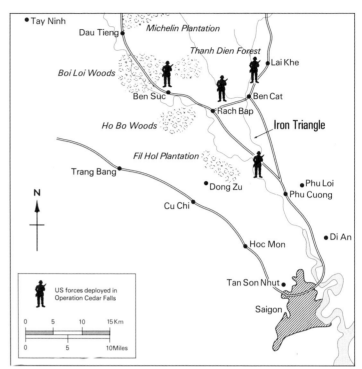

Fighting in the Iron Triangle (*below*): as US troops deploy from a UH-1 Iroquois helicopter, apparently achieving surprise, VC guerrillas move underground into a complex series of tunnels. The failure

The "Iron Triangle" (*right*), showing proximity to the South Vietnamese capital, Saigon. Viet Cong (VC) activity here led to Operation Cedar Falls, in January 1967.

to capture large numbers of VC in Operation Cedar Falls contributed to the growing sense of American frustration in Vietnam.

Division and 196th Light Infantry Brigade took up blocking positions south of the Saigon river by January 8, creating an "anvil" against which the US 1st Infantry Division, 11th Armored Cavalry Regiment and 173rd Airborne Brigade were to "hammer" the enemy. But such a large force was incredibly noisy, warning the communists to go to ground in an elaborate and extensive complex of tunnels beneath the Triangle; thus, although the Americans made full use of their firepower and mobility, they punched thin air. The village of Ben Suc was taken in a helicopter assault and destroyed; the inhabitants were rounded up and the surrounding area levelled by bulldozers and explosives. But no permanent damage was inflicted on the communist infrastructure: instead, the violence of the assault probably ensured the alienation of local people, many of whom drifted back over the next few weeks. By then, the Americans had diverted mainforce units to face NVA activities in the Central Highlands – and the VC had reappeared in the Triangle. Despite American claims that the operation had dealt "a blow from which the VC in this area may never recover", the Iron Triangle was to remain a threat to Saigon right up to the end of the war in 1975.

The fall of South Vietnam to the communists represented a humiliating defeat for the Americans and led, inevitably, to a reassessment of strategy. A special Training and Doctrine Command (TRADOC) was set up to analyse the reasons for the defeat. It came up with a number of factors – that preoccupation with the "mainforce" war against the NVA had diverted attention away from the insurgency in the villages of the South; that the US Army's emphasis on technology and firepower had prevented the sort of face-to-face combat so essential to victory; that US soldiers had lacked the motivation and training needed to cope with the problems of Vietnam – but the overriding feeling, within the army at least, was that it had been asked to fight under extremely restrictive political control. No one argued for an end to restraint, but many commentators pointed out that the US forces had fought well, using their technology, firepower and helicopter mobility to inflict significant defeats on the NVA and VC, which could not then be followed up because of political fears of escalation. In short, the call went out for a strategy which would allow the forces to "fight to win" within general parameters laid down by the politicians.

AirLand Battle – the doctrine

In August 1982 these feelings were translated into a new doctrine known as "AirLand Battle", within which technology had a major role to play. It was based on the presumption that communist aggression was not a thing of the past, but was likely to manifest itself almost anywhere, from Central Europe to Central America, from the Middle East to the Far East, and to take a variety of forms, from full-scale conventional war to local insurgency. The United States therefore needed a flexible strategy, designed to make the most of existing advantages. The most important of these was Emerging Technology (ET), which could be used as a "force multiplier", affording the outnumbered Americans the advantages of surveillance and "deep strike" which would help to redress the balance. If, for example, the communists should choose to commit their forces to an invasion of Western Europe, US tech-

William C Westmoreland (b.1914) was born in South Carolina. He graduated from West Point in 1932 and was commissioned into the artillery. He saw service in North Africa, Sicily and northwest Europe during World War II, after which he specialized in airborne operations. By 1963 he was commanding the US 18th Airborne Corps with the rank of major general. In 1964 he was sent to Saigon to command all US ground forces serving in Vietnam. A firm believer in a strategy of attrition, he conducted large-scale operations along the borders of Vietnam against the North Vietnamese Army (NVA) and in the villages of the South against the Viet Cong (VC), but despite inflicting heavy casualties, his strategy did not work. By early 1968, when the communists mounted attacks throughout South Vietnam during the Tet Offensive and laid siege to Khe Sanh, Westmoreland's views were discredited. He was recalled in mid-1968 and promoted to Chief of Staff of the US Army. He retired in 1972.

South Vietnam, (*left*) showing major towns and the Ho Chi Minh and Sihanouk supply trails. The latter, snaking towards the South through neutral Laos and Cambodia, forced the Americans to keep large numbers of troops in border areas to combat infiltration, leaving the villages of the South poorly protected.

Map labels:
Dong Hoi
North Vietnam
Demilitarized Zone
Khe Sanh
Lang Vei
Quang Tri
Savannakhet
Hué
US marines land March 1965
A Shau
Da Nang
Duy Xuyen
Thailand
Laos
Kham Duc
Quang Ngai
Mekong
Dak To
Kontum
Pleiku
Qui Nhon
Cambodia
Tonle Sap
South Vietnam
Ban Me Thuot
Nha Trang
Dalat
Fish Hook
Phnom Penh
An Loc
Mekong
Tay Ninh
Ben Cat
Ben Suc
Bassac
Iron
Bien Hoa
Chau Doc
Triangle
Saigon
My Tho
Vinh Long
Ben Tre
Can Tho
Mekong Delta
Ca Mau
South China Sea

Main guerrilla actions, 1968 (Tet Offensive)
Ho Chi Minh Trail
Sihanouk Trail
0   40   80 Km
0   40 Miles

nology would monitor the build-up and, once the attack had begun, disrupt and destroy the advancing Warsaw Pact formations in a devastating display of precision and firepower. The aim would be to isolate the "first echelon" of the attacking force from its "second echelon" reserves, thereby creating a battle environment in which the Western allies were not overwhelmingly outnumbered and could "fight to win".

AirLand Battle – the means
This was to be achieved by means of sophisticated surveillance and strike technology, much of which had its origins in Vietnam. In order to monitor enemy advances and to pinpoint appropriate targets deep inside his territory, high-flying TR-1 surveillance aircraft – developments of the U-2 "spy-plane" – would search for radar emissions and, in a technique known as Precision Location Strike System (PLSS), provide precise target information to attack aircraft or long-range ground-based artillery. The former would deploy special "sub-munitions" which, when released, would float down towards their targets, using on-board devices to seek them out before firing a self-forging explosive projectile to destroy them with impressive accuracy. Other devices, such as the Joint Surveillance and Target Attack Radar System (Joint-STARS), would use similar techniques to call down devastating "Ripple fire" from ground-based missiles, while RPVs, "smart" weapons and even robotic defences – unmanned machines equipped with sensors and an attack capability – would add to the chaos. The overall result, according to AirLand Battle doctrine, would be an overwhelming weight of defensive

technology to which the less sophisticated forces of the communist world would be powerless to respond.

In the context of Central Europe, where the forces of the North Atlantic Treaty Organization (NATO) and the communist Warsaw Pact face each other in uneasy peace, AirLand Battle would seem to offer enormous advantages. NATO's current strategy of Flexible Response, designed to counter any Warsaw Pact attack at the appropriate level while threatening to go one stage further if the attack persists, is under serious review, not least because of the dangers inherent in it of early nuclear release – after all, if the Pact should attack using purely conventional weapons, NATO may lack the means to counter it effectively at that level and be forced to "go one stage further", to battlefield nuclear weapons, just to survive. Furthermore, with a superpower agreement to dismantle land-based Intermediate Nuclear Forces (INF) in December 1987, it is even conceivable that NATO will lack the means to carry out its escalation threat, placing the onus firmly on the need for conventional forces which will "fight to win". ET may provide an answer – as early as November 1984, NATO adopted the basic principles of "deep strike" when the doctrine of Follow On Forces Attack (FOFA) was accepted as a long-term planning guideline – but it is no guarantee of success. As recent history shows, the pendulum can swing rapidly from defence to offence: what may seem an advantage now can very quickly be countered and undermined. Conventional war, in short, is as unpredictable as ever: in the end, the only common denominator is the men who fight it. Their motivation, fighting spirit and tactical skills will always be the keys to victory.

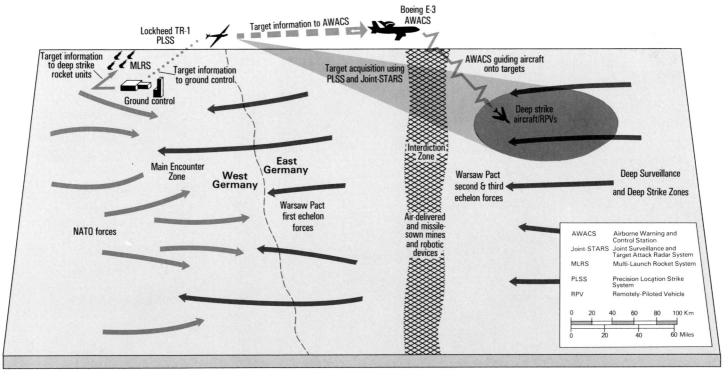

The new US doctrine of Airland Battle, shown here in a projected NATO/Warsaw Pact setting. Introduced in August 1982, it uses high technology to "see deep and strike deep". As enemy first echelon forces attack, defending units will engage them in the "Main Encounter Zone", in this case the Inner German border (IGB). Meanwhile, mines and robotic defences will create an "Interdiction Zone" to prevent enemy second and third echelon forces from reaching the battle area, while high-flying aircraft will monitor their approach, calling in precise air or ground strikes against them in the "Deep Surveillance and Deep Zone".

Bibliography

General

Correlli Barnett *Britain and her Army* (London 1974)

Martin van Creveld *Command in War* (Cambridge 1985; Cambridge MA 1987)
Fighting Power: German Military Performance 1939–45 (Westport Conn. 1982)
Supplying War: Logistics from Wallenstein to Patton (Cambridge 1977; New York NY 1979)

Richard Holmes *Firing Line/Acts of War* (US ed) (London 1984; New York NY 1985)

John Keegan *The Face of Battle* (London 1976; New York NY 1983)
The Mask of Command (London 1987; New York NY 1987)

William H McNeill *The Pursuit of Power: Technology, Armed Force and Society since AD1000* (Oxford 1983; Chicago IL 1982)

Allan R Millett and Peter Maslowski *For the Common Defense: A Military History of the United States of America* (New York, 1984)

Hew Strachan *European Armies and the Conduct of War* (London 1983)

1 The dawn of warfare

Cambridge Ancient History, Vol I Part 2, Vol II Parts 1 & 2 (3rd ed, Cambridge 1980)

John Chadwick *The Mycenaean World* (Cambridge 1976)

L Foxhall and J K Davies (eds) *The Trojan war: Its Historicity and Context* (Bristol 1984)

O R Gurney *The Hittites* (2nd ed, Harmondsworth 1981)

K A Kitchen *Suppiluliuma and the Amarna Pharaohs* (Liverpool 1962)
Pharaoh Triumphant: The Life and Times of Ramesses II (Warminster 1982)

J G MacQueen *The Hittites and their Contemporaries in Ancient Anatolia* (2nd ed, London 1986)

James B Pritchard *Ancient Near Eastern Texts Relating to the Old Testament* (3rd ed, Princeton 1969)

George Roux *Ancient Iraq* (2nd ed, Harmondsworth 1980)

N K Sandars *The Sea Peoples: Warriors of the Ancient Mediterranean* (2nd ed, London 1985)

Yigael Yadin *The Art of Warfare in Biblical Lands* (London 1963)

2 Warriors of Greece and Rome

Arrian *The Campaigns of Alexander* (ed J R Hamilton, London 1981; New York NY 1986)

Brian Caven *The Punic Wars* (London 1980; New York 1980)

Julius Caesar *The Battle for Gaul* (ed/trans Anne and Peter Wiseman, London 1980)

Colin McEvedy *The Penguin Atlas of Ancient History* (Harmondsworth 1968)

H H Scullard *The Elephant in the Greek and Roman World* (London 1974; Ithaca NY 1974)

Thucydides *History of the Peloponnesian War* (trans Rex Warner, Harmondsworth 1968)

John Warry *Warfare in the Classical World* (London 1980; New York NY 1981)

Graham Webster *The Roman Imperial Army* (3rd ed London 1981)

Xenophon *The Persian Expedition* (trans Rex Warner, Harmondsworth 1972)

3 Men of Iron

Jim Bradbury *The Medieval Archer* (Woodbridge 1985)

R Allen Brown *Castles: A History and Guide* (Poole 1980)

Philippe Contamine *War in the Middle Ages* (Oxford 1984)

Colin McEvedy *Penguin Atlas of Medieval History* (Harmondsworth 1961)

R C Smail *Crusading Warfare 1097–1193* (Cambridge 1976)

Malcolm Vale *War and Chivalry* (London 1981)

J F Verbruggen *The Art of Warfare in Western Europe during the Middle Ages* (Amsterdam, Netherlands 1977)

4 Storm from the East

M A Cook (ed) *A History of the Ottoman Empire to 1730* (Cambridge 1976)

Istvan Dienes *The Hungarians Cross the Carpathians* (Hereditas Corvina Press 1972)

John Bagot Glubb *The Great Arab Conquests* (London 1963)

Milos Jankovich *They Rode into Europe* (London 1971)

Otto J Maenchen-Helfen *The World of the Huns* (California 1973)

David Morgan *The Mongols* (Oxford 1986)

Frank Trippet *The First Horsemen* (Time-Life, Amsterdam 1980)

5 War in the Orient

A L Basham *The Wonder that was India* (London 1954; New York NY 1968)

Jack Beeching *The Chinese Opium Wars* (London 1975)

Cambridge History of China vols 3, 10 & 11 (Cambridge 1979–80)

Cambridge History of India vols 1–4 (Cambridge 1922–37)

Cambridge History of Iran vols 1–5 (Cambridge 1968)

Cambridge History of Islam vols 1, 1a, 2, 2a (Cambridge 1980)

H G Creel *The Origins of Statecraft in China* (Chicago 1970)

J W Hall *Japan from Prehistory to Modern Times* (London 1970)

Takashi Hatada *A History of Korea* (Santa Barbara, Calif. 1969)

Clive Irving *Crossroads of Civilization: 3000 Years of Persian History* (London 1979)

J H Longford *Japan* (London 1923)

H G Rawlinson *India: A Short Cultural History* (London 1954)

Percival Spear *Oxford History of India* (3rd ed, Oxford 1958; New York NY 1978)

R Thapar *A History of India*, vol I (London 1966)

S R Turnbull *The Book of the Samurai* (London 1982)

David Walder *The Short Victorious War* (Newton Abbott 1973)

6 "Villainous saltpetre"

Robert Ashton *The English Civil War* (London 1978)

John Childs *Armies and Warfare in Europe 1648–1789* (Manchester 1982)

Christopher Duffy *Siege Warfare: The Fortress in the Early Modern World 1494–1660* (London 1979)

C H Firth *Cromwell's Army* (4th ed, London 1962)

Pieter Geyl *The Revolt of the Netherlands 1555–1609* (London 1958)

Sir Charles Oman *A History of the Art of War in the Sixteenth Century* (London 1937; Ithaca NY 1960)

G Parker *The Army of Flanders and the Spanish Road* (Cambridge 1972; New York NY 1978)

Michael Roberts *Essays in Swedish History* (London 1967)

J W Thompson *The Wars of Religion in France* (London 1957)

C V Wedgwood *The Thirty Years War* (London 1938)

7 The age of the flintlock

David Chandler *The Art of War in the Age of Marlborough* (London 1976)
The Campaigns of Napoleon (London 1967; New York NY 1973)

Christopher Duffy *The Army of Frederick the Great* (Newton Abbott 1974; New York NY 1986)
The Fortress in the Age of Vauban and Frederick the Great (London 1985; New York NY 1988)
Military Experience in the Age of Reason (London 1987)

Don Higginbotham *The War of American Independence* (New York 1971)

J A Houlding *Fit for Service: The Training of the British Army 1715–1795* (Oxford 1981; New York NY 1981)

Robert S Quimby *The Background of Napoleonic Warfare* (New York 1957)

Gunther E Rothenberg *The Art of Warfare in the Age of Napoleon* (London 1977; Bloomington IN 1978)

John Shy *A People Numerous and Armed: Reflections on the Military Struggle for American Independence* (Oxford 1976)

8 The age of factory war

C M von Clausewitz *On War* (Trans Michael Howard and Peter Paret, Princeton 1976)

Gordon A Craig *The Battle of Koniggratz* (London 1965)

William A Frassanito *America's Bloodiest Day: The Battle of Antietam 1862* (New York 1979)

John Gooch *Armies in Europe* (London 1979; New York NY 1980)

Reginald Hargreaves *Red Sun Rising: The Siege of Port Arthur* (London 1962)

Michael Howard *The Franco-Prussian War* (London 1961; New York NY 1981)
The Theory and Practice of War (London 1965)

Thomas Pakenham *The Boer War* (London 1979)

Peter J Parish *The American Civil War* (London 1975; New York NY 1975)

Edward M Spiers *The Army and Society, 1815–1914* (London 1980)

9 The Star of Empire

Carles, Col. P. *Des Millions de Soldats Inconnus* (Paris 1982)

Anthony Clayton *The British Empire as a Superpower* (New York 1986)
France, Soldiers and Africa (London 1988)

John Cloake *Templer, Tiger of Malaya* (London 1985)

T A Heathcote *The Indian Army: The Garrison of British Imperial India* (Newton Abbott 1974)

R Huré *L'Armée d'Afrique 1830–1962* (Paris 1977)

Sir William Jackson *Withdrawal from Empire: A Military View* (London 1986)
Les Troupes de Marine 1622–1984 (Paris 1986)

Philip Mason *A Matter of Honour: An Account of the Indian Army, its Officers and Men* (London 1974)

André Maurois *Marshal Lyautey* (Trans Hamish Miles, London 1931)

Marc Michel *L'Appel à l'Afrique* (Paris 1982)

Boris Mollo *The Indian Army* (Blandford 1981)

Francois Porten de la Morandière *Soldats du Djebel* (Paris 1977)

Peter Young and J P Lawford *History of the British Army* (London 1970)

10 World War I: Advance to Deadlock

Correlli Barnett *The Swordbearers: Studies in Supreme Command in the First War* (London 1963; Bloomington IN 1975)

Fritz Fischer *War of Illusions: German Policies from 1911 to 1914* (London 1975; New York NY 1975)

John Gooch *The Plans of War: The General Staff and British Military Strategy c. 1900–1916* (London 1974)

Paul Kennedy (ed) *The War Plans of the Great Powers* (London 1979; New York 1979)

A J Marder *From the Dreadnought to Scapa Flow* (5 vols, London 1961–69)

G Ritter *The Schlieffen Plan* (New York 1958)

Norman Stone *The Eastern Front 1914–1917* (London 1975; New York NY 1976)

John Terraine *Mons: The Retreat into Victory* (London 1960)

T H E Travers *The Killing Ground* (London 1987)

Barbara W Tuchman *August 1914/The Guns of August* (US ed) (London 1962; New York NY 1962)

11 World War I: Breaking the Fetters

Shelford Bidwell and Dominick Graham *Firepower: British Army Weapons and Theories of War 1904–45* (London 1982)

Vera Brittain *Testament of Youth* (London 1978; New York NY 1980)

Guy Chapman (ed) *Vain Glory* (London 1968)

Paul Fussell *The Great War and Modern Memory* (London 1975)

Bill Gamage *The Broken Years* (Canberra 1974)

Alistair Horne *The Price of Glory: Verdun 1916* (London 1962; New York NY 1979)

Martin Middlebrook *The First Day of the Somme* (London 1971)

John Terraine *Douglas Haig: The Educated Soldier* (London 1963)

Trevor Wilson *The Myriad Faces of War* (Cambridge 1986)

Denis Winter *Death's Men* (London 1978)

12 World War II: Blitzkrieg

Marc Bloch *Strange Defeat* (London 1948)

Brian Bond *British Military Policy between the Two World Wars* (Oxford 1980; New York NY 1980)
France and Belgium, 1939–40 (London 1975)

Wilhelm Deist *The Wehrmacht and German Rearmament* (London 1981)

Jeffery A Gunsberg *Divided and Conquered: The French High Command and the Defeat of the West, 1940* (Westport Conn. 1979)

Alistair Horne *To Lose a Battle: France 1940* (London 1969; New York NY 1979)

Anthony Kemp *The Maginot Line: Myth and Reality* (New York 1979)

Charles Messenger *The Art of Blitzkrieg* (London 1976)

R M Ogorkiewicz *Armoured Forces: A History of Armoured Forces and their Vehicles* (London 1970)

D C Watt *Too Serious a Business: European Armed Forces and the Approach to the Second World War* (London 1975)

13 World War II: The Eastern Front
Omer Bartov *The Eastern Front, 1941–45: German troops and the Barbarisation of Warfare* (London 1985; New York NY 1986)
A Dallin *German Rule in Russia, 1941–45* (London 1981)
John Erickson *The Road to Stalingrad* (London 1975; Boulder CO 1983)
The Road to Berlin (London 1983)
The Soviet High Command: A Military-Political History (London 1962; Boulder CO 1984)
Geoffrey Jukes *Stalingrad* (London 1968)
Erich von Manstein *Lost Victories* (London 1958)
F W von Mellenthin *Panzer Battles* (London 1955)
Albert Seaton *The Russo-German War, 1941–45* (London 1971)
A Werth *Russia at War, 1941–45* (London 1964; New York NY 1984)

14 World War II: The Desert War
Correlli Barnett *The Desert Generals* (2nd ed, London 1986; Bloomington IN 1986)
Michael Carver *Dilemmas of the Desert War* (London 1987; Bloomington IN 1987)
El Alamein (London 1962)
John Ellis *Cassino: The Hollow Victory* (London 1985; New York NY 1984)
Nigel Hamilton *Montgomery: Master of the Battlefield* (London 1983)
W G F Jackson *The North African Campaign, 1940–43* (London 1975)
The Battle for Italy (London 1967)
Eric Newby *Love and War in the Apennines* (London 1971)
Erwin Rommel *The Rommel Papers* (ed B H Liddell-Hart, London 1953)
William Seymour *British Special Forces* (London 1985)

15 World War II: The Pacific War
J Costello *The Pacific War* (London 1985)
W D Dixon *The Battle of the Philippine Sea* (Weybridge 1975)
P S Dull *A Battle History of the Imperial Japanese Navy* (Annapolis, MD 1978)
M Fuchida *Midway: The Battle that Doomed Japan* (London 1957; New York NY 1982)
R Lewin *The Other Ultra: Codes, Cyphers and the Defeat of Japan* (London 1982; New York NY 1983)
G W Prange *At Dawn We Slept* (London 1982; New York NY 1981)
Miracle at Midway (New York 1982)
C G Reynolds *The Fast Carriers* (Melbourne, FL 1978)
R H Spector *Eagle against the Sun* (Harmondsworth 1987)
H P Wilmott *The Barrier and the Javelin* (Annapolis, MD 1983)

Empires in the Balance (Annapolis, MD 1982)
W T Y'blood *Red Sun Setting* (Annapolis, MD 1981)

16 World War II: The end in the West
Carlo d'Este *Decision in Normandy* (London 1983)
Max Hastings *Overlord* (London 1984; New York NY 1984)
John Keegan *Six Armies in Normandy* (London 1982; New York NY 1982)
Harold P Leinbaugh and John D Campbell *The Men of Company K: The Autobiography of a World War II Rifle Company* (New York 1985)
Ronald Lewin *Ultra Goes to War* (London 1978)
Charles B MacDonald *The Battle of the Bulge/A Time for Trumpet* (US ed) (London 1986; New York NY 1984)
R J Overy *The Air War 1939–45* (London 1980)
S W Roskill *The navy at War* (London 1960)
Maurice Tugwell *Airborne to Battle: A History of Airborne Warfare 1918–1971* (London 1971)

17 Strategic bombing
C Bekker *The Luftwaffe War Diaries* (London 1967)
W F Craven and L J Cate (eds) *The Army Air Forces in World War II*, vols 1–5 (Chicago 1949–51)
R H Fredette *The Sky on Fire: The First Battle of Britain* (London 1966)
C V Glines *The Compact History of the US Air Force* (Dallas TX 1963)
A T Harris *Bomber Offensive* (London 1947)
J Killen *A History of the Luftwaffe* (New York 1968)
C E LeMay *Missions with LeMay: My Story* (New York 1965)
R Littauer and N Uphoff (eds) *The Airwar in Indo-China* (Boston, MA 1971)
Sir W Raleigh and H A Jones *The War in the Air*, vols 1–6 and Appendix (Oxford 1922–37)
H Rumpf *The Bombing of Germany* (London 1961)
H St G Saunders *Per Ardua: The Rise of British Air Power 1911–1939* (Oxford 1944)
C Webster and N Frankland *The Strategic Air Offensive against Germany* (London 1961)

18 Guerrilla war
Tom Barry *Guerrilla Days in Ireland* (London 1981)
Geoffrey Fairbairn *Revolutionary Guerrilla Warfare* (London 1974)
Vo Nguyen Giap *People's War, People's Army* (London 1962)
Che Guevara *Guerrilla Warfare* (London 1961)
Chalmers Johnson *Autopsy on People's War* (Los Angeles CA 1974)
Frank Kitson *Low Intensity Operations* (London 1971; Orinda CA 1974)

Walter Lacqueur *Guerrilla: A Historical and Critical Study* (Boston MA 1976)
The Guerrilla Reader (New York NY 1977)
Mao Tse-Tung *On Guerrilla Warfare* (London 1971)
Deneys Reitz *Commando: A Boer Journal of the Boer War* (London 1983)
Charles Townshend *Britain's Civil Wars* (London 1986)
Truong Chinh (ed. Bernard Falls) *Primer for Revolt* (London 1963)

19 Destroyer of worlds
John Baylis, Ken Booth, John Garnett, Phil Williams *Contemporary Strategy*, 2 vols (2nd ed, London 1987)
Nigel Calder *Nuclear Nightmares: An Investigation into Possible Wars* (London 1981)
Christy Campbell *War Facts Now* (London 1982)
Nuclear Facts Now (London 1984)
Lawrence Freedman *Atlas of Global Strategy: War and Peace in the Nuclear Age* (London 1985)
International Institute for Strategic Studies (Annual), "Strategic Survey" and "Military Balance" (London)
Peter Pringle and William Arkin *SIOP: Nuclear War from the Inside* (London 1983)
Gordon Thomas and Max Morgan-Witts *Ruin from the Air: The Atomic Mission to Hiroshima* (London 1978)
E P Thompson (ed) *Star Wars* (London 1983; New York NY 1986)
Andrew Wilson *The Disarmer's Handbook of Military Technology and Organisation* (London 1983)

20 An unquiet peace
M S El Azhary (ed) *The Iran Iraq War* (London 1984)
Frank Barnaby *The Automated Battlefield* (London 1986; New York NY 1986)
Ray Bonds (ed) *The Soviet War Machine* (London 1980; New York NY 1987)
Richard A Gabriel *Operation Peace for Galilee: The Israeli-PLO War in Lebanon* (New York 1984)
Max Hastings and Simon Jenkins *The Battle for the Falklands* (London 1983; New York NY 1983)
Chaim Herzog *The Arab-Israeli Wars: War and Peace in the Middle East* (London 1982)
David Rees *Korea: The Limited War* (London 1964)
Shelby L Stanton *The Rise and Fall of an American Army: US Ground Forces in Vietnam, 1965–73* (California 1985)
Sir Robert Thompson (ed) *War in Peace: An Analysis of Warfare from 1945 to the Present Day* (3rd ed, London 1985; New York NY 1985)
War in Peace partwork (London 1983–85)

Index

Pages quoted may contain more than one reference to the person or subject referred to.

The symbol * means "battle of".

Page numbers in quotation marks " " refer to excerpted sections featuring soldiers' talk; in *italic*, to illustration captions; in **bold**, to maps, plans or diagrams.

Acknowledgments

Abbreviations used are: L, left; R, right; C, centre; T, top; B, bottom;
AAAC – Ronald Sheridan Ancient Art & Architecture Collection, London
BA – Bundesarchiv, Koblenz
BBC – BBC Hulton Picture Library
BL – British Library
BM – British Museum
IWM – Imperial War Museum, London
MARS – Military Archive & Research Services, Lincs.
MH – Michael Holford
NAM – National Army Museum, London
Novosti – Novosti Press Agency, London
PNHP – Peter Newark's Historical Pictures, Bath
PR – Photoresources
RH – Robert Harding Picture Library
RHL – Robert Hunt Library
USN – US Navy
USNA – US National Archives, Washington DC
USAF – US Air Force
V & A – Victoria & Albert Museum

P. 6, IWM; 7, MARS; 8L/R, 10R, AAAC; 11L, Peter Clayton; 12, AAAC; 15TR/BR, PR; 18L, BM; 19T, Ray Gardner; 20B, MARS; 22T, BM/MH; 24L, PR; 24/5, Photo Aerienne Bernard Beaujard; 26TR, MARS; 28R, AAAC; 30L, BL; 30R, Ray Gardner; 33T, MH; 33B, PR; 34R Bibliothèque Nationale, Paris; 35 BL/ET Archive; 36R, Richard Muir; 40,

Mittelalterliches Hausbuch; 43, RH; 47L, Werner Forman Archive; 47R Sonia Halliday Photographs; 48L, Kunsthistorisches Museum, Vienna; 50L, AAAC, 51, BM/RH; 53 70R, Heeresgeschichtliches Museum, Vienna/RH; 57T, Werner Forman Archive; 59T/BR/BL; 60R, BL/RH; 61, MH; 62L, RH; 62R, Pat Hodgeson Library; 63R, RH; 65T, BPCC/Aldus Archive; 65B, ET Archive; 66L, MH; 66BR, MARS; 68, 69B, V & A/MH; 70L, AAAC; 73T, MARS; 73BL, Musée de Strasbourg; 74L, Kunsthalle, Hamburg; 75TR, National Maritime Museum, London; 75BR, 77R, Reproduced by gracious permission of Her Majesty the Queen, The Royal Collection. 77L, Museum of War; 79T, P. Haythornthwaite; 79BL, MARS; 79BR, Aerofilms; 84T, MARS; 86, 87T, Dr. J. Hebbert; 87BL, Musée de la Citadelle/MARS; 87BR, MARS; 89R, V & A/Bridgeman Art Library; 95L, MARS; 95R, Musée du Louvre; 99T, MARS; 100, BBC; 102, PNHP; 104, MARS; 106, NAM/MARS; 110, PNHP; 113T, NA/MARS; 113C, MARS; 116L, Biblioteca Nacional, Madrid/BPCC/Aldus Archive; 116R, NAM/MARS; 117BL, MARS; 118T, BBC; 118BR; 119T, NAM/MARS; 119B, BM; 121TR/BR, BBC; 123L/R, BBC; 125, ECPA, Paris/MARS; 126L, IWM; 126R, RHL; 127, Camera Press; 128L, BBC; 128R, The Keystone Collection; 133, 135L, BA; 135R, IWM; 136T, BBC; 138T, IWM/Bridgeman Art Library; 140T,

RHL; 141L, BA; 141TR/BR, IWM; 142L, RHL; 143, IWM/MARS; 144T, BA; 148BR MARS; 144T, BA; 148BR, MARS; 151T, RHL; 151B, IWM; 153L, MARS; 153TR, ECPA/RHL; 155T, IWM; 156, BA; 157R, IWM; 159, MARS; 162R, BA; 163T, RHL; 164, 165L/TR/BR, IWM; 170L, 171L, 175L/R, RHL; 171R, MARS; 172T, RHL; 173, Helsinki Institute of Military Science/MARS; 177R, BA; 179L/TR, MARS; 179BR, BA; 180L, IWM; 180R, The Keystone Collection; 181TL, MARS; 181BL, BA; 181TR, IWM; 181CR, IWM; 181BR, BA; 185L, MARS; 185R, BA/MARS; 186T, 187R, 189L/TR/BR, Novosti; 190T, MARS; 190BL, 191, 193, 195, Novosti; 196L, IWM; 196R, 197L, Novosti; 202, MARS; 203B, 204T, IWM/MARS; 205, MARS; 206, BA; 211T, RHL; 211B, USN/MARS; 212R, MARS; 213L, USN/MARS; 213R, MARS; 214L, IWM/MARS; 214R, USN/MARS; 216TR, MARS; 217L, Newport News/MARS; 217R, USNA/RHL; 218B, USN/MARS; 220T, 221L/R, 222L, USN/MARS; 222R, IWM/RHL; 223T, MARS; 223BR, Popperfoto; 224, 227TR, The Keystone Collection; 229L, IWM/MARS; 229R, US Army/MARS; 230, IWM; 232TR, The Keystone Collection; 232BR, 236L, IWM; 236R, IWM/MARS; 238, 239T, IWM; 239B, RHL; 240TL/TR, Imperial Tobacco Co.; 241R, 242T, IWM; 243T, MARS; 243BL, IWM/MARS; 246T, USAF/MARS; 247L, MARS; 247R, USAF/MARS; 249L,

Salamander Books; 249TR, MARS; 249BR, USAF/Salamander Books; 252L. Prado Museum/MARS; 252R, Mary Evans Picture Library; 253L, BBC; 253R, Mary Evans Picture Library; 254R, BBC; 255L, Foreign & Commonwealth Office/MARS; 255R, 256L, BBC; 256R, IWM; 257TL, Gamma/Frank Spooner Pictures; 257TR, RTH; 257B, Gamma/Frank Spooner Pictures; 258L, RHL; 258R, SIPA/Rex Features; 259, 260, Salamander Books; 261, Gamma/Frank Spooner Pictures; 262L, SIPA/Rex Features; 262R, UPI/RHL; 263, Gamma/Frank Spooner Pictures; 264T, Camera Press, 264BL, Gamma/Frank Spooner Pictures; 264BR, MARS; 265L, Gamma/Frank Spooner Pictures; 265R, SIPA/Rex Features; 269L/R, RHL; 271L, MARS; 271R, USN; 273, Gamma-Liaison/Frank Spooner Pictures; 274R, MARS; 275, Novosti/MARS; 277L, Rex Features; 277R, Gamma-Liaison/Frank Spooner Pictures; 278, Gamma-Liaison/Frank Spooner Pictures; 279, Marxel-Liaison/Frank Spooner Pictures; 283L/R, RHL; 284TL, Fleet Photographic Unit; 284TR, Salamander Books; 285L, COI; 285TR, SIPA/Rex Features; 286L/R, Camera Press; 287R, Rex Features; 288L, Associated Press; 288R, Camera Press; 289R, Adriano Mordenti/Rex Features; 292R, UPI/RHL.